Ion Channels of Vascular Smooth Muscle Cells and Endothelial Cells

Ion Channels of Vascular Smooth Muscle Cells and Endothelial Cells

Proceedings of the International Society for Heart Research (ISHR) held in Cincinnati, Ohio, May 28 through June 2, 1991.

Editors

Nicholas Sperelakis, PhD
Department of Physiology and Biophysics,
University of Cincinnati College of Medicine,
Cincinnati, Ohio

Hirosi Kuriyama, MD, PhD
Department of Pharmacology,
Faculty of Medicine, Kyushu University,
Fukuoka, Japan

Elsevier
New York • Amsterdam • London • Tokyo

Elsevier Science Publishing Co., Inc.
655 Avenue of the Americas, New York, New York 10010

Sole distributors outside the United States and Canada:
Elsevier Science Publishers B.V.
P.O. Box 211, 1000 AE Amsterdam, The Netherlands

© 1991 by Elsevier Science Publishing Co., Inc.

This book is printed on acid-free paper.

ISBN 0-444-01651-1

Current printing (last digit):
10 9 8 7 6 5 4 3 2 1

Manufactured in the United States of America

TABLE OF CONTENTS

DEDICATION

This book is dedicated to my co-editor and esteemed colleague, Prof. Hirosi Kuriyama, who has been a pioneer in smooth muscle research, including vascular smooth muscle, particularly in the electrophysiological and ion channel aspects.

PREFACE

This book represents part of the proceedings of the annual meeting of the International Society for Heart Research (ISHR) held in Cincinnati, Ohio on May 28 - June 2, 1991. The ISHR meeting covered broad aspects of cardiovascular science, and we decided to publish in this book only one focussed aspect, namely the electrophysiology and ion channels of vascular smooth muscle and vascular endothelial cells. These two cell types are the most important cells comprising the blood vessel wall, and function to regulate the blood flow and its distribution by changing the vascular muscle tone and peripheral resistance. They are also important in vascular pathophysiology and disease, including hypertension and atherosclerosis, and serve as target cells in the therapeutic treatment of hypertension. In order to make this book a comprehensive and up-to-date compilation of the electrophysiology and ion channels of vascular smooth muscle cells and endothelial cells, we also invited a number of leading investigators, who were not present at the ISHR meeting, to contribute articles. Therefore, the reader will find this book to be a timely and thorough discourse on this topic. Articles have been authored by leading researchers from around the world. It has been our pleasure to work with such a distinguished group of scientists.

Nicholas Sperelakis, Ph.D.
Cincinnati, Ohio
August 25, 1991

Hirosi Kuriyama, M.D., Ph.D.
Fukuoka, Japan

ACKNOWLEDGMENTS

The editors wish to acknowledge the dedicated assistance of Rhonda Hentz in preparation of this book.

Support of the following organizations for holding the ISHR-1991/Cincinnati meeting is gratefully acknowledged:

PATRONS

International Society for Heart Research

University of Cincinnati Medical Center

Children's Hospital Medical Center, Cincinnati

Upjohn Co.

Chiquita Brands International

Baxter Healthcare Corp., Blood Substitute Group

Hoffman-LaRoche, Inc.

Procter & Gamble Co.

Tanabe Seiyaku Co., Ltd.

SPONSORS

Alliance Pharmaceutical Corp.

American Heart Association, Basic Science Council

Marion Merrell Dow, Inc.

Sentron Medical, Inc.

CONTRIBUTORS

Boehringer-Ingelheim

Burroughs Wellcome Co.

Miles Laboratories/Bayer AG

SmithKline Beecham Pharmaceuticals

Abbott Laboratories

Eli Lilly & Co.

Glaxo Research Laboratories

Merck & Co., Inc.

Parke-Davis Pharmaceuticals Research Division

Sterling Drug, Inc.

LIST OF CONTRIBUTORS

Abe, Shimako
Division of Molecular Cardiology
Research Institute of Angiocardiology
Faculty of Medicine
Kyushu University
3-1-1 Maidashi
Higashi-ku
Fukuoka 812
Japan

Bkaily, Ghassan
Department of Physiology & Biophysics
Faculty of Medicine
University of Sherbrooke
Sherbrooke, Quebec
Canada J1H 5N4

Bowen, Susan M.
Department of Physiology
University of Nevada
School of Medicine
Reno, NV 89557

Brayden, Joseph E.
Department of Pharmacology and
Vermont Center for Vascular Research
College of Medicine
The University of Vermont
Burlington, VT 05405

Caldwell, W.M.
Department of Physiology
School of Medicine
and Dalton Research Center
University of Missouri
Columbia, MO 65211

Carl, Andreas
Department of Physiology
University of Nevada
School of Medicine
Reno, NV 89557

D'Orleans-Juste, Pedro
Department of Pharmacology
Faculty of Medicine
University of Sherbrooke
Sherbrooke, Quebec
Canada J1H 5N4

DeCoursey, Thomas E.
Department of Physiology
Rush Presbyterian St. Luke's Medical
Center
1653 W. Congress Parkway
Chicago, IL 60612

Economos, Demetri
Department of Physiology & Biophysics
Faculty of Medicine
University of Sherbrooke
Sherbrooke, Quebec
Canada J1H 5N4

Gebremedhin, Debebe
Medical College of Wisconsin
Department of Physiology
8701 Watertown Plank Rd.
Milwaukee, WI 53226

Gelband, Craig H.
Department of Physiology
University of Nevada
School of Medicine
Reno, NV 89557

Goto, Katsutoshi
Institute of Basic Medical Sciences
University of Tsukuba
Tsukuba, Ibaraki 305
Japan

Harder, David R.
Medical College of Wisconsin
Department of Physiology
8701 Watertown Plank Rd.
Milwaukee, WI 53226

Hermsmeyer, Kent
Cardiovascular Research Laboratory
Earle A. Chiles Research Institute
Providence Medical Center and
Departments of Medicine and Cell
Biology and Anatomy
Oregon Health Sciences University
Portland, OR 97213

Hirano, Katsuya
Division of Molecular Cardiology
Research Institute of Angiocardiology
Faculty of Medicine
Kyushu University
3-1-1 Maidashi
Higashi-ku
Fukuoka 812
Japan

Hirano, Mayumi
Division of Molecular Cardiology
Research Institute of Angiocardiology
Faculty of Medicine
Kyushu University
3-1-1 Maidashi
Higashi-ku
Fukuoka 812
Japan

Hume, Joseph R.
Department of Physiology
University of Nevada
School of Medicine
Reno, NV 89557

Humphrey, D.A.
Department of Physiology
School of Medicine
and Dalton Research Center
University of Missouri
Columbia, MO 65211

Inoue, Ryuji
Department of Pharmacology
Kyushu University
Fukuoka 812
Japan

Inoue, Yoshihito
Department of Physiology & Biophysics
University of Cincinnati
College of Medicine
Cincinnati, OH 45267-0576

Ishikawa, Shiro
Fukuoka Children's Hospital
2-5-1 Tohjin-machi
Chuoku
Fukuoka, 810
Japan

Ishikawa, Tomohisa
Department of Physiology
University of Nevada
School of Medicine
Reno, NV 89557

Ito, Mitsuaki
Department of Physiology
School of Medicine
Nagoya University
466, Japan

Kajioka, S.
Department of Pharmacology
Faculty of Medicine
Kyushu University
Fukuoka 812
Japan

Kamouchi, M.
Department of Pharmacology
Faculty of Medicine
Kyushu University
Fukuoka 812
Japan

Kanaide, Hideo
Division of Molecular Cardiology
Research Institute of Angiocardiology
Faculty of Medicine
Kyushu University
3-1-1 Maidashi
Higashi-ku
Fukuoka 812
Japan

Karaki, Hideaki
Department of Veterinary Pharmacology
Faculty of Agriculture
The University of Tokyo
Bunkyo-ku
Tokyo 113
Japan

Kasuya, Yoshitoshi
Institute of Basic Medical Sciences
University of Tsukuba
Tsukuba, Ibaraki 305
Japan

Keef, Kathleen D.
Department of Physiology
University of Nevada
School of Medicine
Reno, NV 89557

Kitamura, Kenji
Department of Pharmacology
Faculty of Medicine
Kyushu University
Fukuoka 812
Japan

Kuriyama, Hirosi
Department of Pharmacology
Faculty of Medicine
Kyushu University
Fukuoka 812
Japan

Ling, Brian N.
Renal Division
Departments of Medicine and Physiology
Emory University
School of Medicine and
Veterans Administration Medical Center
Atlanta, GA 30322

Ma, Yunn-Hwa
Medical College of Wisconsin
Department of Physiology
8701 Watertown Plank Rd.
Milwaukee, WI 53226

Masaki, Tomoh
Institute of Basic Medical Sciences
University of Tsukuba
Tsukuba, Ibaraki 305
Japan

Matsumoto, Toshihiro
Department of Physiology
School of Medicine
Nagoya University
466, Japan

Mironneau, Jean
Laboratoire de Physiologie
Cellulaire et Pharmacologie Moleculaire
Inserm CJF 88-13
Universite de Bordeaux II
33076 Bordeaux
France

Nilius, Bernd
Medical Academy Erfurt
Institut of Medical Physiology
Nordhauser Str. 74
O-5010 Erfurt
Germany

Nozaki, Masahiro
Department of Gynecology & Obstetrics
Faculty of Medicine
Kyushu University 60
3-1-1 Maidashi
Higashi-ku
Fukuoka 812
Japan

O'Neill, W. Charles
Renal Division
Departments of Medicine and Physiology
Emory University
School of Medicine and
Veterans Administration Medical Center
Atlanta, GA 30322

Obye, P.K.
Department of Physiology
School of Medicine
and Dalton Research Center
University of Missouri
Columbia, MO 65211

Ohya, Yusuke
Second Department of Internal Medicine
Kyushu University
Faculty of Medicine
Fukuoka 812
Japan

Oike, M.
Department of Pharmacology
Faculty of Medicine
Kyushu University
Fukuoka 812
Japan

Post, Joseph M.
Department of Physiology
University of Nevada
School of Medicine
Reno, NV 89557

Roman, Richard S.
Medical College of Wisconsin
Department of Physiology
8701 Watertown Plank Rd.
Milwaukee, WI 53226

Sanders, Kenton M.
Department of Physiology
University of Nevada
School of Medicine
Reno, NV 89557

Satoh, Hiroyasu
Department of Physiology & Biophysics
University of Cincinnati
College of Medicine
Cincinnati, OH 45267-0576

Scornik, F.
Department of Molecular Physiology &
Biophysics
Baylor College of Medicine
Houston, TX 77030

Shapiro, Mark S.
Department of Physiology
Rush Presbyterian St. Luke's Medical
Center
1653 W. Congress Parkway
Chicago, IL 60612

Shimamura, Keiichi
Research Institute of Hypertension
Kinki University
Osaka-Sayama
589 Japan

Smith, P.
Department of Physiology
School of Medicine
and Dalton Research Center
University of Missouri
Columbia, MO 65211

Sperelakis, Nicholas
Department of Physiology & Biophysics
University of Cincinnati
College of Medicine
Cincinnati, OH 45267-0576

Stehno-Bittel, L.
Department of Physiology
School of Medicine
and Dalton Research Center
University of Missouri
Columbia, MO 65211

Sturek, Michael
Department of Physiology
School of Medicine
and Dalton Research Center
University of Missouri
Columbia, MO 65211

Sunano, Satoru
Research Institute of Hypertension
Kinki University
Osaka-Sayama
589 Japan

Suzuki, Hikaru
Department of Physiology
Nagoya City University
Medical School
Mizuho-Ku
Nagoya 467
Japan

Syed, Mohsin MD.
Department of Physiology
School of Medicine
Nagoya University
466, Japan

Taira, Norio
Department of Pharmacology
Tohoku University
School of Medicine
Sendai 980
Japan

Teramoto, N.
Department of Pharmacology
Faculty of Medicine
Kyushu University
Fukuoka 812
Japan

Tokuno, Hiroyuki
Department of Physiology
School of Medicine
Nagoya University
466, Japan

Tomita, Tadao
Department of Physiology
School of Medicine
Nagoya University
466, Japan

Toro, L.
Department of Molecular Physiology &
Biophysics
Baylor College of Medicine
Houston, TX 77030

Wagner-Mann, C.
Department of Physiology
School of Medicine
and Dalton Research Center
University of Missouri
Columbia, MO 65211

Wang, Shimin
Department of Physiology & Biophysics
Faculty of Medicine
University of Sherbrooke
Sherbrooke, Quebec
Canada J1H 5N4

Xiong, Zhiling
Department of Pharmacology
Faculty of Medicine
Kyushu University
Fukuoka 812
Japan

Yamamoto, Yoshimichi
Department of Physiology
Nagoya City University
Medical School
Mizuho-Ku
Nagoya 467
Japan

Yanagisawa, Masashi
Department of Pharmacology
Faculty of Medicine
Kyoto University
Kyoto 606
Japan

Yanagisawa, Teruyuki
Department of Pharmacology
Tohoku University
School of Medicine
Sendai 980
Japan

SECTION I

VASCULAR SMOOTH MUSCLE

Chapters 1 - 24

NEUROMUSCULAR TRANSMISSION AT ADRENERGIC NERVE TERMINALS WITH VASCULAR SMOOTH MUSCLE IN GUINEA-PIG MESENTERIC ARTERY

NICHOLAS SPERELAKIS, YOSHIHITO INOUE, MASAHIRO NOZAKI, and SHIRO ISHIKAWA

Department of Physiology and Biophysics, University of Cincinnati College of Medicine, Cincinnati, OH 45267-0576

INTRODUCTION

Almost all blood vessels are innervated by postganglionic sympathetic nerves. The transmitter release from these perivascular nerve terminals is initiated by the action potentials generated in the ganglionic cell body. Under experimental conditions, transmitter release can be evoked by electrical stimulation in isolated blood vessels. The neuroeffector transmission between the adrenergic nerve terminals and vascular smooth muscle (VSM) can be regulated or modulated by a number of factors that act locally on the nerve terminals (prejunctional site) (for review, see 10, 11, 15, 19). However, the situation may differ in other vascular beds. For example, in guinea-pig saphenous artery, it was reported that presynaptic purinergic P_2-receptors could play a physiological role in the autoregulation of neurotransmission (3). Miyahara and Suzuki (13) suggested that the receptor mechanisms for ATP (purinergic-receptor) differed between the pre- and post-junctional membranes, in that $\alpha\beta$-methylene-ATP excited only the post-junctional receptor mechanisms, yet antagonized the facilitating action of ATP on release of NE from the nerve terminals. However, some antihypertensive drugs modulate neurotransmission indirectly with no reduction in the amount of transmitter released (14). Therefore, further studies are necessary to clarify the mechanism of neurotransmission and its importance in vascular smooth muscle.

It has been reported that atriopeptrin-II (20 nM), a peptide related to atrial natriuretic factor, has an inhibitory effect both presynaptically at the nerve terminals and postsynaptically on vascular smooth muscle cells in the guinea-pig saphenous artery (12). However, in the guinea-pig mesenteric and renal arteries, α-HANP (α-human atrial natriuretic polypeptide, up to 10 nM) had no effect on the facilitation (mesenteric artery) or depression (renal artery) of EJP amplitude (2). It was suggested that the main mechanism of α-HANP action was (a) inhibition of Ca release from the cellular stores by reduction in the synthesis of inositol 1, 4, 5-triphosphate (IP_3) and (b) acceleration of Ca extrusion at the sarcolemma by increase in cyclic GMP level.

In rabbit saphenous artery, muscarinic agonists inhibited the EJPs and electrical responses. Komori and Suzuki (8) suggested that this inhibitory action of muscarinic agonists was mainly indirect, i.e., a release of inhibitory substances from the endothelial cells and the inhibition of adrenergic transmission, and that the latter might be more important for vasodilation related to the cholinergic mechanism.

Neuromuscular transmission in blood vessels is important in controlling vascular tone. Some vasoactive substances, such as some hormones and drugs, exert their effects by modulating transmitter release, as well as having direct effects on the vascular smooth muscle (VSM) cells. Our laboratory has recently begun to investigate the underlying mechanisms of neurotransmission, and we summarize some of our results in this article.

PRODUCTION OF EJPs, FACILITATION, AND INITIATION OF APs

Perivascular nerve stimulation with brief impulses evokes excitatory junction potentials (EJPs) in VSM cells. The amplitude of the EJP is thought to reflect the amount of transmitter released, and norepinephrine (NE) and ATP are believed to be the co-neurotransmitters released from the sympathetic nerves (4, 5).

The EJP rises quickly, reflecting the maximum synaptic current, and decays more slowly, as a passive exponential decay, reflecting the R_m C_m (time constant, τ_m) properties of the VSM cell membrane. The duration of the EJP at 50% amplitude (EJP_{50}) is about 500 msec.

Published 1991 by Elsevier Science Publishing Company, Inc.
Ion Channels of Vascular Smooth Muscle Cells and Endothelial Cells
Sperelakis and Kuriyama, Editors

4

The EJPs became successively larger in amplitude upon repetitive nerve stimulation, i.e., there is facilitation of neurotransmitter release. This phenomenon is illustrated in Figures 1 - 4. Facilitation of EJP amplitude does not occur in the renal artery of the guinea pig, but rather fatigue occurs, i.e., depression of EJP amplitude upon repetitive stimulation (2).

Repetitive nerve stimulation at a higher frequency produces temporal summation of the EJPs which can reach the threshold voltage for action potential generation and subsequent contraction of the vascular smooth muscle. This is illustrated in Figures 1 - 2.

Release of transmitter is controlled by the level of excitation at the nerve terminals. In guinea-pig mesenteric artery, EJPs were found to be dependent on external Na^+ and Ca^{2+} ions, and spikes were dependent on external Ca^{2+} ions (20) (Fig. 2). Lowering $[Na]_O$ decreased EJP amplitude, but had no significant effect on spike max dV/dt. In contrast, lowering $[Ca]_O$ produced a significant decrease in both EJP amplitude and spike max dV/dt. Raising $[Ca]_O$ increased both parameters. These results suggest that transmitter release and EJP generation require external Ca^{2+} and Na^+. Intracellular Ca^{2+} may also be important.

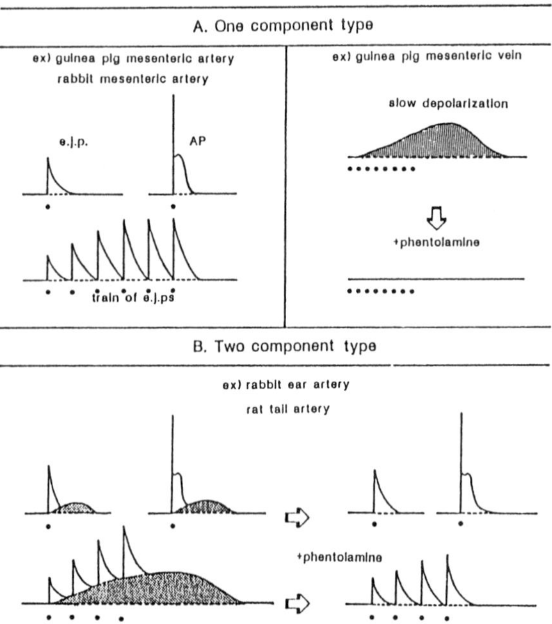

Figure 1. Diagrammatic illustrations of the electrical responses induced by nerve stimulation. **A:** Electrical responses of one-component type. **Left Column:** Fast EJP. Single electrical stimulation of subthreshold intensity induced fact EJP only and when the stimulation is strong enough, action potential is produced following EJP. Train of stimulations induced train of ejps and these EJP shows facilitation. **Right column:** Slow depolarization type. This depolarization was abolished by phentolamine (α-blocker), indicating that this response is mediated by NE released from nerve terminal. **B:** Two-component type. In some vascular smooth muscle (e.g., rabbit ear artery), fast EJPs and following slow depolarization can be seen. Only slow depolarization is abolished by phentolamine. It is well known that these fast EJP was produced by ATP, co-transmitter which is released with NE from nerve terminal.

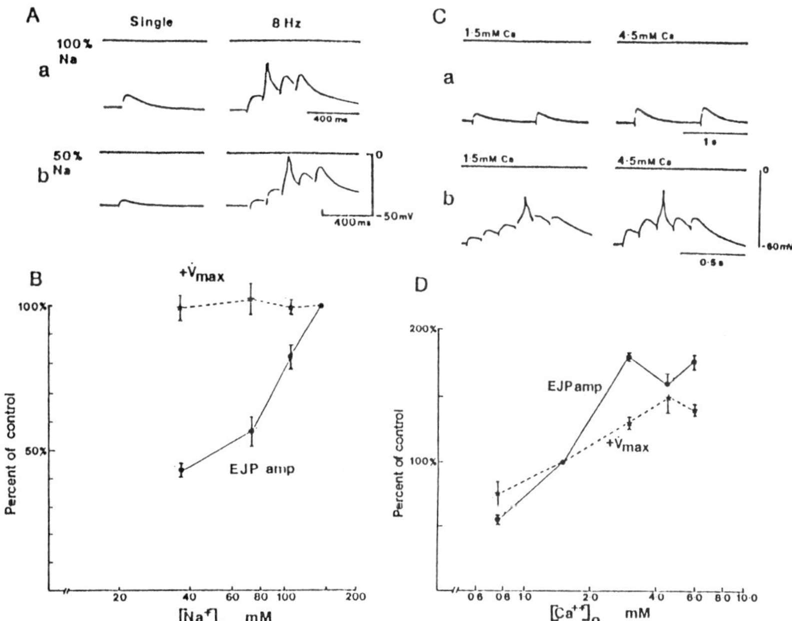

Figure 2. Effects of [Na]$_O$ and [Ca]$_O$ on EJPs and action potentials. **A**: The effects of decreasing external Na$^+$ (replacement with choline + 10^6 g/ml atropline) are shown using data from a single impalement. (a) Control responses to single and 8 Hz stimuli. (b) After 12 min in 50% [Na]$_O$. **B**: The relationship between the external Na concentration (log scale) and EJP amplitude (solid line) and spike max dV/dt (dashed line). **C**: The effects of changing external Ca^{2+} (a and b) from a single cell. **D**: Graph showing the relationship between the external Ca^{2+} concentration (log scale) and EJP amplitude (solid line) and spike max dV/dt (dashed line). Numbers in parentheses are numbers of cells impaled, and numbers of preparations. Each point represents the mean + S.E.M. for paired data. Data were normalized relative to the amplitude of EJPs in each cell, in the normal Na$^+$ or Ca^{2+} solution. Taken from 13.

It has been demonstrated that ω-conotoxin (ω-CTX) irreversibly blocked the EJPs in guinea pig vas deferens at 10-100 nM (1) and inhibited the EJPs in rat small mesenteric arteries at 0.1-3 nM (18). This effect was thought to be due to inhibition of excitation-secretion coupling, possibly mediated by the block of N-type Ca^{2+} channels.

MODULATION OF NEUROTRANSMITTER RELEASE

Various vasoactive substances produce their effects on vascular tissues, in part, by modulating neurotransmission. Adrenergic alpha-2, cholinergic, and purinergic stimulation, and several prostaglandins attenuate transmitter release, presumably by acting through receptors on the nerve terminal. Beta-adrenergic stimulation, in contrast, enhances transmitter release.

Adrenergic Feedback

Norepinephrine. In guinea pig mesenteric artery, norepinephrine decreases the amplitude of the EJPs (Fig. 3). This phenomenon was previously reported to be mediated via presynaptic α_2-adrenoceptors that reduce the amount of neurotransmitter release (9). The mechanism of this action is still unknown.

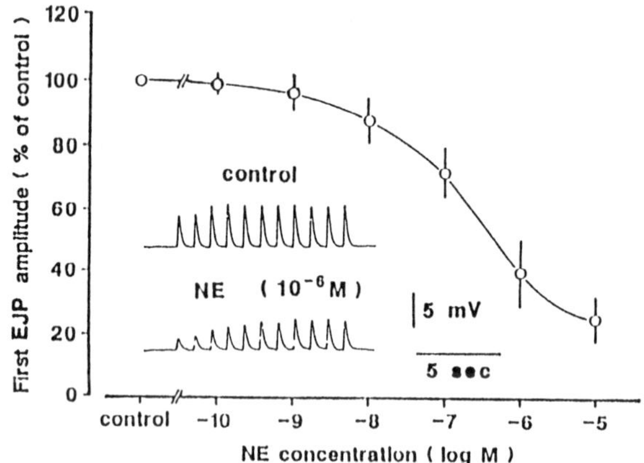

Figure 3. Effect of norepinephrine (NE) on EJPs in guinea-pig mesenteric artery. Concentration-response curves showing effects of various concentration of NE on EJP amplitude. Amplitude of 1st EJP in a train of stimuli was expressed as a fraction of that in control (NE free) EJPs were recorded 5-10 min after application of each concentration of NE. Vertical bars indicate means $\pm$ SE (n = 3-7). **Inset:** Actual traces of EJPs recorded before and after addition of NE (10^{-6} M). Repetitive nerve stimulation was $30\,\mu$sec in duration, 30 V in intensity at 1 Hz for a total of 11 pulses. Modified from 19.

Isoproterenol. We investigated the actions of isoproterenol. Isoproterenol acts on the β-adrenoceptors, and increases intracellular cyclic AMP level. Figure 4 shows that application of $0.1\,\mu$M isoproterenol increased the amplitudes of the first EJP and the mean (11 pulses) EJP. These parameters were increased to $147 \pm 12\%$ (n = 7) and $117 \pm 4\%$ (n = 7) of the control values, respectively (Table 1). The facilitatory action of isoproterenol on EJP amplitude was abolished by $0.3\,\mu$M propranolol (Fig. 4A). Isoproterenol had no effect on the resting potential of the VSM cell. To study the current/voltage relationships, electrotonic potentials were recorded in the presence of $0.3\,\mu$M TTX to block the nerve excitability. The control membrane resistance was 11.8 ± 0.3 MΩ (n = 14) (Fig. 4B). At concentrations up to $1\,\mu$M, isoproterenol had no effect on the current/voltage relationship in the VSM cells (Fig. 4B).

Histamine

We have found a novel type of histamine receptor (H_3) on the perivascular adrenergic nerve terminals of guinea-pig mesenteric artery (7) (Fig. 5). Histamine depresses EJP amplitude. Neither 2-methyl-histamine, an H_1 agonist, nor dimaprit, an H_2 agonist, decreased EJP amplitude, but N-α-methylhistamine, a non-selective histamine derivative and H_3 agonist, mimicked the histamine depression with ten-fold higher potency than histamine itself. These results suggest that an H_3 receptor exists at the adrenergic nerve terminals. This novel class of histamine receptor is pharmacologically distinct from H_1- and H_2-histaminergic receptors. We have found this H_3- receptor on nerve terminals to be much more sensitive to histamine than are H_2-receptors on the VSM membrane, so that they may play an important function in the physiological control of the circulation. Stimulation of H_3-receptors may produce vasodilation by inhibiting the sympathetic tone, i.e., they reduce the contractile influence of the sympathetic nerves on VSM.

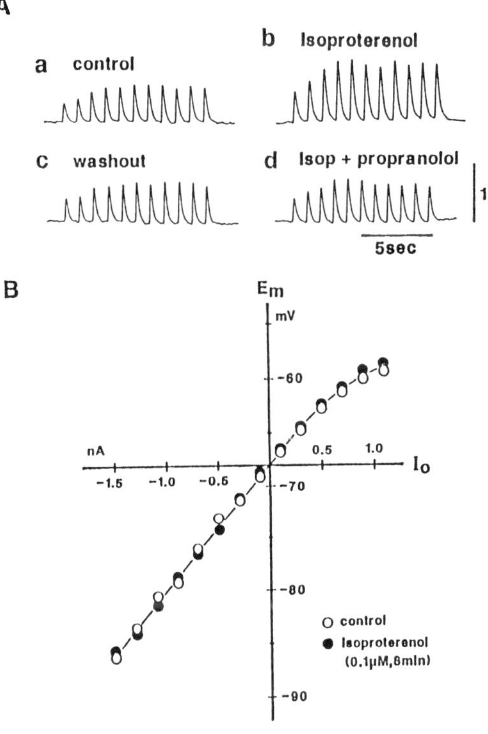

Figure 4. Effects of isoproterenol on EJP amplitude and input resistance of the muscle cells of guinea-pig mesenteric arteries. Repetitive nerve stimulation was 30 μs in duration, 30 V in intensity, at 1 Hz for a total of 11 pulses. **(A)** EJPs were recorded from the same cell before (a) and 8 min after application of 0.1 μM isoproterenol (b). EJPs were also recorded before (c) and 10 min after the combined application of isoproterenol (0.1 μM) and propranolol (0.5 μM) (d). Note that propranolol blocked the isoproterenol potentiation of EJP amplitude. **(B)** Steady-state current-voltage relationship obtained before (control: O) and 8 min after application of 0.1 μM isoproterenol (●). A single microelectrode was used for simultaneously passing current and recording membrane potential by means of a DC bridge circuit. Note that the input resistance of the muscle cell was not altered by isoproterenol. To avoid neuronal effects, 0.3 μM TTX was present throughout the experiment. Taken from 20.

Table 1. Summary of effects of cyclic nucleotides, agents that affect cyclic AMP level, and phorbol ester on EJP amplitude in guinea pig mesenteric artery

	Relative EJP amplitude (%)	
	First EJP	Mean EJP
Isoproterenol (0.1 μM)	147 ± 12 (7)	117 ± 4 (7)
Forskolin (30 μM)	274 ± 7 (4)	139 ± 3 (4)
8-Br-cAMP		
0.1 mM	289 ± 23 (4)	143 ± 12 (4)
1.0 mM	592 ± 20 (3)	183 ± 19 (3)
8-Br-cGMP (1 mM)	266 ± 27 (4)	155 ± 8 (4)
Phorbol ester		
30 nM	132 ± 2* (4)	99 ± 1 (4)
100 nM	194 ± 12* (4)	113 ± 1.5* (4)
300 nM	391 ± 12* (5)	178 ± 10* (5)

The values given are the mean ± SEM (number of observations). The exposure time to PMA was > 40-60 min.
*p < 0.05
Data taken from 18.

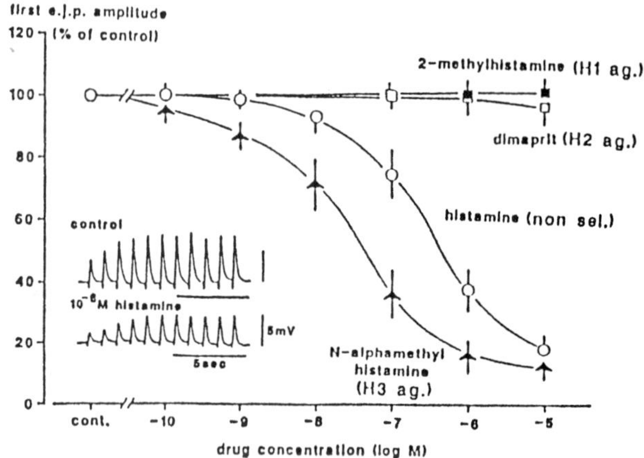

Figure 5. A: Concentration-response curves for the effects of four histamine receptor agonists on neurotransmission in the guinea-pig mesenteric artery. The amplitude of the first e.j.p. in a train of stimuli was expressed as a fraction of that in the control. The e.j.p.s were recorded 5-10 min after application of each agonist. The inset shows the inhibitory effect of histamine on the e.j.p.s. The histamine-induced depression of e.j.p. amplitude was not mimicked by 2-methylhistamine or dimaprit. A non-selective histamine derivative, N-α-methylhistamine, mimicked the histamine depression with tenfold higher potency than histamine istelf. Taken from (7). **B:** Cartoon model for the histamine action on vascular tissue. Norepinephrine (NE) contracts the VSM tissue by means of activation of postjunctional α_1-receptor. NE, ATP, and histamine are contained in high quantities in synaptic nerves and are released by neuronal activation. Released histamine as well as NE inhibits sympathetic tone by interacting with the prejunctional H_3 receptor (in the case of NE, alpha-2). Activation of the postjunctional H_1 receptor induces excitatory effects on VSM tissue (by releasing Ca^{2+} from the sarcoplasmic reticulum, SR, and/or depolarizing the postsynaptic membrane). On the contrary, that of H_2 receptor usually induces inhibitory effects on VSM tissues (by inhibiting Ca^{2+} release from the SR). On endothelium, H_1 receptor is present, and its activation releases the endothelium-dependent relaxing factor (EDRF) and then relaxes the VSM tissue. Under pathophysiological conditions including tissue injury, local release of histamine, mainly from tissue mast cells, may modify the above effects.

Cyclic Nucleotides

In several later studies, we focused on the second messenger systems involved in signal transduction at the adrenergic nerve terminals in guinea-pig mesenteric artery (6). Isoproterenol enhanced the EJP amplitude without modifying the passive membrane properties of the VSM cells. Forskolin (1-10 μM) markedly potentiated the isoproterenol-induced stimulation of EJP amplitude (Table 1). Furthermore, 8-bromo cAMP, a permeable analog of cAMP, enhanced EJP amplitude without changing the resting potential (Table 1; Fig. 6), thus mimicking the effects of isoproterenol and forskolin. 8-bromo cGMP also augmented the EJP amplitude (Table 1), but hyperpolarized the VSM cell membrane by approximately 4 mV (Fig. 6) and decreased the input resistance, presumably by increasing the K^+ conductance. In a second set of experiments, cyclic AMP, not only potentiated EJP amplitude, but also was able to reverse the inhibition of EJP amplitude produced by NE (Table 2).

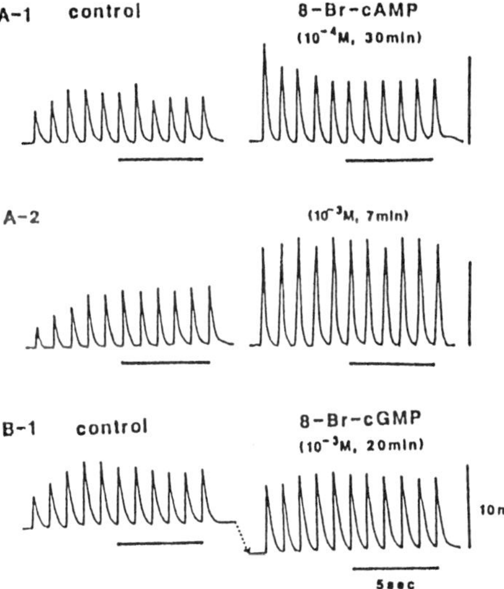

Figure 6. Enhancement of the e.j.p. amplitude by cyclic nucleotide analogs in guinea-pig mesenteric arteries. **A:** Control e.j.p.s were elicited by stimulation of the perivascular nerves at 1 Hz. After addition of 8-bromo cyclic AMP (8-Br-cAMP), the amplitude of the e.j.p.s was increased (A-1: 10^{-4} M, 30 min; A-2: 10^{-3} M, 7 min). **B-1:** Addition of 8-Br-cGMP (10^{-3} M, 20 min) hyperpolarized the membrane and increased the e.j.p. amplitude. Both traces in each row were obtained from the same cell. Taken from 6.

Table 2. Summary of effects of NE and 8-bromo cAMP on EJP amplitude in guinea pig mesenteric artery

	Relative EJP Amplitude (%)	
	First EJP	Mean EJP
NE (1 μM)	53 ± 10 (4)	64 ± 11 (4)
8-bromo cAMP (1 mM)	178 ± 20 (5)	157 ± 23 (5)
8-bromo cAMP (1 mM) + NE (1 μM)	77 ± 12 (4)*	93 ± 14 (4)*
8-bromo cAMP (10 mM)	205 ± 18 (5)	186 ± 21 (5)
8-bromo cAMP (10 mM) +NE (1 μM)	181 ± 21 (4)[ns]	171 ± 19 (4)[ns]

NE: norepinephrine
Mean EJP amplitude is the average of 11 EJPs.
The mean values given are the mean ± SEM (number of observations).
The values are precentages of the control EJP amplitudes.
*p < 0.01 compared to the single drug application immediately above.
[ns] p > 0.05 compared to the single drug application immediately above.
Data taken from Nozaki and Sperelakis, unpublished observations.

Protein Kinase-C

To investigate the possible involvement of protein kinase-C (PK-C) on the nerve terminal, the effects of phorbol esters were examined. Phorbol esters are direct activators of PK-C. Phorbol-12-myristate-13-acetate (PMA), one of the phorbol esters, was applied in normal Krebs solution. After an incubation time of about 60 min, PMA (30, 100, and 300 nM) consistently enhanced the EJP amplitude evoked by nerve stimulation in a concentration-dependent manner (Fig. 7). These facilitatory effects of PMA on EJP amplitude were significant for the first EJP as well as for the averaged EJP amplitude (Table 1).

PMA had no effect on the resting potential of the VSM cell. For example, in 300 nM PMA, the resting potential of the VSM cell was -75.9 ± 0.2 mV (n = 19). The 300 nM PMA solution contained 0.01% ethanol; this concentration of ethanol had no significant effect on the EJP amplitude or resting potential of the VSM cell.

From these results with the cyclic nucleotides and phorbol ester (Table 1), we suggested that three different protein kinase-dependent phosphorylation systems (cAMP/PK-A, cGMP/PK-G, and DAG/PK-C) could be involved in regulating neurotransmitter release from the perivascular nerve terminals. However, the possibility of some action on the postsynaptic VSM cell also exists.

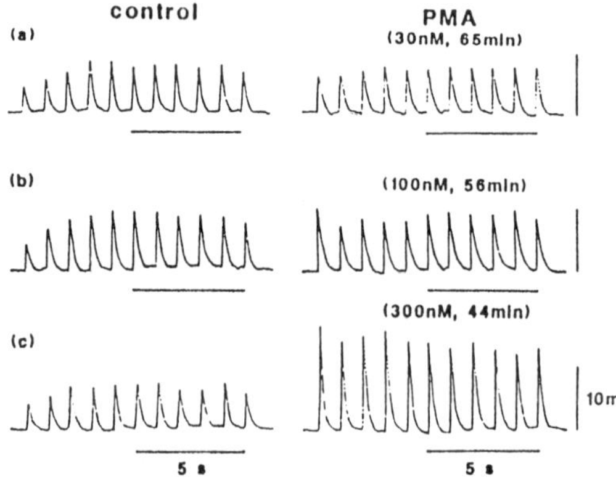

Figure 7. Effects of phorbol-12-myristate-13-acetate (PMA) on the EJPs in guinea pig mesenteric artery. Perivascular nerves were stimulated by a train of rectangular current pulses (25 μs duration, 30 V intensity, 1 Hz frequency, and total of 11 pulses). Each record was obtained from a different cell. EJPs were recorded before (control) and after application of PMA: (a) 30 nM, 65 min; (b) 100 nM, 56 min; and (c) 300 nM, 44 min. Note the marked potentiation of the EJP amplitude (first, and total of the 11) produced by the protein kinase-C activator.

Pertussis Toxin and Cholera Toxin

Pertussis Toxin. We have investigated the possible involvement of GTP-binding protein (G-protein) in neuromuscular transmission at the adrenergic nerve terminal of guinea-pig mesenteric artery (19, 20). Pretreatment of blood vessels with pertussis toxin (PT), a bacterial exotoxin that catalyzes ADP-ribosylation of G-proteins, was used to abolish the effect of G-proteins (G_i or G_i-like protein) at the perivascular nerve terminals (19).

As shown in Figure 8, EJP amplitude was suppressed by histamine in a dose-dependent manner. On the other hand, in VSM preincubated with PT, the inhibition of EJP amplitude by histamine was greatly attenuated. These effects of PT pretreatment were time- and temperature-dependent. Similar results were obtained with NE (not illustrated). In contrast, PT pretreatment had no obvious effect on the postsynaptic membrane: 1) resting potentials, 2) depolarization by NE and ATP, and 3) input resistance. These results suggest that PT-pretreatment has mainly a presynaptic effect. Therefore, the NE and histamine effects on EJP amplitude may be mediated by PT-sensitive G-proteins in the presynaptic nerve terminals of guinea-pig mesenteric artery.

Cholera Toxin. We have also investigated effect of cholera toxin (CTX) on the electrical properties of VSM cells of guinea-pig mesenteric artery (20). CTX is a bacterial exotoxin which ribosylates subunits of the stimulatory G-protein (G_s). EJP amplitude was markedly enhanced by treating isolated blood vessels with CTX (10 μg/ml for 1 hr). The VSM cells became hyperpolarized (control: -68 $\pm$ 2.8 mV, n = 16; CTX: -74.6 $\pm$ 2.1 mV, n = 5), and their input resistances were significantly reduced (control: 12 $\pm$ 0.5 MΩ, n = 6; CTX: 8.2 $\pm$ 0.5 MΩ, n = 6) (Table 3). These effects persisted after washout for 35 min. Addition of G_{M1} ganglioside (5 μg/ml) abolished the CTX effects, indicating that CTX was acting via its normal binding site in the cell membrane, because the ganglioside should inactivate the CTX. The enhancement of EJP amplitude by CTX could not be abolished by injecting currents to depolarize the VSM membrane by 5 mV (to counter the hyperpolarization). CTX also abolished the effects of isoproterenol and 8-bromo-cAMP on the EJPs (Fig. 9 & 10, Table 4). These results suggest that CTX may be exerting its effects by elevating cAMP levels in the nerve terminals and in the VSM cells.

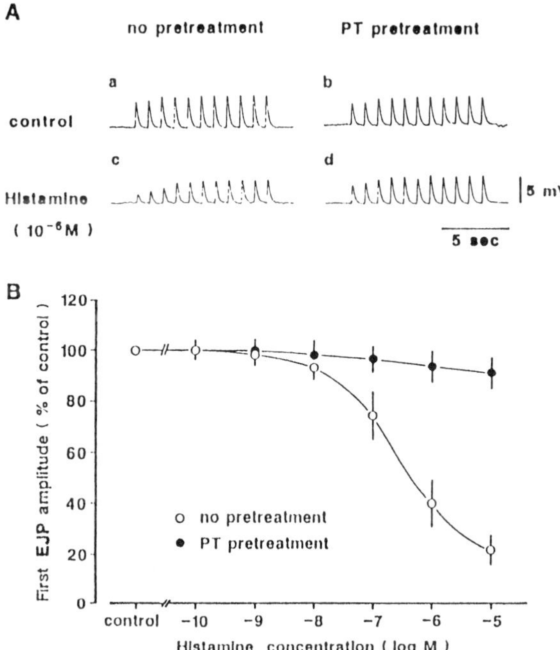

Figure 8. Effect of histamine on excitatory junction potentials (EJPs) in guinea pig mesenteric artery preincubated with pertussis toxin (PT) or without PT. **A:** Actual traces of EJPs recorded before and after addition of histamine (10^{-6} M) to tissues with PT pretreatment (b and d) or without pretreatment (a and c). **B:** Concentration-response curves showing effects of various concentrations of histamine on EJP amplitude. Amplitude of 1st EJP in a train of stimuli was expressed as a fraction of that in control (histamine free). EJPs were recorded 5-10 min after application of each concentration of histamine. Note that PT prevented inhibitory effect of histamine on EJP amplitude. Vertical bars indicate means + SE (n = 3-7).

Table 3. Summary of effects of CTX, G_{M1} ganglioside, isoproterenol and 8-bromo-cAMP on membrane potential and input resistance in guinea-pig mesenteric artery

	Membrane potential (-mV ± SEM)	Input resistance (MΩ ± SEM)
Control	68.7 ± 2.8 (16)	12.1 ± 0.5 (6)
CTX (10 μg/ml)	74.6 ± 2.1 (5)*	8.2 ± 0.5 (4)[a]
CTX (10 μg/ml)+ G_{M1} ganglioside (5 μg/ml)	69.5 ± 3.11 (4)	11.8 ± 0.7 (4)
Isoproterenol (0.1 μM)	70.1 ± 3.5 (6)	12.0 ± 0.9 (4)
8-Bromo-cAMP (10 mM)	69.9 ± 2.8 (4)	11.5 ± 1.2 (3)

*p < 0.01 compared to control.
CTX and G_{M1} ganglioside were applied for 50-65 min.
Isoproterenol and 8-bromo-cAMP were applied for 8-15 min.
TTX (0.3 μM) was present in the bath to inhibit nerve activity.
The values given are the means ± SEM (number of observations).
Data taken from 20.

Table 4. Summary of effects of CTX, isoproterenol and 8-bromo-cAMP on EJP amplitude in guinea-pig mesenteric artery

	Relative EJP amplitude (%)	
	First EJP	Mean EJP
(A) Isoproterenol	195 ± 17 (4)	187 ± 20 (4)
Isoproterenol + CTX	425 ± 25 (4)*	241 ± 23 (4)
CTX	398 ± 42 (3)	212 ± 10 (3)
CTX + Isoproterenol	375 ± 36 (3)	237 ± 19 (3)
(B) 8-bromo-cAMP	225 ± 12 (5)	177 ± 22 (5)
8-Bromo-cAMP + CTX	375 ± 26 (4)	267 ± 31 (4)*
CTX	365 ± 28 (3)	237 ± 20 (3)
CTX + 8-bromo-cAMP	375 ± 26 (3)	267 ± 31 (3)

*$p < 0.01$ compared to the single drug application immediately above.
The drug concentrations were as follows: CTX (10 μg/ml), isoproterenol (0.1 μM), 8-bromo-cAMP (10 mM).
Mean EJP amplitude is the average of 11 EJPs.
The values given are the mean ± SEM (number of observations). The values are percentages of the control EJP amplitudes.
Data taken from 20.

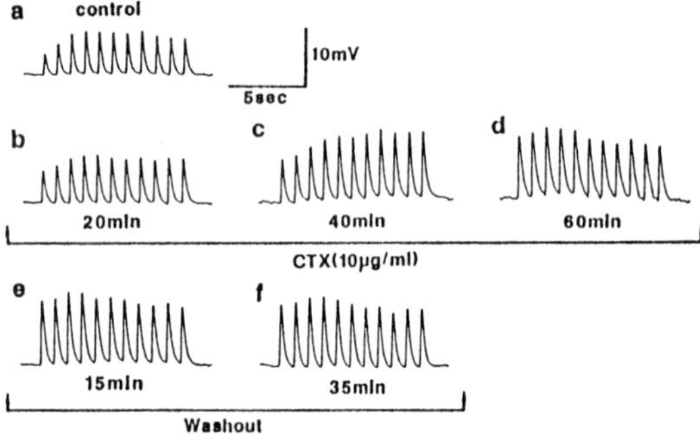

Figure 9. Effect of cholera toxin (CTX) on EJP amplitude recorded from one muscle cell of guinea-pig mesenteric artery. Repetitive nerve stimulation was 30 μs in duration, 30 V in intensity, at 1 Hz for a total of 11 pulses. (a) Control. (b-d) CTX (10 μg/ml) was applied for 1 h. EJP amplitude was potentiated by CTX application. The effects of the CTX appeared in about 40 min (c). (e, f) Washout of the CTX. Note that the potentiation of EJP amplitude persisted after CTX washout.

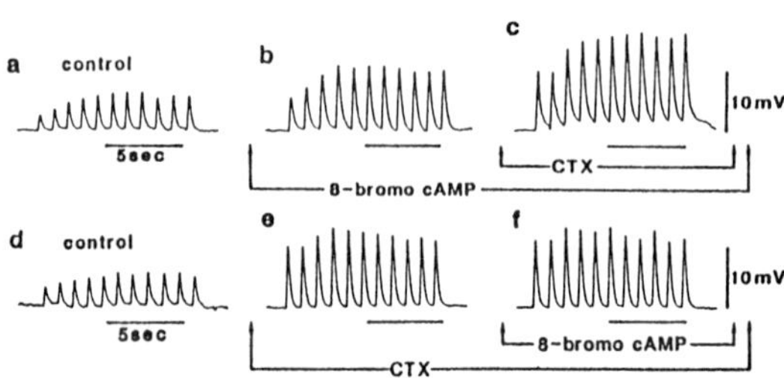

Figure 10. Application of CTX, and 8-bromo-cAMP to guinea-pig mesenteric artery in various combinations. Perivascular nerve stimulation (30 ls in duration, 25 B in intensity) was used. Drug concentrations were as follows: CTX (10 μg/ml), 8-bromo-cAMP (10 mM). EJPs were recorded before (a), 12 min after application of 8-bromo-cAMP (b), and 50 min after the combined addition of 8-bromo-cAMP and CTX (c). d-f: EJPs were recorded before (d), 55 min after application of CTX (e), and 10 min after the combined application of CTX and 8-bromo-cAMP (f).

Endothelin

A preliminary study was done to test whether endothelin had any effect on EJP amplitude in the guinea-pig mesenteric artery preparation. These data are presented in Table 5 (Nozaki & Sperelakis, unpublished observations). As shown, endothelin produced a small depolarization of the VSM cells, from the control value of -69.5 ± 2.3 mV, which was statistically significant at 1×10^{-6} M (to -61.5 ± 4.0 mV) and 5×10^{-6} M (to -60.1 ± 3.7 mV). Despite this depolarization (which would tend to reduce EJP amplitude), endothelin actually increased EJP amplitude, which was statistically significant at 1×10^{-6} M and 5×10^{-6} M (Table 5). For example, mean EJP amplitude was increased by 55% and 49% at 1 μM and 5 μM, respectively. These results suggest that there are functional endothelin receptors, not only on the VSM cells (where depolarization is produced), but also on the adrenergic nerve terminals (where neurotransmitter release is potentiated). An increase in neurotransmitter release is consistent with the N-type Ca channels in the nerve terminals being stimulated by endothelin or by endothelin stimulation of IP_3 production and Ca^{2+} release, resulting in elevated $[Ca]_i$.

Table 5. Effect of endothelin on resting potential and EJP amplitude in guinea pig mesenteric artery

	Resting Potential (mV)	Relative EJP Amplitude (%)	
		First EJP	Mean EJP
Control	-69.5 ± 2.3 (8/6)	100	100
Endothelin			
10^{-9} M	-69.8 ± 3.1 (4/4)	121 ± 23 (3/3)	99
10^{-8} M	-67.9 ± 2.9 (5/4)	118 ± 20 (3/2)	135
10^{-7} M	-64.3 ± 3.5 (5/4)	109 ± 28 (4/3)	121
10^{-6} M	-61.5 ± 4.0 (6/4)	111 ± 19 (3/3)	155*
5×10^{-6} M	-60.1 ± 3.7 (3/3)	151 ± 30 (3/3)	149*

Values in parentheses give: number of cells impaled/number of preps.
* = $p < 0.01$.
Data taken from Nozaki and Sperelakis, unpublished observations.

Table 6. Summary of effects of various receptors and second messengers on EJP amplitude

Receptors	EJP Amplitude
α_2 (NE)	decrease
β_2 (Isop)	increase
M_1 (ACh)	decrease
H_3 (hist)	decrease
Second Messengers	
cAMP (PK-A)	increase
cGMP (PK-G)	increase
PMA (PK-C)	increase

EJP amplitude $\propto$ transmitter release

SUMMARY AND CONCLUSIONS

The results of our studies are summarized in Tables 1 - 6, and our present model for modulation of neurotransmitter release is depicted in Figure 10. EJP amplitude (initial one and mean of 11 EJPs) was used as the index of neurotransmitter released, because most of the agents tested had no significant effect on the electrical properties of the postsynaptic VSM cells, including resting potential and input resistance. The co-transmitters released at the nerve terminals in guinea pig mesenteric artery are ATP and NE. ATP is responsible for the fast component of the EJP and NE is responsible for the slow component (Fig. 1). There is negative feedback regulation of the neurotransmitter release, via receptors on the presynaptic membrane. NE acts on an α_2 receptor and G_i coupling protein (evidenced by the pertussis toxin data) to inhibit neurotransmitter release, as depicted in Figure 11. ATP acts on a P_2 receptor to inhibit release. Histamine acts on an H_3 receptor and G_i coupling protein to inhibit release.

Modulation of neurotransmitter release at adrenegic nerve terminal in guinea-pig mesenteric artery

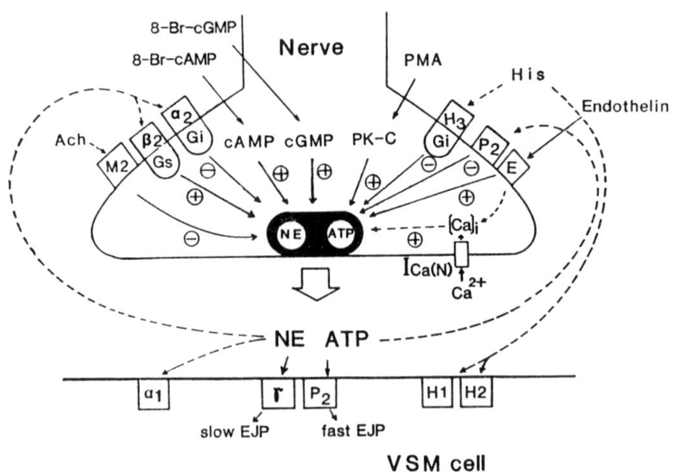

β-adrenoceptor agonists, such as NE and isoproterenol, action the β-adrenoceptors and G_s coupling protein (evidenced by the cholera toxin data) to stimulate transmitter release. This effect is presumably mediated by cAMP elevation. Addition of 8-Br-cAMP and 8-Br-cGMP produced marked stimulation of EJP amplitude due to stimulated transmitter release; 8-Br-cGMP also produced a small hyperpolarization of the postsynaptic VSM cell. Forskolin, a direct activator of adenylate cyclase, also potentiated transmitter release (data not shown). Activation of protein kinase-C by phorbol esters also greatly potentiated neurotransmitter release, and therefore amplitude of the EJPs. Therefore, phosphorylation of some proteins, perhaps in or near the presynaptic membrane, by PK-C, PK-A, and PK-G, potentiate transmitter release.

Endothelin, despite producing some depolarization of the postsynaptic VSM cell, enhanced EJP amplitude and therefore neurotransmitter release. This effect of endothelin may be mediated by a direct or indirect action on the N-type Ca^{2+} channels in the presynaptic membrane to stimulate Ca^{2+} influx and thereby transmitter release. If endothelin produced Ca^{2+} release from the SR, via IP_3 production, this also could account for the enhanced transmitter release.

In conclusion, there is regulation of neurotransmitter release at the adrenergic nerve terminals in vascular smooth muscle, via specific receptors for agonists on the presynaptic membrane. Such regulation includes negative feedback inhibition by the neurotransmitters themselves, including NE (via α_2 receptor), ATP (via P_2 receptor), and histamine (via H_3 receptor). These receptors are functionally connected to G_i coupling proteins for exerting their effects. A positive feedback component occurs with NE activation of β_2 receptors presynaptically, which stimulate cAMP production via a G_s coupling protein. Any agonist that stimulates phospholipase-C (PL-C) and thereby IP_3 and DAG production, would activate PK-C and thereby potentiate transmitter release. Endothelin also potentiates transmitter release, perhaps by IP_3 production and Ca^{2+} release or by directly acting on the voltage-dependent Ca^{2+} channels in the presynaptic membrane.

ACKNOWLEDGEMENTS

This research was supported in part by NIH grant HL-40572. We would like to thank Rhonda S. Hentz for typing this manuscript.

REFERENCES

1. Brock, J.A., T.C. Cunnane, R.J. Evans, and J. Ziogas, Clin. Exp. Pharmacol. and Physiol. 16, 333-339 (1989).
2. Fujii, K., T. Ishimatsu, and H. Kuriyama, J. Physiol. 377, 315-332 (1986).
3. Fujioka and Cheng, Eur. J. Pharmacol. 139, 147-153 (1987).
4. Hirst, G.D.S. and T.D. Neild, J. Physiol. 280, 87-104 (1978).
5. Hirst, G.D.S. and T.D. Neild, J. Physiol. 313, 343-350 (1983).
6. Ishikawa, S. and N. Sperelakis, J. Cardiovasc. Pharmacol. 13, 836-845 (1989).
7. Ishikawa, S. and N. Sperelakis, Nature 327, 158-160 (1987).
8. Komori, K. and H. Suzuki, Circ. Res. 61, 586-593 (1987).
9. Kuriyama, H. and Y. Makita, J. Physiol. 335, 609-627 (1983).
10. Kuriyama, H., Y. Ito, H. Suzuki, K. Kitamura, and T. Ito, Am. J. Physiol. 243, H641-H662 (1982).
11. Langer, S.Z., Pharmacol. Rev. 32, 337-362 (1981).
12. MacKay, M.J. and D.W. Cheung, Can. J. Physiol. Pharmacol. 65, 1988-1990 (1987).
13. Miyahara, H. and H. Suzuki, J. Physiol. 389, 423-440 (1987).
14. Nakarhima, M., Y. Li, N. Seki, and H. Kuriyama, Br. J. Pharmacol. 101, 581-586 (1990).
15. Neild, T.O. and E. Zelcer, Prog. Neurobiol. 19, 141-158 (1982).
16. Nozaki, M. and N. Sperelakis, Am. J. Physiol. 256, H455-H459 (1989).
17. Nozaki, M. and N. Sperelakis, Eur. J. Pharmacol. 197, 57-62 (1991).
18. Pruneau, D. and J.A. Angus, Brit. J. Pharmacol. 100, 180-184 (1990).
19. Vanhoutte, P.M., T.J. Verbeuran, and R.C. Webb, Physiol. Rev. 61, 151-247 (1981).
20. Zelcer, E. and N. Sperelakis, Blood Vessels 19, 301-310 (1982).

ELECTROPHYSIOLOGICAL PROPERTIES OF THE VOLTAGE-DEPENDENT CALCIUM CHANNEL IN VASCULAR SMOOTH MUSCLE CELLS

N. TERAMOTO, M. KAMOUCHI, Z. XIONG, M. OIKE, S. KAJIOKA, K. KITAMURA, and H. KURIYAMA

Department of Pharmacology, Faculty of Medicine, Kyushu University, Fukuoka 812, Japan

INTRODUCTION

It is well known that in various vascular smooth muscle cells cytosolic Ca plays an essential role in the regulation of the contraction-relaxation cycle. The cytosolic Ca is itself regulated by many factors such as influxes of Ca through the membrane, release of Ca from the sarcoplasmic reticulum (SR), extrusion of Ca from the sarcolemma and re-uptake into the SR. In addition, the mitochondria and nucleus also make a contribution to Ca homeostasis in the cytosol. Furthermore, factors from endothelial and neuronal cells also directly or indirectly regulate Ca homeostasis in vascular smooth muscle.

Measurements of the concentration of cytosolic Ca made using the fluorescent dye and Ca-electrode methods suggest that in the resting (relaxing) condition, the amount of free Ca in the cytosol is around 100-150 nM and that this concentration rises to about 1 μM during the maximum activation of smooth muscle cells (24, 27, 42, 105, 113, 127, 130). These values agree well with those estimated from the amplitude of contraction using the skinned muscle procedure (26, 53, 55, 63, 64, 99, 100). Both influxes of Ca through the sarcolemma and release of Ca from the SR require activation of their own Ca channels in the relevant membranes.

The Ca channels responsible for allowing Ca influx and the release of Ca from the SR have each been classified into various subtypes. Influx of Ca occurrs through the voltage-dependent Ca channel, receptor-operated nonselective cation channel (14, 126) and by reversed Na-Ca exchange diffusion (13). The voltage-dependent Ca channel has been further subdivided into an L-subtype (long lasting, high threshold with large conductance and dihydropyridine (DHP)-sensitive) and a T-subtype (transient, low threshold with smaller conductance and DHP-insensitive) (8, 21, 58, 59, 68, 92, 128, 132). The channel responsible for the release of Ca from the SR has been subdivided into a ryanodine (caffeine)-sensitive channel (Ca-induced Ca release-mechanisms-related channel) and a heparin-sensitive channel (inositol-1,4,5-trisphosphate (IP$_3$) sensitive Ca channel) (1, 11, 12, 37, 51, 54, 60, 62, 69, 71, 79, 86, 101, 104, 108, 109, 112, 114, 131,).

As a result of great advances in molecular biology, the amino acid sequences of three types of Ca channel, namely the DHP-sensitive L-subtype and the ryanodine-sensitive and IP$_3$-sensitive channels have been clarified from skeletal muscle and nerve cells (31, 118, 119). It has been clearly shown that in the triad of skeletal muscle, the DHP receptor acts not only as the Ca channel constituents, but also as a voltage sensor which has a key role for voltage-dependent Ca release from the SR (119). In addition, the IP$_3$ receptor for the Ca channel had the same structure as that deduced for P400 (31). However, the features and identity of these receptors in vascular smooth muscle have not yet been clarified.

In this short article, we review the characteristics of the voltage-dependent Ca channels (L- and T- subtypes) distributed on vascular smooth muscle cell membranes and compare them with those distributed on cardiac muscle membranes.

ACTION POTENTIALS IN VASCULAR SMOOTH MUSCLE CELLS

Before we discuss the nature of the macroscopic or unitary currents recorded using whole cell voltage-clamp or patch-clamp procedures, it is worth dwelling briefly on the electrical events recorded using the microelectrode method for a better understanding of the electrical properties of vascular smooth muscle cells. Of vascular smooth muscle cells, only those of the portal vein in many experimental animals produce spontaneous action potentials. The resting membrane potential of the smooth muscle cells of the portal vein was -50 to -60 mV (23). Treatment with Mn, Cd or Ca-free solutions containing EGTA blocked spike generation. For example, action potentials evoked by various procedures in smooth muscle

Published 1991 by Elsevier Science Publishing Company, Inc.
Ion Channels of Vascular Smooth Muscle Cells and Endothelial Cells
Sperelakis and Kuriyama, Editors

cells of the portal vein were blocked by Ca antagonists such as verapamil, diltiazem and dihydropyridine derivatives. At low membrane potentials (more polarized level), smooth muscle cells of the portal vein produced burst discharges between silent periods, while when the membrane potential was high (less polarized level), train discharges occurred. The rate of rise of the action potential was about 10-50 V/sec and the half-spike duration (APD_{50}) was 10-40 msec (23, 73, 74).

In most resistance vessel tissues (mesenteric, tail and ear arteries), the membrane potential was -60 to -70 mV and action potentials occurred on the excitatory junction potential (e.j.p.) evoked by perivascular nerve stimulation (22, 43-48, 75, 76, 115). In many elastic arteries (aorta, pulmonary, carotid and coronary arteries), no action potential was evoked by either perivascular nerve or direct muscle stimulation. However, following pretreatment with K channel blockers, such as tetraethylammonium (TEA) or 4-aminopyridine (4-AP), direct muscle stimulation provoked a graded response or the action potential (36, 39, 74). Therefore, most vascular smooth muscle cells possess voltage-dependent Ca channels. To date, no distribution of the voltage-dependent Ca channel has been reported on the sphincter muscles of the iris, i.e. in high K solution, this muscle tissue produced a contraction which was blocked by atropine via inhibition of the muscarinic receptor activated by acetylcholine (ACh) released from cholinergic nerves. This contraction was not prevented by application of Ca antagonists and, following application of atropine, high concentrations of KCl did not produce contraction (116, 133). In iris smooth muscle, the cytosolic Ca increased due to release of Ca from the SR induced by IP_3 (1).

The e.j.p. generated by perivascular nerve stimulation in resistance vessels is due to activation by released norepinephrine (NE) of the γ-adrenoceptor or by its co-transmitter adenosine triphosphate (ATP) of the purinergic II receptor of post-junctional smooth muscle cells (18, 19, 43-46). In the guinea-pig mesenteric artery, low frequency, repetitive perivascular nerve stimulation (0.5-3 Hz) evoked e.j.ps with no change in the resting membrane potential level, while in the dog mesenteric, rabbit ear and rat tail arteries, such stimulation produced e.j.ps and slow depolarization. The latter, but not the former, was blocked by α-adrenoceptor blockers (either prazosin, an α_1 blocker, or by yohimbine, an α_2 blocker and also by phentolamine, α_1 and α_2 blocker). When NE was applied exogenously, there was an evoked slow depolarization together with spikes and large oscillations of the membrane potential. These potential changes ceased following application of α-adrenoceptor blockers (22, 65, 117). The e.j.p. was evoked mainly by influxes of Na and to some extent Ca. A minute amount of Ca influx may increase the cytosolic Ca and activate some steps of the receptor-GTP-binding protein-phosphatidyl inositol bisphosphate ($PI-P_2$) hydrolysis formation and as a consequence, the cytosolic Ca would be increased by Ca released from the SR by IP_3 (37). ATP, by iontophoretic application, produced a depolarization similar to the e.j.p. and this e.j.p.-like potential was blocked by α-β-methylene ATP, a desensitizer of the purinergic II receptor, as also observed for e.j.p.s (61, 107, 117).

PROPERTIES OF THE VOLTAGE-DEPENDENT CALCIUM CHANNEL IN VASCULAR SMOOTH MUSCLE CELLS

After Hamill et al (35) applied the patch-clamp technique to biological membranes to investigate ionic currents, this procedure was successfully applied to dispersed visceral and vascular smooth muscle cells (7). Using the whole-cell voltage-clamp procedure, isolated Ca (Ba) current was recorded (TEA, Cs with EGTA in the pipette and Na-free TEA with Ca-containing solution in the bath). In many vascular tissues, when depolarizing pulses of various intensities were applied from a holding potential of -60 mV in the above ionic conditions, a current-voltage (I-V) relationship which was a U-shaped curve with the peak value at about 10 mV was obtained. When the Ca in the bath was replaced by Ba, the amplitude of the inward current was markedly enhanced, in a concentration-dependent manner. However, differences between the Ca- and Ba-currents were seen in the inactivation process of the inward-current, i.e. at a holding potential of -60 mV, applied depolarization pulses produced an inward-current with a gradual reduction in amplitude. This decay of the inward-current was much slower in the Ba solution than in the Ca solution. Ca antagonists, as pharmacological tools, or solutions containing only Ca (or Ba) have been used to investigate the macroscopic current measured using the whole-cell voltage-clamp procedure. It is generally accepted that Ca antagonists act on the L-subtype but not on the T-subtype of Ca channel.

Ca antagonists are classified according to their chemical structures. Thus, there are DHPs (nifedipine, nicardipine, nisoldipine, nitrendipine and others), phenylalkyramine (papaverine) derivatives (gallopamil (D600), verapamil and others), benzothiazepine derivatives (Diltiazem and TA-3090), piperazine derivatives (cinarizine and flunarizine) and others (bepridil and others) (28, 34). Following application of Ca antagonists, the inward-current, measured at its peak was consistently inhibited, though the inhibition occurred in a voltage-dependent manner (121-123, 132).

When the effects of four different Ca antagonists (nifedipine or nicardipine, diltiazem, verapamil and flunarizine) were compared on dispersed ileal smooth muscle cells using macroscopic current measurements, a number of effects were observed. Thus, i) The K_d values (IC_{50} value) obtained following applications of nicardipine, diltiazem, verapamil and flunarizine were 24 nM, 1.4 mM, 1.3 mM and 1.4 mM, respectively. This means that DHPs are 100 times more potent than the other antagonists. ii) Nicardipine and flunarizine shifted the steady-state inactivation curve in the hyperpolarizing direction (by more than 10 mV), when the curve was obtained under the following conditions (conditioning pulses of 3 sec duration and various intensities and test pulses of 200 msec duration and 0 mV at holding potentials of -60 mV or -80 mV). iii) When use-dependency was compared by applying various frequencies of stimulation (0.1-1 Hz) in solutions of one the above four agents (200 msec pulse duration, 0 mV intensity, holding potential -60 mV), verapamil and gallopamil showed the strongest use-dependency. iv) An additonal finding was that flunarizine blocked activation of both L- and T-channels (2; and in nerve cells, 3). These Ca antagonists also act on the K channel , and the K_d values for the K-outward current observed following application of nicardipine, diltiazem, verapamil and flunarizine varied only between 5 and 30 μM. Therefore, the ratio between the IC_{50} values for I_K and I_{Ca} was 190 for nicardipine, 21 for diltiazem, 11 for verapamil and 4.1 for flunarizine. This means that nicardipine acts as a more selective Ca channel blocker ,whereas diltiazem, flunarizine and verapamil may inhibit both voltage dependent Ca-channels as welll as K-channels (121-123).

The inward current recorded in the presence of a K channel blocker, had as its main source Ca except in some vascular tissues (pulmonary artery and portal vein), where this current involved Ca and Na. The Na-current occurred with depolarization pulses applied at a holding potential of -80 mV but not at -60 mV. This current had an early onset, was short lasting, and was blocked by tetrodotoxin (K_m value of nM order), but enhanced by chloramine-T (94). In addition, in the myometrium and cultured vascular cells, a tetrodotoxin-less sensitive Na current was obtained (4, 111).

UNITARY CALCIUM CURRENT MEASURED USING PATCH-CLAMP PROCEDURES

More precise classifications of the voltage-dependent Ca channel can be made using the cell-attached or cell-free patch-clamp procedures . Using nerve cells (chick dorsal root ganglia), three different Ca channels have been described, i.e. T-type (transient type; I_{low}), N-type (neither T- nor L-type; I_{middle}) and L-type (long-lasting type; I_{high}). This classification is based on the following findings: (i) The maximum activating potential for T-, N- and L-subtypes was -50 mV, -10mV and -10mV, respectively. (ii) The inactivation potential level for the three subtypes was -100 to -60mV, -100 to -40mV and -60 to -10mV, respectively; (iii) The time constant of inactivation of the three subtypes was 20 - 50 msec for T- and N-types and 700 msec for L-type. (iv) The unitary channel conductance of the three subtypes was 8 - 10pS, 13pS and 25pS, respectively. (v) The K_d values (IC_{50}) for Cd was 7 μM for T- and N-types and 160 μM for L-type, and those for Ni were 280 μM for T-type and 47 μM for N-and L-types; (vi) sensitivities to ω-conotoxin also differed between subtypes; N- and L-subtypes were sensitive to this agent (nerve and cardiac but not smooth muscle) but the T-subtype was not. (vii) L-type but not N-and T-types, was blocked by DHPs (49, 77, 83, 124).

In vascular smooth muscle and in visceral smooth muscle, distributions of the voltage-dependent Ca channel were found to be of only one of these subtypes (L) or of two subtypes (L and T). However, from the conductance of the unitary current and from DHP sensitivity, a different classification into three subtypes could be made, i.e. into 20-30pS ($I_{Ca}L$), 12-15pS ($I_{Ca}M$) and 8PS ($I_{Ca}S$) subtypes in smooth muscle cell membranes as measured using high concentrations of Ba. The guinea-pig aorta (21) contained only the $I_{Ca}M$, while the guinea-pig portal vein (Inoue et al., 1990) possessed the $I_{Ca}L$ and $I_{Ca}M$. Other tissues (rabbit ear artery, 10; dog saphenous vein, 132), possessed the $I_{Ca}M$ and $I_{Ca}S$, while the rabbit basilar artery (92) possessed the $I_{Ca}L$ and $I_{Ca}S$. All $I_{Ca}L$ and $I_{Ca}M$ subtypes recorded from some tissues can be observed at a holding potential of -60mV. To record all $I_{Ca}S$ and $I_{Ca}M$, however, the holding potential needs to be kept more negative than -80mV. When DHPs were applied to the vascular tissues that possess the $I_{Ca}L$ and $I_{Ca}M$, the $I_{Ca}L$ was consistently inhibited. However, inhibition of $I_{Ca}M$ by DHPs was seen in guinea-pig portal vein but not in the basilar artery. In tissues that possessed either $I_{Ca}L$ with $I_{Ca}S$, or $I_{Ca}M$ with $I_{Ca}S$, DHPs inhibited only the $I_{Ca}L$ or the $I_{Ca}M$, respectively.

The $I_{Ca}S$ channel in most vascular cells is insensitive to DHPs and the summated unitary current shows transient activation and rapid inactivation. This channel corresponded to a transient component of the inward current measured using the whole-cell voltage-clamp procedure. While the $I_{Ca}M$ in some tissues behaved as the T-subtype, in others it behaved as the L-subtype (DHP sensitive).

Judging from the unitary current conductances and their sensitivities to various agents, the L subtype recorded in smooth muscle cells seems to possess much the same properties as that in cardiac cells. However, some characteristic differences between cardiac and smooth muscle cells were observed using the voltage- and patch-clamp procedures. Thus:

(a) In cardiac muscle, activation of the β-adrenoceptor produced an increased influx of Ca due to activation of the voltage-dependent Ca channel following increased synthesis of cyclic AMP. Applications of forskolin, dibutyryl cyclic AMP, or isoproterenol increased the opening probability of the channel through activation of cyclic AMP-dependent protein kinase A (A-kinase).   A-kinase phosphorylated the c-terminal of the α_1-subunit of the voltage-dependent L-subtype Ca channel (132). However, cyclic AMP did not modify the Ca channel amplitude in some vascular smooth muscle cells (88, 91) but consistently activated the Ca dependent K channel (I_KL) and hyperpolarized the smooth muscle membrane (72, 88, 98). Recently, it was reported that isoproterenol enhanced the amplitude of the voltage-dependent Ca channel (30). Descrepancies in the observations made using applications of isoproterenol and cyclic AMP need to be clarified.

(b) In vascular smooth muscle, ATP acts from both intra- and extra-cellular sites. From extracellular sites, ATP acts on the purinergic II receptor and also acts directly on the voltage-dependent Ca channel. When intracellular ATP was increased, the voltage-dependent Ca channel was activated and this action was antagonized by AMP-PNP, an unhydrolysable ATP analogue (88). This means that in smooth muscle cells, ATP may or may not act on the voltage-dependent Ca channel through cyclic AMP-dependent phosphorylation of channel proteins, and that ATP itself may act directly on the Ca channel through an unknown mechanism.

(c) In cardiac muscle, papaverine derivatives (the Ca antagonists gallopamil and verapamil) are thought to act from inside the cell membrane and their actions were closely related to the pH of the solution, i.e. intracellular application of these agents is more effective in blocking the voltage-dependent Ca channel (40, 78, 81). However, in smooth muscle, these agents act from extracellular sites and their intracellular perfusion does not attenuate the current amplitude (49, 89).

(d) In the portal vein, when depolarization pulses at 0.5 Hz frequency (0 mV intensity and 300 msec duration) were applied at a holding potential of -80mV, large and small amplitude unitary currents could be recorded with a short open time. When the frequency was reduced from 0.05 Hz to 0.03 Hz, a single channel current with the same amplitude, but a larger mean open time was recorded (58), i.e. transition of the channel state between mode 1 (short-lasting channel opening) and mode 2 (long-lasting channel opening) could be influenced not only by Ca agonists but also by stimulus frequency. Following application of a Ca antagonist, there was a transition of the channel state from mode 1 or 2 to mode 0 (inactivated state; 59). However, in cardiac muscle cells, such frequency dependency of the open time has not yet been reported.

(e) In the rabbit portal vein, inactivation of the Ca current was extremely long and involved several absorbing states. This slow inactivation of Ca channels may be of functional importance in controlling Ca influx. The backward reaction from late inactivated states to earlier ones seems to be voltage-independent (84). Such an action of the Ca channel has not been observed in cardiac muscle.

The above differences in the voltage-dependent Ca channel may suggest that the voltage-dependent Ca channels in smooth muscle and cardiac muscle cells are not of exactly the same nature, presumably due to some differences in the amino acid sequences of the L-type of voltage-dependent Ca channel.

FACTORS MODIFYING THE VOLTAGE-DEPENDENT CALCIUM CHANNEL
Effects of Dihydropyridine Derivatives on the Unitary Ca Current

Most DHPs show a wide spectrum of action from agonistic to antagonistic actions on the Ca channel of vascular smooth muscle cells. (-)-S-Bay-K-8644 increased the open probability and prolonged the duration of the open time of the L-subtype (accelerated the transition from mode 1 to mode 2; 41), while the (+)-R-enantiomer of Bay-K-8644 possessed only a weak antagonistic action. YC 170, CGP 28-392, H 160/51 and Sandoz 202-791 also possessed agonistic actions on the voltage-dependent Ca channel. Ca antagonistic DHPs such as nifedipine, PN200-110, nitrendipine, CV-4093, benidipine and FRC8653 also had some agonistic actions. Even nifedipine, a symmetrical drug, also showed an agonistic action (9, 34, 92). It is likely that most ($\pm$) DHPs possess such dual actions. DHPs with this agonistic action were characterized by esters in 3- and 5-positions of the dihydropyridine ring (an asymmetric carbon atom in the 4-position).

All DHPs Ca antagonists reduced the channel availability without a reduction in the unitary current amplitude (59, 132). The Ca channel shows a voltage-dependent inactivation and DHPs shifted the voltage dependent inactivation of the I_{Ca} to a more hyperpolarized potential level. This action can be explained using the modulated receptor hypothesis by analogy with the action of local anesthetics on the Na channel (16, 41, 59, 66, 67, 102, 121-123). Ca agonists reduced the transition rate from the open to the resting state of the Ca channel (close, open and inactivation states, 102). While the Ca antagonistic action of these agents is thought to be produced by an action on all three states (resting, open and inactivated states), the inactivated state was most strongly affected (38, 41, 70). As these agonistic agents did not modify the unitary current amplitude but increased the current amplitude measured using the whole-cell voltage-clamp procedure, it would appear that Ca agonists accelerate the transition from a non-conducting to a conducting channel state (38). These authors postulated that there are available two closed and two inactivated states, and drugs can act on the Ca channel at any state with a rate constant of the same value which is independent of the membrane potential. Each drug mainly changes the rate constants between the two closed states and also between the two inactivated states (non-modulated receptor hypothesis).

Effects of Second Messengers on the Voltage-Dependent Ca Channel

In vascular smooth muscle cells, second messengers such as Ca, cyclic AMP, cyclic GMP, IP_3 and DG act on ionic channels directly or indirectly. In addition, there is some evidence that GTP-binding proteins also contribute directly to Ca channel activity. Activation of receptors distributed on the sarcolemmal membrane produces two main events: receptor-operated ion channel activation and synthesis of a second messenger. Activation of the receptor-operated ion channel by receptor activations induced a depolarization or a hyperpolarization. The depolarization occurred via increases in the non-selective cation channel or activation of the M current (5, 6, 10, 57, 106). These electrical events occurred independently of the synthesis of a second messenger, and activated the voltage-dependent Ca channel.

Concerning the role of cytosolic Ca, increases in the cytosolic Ca accelerated the inactivation process of the Ca channel in a concentration-dependent manner. As a consequence, the amplitude of the action potential was inhibited. An increase in the concentration of extracellular Ca or Ba increased the open probability of the unitary current and enhanced the macroscopic Ca current.

Cyclic nucleotides, such as cyclic AMP and cyclic GMP, activated their own protein kinases (cyclic AMP or cyclic GMP dependent protein kinase; A- and G-kinase) and modified the ionic channel activity. Using the microelectrode method, applied isoproterenol or nitroprusside hyperpolarized vascular smooth muscle cell membranes (20) and inhibited spontaneously generated action potentials, though isoproterenol enhanced an evoked action potential (30). Atrionatriuretic polypeptide (ANP) synthesized cyclic GMP at the sarcolemma, however, this agent did not produce hyperpolarization (29). On the other hand, other investigators have reported that ANP and 8-bromo cyclic GMP inhibit the action potential (95, 110). In vascular and intestinal smooth muscle, cyclic GMP -synthesized by nitric oxide (NO) from L-arginine as a core substance of endothelium derived relaxing factor (EDRF) released from endothelial cells (52, 96)- produced hyperpolarization and blocked the activation of the voltage-dependent Ca channel. Recently, it was reported that NO may be a neurotransmitter in smooth muscle tissues (17, 32, 120). Thus, cyclic GMP may contribute to the hyperpolarization of the membrane in some tissues. Further experiments would be required on the unitary current for a full explanation of the underlying mechanism.

IP_3 synthesized through activations of various receptors released Ca from the SR via activation of the IP_3-activated Ca channel in the SR and increased the cytosolic Ca: As a consequence, the amplitude of the voltage-dependent Ca current was inhibited due to an acceleration of the inactivation process (90). In addition, inositol 1,3,4,5-tetrakisphosphate (IP_4) did not modify the Ca channel in vascular smooth muscle cells (88).

DG, a co-product with IP_3 through hydrolysis of $PI-P_2$ or PI, activates protein kinase C in the presence of phosphatidylserine and Ca, and plays multiple roles in smooth muscle (85-87, 103). Instead of the unstable substance DG, various derivatives of phorbol esters (tumor promoting agents) are used to activate protein kinase C. Phorbol ester 12,13-diacetate depolarized the smooth muscle membrane, prolonged the action potential duration and finally blocked spike generation (95). On the other hand, it was reported that this agent increased the sustained Ca channel (L-subtype; 25). More detailed experiments are awaited on the role of protein kinase C in vascular smooth muscle.

With regard to the action of second messengers, the actions of GTP-binding protein (G-protein; 33, 125) have been investigated in relation to channel currents (15, 97). The G-protein has been classified into several subtypes, such as G_s, G_i, G_o, or G_t and each subtype is composed of α-, β- and γ-subunits. Individual proteins are also subclassified into types 1 and 2 (G_i1, G_i2 etc) mainly on the basis of the different natures of the α-subunit of the G-protein. In neurons, G_o (brain G-protein) inhibited the voltage-dependent Ca channel, whereas in cardiac muscle (40). G_s (choleratoxin sensitive G-protein) directly enhanced the voltage-dependent Ca channel (56, 132). GTP-γ-S, an unhydrolyzed form of GTP, increased the open probability of the Ca unitary current and the amplitude of I_{Ca} measured using the voltage-clamp procedure (90). Other investigators have reported that GTP-γ-S did not have any effect on I_{Ca}, except when an agonist was applied simultaneously (129). Thus G-protein exhibits a dual action on the voltage-dependent Ca channel in vascular smooth muscle cells, i.e. this protein acts on the synthesis of second messengers, thus modifying the Ca channel indirectly, and also directly modifies the Ca channel through activation of the α-subunit.

As an example, NE synthesizes a second messenger and also acts directly on the Ca channel by lowering the threshold potential needed to evoke the Ca current (82). Presumably, agonists may also have multiple actions, through various receptors and GTP-binding proteins. Oike et al. (93) reported that activation of the H_3-receptor distributed on the sarcolemmal membrane accelerated the voltage-dependent Ca channel in the rabbit saphenous artery. This acceleration of the voltage-dependent Ca channel occurred independently of the depolarization of the membrane, because this action occurred when the holding membrane potential level was fixed under the clamped condition. This activation of the Ca channel by histamine was further accelerated by application of GTP but not of GTP-γ-S, and this action of GTP was not modified by pertussis toxin. Much the same action was observed on application of NE or angiotensin II. Therefore, activation of the receptor induced three different events in smooth muscle cells, viz. receptor activation produced activation of the receptor operated ion channel, synthesized a second messenger and modified the voltage-dependent Ca channel directly (in a manner not related to the receptor-operated depolarization of the membrane) or indirectly (through receptor-operated depolarization). However, individual receptors may not always exhibit the above three events, but either one, two or three events may be involved in individual cells or tissues by activation of various receptors. Moreover, the nature of the known and unknown GTP binding proteins which contribute to these events may also differ.

CONCLUSION

In this chapter, we have discussed the characteristics of the voltage-dependent Ca channel as assessed using the voltage- and patch-clamp procedures. In vascular tissues, at least three different subtypes of Ca channels are heterogenously distributed. The density of these channel distributions differs from one region to another. Some differences in the Ca channels found in smooth muscle and cardiac muscle cells were reviewed. The voltage-dependent Ca channel is directly or indirectly regulated by many factors, such as endothelial cells (endothelial cell derived relaxing and contractile factors), excitatory and inhibitory neurotransmitters and also other putative substances, through depolarization or hyperpolarization of the membrane, the actions of excitatory or inhibitory second messengers or GTP-binding proteins. However, many problems remain to be solved before we can fully understand the voltage-dependent Ca channel in vascular smooth muscle.

REFERENCES

1.   Abdel-Latif, A.A., Pharmacol. Rev. 38, 227-272 (1986).
2.   Akaike, N., H. Kanaide, T. Kuga, M. Nakamura, J. Sadoshima, and H. Tomoike, J. Physiol. 416, 141-160 (1989).
3.   Akaike, N., P.G. Kostyuk, and Y.V. Osipchuk, J. Physiol. 412, 181-195 (1989).
4.   Amedee, T., C. Mironneau, and J. Mironneau, Br. J. Pharmacol. 88, 873-880 (1986).
5.   Benham, C.D., J. Physiol. 419, 689-701 (1989).
6.   Benham, C.D., T.B. Bolton, N.G. Byrne, and W.A. Large, J. Physiol. 387, 473-488 (1987).
7.   Benham, C.D., T.B. Bolton, R.J. Lang, and T. Takewaki, Pflugers Arch. 403, 120-127 (1985).
8.   Benham, C.D., P. Hess, and R.W. Tsien, Circ. Res. 61 Suppl 1, 10-16 (1987).
9.   Becham, M. and M. Schramm, J. Mol. Cell. Cardiol. 19 suppl. 2, 63-75 (1987).
10.   Benham, C.D. and R.W. Tsien, Nature 328, 275-278 (1987).
11.   Berridge, M.J.,   Biochem. J. 220, 345-360 (1984).
12.   Berridge, M.J., Ann. Rev. Biochem. 56, 159-193 (1987).
13.   Blaustein, M.P., Current Topic in Membrane and Transport 34, 289-330 (1989).
14.   Bolton, T.B., Physiol. Rev. 59, 606-718 (1979).
15.   Brown, A.M. and L. Birnbaumer, Am J Physiol. 254, H401-H410 (1988).
16.   Brown, A.M., D.L. Kunze, and A. Yatani, J. Physiol. 379, 495-514 (1986).
17.   Bult, H., G.E. Boeckxstaens, P.A. Pelckmans, F.H. Joraaens, Y.M. Van Maercke, and A.G. Herman, Nature 345, 346-347 (1990).
18.   Burnstock, G., in Handbook of Physiology, The Cardiovascular System (Am. Physiol. Soc. Bethesda, MD 1980) Sect. 2, vol. II, chapt. 19, 567-612.
19.   Burnstock, G., J. Physiol. 313, 1-35 (1981).
20.   Bulbring, E. and T. Tomita, Pharmacol. Rev. 39, 49-96 (1987).
21.   Caffrey, J.M., I.R. Josephson, and A.M. Brown,   Biophys. J. 49, 1237-1242 (1986).
22.   Cheung, D.W., J. Physiol. 328, 461-468 (1982).
23.   Creed, K.E.,   Br. Med. Bull. 35, 243-247 (1982).
24.   DeFeo, T.T. and K.G. Morgan, J. Physiol 369, 269-282 (1985).
25.   Dosemeci. A., R.S. Dhallan, N.M. Cohen, W.J. Lederer, and T.B. Rogers, Circ. Res. 62, 347-357 (1988).
26.   Endo, M., Physiol Rev. 57, 71-108 (1977).
27.   Fay, F.S., M.H. Shleven, W.C. Granger, and S.R. Taylor, Nature 280, 506-508 (1979).
28.   Fleckenstein, A., Med. Res. Rev. 5, 395-425 (1985).
29.   Fujii, K., T. Ishimatsu, and H. Kuriyama, J. Physiol. 377, 315-332 (1986).
30.   Fukumitsu, T., H. Hayashi, H. Tokuno, and T. Tomita, Br. J. Pharmacol. 100, 593-599 (1990).
31.   Furuichi, T., S. Yoshikawa, A. Miyawaki, K. Wada, N. Maeda, and K. Mikoshiba, Nature 342, 32-38 (1989).
32.   Gillespie, J.S., X. Liu, and W. Martin, Br. J. Pharmacol. 98, 1080-1982 (1989).
33.   Gilman, A.G., Ann. Rev. Biochem. 56, 615-649 (1987).
34.   Godfraind, T., R. Miller, and M. Wibo, Pharmacol.Rev. 38, 321-416 (1986).
35.   Hamill, O.P., A. Marty, E. Neher, B. Sakmann, and F. Sigworth, Pflugers Arch. 391, 85-100 (1981).
36.   Hara, Y., K. Kitamura, and H. Kuriyama, Br. J. Pharmacol. 68, 99-106 (1980).
37.   Hashimoto, T., M. Hirata, T. Itoh, Y. Kanmura, and H. Kuriyama, J. Physiol. 370, 605-618 (1986).
38.   Hering, S., D.J. Beech, and T.B. Bolton, Biomed. Biochem. Acta. 467, S657-S661 (1987).
39.   Hermann, A. and A.L.F. Gorman, J. Gen. Physiol. 78, 63-86 (1981).
40.   Hescheler, J., M. Tang, B. Jastorff, and W. Trautwein, Pflugers Arch. 410, 23-29 (1987).
41.   Hess, P., J.B. Lamsman, and R.W. Tsien, Nature 311, 538-544 (1984).
42.   Himpens, B. and A.P. Somlyo, J. Physiol. 395, 507-530 (1988).
43.   Hirst, G.D.S. and F.R. Edwards, Physiol. Rev. 69, 546-604 (1989).
44.   Hirst G.D.S. and T.O. Neild, Nature 283, 767-768 (1980).
45.   Hirst, G.D.S. and T.O. Neild, J. Physiol. 313, 343-350 (1981).
46.   Hirst, G.D.S., T.O. Neild, and G.D. Silverberg, J. Physiol. 328, 351-360 (1982).
47.   Holman, M.E. and A. Surprenant, J. Physiol. 287, 337-351 (1979).
48.   Holman, M.E., and A. Surprenant, Br. J. Pharmacol. 71, 651-661 (1980).
49.   Hume, J.R., J. Pharmacol. Exp. Therap. 234, 134-140 (1985).
50.   Hume, J.R. and N. Leblanc, J. Physiol. 413. 49-73 (1989).
51.   Hwang, K.S. and C. van Breemen, Pflugers Arch. 408, 343-350 (1987).
52.   Ignarro, L.J., Circ. Res. 65, 1-21 (1989).

53. Iino, M., J. Physiol. 320, 449-467 (1981).
54. Iino, M., Biochem. Biophys. Res. Comm. 142, 47-52 (1987).
55. Iino, M., T. Kobayashi, and M. Endo, Biochem.Biophys. Res. Comm. 152, 417-422 (1988).
56. Imoto, Y., A. Yatani, J.P. Reeves, J. Codina, L. Birnbaumer, L. and A.M. Brown, Am. J. Physiol. 255, H722-H728 (1988).
57. Inoue, R., K. Kitamura, and H. Kuriyama, Pflugers Arch. 410, 69-74 (1987).
58. Inoue, Y., M. Oike, K. Nakao, K. Kitamura, and H. Kuriyama, J. Physiol. 423, 171-191 (1990).
59. Inoue, Y., Z. Xiong, K. Kitamura, and H. Kuriyama, Pflgers Arch. 414, 534-542 (1989).
60. Irvine, R.F., Biochem. Soc. Transact. 17, 6-9 (1989).
61. Ishikawa, S., Br. J. Pharmacol. 86, 777-787 (1985).
62. Ito, K., S. Takamura, K. Sato, and J.L. Sutko, Circ. Res. 58, 730-734 (1986).
63. Itoh, T., Y. Kanmura, and H. Kuriyama, J. Physiol. 376, 231-252 (1986).
64. Itoh, T., H. Kuriyama, and H. Suzuki, J. Physiol. 321, 513-535 (1981).
65. Kajiwara, M., K. Kitamura, and H. Kuriyama, J. Physiol. 315, 283-302 (1981) .
66. Kass, R. S., Circ. Res. 61, 1-5 (1987).
67. Kass, R.S. and D.S. Krafte, J.Gen.Physiol. 89, 629-644 (1987).
68. Kawashima, Y. and R. Ochi, J. Physiol. Soc. Japan 49, 369 (1987).
69. Kobayashi, S., A.P. Somlyo, and A.V. Somlyo, Biochem. Biophys. Res. Com. 153, 625-631 (1988).
70. Kokubun, S. and H. Reuter, Proc. Natl. Acad. Sci. USA. 81, 4824-4827 (1984).
71. Komori, S. and T.B. Bolton, J. Physiol. 427, 395-417 (1990).
72. Kume, H., A. Takai, M. Tokuno, and H. Tomita, Nature 341, 152-154 (1989).
73. Kuriyama, H., in, Smooth Muscle, E. Bulbring, A. Brading, A. Jones, and T. Tomita, Eds. (Edward Arnold, London 1970) pp. 366-395.
74. Kuriyama, H., in, Smooth muscles, an assessment of current knowledge, Eds. by E. Bulbring, A.F. Brading, and T. Tomita, eds. (Edward Arnold, London 1981) pp. 171-197.
75. Kuriyama, H. and Y. Makita, J. Physiol. 327, 431-448 (1982).
76. Kuriyama, H. and H. Suzuki, J. Physiol. 317, 383-396 (1981).
77. Lee, K.S. and R.W. Tsien, Nature 302, 790-794 (1983).
78. McDonald, T.F., D. Pelzer, and W. Trautwein, J. Physiol. 325, 217-241 (1984).
79. Meissner, G., J.Biol.Chem. 261, 6300-6306 (1986).
80. Morgan, J.P. and K.G. Mogan, Pflugers Arch. 395, 75-77 (1982).
81. Nawrath, H., R.E. Ten Eick, T.F. McDonald, and W. Trautwein, Circ. Res. 40, 408-414 (1977).
82. Nelson, M.T., N.B. Standen, J.E. Brayden, and J.F. Worley III, Nature 336, 382-385 (1988).
83. Nilius, B., P. Hess, J.B. Lansman, and R.W. Tsien, Nature 316, 443-446 (1985).
84. Nilius, B., K. Kitamura, and H. Kuriyama, J. Physiol. in press (1991).
85. Nishizuka, Y., Nature 308, 693-698 (1984).
86. Nishizuka, Y., Science 233, 305-312 (1986).
87. Nishizuka, Y., Nature 344, 661-665 (1988).
88. Ohya, Y., K. Kitamura, and H. Kuriyama, Pflugers Arch. 408, 465-473 (1987).
89. Ohya, Y., K. Terada, K. Yamaguchi, R. Inoue, K. Okabe, K. Kitamura, M. Hirata, and H. Kuriyama, Pflugers Arch. 412, 382-394 (1988).
90. Ohya, Y. and N. Sperelakis, The Physiologist. 31, 63-67 (1988).
91. Ohya. Y. and N. Sperelakis, Circ. Res. 64, 145-154 (1989).
92. Oike, M., Y. Inoue, K. Kitamura, and H. Kuriyama, Circ. Res. 67, 993-1006 (1990).
93. Oike, M., K. Kitamura, and H. Kuriyama, Jpn. J. Pharmacol. 55, Suppl. 72p. (1991).
94. Okabe, K., K. Kitamura, and H. Kuriyama, Pfluger Arch. 411, 423-428 (1988).
95. Ousterhout, J.M. and N. Sperelakis, Eur. J. Pharmacol. 144, 7-14 (1987).
96. Palmer, R.M., A.G. Ferrige, and S. Moncada, Nature 327, 524-526 (1987).
97. Rosenthal, W., J. Heschler, W. Trautwein, and G. Schultz, FASEB J. 2, 2784-2790 (1988).
98. Sadoshima,J., N. Akaike, H. Tomoike, and M. Nakamura, Am.J.Physiol. 255, H410-418 (1988).
99. Saida, K., J. Gen. Physiol. 80, 191-202 (1982).
100. Saida, K. and Y. Nonomura, J. Gen. Physiol. 72, 1-14 (1978).
101. Sakai, T., K. Terada, K. Kitamura, and H. Kuriyama, Br. J. Pharacol. 95, 1089-1100 (1988).
102. Sanguinetti, M.C. and R.S. Kass, Circ.Res. 55, 336-348 (1984).
103. Shearman, N.S., K. Sekiguchi, and Y. Nishizuka, Pharmacol. Rev. 41, 211-237 (1989).
104. Shears, S.B., Biochem. J. 260, 313-324 (1989).
105. Shogakiuchi, Y., H. Kanaide, S. Kobayashi, J. Nishimura, and M. Nakamura, Biochem. Biophys. Res. Comm. 135, 9-15 (1986).
106. Sims, S.M., J.J. Singer, and J.V. Walsh, Jr., J. Physiol. 367, 503-529 (1985).
107. Sneddon, P. and G. Burnstock, Eur. J. Pharmacol. 100, 85-90 (1984).
108. Somlyo, A.V., M. Bond, A.P. Somlyo, and A. Scarpa, Proc. Natl. Acad. Sci. USA. 82, 5231-5235 (1985).

109. Somlyo, A.P., J.W. Walker, Y.E. Goldman, D.R. Trenthan, S. Kobayashi, T. Kitazawa, and A.V. Somlyo, Phil. Trans. Roy. Soc. Lond., Series 13-Biol. Sci. 320, 399-414 (1988).

110. Sperelakis, N. and Y. Ohya, in Frontiers in Smooth Muscles. N. Sperelakis and J.D. Woods, Eds. (Wiley-Liss, New York 1990).

111. Sturek, M. and K. Hermsmeyer, Science 233, 475-478 (1987).

112. Suematsu, E., M. Hirata, T. Hashimoto, and H. Kuriyama, Biochem. Biophys. Res. Comm. 120, 481-485 (1984).

113. Sumimoto, K. and H. Kuriyama, Pflugers Arch. 406, 173-180 (1987).

114. Sutko, J.L., K. Ito, and J.L. Kenyon, Fed. Proc. 44, 2984-2988 (1985).

115. Suzuki, H., J. Physiol. 321, 495-512 (1981).

116. Suzuki, R., Br. J. Pharmacol. 78, 591-597 (1983).

117. Suzuki, H., J. Physiol. 359, 401-415 (1985).

118. Takeshima, H., S. Nishimura, T. Matsumoto, H. Ishida, K. Kangawa, N. Minamino, H. Matsuo, M. Ueda, M. Hanaoka, T. Hirose, and S. Numa, Nature 339, 439-445 (1989).

119. Tanabe, T., K.G. Beam, J.A. Powell, and S. Numa, Nature 336, 134-139 (1988).

120. Tare, M., H.C. Parkington, H.A. Coleman, T.O. Neild, and G.J. Dusting, Nature 346, 69-71 (1990).

121. Terada, K., K. Kitamura, and H. Kuriyama, Pflugers Arch. 408, 552-557 (1987).

122. Terada, K., K. Nakao, K. Okabe, K. Kitamura, and H. Kuriyama, Br. J. Pharmac. 92, 615-625 (1987).

123. Terada, K., Y. Ohya, K. Kitamura, and H. Kuriyama, J. Pharmac. Exp. Therap. 240, 978-983 (1987).

124. Tsien, R.W., P. Hess, E.W. McCleskey, and R.L. Rosenberg, Ann. Rev. Biophysic. Chem. 16, 265-290 (1987).

125. Ui, M, TIPS. 5, 277-279 (1984).

126. Van Breemen, C., P. Aaronson, and Loutzenhiser, Pharmacol. Rev. 30, 167-208 (1979).

127. Williams, D.A., K.E. Fogarty, R.Y. Tsien, and F.S. Fay, Nature 318, 558-561 (1985).

128. Worley III J.F., J.W. Deitmer, and M.T. Nelson, Proc. Nat. Acad. Sci., U.S.A. 83, 5746-5750 (1986).

129. Xiong, Z.L., K. Kitamura, and H. Kuriyama, J. Physiol. (in press) (1991).

130. Yamaguchi, H., Cell Calcium 7, 203-219 (1986).

131. Yamamoto H. and C Van Breemen, Biochem. Biophys. Res. Comm. 130(1), 270-274 (1985).

132. Yatani, A., C.L. Seidel, J. Allen, and A.M. Brown, Circ. Res. 60, 523-533 (1987).

133. Yoshitomi, T. and Y. Ito, Invest. Ophthalmol. Vis. Sci. 27, 83-91 (1986).

REGULATION OF CALCIUM SLOW CHANNELS IN VASCULAR SMOOTH MUSCLE CELLS

NICHOLAS SPERELAKIS and YUSUKE OHYA[*]

Department of Physiology & Biophysics, University of Cincinnati, College of Medicine, Cincinnati, Ohio and Second Department of Internal Medicine, Kyushu University, Faculty of Medicine, Fukuoka 812, Japan

INTRODUCTION

Considerable attention during the past few years has been given to phosphorylation of ion channels as a means whereby the activity of the ion channels can be regulated. This article will cover the evidence that cyclic nucleotides regulate the Ca^{2+} influx into vascular smooth muscle (VSM) cells. This regulation is presumably mediated by phosphorylation(s) of the Ca^{2+} slow channel protein and/or of associated regulatory protein(s). In some VSM cells, phosphorylation by cAMP-PK or cGMP-PK inhibits the Ca^{2+} slow channel activity and thereby produces vasodilation, whereas phosphorylation by PK-C stimulates the Ca^{2+} slow channel activity and produces vasoconstriction.

In myocardial cells, such phosphorylation (Fig. 1) presumably (a) increases the number of Ca^{2+} slow channels available for voltage activation during the action potential (AP); (b) increases the probability of their opening, and (c) increases their mean open time. A greater density of available Ca^{2+} channels increases Ca^{2+} influx and inward Ca^{2+} slow current (I_{si}) during the AP, and so increases the force of contraction of the heart. Because of the relationship between cAMP and the number of available slow channels, and because of the dependence of the functioning of the slow channels on metabolic energy, it was postulated that a membrane protein must be phosphorylated in order for the slow channel to become available for voltage activation (26, 32, 35, 38). Elevation of cAMP by a positive inotropic agent activates a cAMP-dependent protein kinase (cA-PK), which phosphorylates the slow channel protein itself or a contiguous regulatory type of protein (Fig. 1). Phosphorylation could make the slow channel available for activation by a conformational change that either allowed the activation gate to be opened upon depolarization or effectively opened or increased the diameter of the water-filled pore so that Ca^{2+} could pass through. The phosphorylated form of the slow channel is the active (operational) form, and the dephosphorylated form is the inactive (inoperative) form which is electrically silent. Thus, phosphorylation increases the probability of channel opening with depolarization. An equilibrium would exist between the phosphorylated and dephosphorylated forms of the slow channels. Thus, agents that elevate the cAMP level would increase the fraction of the slow channels that are in the phosphorylated form, and hence readily available for voltage activation (Table 1).

TABLE 1

Comparison of Properties of Ca^{2+} Slow Channels in

Vascular Smooth Muscle Cells With Those of Myocardial Cells

	Myocardial Cells	VSM Cells
ATP	stim.	stim.
cAMP	stim.	inhib.
cGMP	inhib.	inhib.
acidosis	inhib.	?
Ca antagonist drugs	inhib.	inhib.
Ca agonist drugs	stim.	stim.

Published 1991 by Elsevier Science Publishing Company, Inc.
Ion Channels of Vascular Smooth Muscle Cells and Endothelial Cells
Sperelakis and Kuriyama, Editors

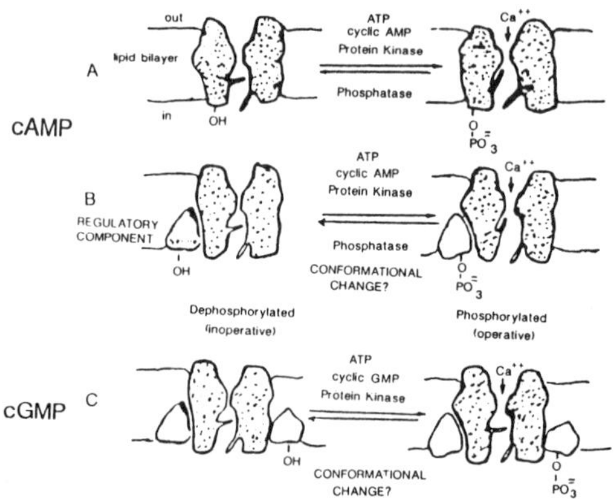

Figure 1. Schematic model for a Ca^{2+} slow channel in myocardial cell membrane in two hypothetical forms: dephosphorylated (or electrically silent) form (left diagrams) and phosphorylated form (right diagram). The two gates associated with the channel, an activation gate and an inactivation gate, are kinetically much slower than those of the fast Na^+ channel. The phosphorylation hypothesis states that a protein constituent of the slow channel itself (A) or a regulatory protein associated with the slow channel (B) must be phosphorylated in order for the channel to be in a state available for voltage activation. Phosphorylation occurs by a cAMP-dependent protein kinase (PK-A) in the presence of ATP. Presumably, a serine or threonine residue in the protein becomes phosphorylated. Phosphorylation may produce a conformational change that effectively allows the channel gates to operate. The slow channel (or an associated regulatory protein) may also be phosphorylated by a cGMP-PK (C), thus mediating the inhibitory effects of cGMP on the slow Ca^{2+} channel. (Modified from 35.)

Besides the standard slow Ca^{2+} channel, a fast-type of Ca^{2+} channel has been found in cardiac muscle and VSM cells on the basis of kinetics (3, 20). These fast Ca^{2+} channels are much more rapidly inactivated than the slow Ca^{2+} channels, are active over a more negative voltage range, and are little affected by cAMP or Ca^{2+} antagonists (Table 2). Their function is as yet unclear.

TABLE 2

Summary of Major Differences Between the Slow (L-Type) and Fast (T-Type) Ca^{2+} Channels

Properties	Ca^{2+} Channels	
	Slow (L-type)	Fast (T-type)
Duration of current	Long-lasting (sustained)	Transient
Inactivation kinetics	Slower	Faster
Activation kinetics	Slower	Faster
Threshold	High (ca. -30 mV)	Low (ca. -50 mV)
Half-inactivation potential	ca. -20 mV	ca. -50 mV
Single-channel conductance	High (18-26 pS)	Low (8-10 pS)
Regulated by cAMP and cGMP	Yes	No
Regulated by phosphorylation	Yes	No
Blocked by Ca^{2+} antagonist drugs	Yes	No (slight)
Opened by Ca^{2+} agonist drugs	Yes	No
Permeation by Me^{2+}	Ba > Ca	Ba $\simeq$ Ca
Inactivation by $[Ca]_i$	Yes	Slight (?)
Recordings in isolated patches	Runs down	Rel. stable

REGULATION OF Ca^{2+} SLOW CHANNELS BY CYCLIC NUCLEOTIDES

Figure 2 depicts some of the types of ion channels in VSM cells, including the two types of Ca^{2+} channels: slow (L-type) and fast (T-type). The properties of these two types of Ca^{2+} channels are rather similar to those in myocardial cells, with the exception of the regulatory mechanisms (Table 2).

Both cAMP and cGMP elevation have been implicated in the relaxation of smooth muscle tissue in response to some vasodilators, such as EDRF, ANF, and nitroprusside. We examined the effects of cyclic nucleotide analogs and related agents on the electrical properties of cultured rat aortic reaggregates (27). Agents that activate adenylate cyclase, such as isoproterenol and forskolin, depressed or abolished the Ca^{2+}-dependent APs and hyperpolarized the membrane. These effects are mimicked by membrane permeable analogs of cAMP (dibutyryl or 8-bromo cAMP) (Fig. 3). 8-Bromo-cGMP also depressed or abolished the APs. Synthetic atrial natriuretic factor (ANF), which activates the membrane-bound guanylate cyclase, had inhibitory effects, whereas nitroprusside, which stimulates only cytosolic G-cyclase, had little effect. These results are summarized in Table 3. They suggest that cAMP and cGMP decrease the inward Ca^{2+} current and/or increase an outward K$^+$ current in the VSM cells.

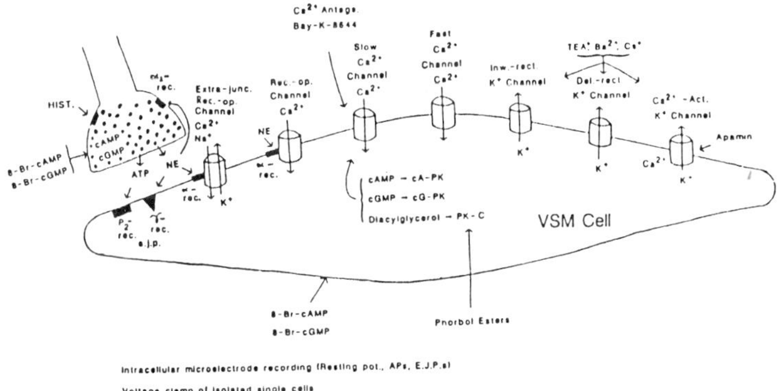

Figure 2. Diagrammatic representation of various ion channels in a vascular smooth muscle (VSM) cell and a number of agents that may either activate or inhibit these channels. Depicted are three different types of Ca^{2+} channels (fast, slow, and receptor-operated) and a nonselective ion channel (extrajunctional, receptor-operated), which allows Ca^{2+}, Na$^+$, and K$^+$ to pass through. The voltage-dependent Ca^{2+} slow channels are blocked by Ca^{2+} antagonists and enhanced by Ca^{2+} agonists (Bay-K-8644). The three different K$^+$ channels (inward rectifier, delayed rectifier, and Ca^{2+} activated) are blocked by TEA$^+$, Ba^{2+}, and Cs$^+$. Also shown is an adrenergic nerve terminal from which norepinephrine (NE) and ATP are released to activate α- or γ-receptors and purinergic (P$_2$) receptors, respectively, on the postsynaptic membrane. Release of neurotransmitters is modulated by cyclic nucleotides and phorbol esters.

In voltage-clamp experiments on cultured single cells from rabbit aorta, the dibutyryl analogs of cAMP and cGMP suppressed the inward Ba^{2+} current carried through the Ca^{2+} slow channels (6). The 8-bromo analogs of cAMP and cGMP and nitroprusside also enhanced the delayed-rectifier outward K$^+$ current (6). Thus, these effects of cyclic nucleotides on the Ca^{2+} and K$^+$ channels can explain their inhibitory effects on the APs.

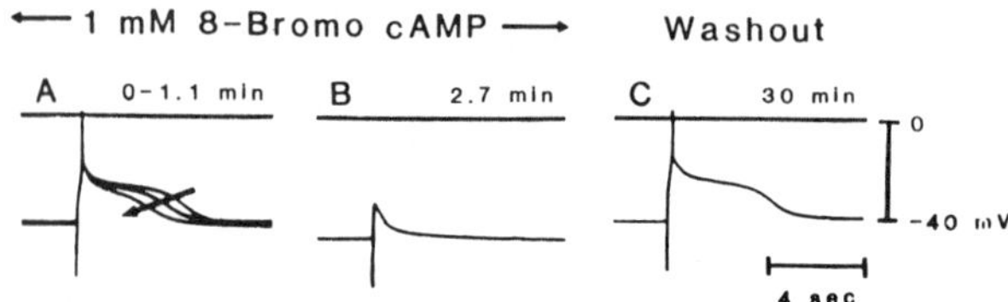

Figure 3. Depression and abolition of TEA-induced APs APs in cultured rat aortic smooth muscle cells by 8-bromo cAMP. **A:** Superimposed traces of a control AP (in 15 mM TEA) and depression of the AP within 1.1 min. after addition of 1 mM 8-bromo-cAMP. **B:** Abolition of the AP after 2.7 min. **C:** Recovery of the AP upon washout for 30 min. (Records A-C were from the same cell.) The stimulation frequency was 0.04 Hz. (Data taken from 27.)

TABLE 3. Summary of the Effects of Cyclic Nucleotides and Related Agents on the Action Potentials of Cultured Rat Aortic Cells

Compound	Effect on APs
8-Bromo cyclic AMP (10^{-3} M)	Abolished
Dibutyryl cyclic AMP (10^{-3} M)	Abolished
Isoproterenol (10^{-6} M)	Abolished
Forskolin (10^{-6} M)	Abolished
8-Bromo cGMP (10^{-4} - 10^{-3} M)	Abolished or depressed
Nitroprusside (10^{-6} - 10^{-5} M)	Depressed slightly
Atrial Natriuretic Factor (10^{-8} - 10^{-7} M)	Abolished

Taken from Ousterhout and Sperelakis, 1987.

We propose a phosphorylation model for regulation of Ca^{2+} slow channels in VSM cells, similar to that for myocardial cells, except in the case of VSM cells, cAMP and cGMP have the same effect, namely both inhibit the Ca^{2+} slow channels (Fig. 4). A second regulatory protein, when phosphorylated by protein kinase C (PK-C), stimulates the Ca^{2+} slow channels, as depicted (Fig. 4).

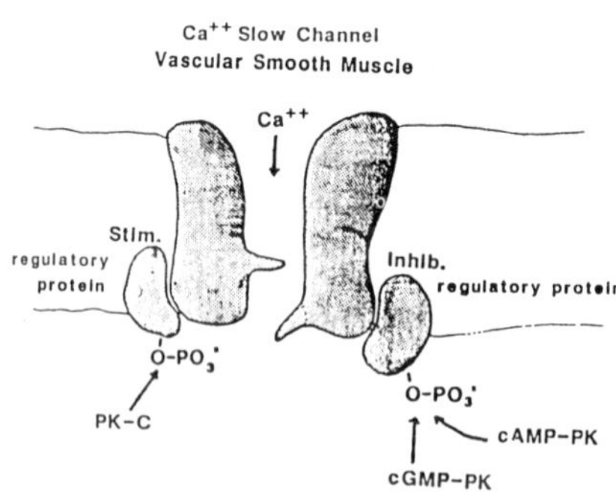

Figure 4. Cartoon model of a Ca^{2+} slow channel protein imbedded in the lipid bilayer of the cell membrane of a vascular smooth muscle cell. Depicted are two regulatory proteins, inhibitory and stimulatory, associated with the channel protein. As shown, both regulatory proteins have phosphorylation sites which affect the function of the regulatory proteins. Phosphorylation of the inhibitory protein by cAMP-protein kinase (PK-A) or by cGMP-PK (PK-G) inhibits the function of the ion channel, e.g., decreases probability of channel opening at depolarized potentials. Phosphorylation of the stimulatory protein by PK-C stimulates the Ca^{2+} slow channel, e.g., increases probability of opening. Thus, channel function is regulated by cyclic nucleotides and by diacylglycerol.

Figures 5 and 6 summarize the roles played by cyclic nucleotides and PK-C in regulating the intracellular Ca^{2+} concentration ($[Ca]_i$) in VSM cells. Activation of A-cyclase or G-cyclase by appropriate membrane receptors (e.g., beta-adrenergic, prostacyclin, ANP) and G coupling proteins or directly (by agents like nitroprusside, nitric oxide free radical, or forskolin) produces elevation of cAMP and cGMP and activation of cA-PK and cG-PK. These kinases can phosphorylate the Ca^{2+} slow channel (to inhibit channel opening) and the K^+ channels (to stimulate channel opening). Both mechanisms would inhibit the Ca^{2+}-dependent APs, and thereby diminish Ca^{2+} influx and lower $[Ca]_i$. cG-PK and cA-PK also phosphorylate the sarcolemmal Ca^{2+}-ATPase and stimulate the Ca^{2+} pump in the surface membrane, thus lowering $[Ca]_i$. Lowering $[Ca]_i$ inhibits contraction and produces vasodilation.

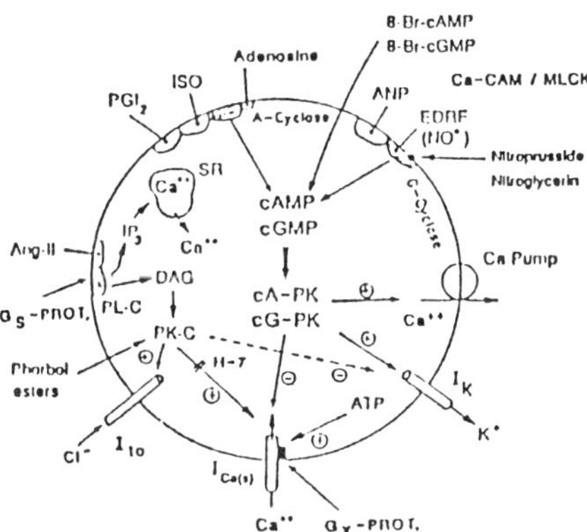

Figure 5. Diagram of a vascular smooth muscle cell to illustrate one mechanism whereby elevation of cyclic nucleotides can lead to vasodilation. Depicted are membrane receptors for stimulation of adenylate cyclase and guanylate cyclase, resulting in elevation of either cAMP or cGMP. As shown, cAMP activates PK-A and cGMP activates PK-G, which phosphorylate at least two types of ion channels and a Ca pump in the sarcolemma. Phosphorylation of the Ca^{2+} slow channel (or an associated inhibitory-type regulatory protein) produces inhibition of the channel, e.g., decreases probability of opening at depolarized potentials. Also depicted is the fact that phosphorylation by PK-C stimulates the Ca^{2+} slow channel. The delayed rectifier K^+ channel is stimulated by phosphorylation with cA-PK and cG-PK, as illustrated. Also depicted is stimulation of the sarcolemmal Ca^{2+} pump by cA-PK and cG-PK.

Activation of PL-C by appropriate membrane receptors (e.g., Ang-II receptor) and G coupling proteins stimulates PI turnover with IP_3 and DAG production. DAG activates PK-C to phosphorylate the Ca^{2+} slow channels and stimulate Ca^{2+} influx and raise $[Ca]_i$. Phorbol esters act, like DAG, to directly activate PK-C. IP_3 acts on the SR to release Ca^{2+} (via the Ca-release channels) and elevate $[Ca]_i$. Raising $[Ca]_i$ potentiates contraction and produces vasoconstriction.

The cyclic nucleotides also phosphorylate the myosin light-chain kinase (MLCK), which diminishes its Ca^{2+} sensitivity. In addition, the cyclic nucleotides have been reported to inhibit phospholipase C (PL-C), and therefore production of IP_3 and diacylglycerol (DAG) is inhibited. These factors would contribute to the vasodilation.

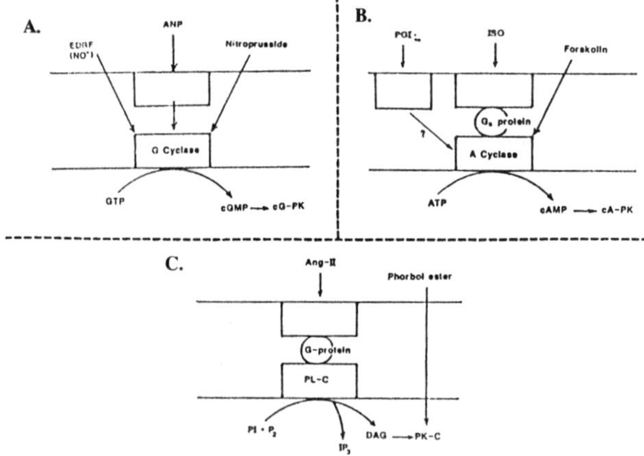

Figure 6. Diagram of expanded regions of the sarcolemma of a VSM cell to better illustrate the relationship between various membrane receptors and the cyclases and phospholipase C (PL-C). **A:** A receptor for atrial natriuretic peptide (ANP) is depicted, along with the guanylate (G) cyclase. Stimulation of G-cyclase by ANP, or directly by nitroprusside or EDRF (presumably nitric oxide free radical) elevates cGMP and activates cG-PK. **B:** Receptors for prostacyclin (PGI$_2$) and isoproterenol (beta-adrenergic agonist) are depicted, along with the adenylate (A) cyclase. The beta receptor is coupled to the A-cyclase by a GTP-binding protein (G$_s$). Activation of A-cyclase by agonists, or directly by forskolin, elevates cAMP level and thereby activates cA-PK. **C:** Depicted is the angiotensin-II receptor, coupled by a G-protein to PL-C. Stimulation of PL-C stimulates PI turnover and leads to production of IP$_3$ and DAG. DAG activates PK-C which phosphorylates several proteins including the Ca^{2+} slow channel. Phorbol esters directly stimulate PK-C (like DAG). Inositol trisphosphate (IP$_3$) causes Ca^{2+} release from the SR by activation of Ca release channels.

REGULATION OF Ca^{2+} CHANNELS BY ATP

Two distinct types of voltage-sensitive Ca^{2+} channels are found in VSM cells (3, 13, 29, 41), as in myocardial cells. The two types of Ca^{2+} currents can be distinguished by differences in their kinetics, voltage ranges of activation and inactivation, and sensitivities to pharmacological agents (Table 2). The slow (sustained) Ca^{2+} channel is sensitive to dihydropyridine Ca^{2+} agonists and antagonists, prefers Ba^{2+} over Ca^{2+} as charge carrier, and activates at more positive membrane potentials (high threshold) than the fast (transient) Ca^{2+} channel. The fast Ca^{2+} channel activates at more negative membrane potentials (low threshold), has an equal preference for Ca^{2+} and Ba^{2+}, and is insensitive to the dihydropyridines.

We examined the properties of the two types of Ca^{2+} channels in freshly isolated cells from guinea-pig small mesenteric arteries (resistance vessels), using the whole-cell voltage clamp technique (22). Fast and slow Ca^{2+} channel currents were recorded, and basal characteristics of these channels were the same as previously reported (Fig. 7). Injection of ATP (0.3-5 mM), using an intracellular perfusion method, modified only the slow current, but did not affect the fast current (Fig. 8). When the production of ATP was inhibited by adding cyanide and 2-deoxy-D-glucose to the bath (in absence of glucose), the Ca^{2+} slow current was abolished within 10 min (Table 1), whereas the fast Ca^{2+} current was only slightly affected. These results indicate that only the slow Ca^{2+} channel is metabolically dependent. The ATP dependence of the activity of the Ca^{2+} slow channels was also shown at the single-channel level (Ohya & Sperelakis, 1989b) (Fig. 9).

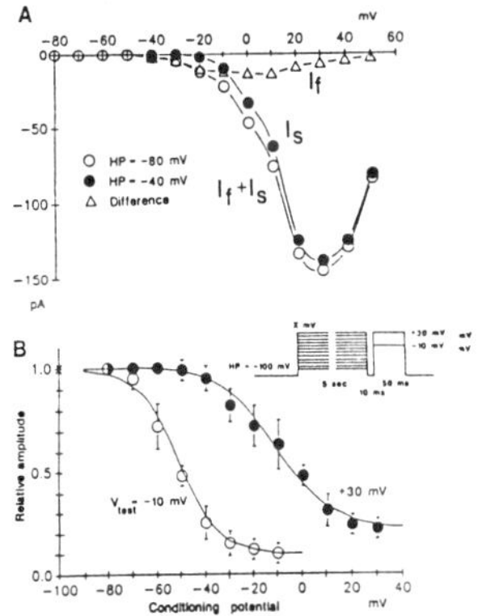

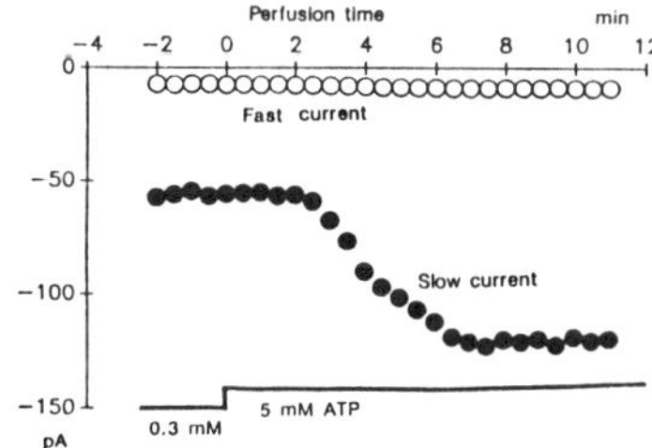

Figure 7. Two types of Ca^{2+} channel currents recorded from freshly-isolated single smooth muscle cells of guinea-pig mesenteric arteries. **A:** I/V curves of the peak current amplitudes obtained from HPs of -80 mV (unfilled circles) and -40 mV (filled circles), and the difference curve between these two currents (unfilled triangles). **B:** Steady-state inactivation curves for the fast current (unfilled circles) and slow current (filled circles. Various levels of conditioning pulses (5 sec) were applied before application of the test pulse (to -10 mV for the fast current, and to +30 mV for the slow current); 10 msec intervals were used between the conditioning pulse and test pulse. The HP was kept at -100 mV. Data are shown as mean $\pm$ SD (n = 3-5). Continuous curves were drawn by fitting the data to a Boltzmann distribution. The pipette contained high Cs^+ and 5 mM ATP, and the bath contained isotonic Ba^{2+} solution. (Data taken from 23 .)

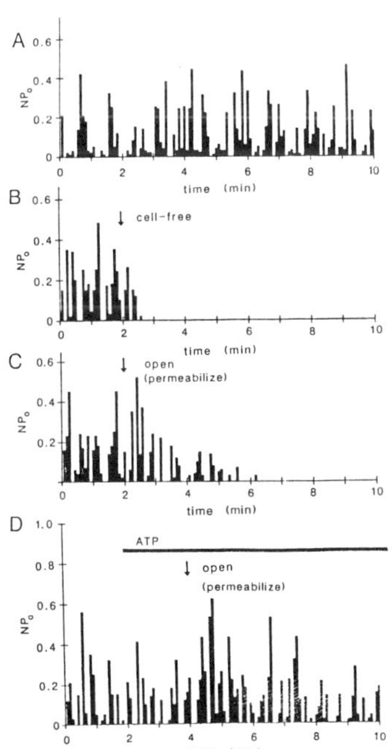

Figure 8. Effects of intracellularly-perfused ATP on the separated fast and slow Ca^{2+} channel currents. The concentration of ATP was increased from 0.3 mM to 5 mM using the intracellular perfusion technique. The time after beginning of the perfusion is indicated on the abscissa. The slow current was isolated by using a HP of -40 mV to voltage-inactive the fast channels; command potentials to +30 mV were applied every 30 sec. The fast current was recorded using a HP of -80 mV and a command potential to -20 mV. The peak magnitudes of the fast (open circles) and slow (filled circles) Ca^{2+} channel currents were plotted as a function of time following perfusion with 5 mM ATP. The fast and slow currents were obtained from two different cells. (Data taken from 23.)

Figure 9. Histogram of Ca^{2+} slow channel activity to show changes over time in cell-attached patch (A), isolated inside-out patch (cell-free) (B), and cell-attached patch with cell opened distally without ATP (C) and with ATP in the bath (D). The values of NP_O (ordinates) were calculated for each sweep from 5 to 100 ms after onset of the depolarizing pulse from a holding potential of -70 mV to 0 mV (at 0.2 Hz), and were plotted against time (abscissa). In B, the channel activity was recorded in the cell-attached configuration initially, and, at the arrow, the patch was excised (isolated patch). In C, the cell-attached configuration was used, and, at the arrow, one end of the cell was mechanically disrupted by using another electrode to permeabilize the membrane. In D, 5 mM ATP was applied to the bath solution before the membrane was permeabilized (at the arrow). The pipette contained Ba^{2+} (100 mM) and the bath was K^+-rich. (Data taken from 23.)

AMP-PNP and cAMP could not substitute for ATP in sustaining the slow Ca^{2+} current in intestinal smooth muscle cells (25). Since studies on guinea-pig portal vein showed that AMP-PNP did not have similar effects as ATP, but ATP-γ-S did (23), ATP may act via phosphorylation of ion channels or associated proteins. Alternatively, there may be an obligatory binding site for ATP on the inner surface of the channel which must be occupied for channel activity.

MODULATION OF Ca^{2+} CHANNELS BY AGONISTS

Agonists, such as norepinephrine (NE) or angiotensin II (A-II), may induce contraction of VSM via several mechanisms, including (a) stimulation of Ca^{2+} influx through receptor-operated channels (ROCs); (b) stimulation of Ca^{2+} entry through voltage-dependent Ca^{2+} channels, opened indirectly by the depolarization resulting from an increase in membrane conductance for other ions; and (c) release of intracellular Ca^{2+} from store sites (for review: 7, 17, 19, 34).

It is an attractive hypothesis that some agonists (NE, AII, etc.) may modify the voltage-dependent Ca^{2+} channels of VSM cells, as occurs in cardiac cells by beta-adrenergic agonists. The reported results have been contradictory, perhaps due to the use of different protocols and experimental conditions. For example, one group (1) reported that NE enhanced the Ca^{2+} channel current in rabbit ear artery, whereas another group (11) reported that NE inhibited the Ca^{2+} channel. Pacaud et al. (29) reported that NE enhanced the fast-type of Ca^{2+} channel, but inhibited the slow-type in cultured rat portal vein cells. Others could not observe a significant change in mesenteric artery and dog saphenous vein (3, 41). Agonists, such as NE may affect the ion channel activity in VSM cells via a number of different mechanisms, some direct and some indirect. Further, some intracellular factor which regulates the response to agonists might be lost by perfusion of the cell during whole-cell voltage clamp.

We previously examined the effects of A-II on the electrical activity of cultured VSM cells from rat aorta (42). A-II, applied by bolus perfusion, elicited a transient depolarization of up to 30 mV, which sometimes triggered an AP (Fig. 10, top). The A-II-induced depolarization disappeared in Na^+-free solution, suggesting that it was primarily due to an increase in Na^+ conductance, with resulting Na^+ influx. Continuous superfusion of cultured aortic reaggregates with A-II (10^{-6} M) also produced depolarization, which triggered a long-lasting AP (18). The membrane resistance was decreased by A-II, consistent with an increase in conductance for an inward depolarizing current. Low concentrations of A-II (10^{-9} M) caused a marked prolongation of the Ca^{2+}-dependent APs without depolarizing the membrane (Fig. 10, bottom).

POSSIBLE MECHANISMS FOR AGONIST MODULATION OF Ca^{2+} CHANNEL FUNCTION
Regulation by C-kinase

Many neurotransmitters and hormones, whose action involve Ca^{2+} mobilization, promote the breakdown of membrane phospholipids and the formation of two putative second messengers, inositol-1,4,5-trisphosphate (IP_3) and 1,2-diacylglycerol (DAG). One of the major roles of IP_3 is to release Ca^{2+} from intracellular stores (see review: 4), including the SR of VSM cells (33, 37). It has been suggested (30) that activation of PK-C may be involved in agonist-induced Ca^{2+} influx, which is responsible for the tonic phase of smooth muscle contraction. The phorbol esters, which activate PK-C, produce a slowly-developing sustained contracture of VSM that is partially dependent on $[Ca]_o$ (9, 10, 14, 16).

The whole-cell voltage clamp technique was used to examine the effects of A-II on Ca^{2+} and K^+ channels in cultured VSM cells from rabbit aorta (5). When the bath solution contained 50 mM Ba^{2+} (with TEA and 4-AP to block the outward current), depolarizing voltage steps from a holding potential of -80 mV elicited a small inward current. Addition of A-II caused a substantial increase in this current at all depolarizing steps, and an A-II antagonist blocked this effect. A-II (1-5 x 10^{-8} M) suppressed the delayed-rectifier outward K^+ current. These effects of A-II on the Ca^{2+} and K^+ channel currents can explain the depolarization and AP prolongation found in cultured rat aortic cells. These observations suggest that agonists control vascular tone, at least in part, by modifying the Ca^{2+} channel and K^+ channel activities, and thereby changing $[Ca]_i$.

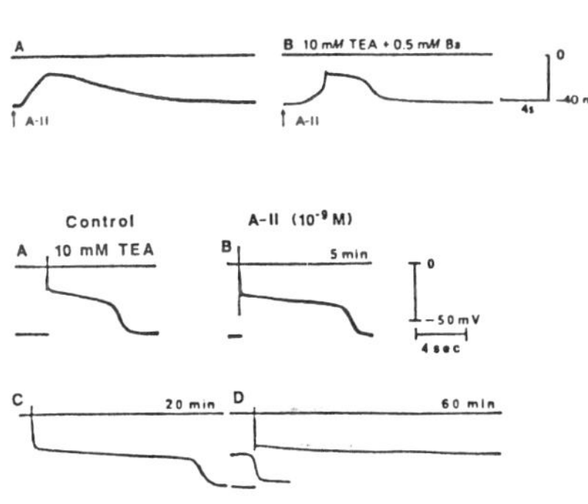

Figure 10. Effects of angiotensin (A-II) on the membrane potential and TEA-induced action potentials (APs) in cultured smooth muscle cells from rat aorta. **Top:** Bolus exposure (SHR preparations). **A:** A bolus of A-II (20 ml of 10^{-6} M) rapidly depolarized the cell. **B:** In the presence of 10 mM TEA and 0.5 mM Ba^2, an A-II bolus depolarized and induced an AP. (Taken from Zelcer & Sperelakis, 1981.) **Bottom:** Steady-state exposure. **A:** Control AP induced by electrical stimulation in the presence of 10 mM TEA. **B-D:** Bath addition of 10^{-9} M A-II (continuous exposure) progressively increased the AP duration at 5 min (B), 20 min (C), and 60 min (D). All records were from the same impalement. (Data taken from 18.)

We also demonstrated that A-II greatly stimulated the Ca^{2+} slow channel current in isolated single cells from guinea pig portal vein (24) (Fig. 11). This effect of A-II was mediated via a G-protein, based on inhibition of the A-II effect by GDP-B-S and stimulation of the basal I_{Ca} by GTP-γ-S. However, this G-protein was not sensitive to pertussis toxin or to cholera toxin, hence is not G_i-like or G_s-like.

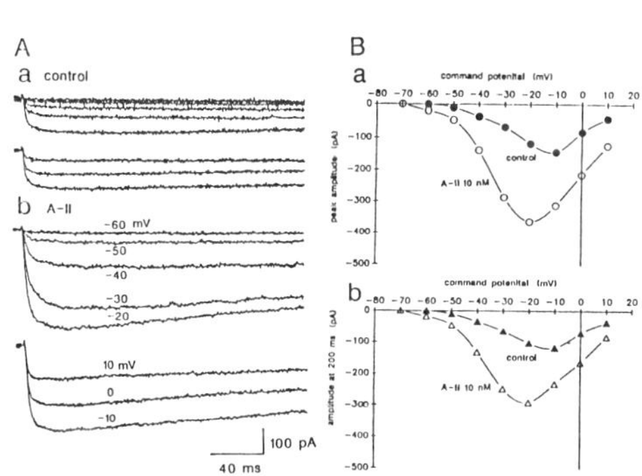

Figure 11. Effects of angiotensin-II (A-II) on the Ca^{2+} channel current recorded from guinea-pig portal vein. **A:** Ba^{2+} current recorded before (a) and 3 min after application of 10 nM A-II. Step command pulses were applied over the voltage range of -60 mV to 10 mV from a holding potential (HP) of -90 mV. **B:** Peak current amplitude (a) and current amplitude measured at 200 ms (b), measured before (filled symbols) and after (unfilled symbols) 10 nM A-II application, were plotted against the command potentials. Data in A and B were collected from the same cell. Bath solution contained 2 mM Ba^{2+} as the charge carrier. Pipette solution contained high Cs with 5 mM ATP and 0.1 mM GTP. (A), (B), and (C) were taken from three different cells.

To examine a possible role of PK-C in the modulation of electrical activity in VSM cells, the effects of a phorbol ester, phorbol 12,13-diacetate (PDA), were determined on the resting potential and APs in cultured aortic reaggregates (28). Superfusion with PDA (0.1-5 μM) produced a gradual depolarization and allowed the appearance of an active (plateau-type) AP response to electrical stimulation (Fig. 12). This response was abolished by verapamil. When APs were elicited in the presence of 15 mM TEA, PDA also depolarized and prolonged the AP duration. However, after prolonged exposure (20-30 min) to higher concentrations of PDA, the APs became depressed, perhaps due to depolarization-induced inactivation of the Ca^{2+} channels.

36

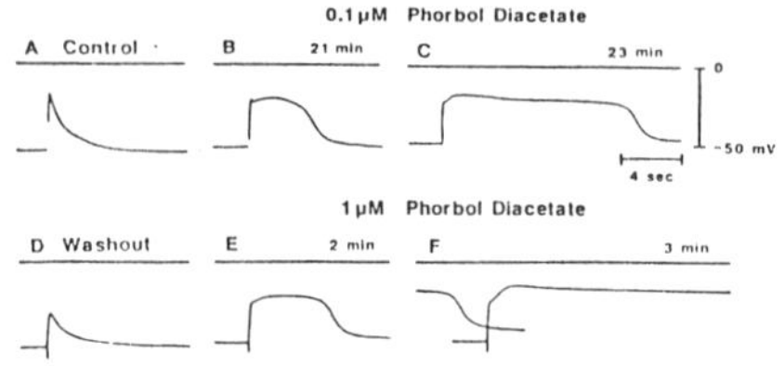

Figure 12. Induction of APs by phorbol diacetate in cultured rat aortic cells. **A:** Control record showing a lack of response to electrical stimulation in normal physiological salt solution. **B:** Induction of an AP after exposure to 10^{-7} M phorbol diacetate for 21 min. **C:** Further prolongation of the plateau component (23 min). **D:** Lack of response to stimulation after washout for 2.5 hr. **E:** Induction of the AP after exposure to 10^{-6} M phorbol diacetate for 2 min. **F:** Further prolongation of the plateau component (repolarization shown in subsequent sweep superimposed). (Ousterhout & Sperelakis, unpublished observations.)

Thus, the effects of phorbol esters on the ionic currents in VSM cells may be complex. The depolarization produced by the phorbol esters could be due to inhibition of K^+ channels, and the induction of APs could result from this plus the activation of a Ca^{2+} conductance, possibly mediated by phosphorylation. For example, in cultured aortic (A7r5 cell line) cells, phorbol esters increased the slow (sustained) Ca^{2+} channel current (12). Werz and MacDonald (39) observed dual effects of phorbol esters on the Ca^{2+}-dependent APs in cultured neurons, and suggested that either Ca^{2+} or K^+ conductances could be depressed depending on the membrane potential. Our observations suggest that C-kinase may be involved in the regulation of ion channels in VSM cells.

Direct Regulation by G-Protein

Direct regulation of ionic channels by GTP-binding protein (G-protein) has been identified in several tissues (for review: 8, 31). In some neurons, G-protein (G_0) is thought to mediate the modulation (inhibition) of Ca^{2+} channels produced by receptor activation (15). In cardiac cells, G-protein (G_s) was reported to directly enhance the activity of Ca^{2+} channels (40).

We examined the effects of GTP-γ-S on Ca^{2+} channels in freshly-isolated single cells from guinea-pig portal vein (21). Single Ca^{2+} channels (conductance of 18-22 pS) were recorded in cell-attached patch configuration with 100 mM Ba^{2+} and Bay-K-8644 in the pipette and high K^+ solution in the bath. After permeabilization of one end of the cell, GTP-γ-S (0.1 mM) applied to the bath (access into the cell interior) enhanced the channel activity. This observation suggests that a G-protein may be one of the factors regulating Ca^{2+} channels in VSM cells also. However, it is not known whether G-protein acts directly or indirectly on the Ca^{2+} channel.

CONCLUSIONS

Vascular tone is regulated by a variety of neurotransmitters, vasoactive hormones and autacoids, and vasoactive drugs. These actions are mediated, at least in part, by actions on the membrane ion channels, exerted either directly or indirectly. In this section, we described evidence that three different protein kinase systems (protein kinases A and G, and C-kinase) act on and modulate the Ca^{2+} slow channels in VSM cells, and that ATP is required for activity of these channels (Fig. 13).

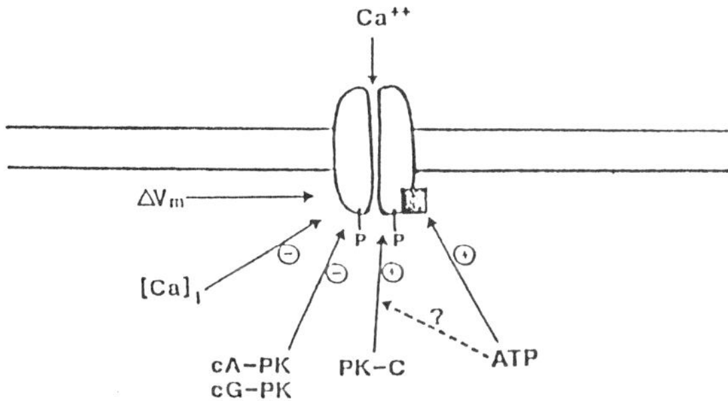

Figure 13. Diagram depicting some of the factors that regulate the Ca^{2+} slow channels in VSM cells. As shown, phosphorylation of a site on the channel by cA-PK or cG-PK inhibits channel activity, whereas phosphorylation of a second site by PK-C stimulates channel activity. Therefore, agents that elevate cAMP or cGMP inhibit the channels, and therefore decrease Ca^{2+} influx, and produce vasodilation. Agonists that increase DAG production stimulate the channels, and therefore increase Ca^{2+} influx, and produce vasoconstriction. As shown, ATP is required for activity of the channels ($K_{0.5}$ of ca 0.3 mM). There may be a binding site for ATP that must be occupied, or ATP may be acting via phosphorylation. Ohya, Kitamura and Kuriyama (25) had shown that internal Ca^{2+} inhibits channel activity.

ACKNOWLEDGEMENTS

These research projects were supported by N.I.H. Grant HL-22619. We would like to thank Rhonda S. Hentz for typing this manuscript, and Anthony Sperelakis for making the figures.

REFERENCES

1.   Aaronson, P.I., Benham, C.D., Bolton, T.B., Hess, P., Lang, R.J., and Tsien, F.W., J. Physiol. _377_, 36 (1986).
2.   Bean, B., J. Gen. Physiol. _86_, 1-30 (1985).
3.   Bean, B.P., Sturek, M., Puga, A., and Hermsmeyer, K., Circ. Res. _59_, 229-235 (1986).
4.   Berridge, M.J. and Irvine, R.F., Nature _312_, 315-321 (1984).
5.   Bkaily, G., Peyrow, M., Sculptoreanu, A., Jacques, D., Chahine, M., Regoli, D., and Sperelakis, N., Pflugers Arch. _412_, 448-450 (1988).
6.   Bkaily, G., Yamamoto, T., Peyrow, M., Sculptoreanu, A., Jacques, D., and Sperelakis, N., Mol. Cell. Biochem. _80_, 59-72 (1988).
7.   Bolton, T.B., Physiol. Rev. _59_, 606-718 (1979).
8.   Brown, A.M. and Birnbaumer, L., Am. J. Physiol. _254_, H401-H410 (1988).
9.   Chiu, A.T., Bozarth, J.M., Forsyth, M.S., and Timmermans, P.B.M.W.M., J. Pharmacol. Exp. Therap. _242_, 934 (1987).
10. Danthuluri, N.R. and Deth, R.D., Eur. J. Pharmacol. _125_, 1103 (1986).
11. Droogmans, G., Declerck, I., and Casteel, R., Pflugers Arch. _409_, 7-12 (1987).
12. Fish, R.D., Sperti, G., Colucci, W.S., and Clapham, D.E., Circ. Res. _62_, 1049 (1988).
13. Friedman, M.E., Suarez-Kurtz, G., Karzorowski, G.J., Katz, G.M., an Reuben, J.P., Am. J. Physiol. _198_, H699-H703 (1986).
14. Gleason, M.M. and Flaim, S.F., Biochem. Biophysi. Res. Comm. _138_, 1362 (1986).
15. Hescheler, J., Kameyama, M., Trautwein, W., Mieskes, G., and Soling, H.D., Eur. J. Biochem. _165_, 261-266 (1987).
16. Itoh, H. and Lederis, K., Am. J. Physiol. _252_, C244 (1987).
17. Johansson, B. and Somlyo, A.P., in Handbook of Physiology, Sect. 2, The Cardiovascular System, Vol. 2, (Am. Physiol. Soc., Bethesda 1980) pp. 301-323.
18. Johns, D.W. and Sperelakis, N., Eur. J. Pharmacol. _187_, 183-191 (1990).
19. Kuriyama, H., Ito, Y., Suzuki, H., Kitamura, T., and Itoh, T., Am. J. Physiol. _243_, H641-H662 (1982).

20. Nilius, B., Hess, P., Lansman, J.B., and Tsien, R.W., Nature 316, 443-446 (1985).

21. Ohya, Y. and Sperelakis, The Physiologist 31, A88 (1988).

22. Ohya, Y. and Sperelakis, N., Circ. Res. 64, 145-154 (1989).

23. Ohya, Y. and Sperelakis, N., Pflugers Arch. 414, 257-264 (1989).

24. Ohya, Y and Sperelakis, N., Circ. Res. 68, 763-777 (1991).

25. Ohya, Y., Kitamura, K., and Kuriyama, H., Pflugers Arch. 408, 465-473 (1987).

26. Osterrieder, W., Brum, G., Hescheler, J., Trautwein, W., Flockerzi, V., and Hofmann, F., Nature 298, 576-578 (1982).

27. Ousterhout, J.M. and Sperelakis, N., Eur. J. Pharmacol. 144, 7-14 (1987).

28. Ousterhout, J.M. and Sperelakis, N., (Unpublished observations).

29. Pacaud, P., Lorrand, G., Mironneau, C., and Mironneau, J., Pflugers Arch. 410, 557-559 (1987).

30. Rasmussen, H., Takuwa, Y., and Park, S., FASEB J. 1, 177 (1987).

31. Rosenthal, W., Hescheler, J., Trautwein., W., and Schultz, G., J. Molec. Cell. Cardiol. 3, 271-286 (1971).

33. Somlyo, A.V., Bond., M., Somlyo, A.P., and Scarps, A., Proc. Natl. Acad. Sci. 82, 5231 (1985).

34. Sperelakis, N. and Ohya, Y., in Physiology and Pathophysiology of the Heart, 2nd edition., (Kluwer Academic Press., Boston. 1989) pp. 773-811.

35. Sperelakis, N. and Schneider, J.A., Am. J. Cardiol. 37, 1079-1085 (1976).

36. Sperelakis, N., N. Tohse, and Y. Ohya, (in press) (1991).

37. Suematsu, E., Hirata, M., Sasaguri, T., Hashimoto., T., and Kuriyama, H., Comp. Biochem. Physiol. 82A, 645-649 (1985).

38. Tsien, R.W., Giles, W., and Greengard, P., Nature 240, 181-183 (1972).

39. Werz, M.A. and MacDonald, R.L., J. Neurosci. 7, 1639-1647 (1987).

40. Yatani, A., Condina, J., Imoto, Y., Reeves, J.P., Birnbaumer, L., and Brown, A.M., Science 238, 1288-1292 (1987).

41. Yatani, A., Seidel, C.L., Allen, J., and Brown, A.M., Circ. Res. 60, 523-533 (1987).

42. Zelcer, E. and Sperelakis, N., Blood Vessels 18, 263-279 (1981).

AGONIST MODULATION OF VOLTAGE-DEPENDENT CALCIUM CHANNELS IN VASCULAR SMOOTH MUSCLES

YUSUKE OHYA* and NICHOLAS SPERELAKIS

*Second Department of Internal Medicine, Kyushu University, Faculty of Medicine, Fukuoka 812, Japan and Department of Physiology & Biophysics, University of Cincinnati, College of Medicine, Cincinnati OH 45267-0576,USA

INTRODUCTION

Recent progress in voltage clamp experiments facilitates our understanding of regulation of the voltage-dependent Ca^{2+} channels in vascular smooth muscles. In the present article, we review the agonist modulation of voltage-dependent Ca^{2+} channels based on recent results from our laboratories.

MODULATION OF CA^{2+} CHANNELS BY AGONISTS

Agonists, such as norepinephrine(NE) or angiotensin II (Ang-II) increase intracellular Ca^{2+} concentration in vascular smooth muscles via several mechanisms, including (a) stimulation of Ca^{2+} influx through receptor-operated channels, (b) activation of voltage-dependent Ca^{2+} channels, opened indirectly by the depolarization resulting from increases in membrane conductances for various ions, and (c) release of Ca^{2+} from intracellular store sites (for review; 6, 23, 27, 44).

Agonists may modify the voltage-dependent Ca^{2+} channels of vascular smooth muscle cells, as occurs in cardiac cells by beta-adrenergic agonists. However, one group (3) has reported that NE enhances the Ca^{2+} channel current in rabbit ear artery, whereas another group (12) has reported that it is inhibited. It was reported that NE enhanced the fast-type Ca^{2+} channels in cultured rat portal vein cells (38). Others observed no significant changes in rat mesenteric artery and dog saphenous vein (2, 51). Some of the variability in these findings may be due to the use of different protocols and experimental conditions. However, it is likely that agonists affect the ion channel activity via a number of different mechanisms. Furthermore, there might be some factor which regulates the response to agonists, which is lost from the cell by perfusion under whole-cell voltage clamp configuration. Ca^{2+}-induced Ca^{2+} channel inactivation also may modify agonist action on the Ca^{2+} channels. We have previously reported that this mechanism was present in vascular muscles, as in cardiac muscles and neurons, and that the threshold concentration of intracellular Ca^{2+} required to inactivate the Ca^{2+} channels was lower than in cardiac muscle (31). Agonists trigger Ca^{2+} release from the store sites, and this released Ca^{2+} may inhibit the Ca^{2+} channel activation. The Ca^{2+}-induced Ca^{2+} inactivation mechanism, along with possible unknown factors, makes agonist action on Ca^{2+} channels complicated.

ACTION OF ANG-II ON THE ELECTRICAL PROPERTIES OF VASCULAR SMOOTH MUSCLE CELLS

Ang-II is a potent vasoconstrictive peptide, and is known to enhance membrane excitation. We previously examined the effects of Ang-II on the membrane potentials of cultured vascular smooth muscle cells from rat aorta (52). Ang-II (10^{-6} M), applied by bolus perfusion or continuous superfusion, elicited a depolarization, which sometimes triggered an action potential. The Ang-II-induced depolarization disappeared in Na^+-free solution, suggesting that it was primarily due to an increase in Na^+ conductance. Low concentrations of Ang-II (10^{-9} M) also caused a marked prolongation of the Ca^{2+}-dependent action potentials (TEA-induced) without depolarizing the membrane (24). Similar observations were reported in rat portal vein, in which Ang-II enhanced the Ca^{2+}-dependent action potentials with or without preceeding depolarization (46).

The effects of Ang-II on Ca^{2+} and K^+ channels of cultured rabbit aortic cells was examined by the whole-cell voltage clamp technique (5). TEA and 4-AP were used to

40

block the outward K^+ currents, and Ba^{2+} (50 nM) was used as the charge carrier. Depolarizing voltage steps, from a HP of -80 mV, elicited a small inward Ca^{2+} channel current. Ang-II substantially increased this current at all depolarizing steps. The Ang-II antagonist, [Leu-8]-Ang-II, blocked the effect of Ang-II. In physiological bath solution, depolarizing pulses evoked a delayed outward K^+ current. Superfusion with Ang-II (1-5 x 10^{-8} M) suppressed the delayed outward current; [Leu-8]-Ang-II inhibited the effect of Ang-II. These effects of Ang-II on the Ca^{2+}-channel and K^+-channel currents can explain the previous observations that Ang-II depolarized the membrane and prolonged the action potential duration.

Enhancement of the Ca^{2+} channel current by Ang-II was further clarified in guinea-pig portal vein (Fig.1) (34). The pipette solution contained high Cs^+ to inhibit K^+ currents and isolate the Ca^{2+}-channel current. Ba^{2+} (2 mM) was in the bath solution as a charge carrier for the Ca^{2+} channel. To ensure consistent results, the conditions we used were as follows: 1) We used Ba^{2+} ion as a charge carrier to record the Ca^{2+}-channel currents, and the single cells were stored in the Ca^{2+}-free solution before experimentation, to avoid Ca^{2+} loading of the store sites. 2) The pipettes used had a relatively small tip diameter to avoid loosing unknown intracellular factors. 3) Ang-II was applied in the bath solution after the same preincubation time for stabilizing the intracellular content by the pipette solution (after making whole-cell configuration). Ang-II enhanced the Ca^{2+}-channel currents in about 90 % of the cells used. The threshold concentration of Ang-II to enhance the current was about 10^{-10} M. Half-enhancement of the maximal response was observed at about 10^{-8} M.

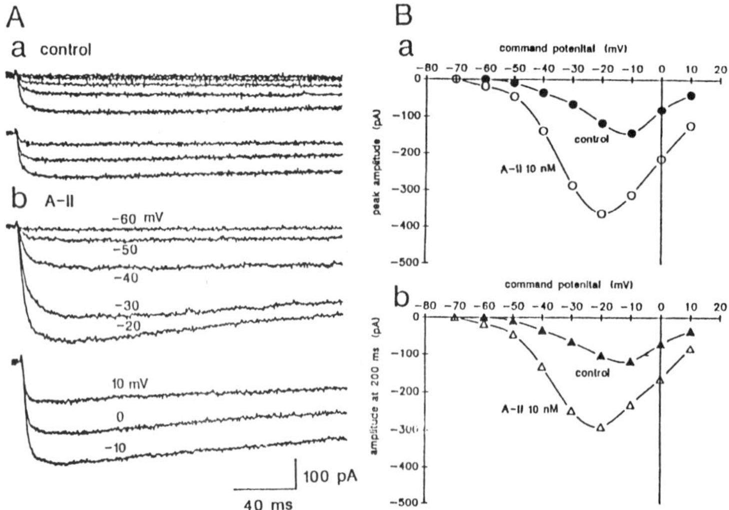

Figure 1. Effects of angiotensin-II (Ang-II) on the Ca^{2+} channel current recorded from single smooth muscle cells isolated from guinea-pig portal vein. **A:** Ba^{2+} current recorded before (a) and 3 min after (b) application of 10 nM Ang-II. Step command pulses were applied over the voltage range of -60 mV to 10 mV from a holding potential (HP) of -90 mV. **B:** Peak current amplitude (a; circles) and amplitude measured at 200 ms (b; triangles), before (filled symbols) and after (unfilled symbols) Ang-II application, were plotted against the command potentials. Data in A and B were collected from the same cell. Bath solution contained 2 mM Ba^{2+} as the charge carrier. Pipette solution contained high Cs^+, with 5 mM ATP and 0.1 mM GTP.

Besides enhancement of the amplitude, Ang-II also shifted the current/voltage relationship in the negative direction by 5-15 mV. With 2 mM Ba^{2+}, threshold was between -60 mV and -50 mV, and maximum amplitude was observed around -10 mV; with 2 mM Ca^{2+}, threshold was about -40 mV, and maximum amplitude was at about 0 mV. If simple extrapolation is applied, in physiological condition (2 mM Ca^{2+}), the threshold potential should be shifted to be between -55 and -45 mV. This value is similar to the resting potential reported for this tissue. This could explain why Ang-II enhanced spike activity without depolarization (24, 46). Furthermore, the shift of the threshold potential by agonists may contract the vascular smooth muscle, even without membrane depolarization.

We also observed a secondary effect of Ang-II on the Ca^{2+} channels, namely an inhibitory effect. After the Ca^{2+} channels were stimulated, the current decreased gradually. This phenomenon was most evident with high concentrations of Ang-II. This is not due to the Ca^{2+}-induced Ca^{2+} channel inactivation, because we used Ba^{2+} ion, instead of Ca^{2+}, in the bath solution as the charge carrier.

POSSIBLE INVOLVEMENT OF GTP-BINDING PROTEIN (G-PROTEIN) IN ANG-II ACTION

Involvement of G protein in the regulation of Ca^{2+} channels by Ang-II in vascular smooth muscle cells was investigated by whole-cell voltage clamp method (34). The analogues of GTP, GDP-β-S and GTP-γ-S, are considered to be good tools to examine the involvement of G-protein in the receptor coupling to various effectors (16). Single cells from guinea pig portal vein were used. Either GTP (0.1 mM), GDP-β-S (0.3 mM or 1 mM) or GTP-γ-S (0.3 mM or 1 mM) was contained in the pipette solution for diffusion into the cell interior. Ang-II (10 nM) was applied to the cell after a short (1-3 min) or long (5-8 min) incubation. To assess the effects of GTP-γ-S or GDP-β-S on the Ang-II action, the relative current amplitude after Ang-II application (Ang-II stimulated current) with GTP-γ-S or GDP-β-S in the pipette solution was compared with the control. Control data were obtained with 0.1 mM GTP. Effects on GTP-γ-S on the unstimulated Ca^{2+} channel current were also examined. GTP-γ-S is considered to: 1) substitute for endogenous GTP in the binding site of the G-protein, 2) be a poor substrate for the GTPase, causing its effect to be persistent, and 3) cause slowly increasing activation of G-protein in the absence of agonist stimulation (16). In electrophysiological studies, persisting and strong effects of GTP-γ-S in the activation of G_k channel in atrial cells and in the inhibition of the Ca^{2+} currents of neuronal cells were observed (17, 19, 50).

In our experiments of guinea pig portal vein, GTP-q-S (0.3 mM and 1 mM) produced a slowly progressive increase in the basal current amplitude, and inhibited the subsequent Ang-II action (Fig. 2), perhaps due, at least in part, to substitute for endogenous GTP in the G-protein and to the increased basal level (Fig. 2A). The slow increase in the basal amplitude might be due to a slow increase in degree of activation of G-protein.

GDP-β-S is a competitive inhibitor of GTP, binding to the G-proteins (16). There are several reports with electrophysiological studies showing that intracellular GDP-β-S blocked the agonist-dependent responses (11, 17-19). In our observations, GDP-β-S suppressed the effect of Ang-II, dose-dependently and time-dependently (Fig. 3). Therefore, this result, together with those of GTP-c-S, provides positive evidence of the involvement of G-protein.

Pertussis toxin (PT) is the bacterial toxin which ribosylates the subunit of G_i or G_i-like protein, and consequently inhibits the action of this G-protein (see 10). Preincubation of muscle tissues with pertussis toxin (PT; 10^{-6} g/ml) (for up to 6 hr at 36° C) or intracellular application of preactivated PT (10^{-6} g/ml) with NAD (10^{-3} M) did not inhibit the Ang-II action (Fig. 4). In biochemical studies and voltage clamp studies, these procedures were demonstrated to be effective in ribosylating the G-protein in cardiac tissues (26) and vascular tissues (20, 41). To confirm our results, we treated the muscle tissues by PT (10^{-6} M) for 24 hr at 36° C, and measured the mechanical force produced by high K^+ and Ang-II. There was no significant difference between control tissue (incubation without PT) and PT-pretreated tissue in the contraction induced by 10 nM Ang-II (Y. Ohya, unpublished observations).

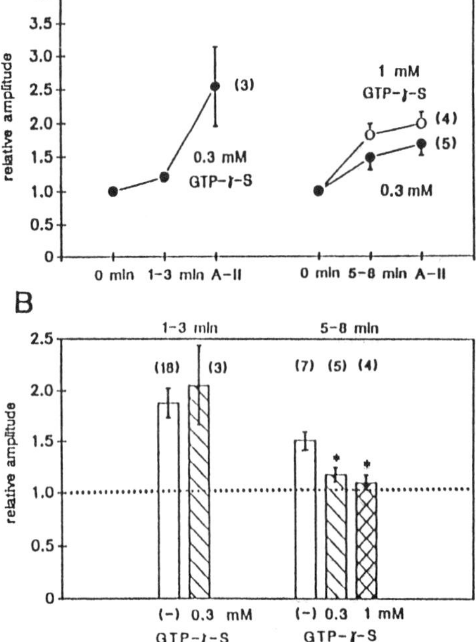

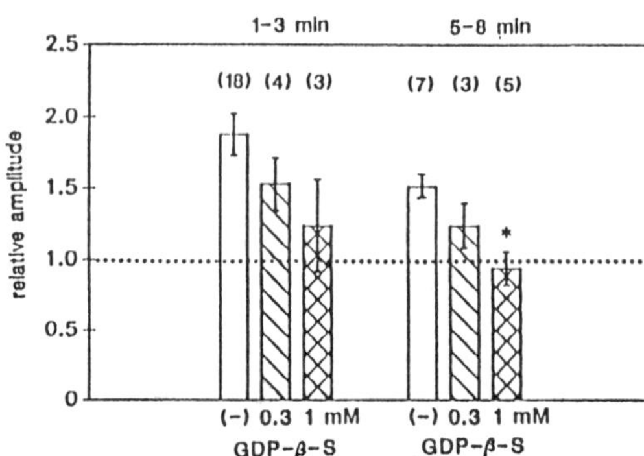

Figure 2. Relative amplitudes of the Ba^{2+} current affected by 10 nM A-II with and without presence of GDP-β-S. The basal current amplitude before application of A-II was normalized to 1.0 (horizontal dotted line). Pipette solution contained either 0.1 mM GTP as control (--) or GDP-β-S at 0.3 or 1 mM. Incubation with the pipette solution was either short (1-3 minutes) or long (5-8 minutes). The number of cells for each condition is indicated in parentheses. *Statistical significance compared with control.

Figure 3. Effect of GDP-γ-S on the action of angiotensin II (A-II) on the Ca^{2+} channel current. Ba^{2+} current was recorded before and after application of 10 nM A-II. Currents were evoked by a command potential to -20 mV from a holding potential of -90 mV. Effect of incubation with GTP-γ-S was recorded 1-8 min after cell dialysis with the pipette solution containing 0.3 mM or 1.0 mM GTP-γ-S. A-II was applied to the bath solution and stimulated the current. Control current indicates the basal current recorded just after the cell dialysis was started. **Panel A:** Graphic plot of all data: current amplitude recorded just after the cell dialysis was started (0 min), after short incubation (1-3 min), after long incubation (5-8 min), and after application of 10 nM A-II. Concentrations of GTP-γ-S used were 0.3 mM (filled circles) and 1 mM (open circles). **Panel B:** Summary of the effects of GTP-γ-S on A-II on action. Relative current amplitudes enhanced by A-II were normalized and plotted against incubation with and without GTP-γ-S. Current amplitude just before application of A-II was normalized to 1.0. Pipette solution contained either 0.1 mM GTP as control (--) or GTP-γ-S at 0.3 or 1 mM. Exposure to GTP-γ-S was either short (1-3 min) or long (5-8 min). The horizontal dotted line indicates the basal (unstimulated) Ca^{2+} channel current. The number of cells for each condition is indicated in parentheses. *Statistical significance compared with control.

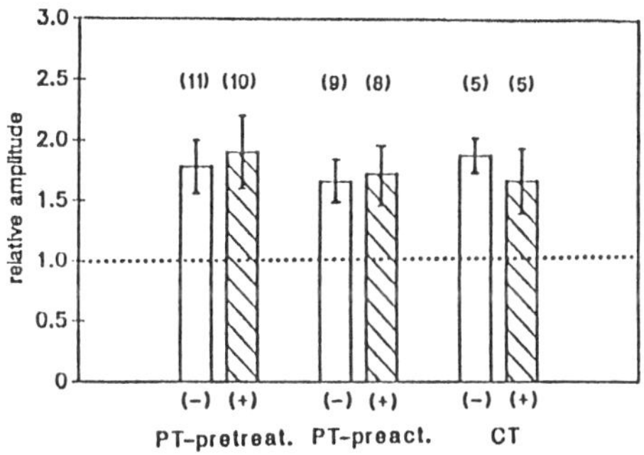

Figure 4. Lack of effect of pertussis toxin (PT) and cholera toxin (CT) treatments on the action of angiotensin II (A-II) on the Ca^{2+} channel current. Graphic summary of the data on toxins. Relative current amplitudes stimulated by 10 nM A-II with (+) and without (-) toxin treatments are plotted. The horizontal dotted line indicates the basal (unstimulated) Ca^{2+} channel currents. As can be seen, treatment with PT (pretreated and preactivated) or CT had no significant effect on the A-II stimulation of the basal Ca^{2+} current.

Effect of PT treatment on the contraction and Ca^{2+}-transient (using Ca^{2+}-sensitive dyes) measured from vascular smooth muscles in response to agonists (such as NE, serotonin or Ang-II) were controversial. In dog and rabbit mesenteric artery, PT treatment abolished the increase in Ca^{2+} measured by Fura-2 and mechanical force produced by NE and Ang-II (42). Inhibition of the Ca^{2+} transient was also reported for cultured rat aortic muscle in response to NE, Ang-II, and serotonin (8, 25). However, the inhibition was only partial (8, 25). Furthermore, even in the same tissue (rat aortic cultured cells), others reported PT-treatment did not affect the Ca^{2+} increase produced by Ang-II (47). Others also reported that PT pretreatment had little effect on the contraction of porcine coronary artery induced by acetylcholine (41). In an electrophysiological study, Nozaki & Sperelakis (30) reported that PT treatment of guinea-pig mesenteric artery did not change resting potential, input resistance, or depolarization induced by NE or ATP. These reports suggest (a) that there may be regional and/or species differences in the PT effects, and (b) that the Ca^{2+} increase and contraction produced by agonists may be mediated by several mechanisms that are both PT-sensitive and PT-insensitive.

Another important toxin for investigation of the involvement of GTP-binding protein in the agonist response is cholera toxin (CT). CT ribosylates G_S protein (16, 20). In vascular smooth muscle, CT was reported to inhibit agonist-induced contraction, probably due to increasing cytosol cAMP level (1, 37). In our experiments, CT did not change the Ang-II response, suggesting that G_S does not have an important role in this system (Fig. 4). It was reported that G_S protein had a direct stimulatory action on the Ca^{2+} channels of cardiac muscles (7). These results suggest that Ang-II-induced stimulation of Ca^{2+} channels in vascular smooth muscle cells may be mediated by a G-protein which is not PT-sensitive or CT-sensitive.

POSSIBLE MECHANISMS UNDERLYING AGONIST MODULATION OF CA^{2+} CHANNELS
Regulation by C-kinase

Neurotransmitters and hormones promote the breakdown of membrane phospholipids and the formation of two putative second messengers, inositol-1,4,5-triphosphate (IP_3) and 1,2-diacylglycerol (DAG) by activating phospholipase C. (reviewed in 29). Phospholipase C is known to be modified by G-protein. Therefore, products from this phospholipid-breakdown may be the candidates to explain the G-protein-mediated Ang-II action. IP3 is known to release Ca^{2+} from intracellular stores (reviewed in 4), including the SR of vascular smooth muscle (43, 45). Recent studies also suggest an important role of IP_4 in regulating Ca^{2+} influx (21). We have previously reported that neither IP_3 or IP_4, perfused into the cell, modify the Ca^{2+} channel current directly, suggesting that these two phospholipid metabolites were not candidates for modifying the Ca^{2+} channels in vascular smooth muscles (35).

Activation of protein kinase C has been suggested (39) to be involved in agonist-induced Ca^{2+} influx, responsible for the tonic phase of smooth muscle contraction. Phorbol esters, which activate protein kinase C (PK-C), produce a slowly developing sustained contracture of vascular muscle that is dependent at least partially on $[Ca]_o$ (9, 15, 22). To examine a possible role of PK-C in modulation of electrical activity of vascular smooth muscle cells, the effects of phobol 12,13-diacetate (PDA) were determined on the membrane potential and action potentials (APs) in cultured aortic reaggregates (36). Superfusion with PDA ($0.5 - 1 \times 10^{-3}$ M) produced a gradual depolarization and, in most cases, also allowed the appearance of an active AP (plateau-type) in response to electrical stimulation. This AP response was abolished by verapamil. When APs were elicited in the presence of TEA, PDA also depolarized and, at first, prolonged the AP duration. After prolonged exposure to PDA (20-30 min), the APs became depressed or abolished. Therefore, phorbol esters had several effects on membrane excitability, including depolarization accompanied by induction or enhancement of Ca^{2+}-dependent APs, followed by depression of the APs. The depolarization produced by the phorbol esters could be due to inhibition of K^+ channels, and the induction of APs could result from the depression of a K^+ conductance and/or the activation of a Ca^{2+} conductance. For example, in cultured aortic cells (A7r5 cell line) (14) and neonatal rat cardiac myocytes, phorbol esters increased the slow (sustained) Ca^{2+} channel current (13). The subsequent depression of the APs could be due, in part, to depolarization-induced inactivation of the Ca^{2+} channels, or to a more direct inhibition of the Ca^{2+} channels. Such dual actions of phorbol esters were reported in cultured neurons and cardiac myocytes (28, 48). Thus, C-kinase might be involved in the agonist actions on Ca^{2+} channels. However, since several reports (including personal communication) suggested that C-kinase did not have stimulatory effects, further study is required to clarify the action of C-kinase.

Direct Regulation by G-protein

Direct regulation of ionic channels by G-protein has been reported in several tissues (reviewed in 7, 40). In some neurons, G-protein (G_o) mediates the inhibition of Ca^{2+} channels produced by receptor activation by agonists (17). In cardiac cells, G-protein (G_s) was reported to directly enhance the activity of Ca^{2+} channels (50).

Effects of GTP-q-S on single Ca^{2+} channels were examined in single cells isolated from guinea pig portal vein (32). The activity of single Ca^{2+} channels (conductance of 18-22 pS) was recorded in cell-attached patch configuration with 100 mM Ba^{2+} and Bay-K-8644 in the pipette and high K^+ solution in the bath. One end of the cell was disrupted mechanically by another electrode during the single channel recording in the cell-attached configuration, so that the cell interior communicated with the bath solution (open-cell-attached configuration; see 33). In this configuration, chemicals applied to the bath solution can diffuse into the cell. Conversely, intracellular factors may diffuse out of the cell. Run-down of the single Ca^{2+} channel activity was prevented by using 1 mM Bay-K-8644. After obtaining the control records, 0.1 mM GTP-γ-S was applied inside the cell. GTP-γ-S, which activates the G-protein without agonist stimulation, enhanced the Ca^{2+} channel activity in about 30% of the cells. Figure 5 shows results obtained from such an experiment. As can be seen, the single-channel activity of the Ca^{2+} channels was stimulated. Single channel amplitude was not affected. These results are consistent with the possibility that a G-protein may directly regulate Ca^{2+} channels in vascular smooth muscle cells. However, the failure in about 70% of the cells may also suggest that the regulation of voltage dependent Ca^{2+} channels in vascular smooth muscle has multiple factors.

In conclusion, Ang-II contracts vascular smooth muscle, at least in part, by stimulating Ca^{2+} influx via voltage-dependent Ca^{2+} channels. Stimulation of the Ca^{2+} influx could occur with or without depolarization, because the threshold potential was shifted in a negative direction. Ang-II action on the Ca^{2+} channels may be mediated by G-protein, which is insensitive to both pertussis toxin and cholera toxin. The underlying mechanisms of Ang-II action are still not fully understood. Involvement of C-kinase-dependent phosphorylation or a direct modulation of the Ca^{2+} channel by G-protein should be further clarified.

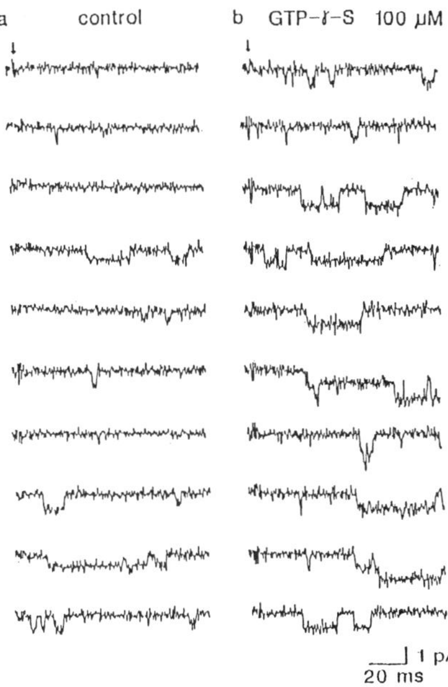

Figure 5. Effects of GTP-γ-S on the Ca^{2+} channel activity in a single cell isolated from guinea pig portal vein, using the open-cell-attached configuration with presence of 1 mM Bay-K-8644. After one end of the cell was disrupted in the cell-attached configuration, control recordings were obtained (a). Bath application of 0.1 mM GTP-γ-S (access into the cell interior) enhanced the channel openings (b).

REFERENCES

1. Asano, M., K. Masuzawa, and T. Matsuda, Br. J. Pharmacol. 95:241-251(1988).
2. Bean, B.P., M. Sturek, A. Puga, and K. Hermsmeyer, Circ. Res. 59:229-235 (1986).
3. Benham, C.D. and R.W. Tsien, J. Physiol. 404:767-784 (1988).
4. Berridge, M.J., and R.F. Irvine, Nature 312:315-321 (1984).
5. Bkaily, G., M. Peyrow, A. Sculptoreanu, D. Jacques, M. Chahine, D. Regoli, and N. Sperelakis, Pflugers Arch 412:448-450 (1988).
6. Bolton, T.B., Physiol.Rev. 59 , 606-718 (1979).
7. Brown, A.M., and L. Birnbaumer, Am. J. Physiol. 254:H401-410 (1988).
8. Bruns, C., and D. Marme, FEBS Letters 212:40-44 (1987).
9. Danthuluri, N.R., and R.C. Deth, Eur.J.Pharmacol. 125:1103-1107 (1986).
10. Dolphin, A.C., Trends in Neurosci. 10:53-57 (1987).
11. Dolphin, A.C., and R.H. Scott, J.Physiol. 86:1-17 (1987)
12. Droogmans, G., I. Declerck, and R. Casteels, Pflugers Arch. 409:7-122 (1987).
13. Dosemeci, A., R.S. Dhalla, N.M. Cohen, W.J. Lederer, and T.B. Roger, Circ. Res. 62:347-355 (1988).
14. Fish, R.D., G. Sperti, W.S. Colucci, and D.E. Clapham, Circ. Res. 62:1049-1052 (1988).
15. Gleason, M.M., and S.F. Flaim, Biochem. Biophys. Res. Comm. 138:1362-1365 (1986).
16. Gilman, A.G., Ann. Rev. Biochem. 56:615-649 (1987).
17. Hescheler, J., W. Rosenthal, W. Trautwein, and G. Sculots, Nature 325:445-447 (1987).
18. Holz IV, G.G., S.G. Rane, and K. Dunlap, Nature 319:670-672 (1986).
19. Ikeda, S.R., and G.G. Schofield, J.Physiol. 409:221-240 (1989).
20. Inoue, R. and G. Isenberg, Am. J. Physiol. 258:C1173-C1178 (1990).
21. Irvine, R.F., and R.M. Moor, Biochem. J. 240: 917-920 (1986).
22. Itoh, H., and K. Lederis, Am. J. Physiol. 252:C244-C249 (1987).
23. Johansson, B. and A.P. Somlyo, in: Handbook of Physiology, Sect 2: Cardiovascular System, Vol. 2, D.F. Bohr, A.P. Somlyo, H.V. Sparks, eds. (American Physiol. Soc., Bethesda 1980) pp. 301-323.
24. Johns, D.W. and N. Sperelakis, Circulation 66:204 (1982).
25. Kanaide, H., T. Mastumoto, and M. Nakamura, Biochem. Biophys. Res. Com. 140:195-203 (1986).

26. Kurachi, Y., T. Nakajima, and T. Sugimoto, Pflugers Arch. 407:264-274 (1986).
27. Kuriyama, H., Y. Ito, H. Suzuki, T. Kitamura, and T. Itoh, Am. J. Physiol. 243: H641-662 (1982).
28. Lacerda, A.E., D. Rampe, and A.M. Brown, Nature 335:249-251 (1988).
29. Nishizuka, Y., Nature 308:639-698 (1984).
30. Nozaki, M., and N. Sperelakis, Am. J. Physiol. 256:H455-H459 (1989).
31. Ohya, Y., K. Kitamura, and H. Kuriyama, Circ.Res. 62:375-383 (1988).
32. Ohya, Y. and N. Sperelakis, Physiologist 31:A88 (1988).
33. Ohya, Y. and N. Sperelakis, Pflugers Arch. 414:257-264 (1989).
34. Ohya, Y. and N. Sperelakis, Circ. Res. 68:763-771 (1991).
35. Ohya, Y., K. Terada, K. Yamaguchi, R. Inoue, K. Okabe, K. Kitamura, M. Hirata, and H. Kuriyama, Pflugers Archiv. 412: 382-389 (1988).
36. Ousterhout, J.M., and N. Sperelakis, unpublished observations.
37. Ousterhout, J.K., and O.S. Steinsland, Life Sci. 28:2687-2695 (1981).
38. Pacaud, P., G. Lorrand, C. Mironneau, and J. Mironneau, Pflugers Archiv. 410:557-559 (1987).
39. Rasmussen, H., Y. Takuwa and S. Parks, FASEB J. 1:177-180 (1987).
40. Rosenthal, W., J. Hescheler, W. Trautwein, and G. Schultz, FASEB J. 2:2784-2790 (1988).
41. Sasaguri, T., M. Hirata, T. Itoh, T. Koga, and H. Kuriyama, Biochem.J. 239:567-574 (1986).
42. Satoh, S., T. Itoh, and S. Kuriyama, Pflugers Arch. 410:132-138 (1987).
43. Somlyo, A.V., M. Bond, and A.P. Somlyo, Proc. Natl. Acad. Sci. 82:5231-5233 (1985).
44. Sperelakis, N. and Y. Ohya, in: Physiology and Pathophysiology of the Heart, 2nd edition, N. Sperelakis, ed. (Kluwer Academic Press, Boston 1988) pp.773-811.
45. Suematu, E., M. Hirata, T. Sasaguri, T. Hashimoto, and H. Kuriyama, Com. Biochem. Physiol. 82A:645-649 (1985).
46. Takata, Y. and H. Kuriyama, Jpn. J. Pharmacol. 29:639-651 (1979).
47. Tawada, Y., K.I. Furukawa, and M. Shigekawa, J. Biochem. 102:1499-1509 (1987).
48. Wertz, M.A. and R.L. MacDonald, J. Neurosci. 7:1639-1647 (1987).
49. Yatani, A., J. Cordina, A.M. Brown, and L. Birnbaumer, Science 235:207-211 (1987).
50. Yatani, A., J. Cordina, Y. Imoto, J.P. Reeves, L. Birnbaumer, and A.M. Brown, Science 238:1288-1292 (1987).
51. Yatani, A., C.L. Siegel, J. Allen, and A.M. Brown, Circ. Res. 60:523-533 (1987).
52. Zelcer, E. and N. Sperelakis, Blood Vessels 18:263-279 (1981).

NORADRENALINE MODULATION OF IONIC CHANNELS IN VASCULAR SMOOTH MUSCLE CELLS

JEAN MIRONNEAU

Laboratoire de Physiologie Cellulaire et Pharmacologie Moléculaire, INSERM CJF 88-13, Université de Bordeaux II, 33076 Bordeaux, France

Abstract

The effects of noradrenaline (NA) were studied on smooth muscle of portal vein using both the patch-clamp technique and radioligand binding experiments. NA induced activation of chloride current as a consequence of a rise in cytoplasmic Ca^{2+} depending of release of Ca^{2+} from intracellular stores. NA stimulated also the voltage-dependent Ca^{2+} current by increasing the probability of Ca^{2+} channel being open. The transduction mechanism between alpha-1-adrenoceptors and ionic channels involved activation of phospholipase C by a pertussis toxin-insensitive G-protein and production of both IP_3 which depleted Ca^{2+}-stores and DAG which activated protein kinase C. Both NA and phorbol esters caused an increase in (+) - [^{3}H]isradipine binding affinity to Ca^{2+} channels without change in the maximal binding capacity. This stimulatory effect was dependent on the presence of both ATP and Mg^{2+} in the incubation medium suggesting that regulation of dihydropyridine binding sites by protein kinase C might involve a phosphorylation process.

INTRODUCTION

In vascular smooth muscles, noradrenaline (NA) application produces contractions that can be dependent on an increase in Ca^{2+} influx through voltage-dependent Ca^{2+} channels, and a release of Ca^{2+} from intracellular stores (5, 12, 14). The Na-induced contraction is generally associated with a membrane depolarization. Recent studies concerning the ionic mechanisms underlying the electrophysiological response to NA have suggested that the NA-induced depolarization can be mediated by an increase in Cl^- conductance (3, 17), the opening of non-specific cation channels (1, 3), or a decrease in K^+ conductance (19).

In the present study, we have investigated the effects of NA in portal vein smooth muscle in order to characterize: (1) the type of adrenoceptors involved in the NA action; (2) the ionic channels responsible for both NA-induced depolarization and activation of contraction; (3) the transduction mechanism involved between adrenoceptors and ionic channels. In addition, we tested the role of the different ionic mechanisms previously identified in order to account for the contractile effects of NA in intact portal vein strips.

BINDING OF [^{3}H]PRAZOSIN TO RAT PORTAL VEIN STRIPS

In veins, binding studies have frequently revealed the existence of alpha-1 and alpha-2 adrenoceptors. However, in portal veins the alpha-1-receptors predominate and selective alpha-1-agonists have been demonstrated to cause contractions (18).

Figure 1 shows the saturation curve of specific binding for [^{3}H]prazosin to intact strips of rat portal vein. Scatchard analysis indicated the existence of a single class of specific binding sites, characterized by a dissociation constant of about 100 pM and a maximal binding capacity of about 14 fmol/mg wet weight. It should be noted that similar values of K_D (50-100 pM) were obtained for the binding of prazosin to isolated plasma membranes of portal veins.

Published 1991 by Elsevier Science Publishing Company, Inc.
Ion Channels of Vascular Smooth Muscle Cells and Endothelial Cells
Sperelakis and Kuriyama, Editors

48

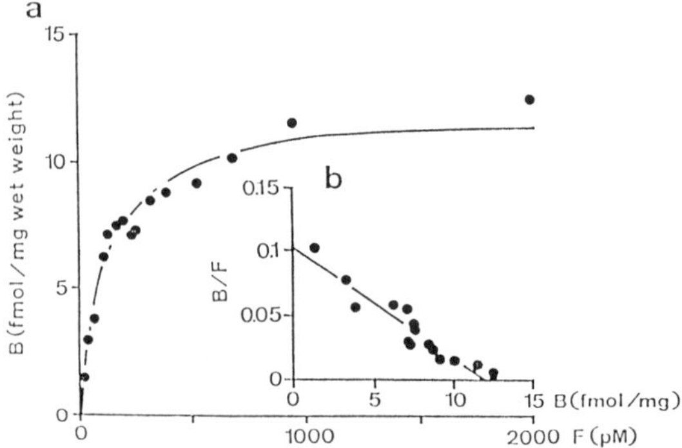

Figure 1. Equilibrium binding of [^{3}H]prazosin in rat portal vein strips. (a) Saturation binding experiments were carried out by incubating segments of portal vein (0.5-1 mg wet wt) with increasing concentrations of [^{3}H]prazosin in 5.6 mM K$^+$-containing solution for 60 min at 37° C. Specific binding was defined as the binding displaceable by 2 μM unlabelled prazosin. (b) Scatchard analysis of specific binding values was done with a nonlinear least-square programme. B/F, bound/free.

ACTIVATION OF CHLORIDE CURRENT BY NORADRENALINE

Direct evidence that Na-activated depolarizing current was obtained by micro-ejections of 10 μM NA on cells held at -70 mV. A similar response was obtained with phenylephrine but not with clonidine. With depolarizations, the NA-evoked current decreased in amplitude, was null near zero mV and reversed at positive membrane potentials. The NA-activated current was selectively blocked by prazosin (1 μM) and phentolamine (1 μM), while it remained unaffected in the presence of yohimbine (1 μM) and propranolol (10 μM). In this experiment, the external Cl$^-$ concentration was 146 mM and the internal concentration was 130 mM, so that the equilibrium potential for Cl$^-$ ions was estimated to be -3 mV. In addition, CsCl was used instead of KCl in both external and internal solutions in order to block K$^+$ currents. The current-voltage relationship obtained by plotting the peak current induced by NA against membrane potential was linear. The reversal potential was obtained at 0 mV was close to the calculated E$_{C1}$.

In order to confirm to the chloride nature of the NA-induced Cl$^-$ current, the reversal potential was determined with various chloride concentrations in the bath and pipette solutions. The mean reversal potentials obtained in 5 different chloride solutions as indicated in Figure 2 are plotted in ordinate against the calculated equilibrium potential for Cl$^-$ ions in abscissa. The experimental points are closely distributed along a straight line with a slope of 1, as expected for a pure chloride conductance. The absence of any variation in reversal potential by replacing internal Cs$^+$ by Na$^+$ ions suggests that it is unlikely that Na opens a cation channel in smooth muscle cells from rat portal vein.

It has been noted that in vascular smooth muscle the intracellular Cl$^-$ concentration is estimated to be about 60 mM, so that the reversal potential for the Cl$^-$ current can be calculated to be about -20 mV. The role of Ca^{2+} ions in activation of Cl$^-$ channels was examined first by comparing the NA-induced response in the presence of Ba^{2+} and Ca^{2+} ions. No responses were observed when the cells are perfused for 20 min in 5 mM Ba^{2+} solution. After replacement of Ba^{2+} with Ca^{2+} for 5 min, inward currents could be evoked by NA applications. Similarly, when the pipette solution contained 10 mM EGTA in order to reduce the cytoplasmic Ca^{2+}-concentration in the nanomolar concentration range, the responses to NA were never observed, even in the presence of 5 mM external Ca^{2+}. In contrast, the NA-induced current was not affected in the presence of 3 mM Co^{2+} applied for 10 min to block voltage-dependent Ca^{2+} channels. Caffeine (10 mM) applied to the bathing solution also induced a transient increase in Cl$^-$ current but the NA-induced response was lost after application of caffeine. These results suggests that the Cl$^-$

current is activated by a release of intracellular Ca^{2+} and the caffeine and NA acts on the same intracellular Ca^{2+} stores.

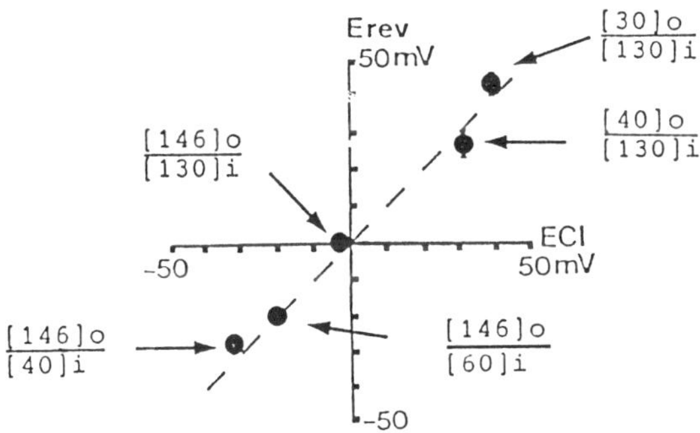

Figure 2. Effects of chloride substitution on the noradrenaline-induced currents measured in single cells of rat portal vein with the whole-cell mode of the patch-clamp technique. Reversal potential (E_{rev}) of the NA-induced current was plotted as a function of the calculated equilibrium potentail for chloride ions (E_{Cl}) for reference chloride gradient (146 mM): 130 mM, low external chloride (40 mM or 30 mM). Each point is the mean of 3-15 cells; vertical lines show s.e. mean and are often smaller than the symbol size. Dashed line predicts the reversal potential for a pure chloride conductance.

Activation of Ca^{2+}-dependent Cl^- current supports the NA-induced depolarization as E_{Cl} in physiological solutions is estimated to be -20 mV, a value different from the resting potential (-60 mV). In K^+-containing solutions but in the presence of 2.5 mM Co^{2+} to inhibit voltage-dependent Ca^{2+} channels, applications of NA produced a large depolarization reaching 30-40 mV when the cell was held at -60 mV. At depolarization membrane potential, the Na-induced response was composed of an early hyperpolarization corresponding to activation of Ca^{2+}-dependent K^+ channels, followed by a depolarization which clearly reached the threshold for activation of voltage-dependent Ca^{2+} channels. It could be observed that the delay between NA application and onset of Cl^- current was about 1-2 s and that maximal depolarization was reached within 2-3 s. In cells showing spontaneous spikes, the first spike appeared 3-4 s after NA application, and the increase in spike amplitude was observed only after 5-10 s. These observations indicate that activation of Cl^- channels clearly precedes activation of Ca^{2+} channels.

Ca^{2+}-activated chloride current can be inhibited by different drugs which are known to interfere with Cl^- movements. For example, anthracene-9-carboxylic acid (5 mM) completely inhibited the Cl^- current activated by caffeine and the effect was completely reversible within 4-5 min (Baron, Pacaud, Loirand, Mironneau & Mironneau, unpublished results).

MODULATION OF VOLTAGE-DEPENDENT CALCIUM CHANNELS BY NORADRENALINE

In portal vein smooth muscle cells, two types of Ca^{2+} channel have been separated (9, 11) which resemble the T-type and L-type Ca^{2+} channels described in other excitable cells. When the holding potential was held at -70 mV, single channel current typified by brief openings of about -1 pA in amplitude were recorded. At more depolarized test potentials, a second type of channel activity was seen with long-lasting openings of about -1.5 pA in amplitude. The slope conductance was 8 pS for the small-conductance channel and 17 pS for the large conductance channel.

Single-channel current recordings provide the most direct approach to identifying the NA-sensitive Ca^{2+} channels and characterizing the modulatory effects of NA.

NA did not change the amplitude of the single channel current recorded in a 90 mM Ba^{2+} solution and did not increase the number of functional channels in membrane patches. NA increased the probability of Ca^{2+} channel being open. This is seen as an increase in the time during which the channel is opened and an increase of the sweep number with openings. It may be noted that in cell-patches where Ca^{2+} channels regularly open during depolarizing pulses, openness of both types of Ca^{2+} channels is undetectable at -50 mV in the presence of NA, for recording times as long as 2 hours in the same cell. Therefore, it is unlikely that NA opens Ca^{2+} channels at resting membrane potential.

This observation is complemented by the fact that NA increased the whole-cell current, in 5 mM external Ba^{2+}, with no apparent change in threshold potential, potential for maximal current and reversal potential. Response to NA was not affected by a pre-exposure of cells to $1 \mu M$ propranolol or 5 mM anthracene-9 carboxylic acid, a blocker of Cl^- current. In contrast, the effect of NA was blocked by phentolamine and prazosin (1 μM) indicating that it was related to activation of alpha 1-adrenoceptors.

TRANSDUCTION MECHANISM

To characterize the transduction mechanism for NA-induced increase in Ca^{2+} and Cl^- currents, we examined: (1) the effects of predicted second messengers, such as cAMP, and arachidonic acid; (2) the effects of substances which replicate the action of second messengers, for example phorbol esters which activate protein kinase C; and (3) the effects of substances stimulating or inhibiting G-protein activity.

Inclusion of 1 mM arachidonic acid as well as 0.3 mM cAMP in the pipette solution had no stimulation effect on Ca^{2+} channel current. In contrast, external application of phorbol dibutyrate (PDBu) stimulated the Ca^{2+}-channel current. Therefore, the most plausible transduction process may be the activation of phospholipase C (PLC) leading to generation of diacylglycerol (DAG) and inositol 1,4,5-trisphosphate (IP_3). DAG is the natural activator of protein kinase C and IP_3 releases Ca^{2+} channels through the same transduction process as their stimulatory effects were not additive. PDBu (0.1 μM) approximately doubled the Ca^{2+}-channel current but a further application of 10 μM NA was ineffective. Similarly, when the Ca^{2+} channel current was stimulated by NA, PDBu was ineffective or produced a slight inhibition of current. Furthermore, when the inactive form of phorbol ester, phorbol diacetate, was externally applied, the inward current was not affected. These observations support the idea that the transduction pathway involves activation of protein kinase C.

To determine whether a G-protein was involved in the stimulation of phospholipase C activity, we tested the effect of GTP and GDP analogues.

GDP-β-S, a stable analogue of GDP, acts as a competitive inhibitor of GTP on G-protein activation, and thereby traps the G-protein in an inactivated state. Internal application of 1 mM GDP-β-S suppressed the stimulatory effect of NA on Ca^{2+} channel current, but PDBu still induced an increase in current.

In contrast, when GTP-γ-S is applied in the pipette solution, the G-protein is trapped in the activated state. Therefore, the Ca^{2+} channel current was increased in a concentration-dependent manner by GTP-γ-S while the NA-induced stimulation of current was progressively reduced to zero. With 1 mM of GTP-γ-S NA failed to enhance the Ca^{2+} channel current suggesting that the transduction mechanism was fully activated. A similar stimulation of Ca^{2+} channels was obtained by using a combination of aluminium (10^{-5} M $AlCl_3$) and fluoride (10^{-2} M NaF) which was believed to stimulate G-proteins by mimicking the effects of GTP-γ-S on the alpha-subunit. In contrast, incubation of the cells in the presence of 10 $\mu g/ml$ of pertussis toxin for 3-30 hours had no effect on Ca^{2+}-channel current (10).

Effects of GDP-β-S and GTP-γ-S were also studied on the Ca^{2+}-activated Cl^- current. In a cell where Cl^- current could be activated by Ca^{2+} influx, internal application of 1 mM GDP-β-S suppressed the NA-induced response while the caffeine-induced response was maintained. As caffeine is still effective in the presence of intracellular GDP-β-S, these results suggest that the blockade of the NA-induced response is not due to a depletion of the internal Ca^{2+} stores or a direct inhibition of Cl^- channels. At high concentrations (1 mM), GTP-γ-S maximally activated the Cl^- current. Further applications of NA and caffeine failed to induce any responses indicating that the transduction way was fully activated in the presence of high concentrations of GTP-γ-S, but also that intracellular Ca^{2+} stores had been completely emptied.

In control conditions, the NA-activated Cl⁻ current increased in a concentration-dependent manner. The concentration required to produce 50% of the maximal response was 0.1 μM. When 1 μM GTP-γ-S was added to the pipette solution there was no activation of Cl⁻ current but a subthreshold concentration of NA (10 nM) induced a quasi-maximal Cl⁻ current. These results suggest that the sensitivity to NA is strongly dependent on the activity of the transduction mechanism.

Taken together, these results suggest that the NA activation of both Ca^{2+} and Cl⁻ currents is mediated through the accelerated activity of phospholipase C and production of both IP₃ and DAG. Control of the phospholipase C involves a pertussis toxin insensitive-G protein.

Therefore, two important points were investigated: (1) the direct involvement of phosphatidylinositides as substrates for PLC activity; (2) the possible phosphorylation of Ca^{2+} channels or of some regulatory proteins during neuromediator activation.

In order to demonstrate the involvement of phosphatidylinositides in the generation of second messengers, one possibility consisted to use antibodies raised against phosphatidylinositides. Recently, autoantibodies directed against phosphatidylinositides have been identified in highly diluted sera of patients with malignant tumors (leukemia, breast cancer or gastrointestinal cancer). Immunochemical analysis has revealed high levels of autoantibodies with high avidity and specificity for phosphatidylinositol (Loirand, Faiderbe, Baron, Geffard, Mironneau, unpublished data). Displacement curves were established by competition experiments in an ELISA system for the following compounds: phosphatidylinositol (PI), phosphatidylinositol 4-monophosphate (PIP), phosphatidylinositol 4,5-biphosphate (PIP₂), inositol and phosphatidylserine. The best avidity was obtained for PI (half-displacement in the nanomolar range) but these autoantibodies also recognized PIP and PIP₂. Inositol was 100-fold less recognized. Non-significant displacement was observed with phosphatidylserine indicating that the autoantibodies were specific for phosphatidylinositides.

When these autoantibodies were added to the pipette solution at concentrations of 1.2 μg IgG/ml the number of cells that displayed a NA-induced Cl⁻ current decreased from 17 to 4, while the mean amplitude of current decreased from 300 pA to 40 pA. Similar results were obtained on the NA-induced enhancement of the Ca^{2+} current. In contrast, when IgG fractions from healthy patients or IgG fractions incubated with a 10^{-9} M solution of phosphatidylinositol were intracellularly applied, the NA-induced responses were not modified.

Autoantibodies did not interact with Ca^{2+} and Cl⁻ channels as well as the stimulation of these currents induced by phorbol ester and caffeine. They did not affect protein kinase C or the intracellular Ca^{2+} stores. Therefore, these results support the idea that autoantibodies against phosphatidylinositides block the transduction pathway before hydrolysis of PIP₂ by PLC.

EFFECTS OF NORADRENALINE AND PROTEIN KINASE C MODULATORS ON (+) - [³H]ISRADIPINE BINDING

It is now well established that specific binding sites for different Ca^{2+} channel antagonists can be identified on the alpha-subunit of the Ca^{2+} channel protein (4, 8).

The binding site for dihydropyridines in intact strips of vascular smooth muscle had a high-affinity (about 0.1 nM) and was modulated by the membrane potential such as depolarization increased the affinity of dihydropyridines to the specific binding site (7, 13). This property has been demonstrated in intact strips or in cell suspensions which can be depolarized during incubation in high K^+ solutions.

Isradipine specific DHP binding in intact venous strips was increased in a concentration-dependent manner by NA. The maximal effect was obtained with 10 μM NA and corresponded to about 80% of the binding of isradipine in the absence of NA. The increase in isradipine specific binding was suppressed in the presence of 1 μM prazosin, but remained unaffected when the strips were preincubated in the presence of 10 μg/ml pertussis toxin.

52

Phorbol esters increased isradipine specific binding in a manner similar to NA and the maximal effect was obtained with 1 μM PDBu. The stimulatory effect of PDBu was suppressed in the presence of 1-10 μM h7, an inhibitor of protein kinase C. H7 by itself had no effect on isradipine specific binding. Therefore, isradipine binding to Ca^{2+} channels in intact vascular strips was similarly modulated by both NA and phorbol esters suggesting a possible phosphorylation-dependent mechanism through activation of protein kinase C.

In order to confirm a possible phosphorylation process in the increase in isradipine binding, we studied the effects of phorbol esters in microsomes of portal vein smooth muscle. When PDBu was added in the presence of 10 μM ATP and 1 mM Mg^{2+}, isradipine specific binding was notably increased. This enhancement was not obtained when PDBu, GTP-γ-S or NA were added alone. Thus, these results clearly indicate that the regulation of dihydropyridine binding sites by protein kinase C may involve a phosphorylation. These results are in good agreement with recent data showing that intracellularly-perfused ATP increases the L-type Ca^{2+} current in a concentration-dependent manner (16).

To elucidate the effect of NA and PDBu on Ca^{2+} channels, Scatchard analysis of saturation experiments were carried out in the absence and in the presence of either NA or PDBu. Both PDBu and Na increased the affinity of isradipine to its binding sites without affecting the maximal binding capacity. These results are in good conductance with the observation that NA activation does not increase the number of Ca^{2+} channels but stimulates the open-state probability of these channels.

NORADRENALINE-INDUCED CONTRACTION

Figure 3 summarizes the role of the ionic channels which are believed to be involved in NA activation of vascular smooth muscle cells. Applications of 10 μM NA induced a complex contraction in which a large transient component was followed by a sustained plateau during NA application. In the presence of 5 mM anthracene-9-carboxylic acid (which completely blocked Cl^- channels) the peak of NA contraction was reduced by 66 $\pm$ 7% (n = 4) while the plateau component was suppressed. A similar inhibition of NA-induced contraction (71 $\pm$ 9%, n = 5) was observed in the presence of 10 nM isradipine (Pacaud, Loirand, Baron, Mironneau, Mironneau, unpublished data). The NA-induced contraction which was resistant to anthracene-9-carboxylic acid and isradipine was similar to that obtained when external Ca^{2+} was suppressed (6). This component of contraction is related to Ca^{2+} release from intracellular stores.

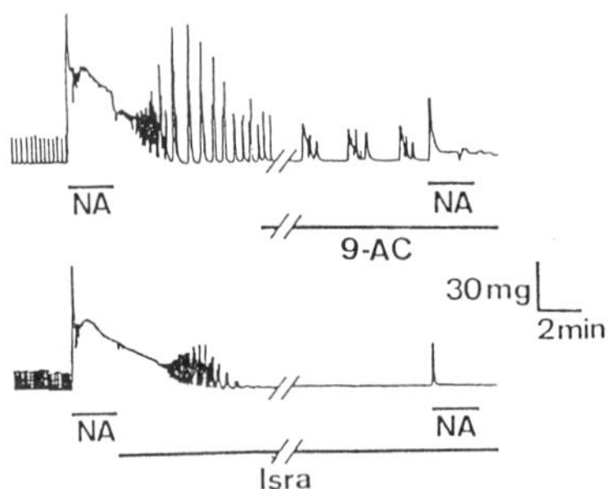

Figure 3. Effects of Ca^{2+} channel and Cl^- channel blockers on NA-induced contractions in intact strips of rat portal vein. Contractins were induced by 10 μM NA in control conditions and in the presence of 5 mM anthracene-9-carboxylic acid (9-AC, a) or 10 nM isradipine (b).

CONCLUSIONS

These data demonstrate that in single cells of rat portal vein the responses to noradrenaline (NA) are mediated by alpha 1-adrenoceptors. NA activates at least two separate membrane conductances. First, NA evokes a Cl^- current which leads to membrane depolarization. The Cl^- current is mediated by intracellular Ca^{2+} release from intracellular stores. This is consistent with the observation that caffeine stimulates the Cl^-

current in a manner similar to NA (3, 17). Secondly, NA appears to enhance voltage-dependent Ca^{2+} channel current by increasing the open-state probability of single Ca^{2+} channels. However, NA does not open Ca^{2+} channels when the membrane patch potential is held at -50 mV which is about the resting potential in physiological conditions. These observations do not support the idea that NA can shift the steady-state activation curve to more negative membrane potentials and in turn increase Ca^{2+} channel current for a given membrane potential (15).

The transduction mechanisms for NA-induced increase in Ca^{2+} channel current and internal Ca^{2+} release involve a G-protein. In view of the pertussis toxin-insensitivity of G-protein and the blockade of NA action by antibodies raised against phosphatidylinositides, the most plausible transduction pathway is the activation of phosphatidylinositide hydrolysis by phospholipase C leading to generation of two second messengers: inositol 1, 4, 5-trisphosphate (IP_3) and diacylglycerol (DAG) (2). As the effects of NA and phorbol esters (PDBu) on Ca^{2+} current are not additive, our results suggest that modulation of Ca^{2+} channels by NA may involve the generation of DAG and activation of protein kinase C.

Both NA and phorbol esters (activators of protein kinase C) cause an increase in the affinity of (+) - [^{3}H]isradipine binding to Ca^{2+} channels in intact strips without any changes in the maximal binding capacity. These results are in concordance with electrophysiological data which show that NA increases the open-state probability of Ca^{2+} channels without modifying neither the number of functional channels nor the amplitude of the single-channel current. Increase in isradipine affinity to Ca^{2+} channels is associated with direct activation of protein kinase C as well as activation of alpha 1-adrenoceptors by NA. In venous membranes, the phorbol ester-induced increase in isradipine affinity is only obtained when the incubation medium contains both ATP and Mg^{2+} suggesting that modulation by phorbol ester involves a phosphorylation process. Whether this phosphorylation directly affects Ca^{2+} channels or an intermediary intracellular protein remains to be determined. The effects of phorbol esters on isradipine binding are independent of the membrane potential and blocked by H7, a protein kinase C inhibitor. These observations are therefore consistent with the idea that protein kinase C may increase the affinity of dihydropyridine binding to Ca^{2+} channels independently of a separated modulation by membrane potential (7, 13).

Importance of both Cl^- and Ca^{2+} channels in NA-induced contraction has been directly determined by inhibiting Cl^- or Ca^{2+} channels with specific inhibitors. These results confirm that membrane depolarization induced by activation of Ca^{2+}-activated Cl^- channels is a prerequisite to opening of voltage-dependent Ca^{2+} channels and further stimulation of the channels in response to NA.

ACKNOWLEDGEMENTS

This work was supported by grants from Institut National de la Santé de la Recherche Médicale, Ministére de la Recherche et de la Technologie, Région Aquitaine, Fondation pour la Recherche Médicale and Association Francaise contre les Myopathies, France. We thank Ms. N. Biendon for excellent assistance.

REFERENCES

1.   Amde, T., W.A. Large, and Q. Wang, J. Physiol. London. <u>428</u>, 501-516 (1990).

2.   Berridge, M.J., Biochem. J. <u>220</u>, 345-360 (1984).

3.   Byrne, N.G. and W. A. Large, J. Physiol. London, <u>404</u>, 557-573 (1988).

4.   Catterall, W.A., Science <u>424</u>, 50-61 (1988).

5.   Dacquet, C., C. Mironneau, and J. Mironneau, Br. J. Pharmacol. <u>92</u>, 203-211 (1987).

6.   Dacquet, C., G. Loirand, C. Mironneau, J. Mironneau, and P. Pacaud, Br. J. Pharmacol. <u>92</u>, 535-544 (1987).

7.   Dacquet, C., G. Loirand, L. Rakotoarisoa, C. Mironneau, and J. Mironneau, Br. J. Pharmacol. <u>97</u>, 256-262 (1989).

8.   Hosey, M.M. and M. Lazdunski, J. Membr. Biol. <u>104</u>, 81-105 (1988).

9.   Loirand, G., P. Pacaud, C. Mironneau, and J. Mironneau, J. Physiol. Lond. <u>412</u>, 333-349 (1989).

10.   Loirand, G., P. Pacaud, C. Mironneau, and J. Mironneau, J. Physiol. Lond. <u>428</u>, 517-529 (1990).

11. Loirand,G., P. Pacaud, C. Mironneau, and J. Mironneau, Pflugers Arch. $\underline{407}$, 566-568 (1986).
12. Mironneau, J. and Y.M. Gargouil, Eur. J. Pharmacol. $\underline{57}$, 57-67 (1979).
13. Morel, N. and T. Godfraind, J. Pharmacol. Exp. Ther. $\underline{234}$, 711-715 (1987).
14. Nanjo, T., Br. J. Pharmacol. $\underline{81}$, 427-440 (1984).
15. Nelson, M.T., J.B. Patlak, J.F. Worley, and N.B. Standen, Am. J. Physiol. $\underline{259}$, C3-C18 (1990).
16. Ohya Y. and N. Sperelakis, Circul. Res. $\underline{64}$, 145-154 (1989).
17. Pacaud, P., G. Loirand, C. Mirronneau, and J. Mironneau, Pflugers Arch. $\underline{413}$, 629-636 (1989).
18. Shi, A.G.. C.Y. Kwan, and E.E. Daniel, J. Pharmacol. Exp. Ther. $\underline{250}$, 1119-1124 (1989).
19. Suzuki, H., J. Physiol. London $\underline{321}$, 495-512 (1982).

CALCIUM AND POTASSIUM CURRENTS IN CULTURED RAT AORTIC VASCULAR SMOOTH MUSCLE CELL LINES

HIROYASU SATOH and NICHOLAS SPERELAKIS

Department of Physiology and Biophysics, University of Cincinnati, College of Medicine, Cincinnati, Ohio 45267, USA

INTRODUCTION

Contractility of vascular smooth muscle (VSM) is modulated by intracellular effectors through a variety of membrane receptors. The contractile state of VSM cells is dependent on the concentration of intracellular free Ca^{2+} ($[Ca]_i$), which acts as a second messenger to activate the contractile myofilaments. $[Ca]_i$ is elevated by (a) Ca^{2+} influx across the plasma membrane, and (b) Ca^{2+} release from internal stores, primarily the sarcoplasmic reticulum (SR). Ca^{2+} release from SR is brought about: (a) by the Ca^{2+} influx across the plasma membrane by a Ca^{2+}-induced Ca^{2+} release process, and (b) by production of inositol 1, 4, 5-triphosphate (IP_3), resulting from activation of membrane receptors. Caffeine and ryanodine act on the Ca^{2+} release channels of the SR membrane to bring about Ca^{2+} release and resulting depletion of the Ca^{2+} stores in the SR. Thus, VSM cells can contract even in the absence of extracellular Ca^{2+}, or without an associated change in membrane potential ('pharmacomechanical coupling') (4, 5, 13).

The cellular effectors detect the levels of the cyclic nucleotides, and their physiological functionings are affected by phosphorylation with cyclic AMP (cAMP)-dependent or cyclic GMP (cGMP)-dependent protein kinase. The effect of cyclic nucleotides on $[Ca]_i$ may be one key regulator of vascular tone. The contraction of VSM cells is influenced by cyclic nucleotides and PK-C. Both cAMP and cGMP relax, whereas PK-C stimulates the contraction of VSM cells (15). The cAMP accumulation is associated with vascular relaxation through β-adrenoceptor stimulation by catecholamines. Cholinergic-mediated vasodilation, accompanied with hyperpolarization, was associated with cGMP accumulation (3, 14). ACh can not only accumulate cGMP through muscarinic receptors, but also can stimulate phosphatidyl inositol (PI) turnover, and produce IP_3 and diacylglycerol (DAG) for stimulation of protein kinase C (PK-C)(19). By the latter mechanism, ACh would tend to contract the VSM cells.

Much remains to be investigated on the electrical properties and ion channels of VSM cells, because it is difficult to obtain satisfactory whole-cell voltage clamp from the small vascular smooth myocytes (size: ca. 5 x 200 μm). The cells are difficult to separate because they are covered with tough conective tissue. In addition, there are many regional differences in properties of the cells. This article focuses on the slow Ca^{2+} current and the Ca^{2+}-activated K^+ current in cultured rat aortic VSM cells, and the effects of isoprenaline, carbachol, and phorbol ester (for stimulation of PK-C) on the Ca^{2+} current.

METHODS
Cell Culture and Preparation

Stable VSM cell lines (A7r5 and A10), derived from rat aorta and purchased from Americal Type Tissue Culture Collection (Bethesda, MD, USA), were used. These cell lines present many characteristics of adult aortic VSM cells, including a well-developed rough endoplasmic reticulum, dispersed contractile filaments, and muscle-type creatine phosphokinase isozyme (12, 21). The cells were transferred to glass cover-slips (placed in 35 mm petri dishes) and bathed in tissue culture medium (GIBCO, Grand Island, NY, USA) containing 10% fetal calf serum. The cells were incubated at 37° C in a humidified atmosphere under 5% CO_2 and 95% air.

Electrical Measurements

The membrane currents were measured by the whole-cell recording technique of the patch-clamp method (11). An Axopatch patch-clamp amplifier (Axon Instruments,

Published 1991 by Elsevier Science Publishing Company, Inc.
Ion Channels of Vascular Smooth Muscle Cells and Endothelial Cells
Sperelakis and Kuriyama, Editors

Burlingame, CA, USA) was used, and the data were stored and analyzed on an IBM-AT microcomputer using the PCLAMP analysis program. Test pulses were applied once every 20 sec for I_{Ca} and 30 sec for I_K. Current traces were filtered at a cut-off frequency of 2 KHz for plotting. All values are given as mean $\pm$ SEM. All experiments were performed at room temperature (22° C).

Experimental Solutions

Slow Ca^{2+} Current (I_{Ca}). The composition of modified Tyrode solution used was (in mM): NaCl 110, KCl 5.4, CaCl$_2$ 20, MgCl$_2$ 1, NaH$_2$PO$_4$ 0.3, glucose 5, and HEPES 5 (pH 7.4)(Fig. 1). High Ca^{2+} concentration (20 mM) was used to facilitate the measuring of the I_{Ca}. TEA (10 mM) and 4-aminopyridine (4-AP)(2 mM) were also added to block outward K^+ currents. The solution used in the pipette was composed of (mM): CsOH 110, CsCl 20, MgCl$_2$ 1, MgATP 5, EGTA 10, creatine phosphate 5, aspartic acid 100, and HEPES 5 (adjusted to pH 7.2 with HCl).

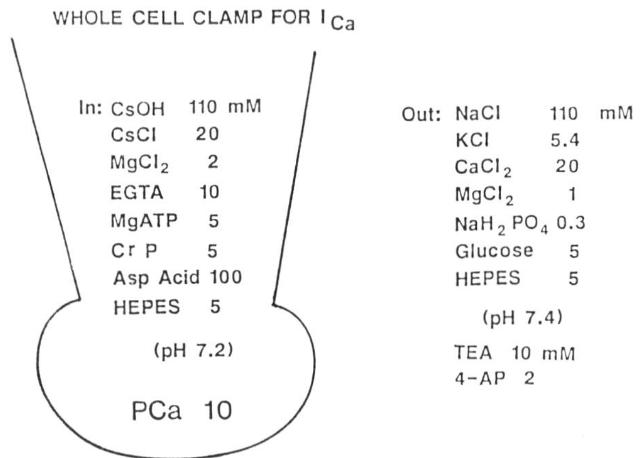

Figure 1. Composition of the pipette solution and bath solution used for measuring I_{Ca}. To block the outward currents, K^+ in the pipette solution was replaced by Cs^+, and TEA was added to the bath solutions. High Ca^{2+} (20 mM) was used to facilitate the amplitude of I_{Ca}. CrP: creatine phosphate. Asp Acid: aspartic acid. TEA: tetraethyl anmonium. 4-AP: 4-aminopyridine.

Outward K^+ Current. Na^+, K^+ and Ca^{2+} were replaced by choline$^+$ to avoid the interference of fast Na^+ current, Ca^{2+} current, Na-Ca exchange current, and Na-K pump current; atropine (3 μM) was added to avoid the pharmacological effects of choline$^+$ (Fig. 2). The composition of the bath solution was as follows (in mM): choline-Cl 144, MgCl$_2$ 1.0, NaH$_2$PO$_4$ 0.3, glucose 5, and HEPES 5 (pH 7.4). The pipette solution contained (in mM): KOH 110, KCl 20, MgCl$_2$ 2, EGTA 10, MgATP 5, creatine phosphate 5, aspartic acid 100, and HEPES 5 (adjusted to pH 7.2). Two Ca^{2+} concentrations (pCa 10 and pCa 7) in the pipette solutions were used to examine the Ca^{2+}-activated outward current ($I_{K(Ca)}$), according to the calculation of Fabiato and Fabiato (6) and the correction of Rink and Tsien (20). The outward K^+ current at pCa 10 was subtracted from that at pCa 7, and the difference current was considered as $I_{K(Ca)}$.

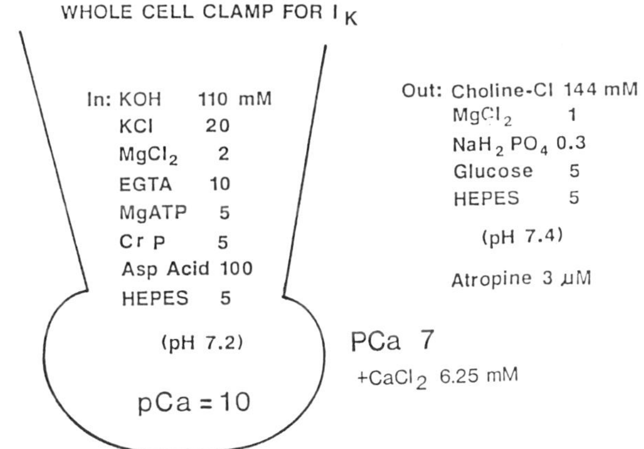

Figure 2. Composition of the pipette solution and bath solution used for measuring I_K. The pipette solution was adjusted to pCa 10. Na^+, K^+ and Ca^{2+} in the bath solution were replaced by choline$^+$ to block I_{Na}, I_{Ca}, Na-Ca exchange, and Na-K pump. Atropine was added to the bath solution to block the pharmacological effects of choline$^+$. CrP: creatine phosphate. Asp acid: aspartic acid.

RESULTS AND DISCUSSION
Ca^{2+} Current

Basal Current. Voltage-dependent Ca^{2+} channels are a major pathway of Ca^{2+} ion entry across the cell membrane. In most VSM cells, the action potential is considered to be due to an activation of the voltage-dependent Ca^{2+} channels. Two types (L and T) of calcium channels have been reported to be present in the A7r5 VSM cells, as in cardiac muscle cells (7). These are distinguished by differences in (a) their voltage-dependency of activation and inactivation, (b) their sensitivity to dihydropyridine blockers, (c) their single-channel conductance, and (d) their permeability to various divalent cations.

In these experiments, the holding potential (HP) was held at -60 mV to prevent run-down of I_{Ca} (L-type Ca^{2+} channels). It has been commonly recognized that run-down is minimized at more negative holding potentials. Extracellular Ca^{2+} concentration ($[Ca]_o$) was 20 mM. The measured capacitance was 12.4 $\pm$ 1.1 pF (n = 19). The peak current of I_{Ca} occurred at +10 to +20 mV, and the current density was 12.8 $\pm$ 2.1 pA/pF (n = 15)(Fig. 3). The threshold potential was -20 to -30 mV, and the reversal potential was about +70 mV. This I_{Ca} current was recognized as L-type Ca^{2+} current (a) by blockade with nifedipine, (b) by its slow inactivation, (c) by the high threshold voltage (ca. -25 mV), and (d) by the high voltage at peak current occurred (ca. +10 mV). Comparative characteristics of the L-type I_{Ca} current in the A7r5 and A10 cell lines are summarized in Table 1.

Isoprenaline Effects on I_{Ca}

Isoprenaline (ISO) binds to β-adrenoceptors and accumulates cAMP by activation of adenylate cyclase through stimulatory G-binding protein (8). The Ca^{2+} channel proteins are phosphorylated by cAMP-dependent protein kinase, resulting in stimulation of channel activity in myocardial cells; both cyclic nucleotides inhibit Ca^{2+} channel activity in some VSM cells (23).

In the A7r5 cell line, ISO (1 μM) inhibited the I_{Ca} current only by about 6% (n = 5), and at 5 μM, ISO decreased I_{Ca} by 20% (n = 1)(Fig. 3A). Test pulses for 200 ms were applied to -30 to +70 mV from the HP of -60 mV. External solution contained high Ca^{2+} (20 mM), and $[Ca]_i$ was pCa 10. The current/voltage relationships for I_{Ca} in the absence and presence of ISO (1-5 μM) are shown in Fig. 3B.

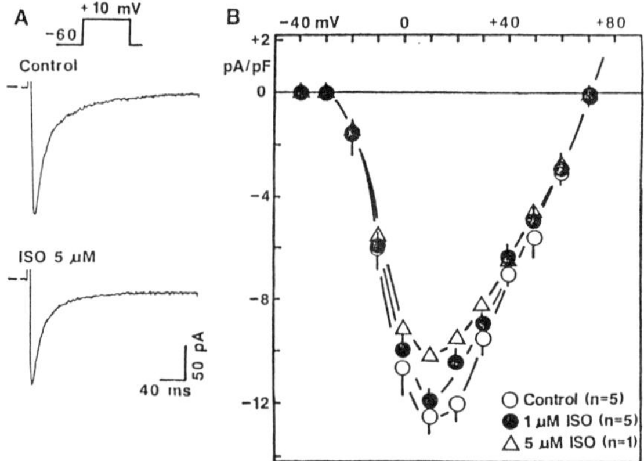

Figure 3. Measurement of I_{Ca} in a cultured A7r5 cell and the effect of isoprenaline. The threshold potential was -20 to -30 mV, and the peak amplitude of I_{Ca} was 150 to 180 pA at +10 to +20 mV, as shown in panel B (control = open circles). The reversal potential was +70 mV. Holding potential was -60 mV. **A:** Current traces in absence and presence of 5 µM isoprenaline (ISO). Test pulses were applied to +10 mV. The short line at the left of the current records represents the zero current level. **B:** Current/voltage relationship for I_{Ca} in the absence (control; open circles, n = 5) and presence of isoprenaline at 1 µM (filled circles, n = 5) and at 5 µM (triangles, n = 1). The values plotted are mean ± SEM.

Table 1. L-Type Calcium Current in Cultured Rat Aortic Vascular Smooth Muscle Cell Lines

	$[Ca]_o$ or $[Ba]_o$ (mM)	[EGTA] in pipette (mM)	Threshold potential (mV)	Potential at peak I_{Ca} (mV)	Peak current amplitude (pA)	E_{rev} (mV)
A7r5 cell line						
Present data	20 Ca	10	-20-30	+10-20	150-180 [a]	+70
Fish et al. (1988)	20 Ba	10	-20	+10	106 [b]	- -
Tamura et al. (1991)	1.8 Ca	0	-20	+10	90	- -
A10 cell line						
Friedman et al. (1986)	10 Ca	5	-30	+10	25	+70
	10 Ba	5	-30	+10	145	+70

E_{rev}: reversal potential.
a : Current density was -13 pA/pF.
b : Current density was -2.8 pA/pF.

Consistent results about the effects of catecholamines on I_{Ca} have not been obtained. Ohya et al. (16, 17) have reported that in intestinal and portal venous cells, cAMP-dependent phosphorylation did not play an important role in the regulation of Ca^{2+} channels. Intracellular or extracellular application of cAMP had no effect on I_{Ca}. Droogmans et al. (5) reported that ISO (10 μM) inhibited or had no effect in rabbit ear artery. Sperelakis and colleagues (2, 15) showed that the dibutyryl analogs of cAMP suppressed the slow inward Ba^{2+} current in single rabbit aortic VSM cells. On the other hand, Fukumitsu et al. (10) showed that ISO (1-5 μM) actually increased I_{Ca} in porcine coronary artery.

Although ISO could inhibit I_{Ca} secondarily due to elevated $[Ca]_i$ (because of release from SR), in the present experiments, $[Ca]_i$ was fixed at pCa 10 (10 mM EGTA). Therefore, these results indicate that, in the A7r5 cell line, ISO inhibits I_{Ca} by a small amount even at low $[Ca]_i$.

Carbachol Effects

ACh elevates cGMP through muscarinic receptors, and can also activate PK-C by the production of diacylglycerol (DAG) due to increased PI turnover. An agonist specifically binds to a cell surface receptor, which then activates a phospholipase C for PI located on the cytosolic face of the plasma membrane. The active enzyme produces phosphatidyl-inositol-4, 5-bisphosphate (PIP$_2$), and subsequently inositol-1, 4, 5-triphosphate (IP$_3$) and DAG. IP$_3$ mobilizes Ca^{2+} stored in the SR (1, 29). Diacylglycerol activates PK-C by stabilizing its insertion into the membrane. Activated PK-C phosphorylates substrate proteins, and thereby the physiological responses are elicited (1, 24).

In the present experiments, carbachol (0.1 μM) decreased the I_{Ca} current by about 14% (n = 5) (Fig. 4). In one experiment, 1 μM carbachol depressed I_{Ca} by 22%. The current/voltage curves for I_{Ca} are summarized in Fig. 4B. In frog visceral smooth muscle cells, ACh (50 μM) increased I_{Ca}, possibly by activating PK-C (27). Since cGMP relaxes the VSM cells, cGMP elevation induced by carbachol may depress I_{Ca}. Actually, the dibutyryl analogs of cGMP depressed I_{Ba} in rabbit aortic VSM cells (2). These results suggest that in the cultured A7r5 cell line, the effect of carbachol-induced cGMP may predominate, as compared with the effect of PK-C stimulation by carbachol (via DAG).

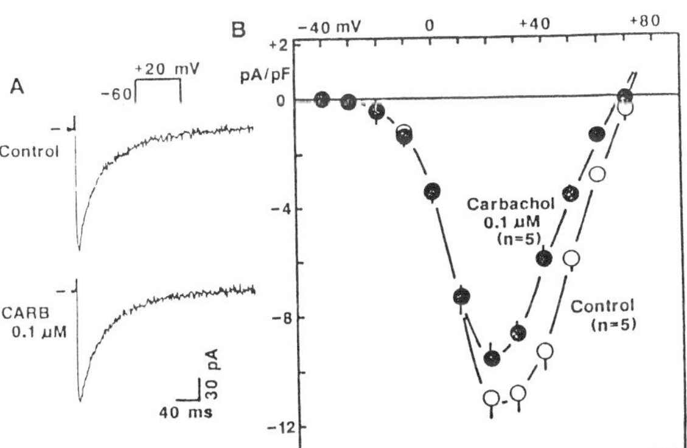

Figure 4. Depression of I_{Ca} in the A7r5 cell line by carbachol. **A:** Current traces in control and in 0.1 μM carbachol (CARB). Test pulse was applied to +20 mV from -60 mV of the holding potential. The short line at the left of the current records represents the zero current level. **B:** Current-voltage relationship for I_{Ca} in the absence (open circles, n = 5) and presence of 0.1 μM carbachol (filled circles, n = 5). The values are represented as mean $\pm$ SEM.

PK-C Stimulation With Phorbol Ester

Phorbol ester, 4-β-phorbol-12, 13-dibutyrate (PDB), concentrations ranging from 10 nM to 1 μM, produced very little or no increase in I_{Ca} (Fig. 5A). PDB is a potent stimulator of PK-C among phorbol esters (22). Current/voltage relationships for I_{Ca} in control (open circles, n = 14) and PDB (1 μM)(filled circles, n = 11) are shown in Fig. 5B.

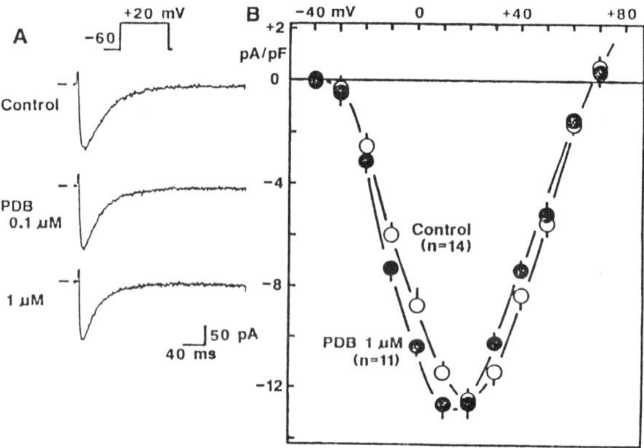

Figure 5. Lack of effect of phorbol ester (4-β-phorbol-12,13-dibutyrate; PDB) on I_{Ca} in the A7r5 cell line. **A:** Current traces in the absence and presence of PDB (0.1-1 μM). The holding potential was -60 mV. Test pulse was applied to +20 mV. The short line at the left of the current records represents the zero current level. **B:** Current/voltage curves for I_{Ca} in control (open circles, n = 14) and in 1 μM PDB (filled circles, n = 11). The values are represented as mean $\pm$ SEM.

It has been reported that phorbol esters (TPA and PDB) enhanced I_{Ca} in frog visceral smooth muscle cells (27) and in the A7r5 cell line (7). In guinea pig portal vein cells, angiotensin-II (Ang-II) enhanced I_{Ca}, presumably via stimulation of PK-C (18). In the A7r5 cell line used in the present experiments, however, carbachol inhibited I_{Ca}, and PDB did not affect I_{Ca}. One possibility to explain the lack of effect of phorbol ester is that the basal PK-C activity in the cultured cells may have been already maximum before application of phorbol ester, resulting in no effect of added phorbol ester. Another possibility is that, since our experiments were performed only at room temperature (22° C), the effect of phorbol ester may have been masked, because the regulation of ionic channels by protein kinase is temperature-dependent (28).

When PDB (1 μM) was added to bath solution in the presence of ISO (1 μM), Ca^{2+} current was further reduced (Fig. 6). That is, phorbol ester (which by itself did not affect I_{Ca}) slightly potentiated the small inhibitory effect of ISO. Sugden et al. (25) have reported that application of phorbol esters in the presence of ISO potentiated the cAMP accumulation by the β-adrenergic stimulation. In rabbit sino-atrial (SA) nodal cells, the phorbol ester TPA increased I_{Ca} (22). Addition of 0.1 μM ISO enhanced I_{Ca}, and simultaneously induced a transient inward current (I_{ti}) due to calcium overload. The calcium overload results from the combined I_{Ca} stimulating effects of cAMP elevation (due to ISO) and PK-C activation (due to TPA).

In the present study, it was found that ISO inhibits I_{Ca} to a small degree, presumably by elevation of cAMP. Phorbol ester may potentiate the cAMP elevation (25), thereby resulting in further inhibition of I_{Ca}.

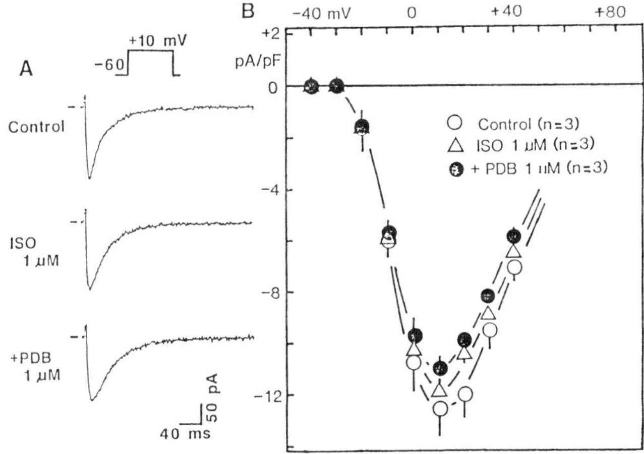

Figure 6. Slight potentiation by phorbol ester of the small depression in I_{Ca} produced by ISO. **A:** Current traces in control, in ISO (1 µM), and ISO (1 µM) plus PDB (1 µM). The holding potential was -60 mV. Test pulses were applied to +10 mV. The short line at the left of the current records represents the zero current level. **B:** Current/voltage curves for I_{Ca} in control (open circles, n = 3), in 1 mM ISO (triangles, n = 3), and in 1 µM ISO plus 1 µM PDB (filled circles, n = 3). The data points are mean ± SEM.

Outward K$^+$ Current

Outward K$^+$ currents play an important role of regulation of VSM cells, as well as the Ca^{2+} current. Blockade of K$^+$ channels depolarizes the membrane, and Ca^{2+}-dependent spikes are elicited more easily. Aortic VSM cells possess a delayed rectifier outward current ($I_{K(del)}$), and a current regulated by [Ca]$_i$, which is a Ca^{2+}-activated outward K$^+$ current ($I_{K(Ca)}$).

In the present experiments, the A10 cell lines were used, and Na$^+$, K$^+$ and Ca^{2+} in the bath solution were replaced by choline$^+$ to avoid the fast Na$^+$ current, the Ca^{2+} current, the Na-Ca exchange current, and the Na-K pump current.

Dependency on [Ca]$_i$. To examine whether the voltage-dependent outward K$^+$ current is stimulated by an increase in [Ca]$_i$ in A10 cells, we chose two [Ca]$_i$ levels: pCa 10 and pCa 7.

At low [Ca]$_i$ (pCa 10) in the pipette solution, $I_{K(Ca)}$ presumably should not be activated (Fig. 7A). At pCa 7, the amplitude of the outward K$^+$ current was potentiated (Fig. 7B). The HP was -40 mV. Test pulses (1 s) were applied from -20 to +70 mV in increment of 10 mV. The capacitance of the A10 cells was 8 ± 2 pF (n = 15). Maximum current density at +70 mV was about 11 pA/pF at pCa 10 in the pipette, and was about 28 pA/pF at pCa 7. The family of current traces at pCa 10 (A) and at pCa 7 (B) are given in Fig. 7A-B. The mean current/voltage relationships (current density) for the outward K$^+$ currents at pCa 10 (open circles, n = 8) and pCa 7 (filled circles, n = 8) are summarized in Fig. 7C. The difference current/voltage curve is also plotted (triangles), by subtracting the pCa 10 curve from the pCa 7 curve; this difference curve represents the Ca^{2+}-activated I_K ($I_{K(Ca)}$). The $I_{(KCa)}$ current density was 16.4 pA/pF at +70 mV. These results indicate that an outward current in aortic VSM cells can be activated by elevation of [Ca]$_i$ to 1 x 10$^-$ M.

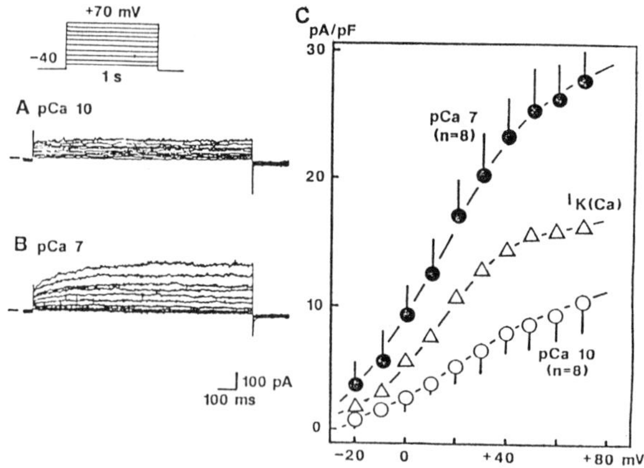

Figure 7. Dependency of outward K^+ current on [Ca]$_i$ in the A10 cell line. **A:** Superimposed outward current traces at pCa 10. Test pulses for 1 s were applied between -20 mV and +70 mV from a holding potential of -40 mV. **B:** Current traces at pCa 7. **C:** Mean current/voltage curves at pCa 10 (open circles, n = 8) and pCa 7 (filled circles, n = 8), and the difference current $I_{K(Ca)}$(triangles). The values are represented as mean $\pm$ SEM. The short line at the left of the current records in A-B represents the zero current level.

SUMMARY AND CONCLUSIONS

The present experiments were performed on single cultured rat aortic VSM cells by means of the patch electrode whole-cell voltage-clamp technique at room temperature (22° C). With 20 mM Ca^{2+} in the bath solution and 10 mM EGTA in the pipette solution, a slow I_{Ca} (L-type) current was observed in the A7r5 cell line. The peak current amplitude was 150 - 180 pA at +10 - 20 mV. The current density was about 13 pA/pF. The threshold potential was -20 - 30 mV, and the reversal potential was +70 mV (see Table 1). These results are generally consistent with recent reports (7, 9, 26).

Modulations of voltage-sensitive Ca^{2+} current by ISO, carbachol, and phorbol ester were examined. The addition of ISO for activation of the β-adrenoceptors and consequent elevation of cAMP had only little depressant effect on I_{Ca}. cGMP, produced secondarily by stimulation of muscarinic receptors, inhibited I_{Ca}. Phorbol ester (for PK-C stimulation) had very little or no effect on I_{Ca} in the present experiments, for possible reasons discussed above.

An outward K^+ current was activated in the A10 cell line by an increase in [Ca]$_i$ to 1×10^{-7} M from pCa 10. The outward K^+ current at pCa 10 was subtracted from the outward current at pCa 7 to give $I_{K(Ca)}$.

REFERENCES

1.   Berridge, M.J., Biol. Chem. Hoppe-Seyler. 367, 447-456 (1986).
2.   Bkaily, G., M. Peyrow, T. Yamamoto, A. Sculptoreanu, D. Jacques, and N. Sperelakis, Mol. Cell. Biochem. 80, 59-72 (1988).
3.   Bolton, T.B, R.J. Lang, and T. Takewaki, J. Physiol. 351, 549-572 (1984).
4.   Droogmans, G., L. Raeymaekers, and R.J. Casteels, Gen. Physiol. 70, 129-148 (1977).
5.   Droogmans, G., L. Raeymaekers, and R. Casteels, Pflugers Arch. 409, 7-12 (1987).
6.   Fabiato, A. and F. Fabiato, J. Physiol. (Lond) 75, 463-505 (1979).
7.   Fish, R.D., G. Sperti, W.S. Colucci, and D.E. Clapham, Circ. Res. 62, 1049-1054 (1988).

8.   Gilman, A.G., Ann. Rev. Biochem. 56, 615-649 (1987).

9.   Friedman, M.E., G.S. Kurtz, G.J. Kaczorowski, G.M. Katz, and J.P. Reuben, Am. J. Physiol. 250, H699-H703 (1986).

10.   Fukumitsu, T., H. Hayashi, H. Tokuno, and T. Tomita, Br. J. Pharmacol. 100, 593-599 (1990).

11.   Hamill, O.P., A. Marty, E. Neher, B. Sakmann, and F.J. Sigworth, Pfluger Arch. 391, 81-100 (1981).

12.   Kimes, B.W. and B.L. Brandt, Exp. Cell. Res. 98, 349-366 (1976).

13.   Kuriyama, H., Y. Ito, H. Suzuki, K. Kitamura, and T. Ito, Am J. Physiol. 243, H641-H662 (1982).

14.   Kuriyama, H. and H. Suzuki, Br. J. Pharmacol. 64, 493-501 (1978).

15.   Ousterhout, J.M. and N. Sperelakis, Eur. J. Pharmacol. 144, 7-14 (1987).

16.   Ohya, Y., K. Kitamura, and H. Kuriyama, Pfluger Arch. 408, 465-473 (1987).

17.   Ohya, Y., K. Kitamura, and H. Kuriyama, Circ. Res. 62, 375-383 (1988).

18.   Ohya, Y. and N. Sperelakis, Circ. Res. 68, 763-771 (1991).

19.   Rasmussen, H. and P.Q. Barrett, Physiol. Rev. 64, 938-984 (1984).

20.   Rink, T.J. and R.Y. Tsien, Biochimica. Biophysica. Acta. 599, 623-638 (1980).

21.   Ruegg, U.T., V.M. Doyle, J. Zuber, and R.P. Hof, Biochem. Biophy. Res. Commun. 130, 447-453 (1985).

22.   Satoh, H. and K. Hashimoto, Naunyn-Schmiedeberg's Arch. Pharmacol. 337, 308-315 (1988).

23.   Sperelakis, N., Mol. Cell. Biochem. 99, 97-109 (1990).

24.   Streb, H., R.F. Irvine, M.J. Berridge, and Schulz, I., Nature 306, 67-69 (1983).

25.   Sugden, D., J. Vanecek, D.C. Klein, T.P. Thomas, and W.B. Anderson, Nature 314, 359-361 (1985).

26.   Tamura, T., Saigusa, A., and S. Kokubun, Naunyn-Schmiedeberg's Arch. Pharmacol. 343, 405-410 (1991).

27.   Vivaudou, M.B., L.H. Clapp, J.V. Walsh, and J.J. Singer, FASEB J. 2, 2497-2504 (1988).

28.   Walsh, K.B. and R.S. Kass, Science 242, 67-69 (1988).

29.   Williamson, J.R., R.H. Cooper, S.K. Joseph, and A.P. Thomas, Am. J. Physiol. 248, C203-C216 (1985).

MODULATION OF ION CHANNELS BY CALCIUM RELEASE IN CORONARY ARTERY SMOOTH MUSCLE

M. STUREK, L. STEHNO-BITTEL, and P. OBYE

Department of Physiology, School of Medicine, and Dalton Research Center, University of Missouri, Columbia, Missouri 65211

INTRODUCTION

Intracellular free Ca^{2+} is involved in regulation of a variety of events occurring in vascular smooth muscle cells. Commonly, Ca^{2+} is considered an intracellular messenger transducing extracellular information to organelles and myofilaments. However, intracellular Ca^{2+} may also affect sarcolemmal ion currents in vascular smooth muscle cells. The interaction of subsarcolemmal Ca^{2+} and ion channels is the focus of this manuscript.

Indirect evidence for intracellular Ca^{2+} modulation of ion currents is provided by work with vasoconstrictors such as histamine, serotonin, and acetylcholine, which initiate coronary artery smooth muscle cell contraction by release of Ca^{2+} from the sarcoplasmic reticulum (SR) (16, 17, 23). Histamine (12) and acetylcholine (10, 15) cause hyperpolarization of coronary artery smooth muscle, suggesting that receptor occupation and intracellular Ca^{2+} release by these ligands is associated with alterations in ionic currents (31, 32) that are not dependent on membrane depolarization for gating. In addition, Kitamura and Kuriyama (15) have found that acetylcholine caused a selective increase in coronary smooth muscle K^+ permeability through Ca^{2+}-activated K^+ channels (10). Our hypothesis was that both the inward and outward ion currents in coronary artery smooth muscle may be modulated by a common messenger, intracellular free Ca^{2+}. Specifically, we investigated the effects of Ca^{2+} released by the sarcoplasmic reticulum (SR) on sarcolemmal currents.

Through the Ca^{2+} pathways shown in Fig. 1 free intracellular Ca^{2+}, including Ca^{2+} in the subsarcolemmal region (Ca_{is}), is tightly regulated in resting cells. High Ca^{2+} concentrations in the bulk myoplasm (Ca_{im}) of the cell induce contraction via a Ca^{2+}-calmodulin cascade. Thus, in resting cells bulk myoplasmic Ca^{2+} concentrations must be kept low. This is accomplished through Ca^{2+} efflux via the sarcolemmal Ca^{2+} pump and Na^+-Ca^{2+} exchanger and Ca^{2+} sequestration by the SR Ca^{2+} pump. Ca^{2+} may enter the cell through voltage-gated Ca^{2+} channels and be sequestered by the SR, thus filling the SR Ca^{2+} store (37, 42).

Casteels and Droogmans (6) and van Breemen (20) proposed that in addition to the well-known role of the SR in agonist-induced Ca^{2+} release, the SR also buffers myoplasmic free Ca^{2+} in smooth muscle cells. This hypothesis was further refined by van Breemen and coworkers and is known as the superficial buffer barrier hypothesis (26, 45). A major prediction of the hypothesis was that a fraction of the Ca^{2+} entering the cell through the sarcolemma would be preferentially sequestered by the SR, thereby buffering (attenuating) the increase in Ca^{2+} concentration in the bulk myoplasm. Such buffering of Ca^{2+} would serve an immediate role in attenuating contraction. Several studies have provided support for this "protective aspect" of the buffer barrier mechanism (20, 33, 37, 42). However, the SR may become filled to capacity with Ca^{2+} by this avid sequestration and subsequently release the Ca^{2+} in response to agonists, such as endothelin, whose primary mechanism of action is on the SR of vascular smooth muscle cells (47,48). A second major prediction of the SR buffer barrier hypothesis is that the SR, once Ca^{2+}-filled, would release Ca^{2+} preferentially toward the sarcolemma for extrusion from the cell without increasing free myoplasmic Ca^{2+} (Ca^{2+} surrounding the myofilaments in the bulk myoplasm) (45). In doing so the SR "unloads" its Ca^{2+} store (37,38). One implication of this mechanism that is most relevant to this chapter is that the Ca^{2+} released from the SR could profoundly affect sarcolemmal ion currents via effects on Ca^{2+}-activated ion channels.

Three classifications of sarcolemmal ion channels are known to be sensitive to free subsarcolemmal Ca^{2+} concentrations (Fig. 1): 1) Ca^{2+}-activated K^+ channels; 2) Ca^{2+}-activated non-selective cation channels; 3) Ca^{2+}-activated Cl^- channels (22). Activity

Published 1991 by Elsevier Science Publishing Company, Inc.
Ion Channels of Vascular Smooth Muscle Cells and Endothelial Cells
Sperelakis and Kuriyama, Editors

(probability of the channel opening) appears to increase in the presence of high subsarcolemmal Ca^{2+} concentrations in each of the three types of channels. The purpose of this chapter is to briefly review the literature and our evidence that sarcolemmal Ca^{2+} concentrations affect unitary and whole-cell ion currents in vascular smooth muscle. Primarily, we will review the role of Ca^{2+} release from the SR in modulating sarcolemmal ion currents. Specifically, the types of Ca^{2+} release are: 1) localized, bolus release initiated by agonists or occurring spontaneously toward the sarcolemma and; 2) constant, uniform (not bolus) release from the SR toward the sarcolemma.

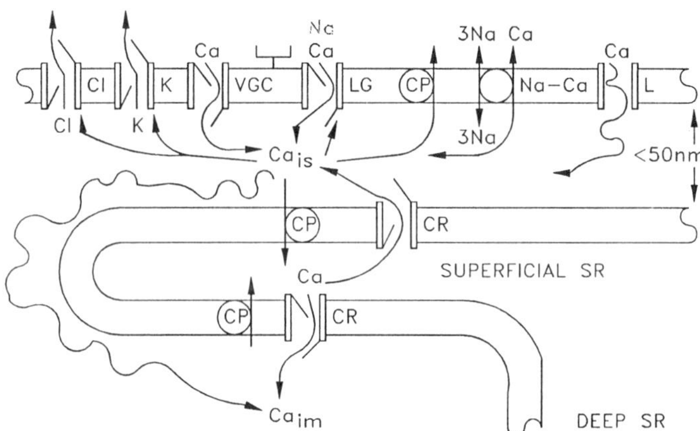

Figure 1. Calcium pathways in the sarcolemma and sarcoplasmic reticulum (SR). Ca^{2+} enters cells through voltage- gated Ca^{2+} channels (VGC), ligand-gated Ca^{2+} channels (LG) and a passive influx of Ca^{2+} through leak channels (L). Calcium is removed from the cell via the sarcolemmal Ca^{2+} pump (CP) and the Na^{+}-Ca^{2+} exchanger (Na-Ca). The SR sequesters Ca^{2+} via a Ca^{2+} pump (CP) and releases Ca^{2+} through Ca^{2+}-activated Ca^{2+} release channels or inositol trisphosphate-sensitive channels (both indicated as CR). Ca^{2+} activates sarcolemmal Cl^{-}, K^{+} channels, and Ca^{2+}-activated non-selective cation channels (indicated as LG).

MATERIALS AND METHODS

Cell dispersion, fura-2 measurements, and patch-clamp methods have been described previously (37, 38, 47) and are presented in greater detail in chapters by Sturek et al. on instrumentation (41) and physiological methods (43). Briefly, coronary artery smooth muscle cells from porcine and bovine hearts were enzymatically dispersed by methods similar to those described by Washaw et al. (47). When intracellular Ca^{2+} measurements were required prior to dialysis during whole-cell recordings, cells were loaded with the membrane permeant intracellular Ca^{2+} indicator, fura-2, by incubating the cells in 2.5 μM fura-2 for 15 to 30 min at 37° C in physiological buffer. The buffer solution, composed of (in mM): 138 NaCl, 2 $CaCl_2$, 5 KCl, 1 $MgCl_2$, 10 HEPES, 10 glucose, pH 7.4, was superfused over cells during experiments unless otherwise stated. Drugs were added to the buffer solution and time of drug application is indicated by horizontal lines in all figures.

Whole-cell and inside-out excised patch configurations were used to record ionic currents from single cells. Fire-polished micropipettes (4-8 MΩ resistance) were sealed against the cell membrane with suction. After obtaining a GΩ seal the measurement aperture was moved to mask off the pipette tip from the measurement area of the photomultiplier tube. After formation of a gigaseal ionic currents were measured in the cell-attached mode, or the membrane under the pipette was ruptured with additional suction for whole-cell recordings allowing dialysis of the pipette solution into the cells. Unless otherwise indicated the pipette solution for whole-cell recordings included (in mM): 135 KCl, 10 NaCl, 1 $MgCl_2$, 2 MgATP, 0.5 Tris GTP, 0.2 EGTA, 0.1 fura-2 pentapotassium salt, pH to 7.1. Calculated equilibrium potentials, given the whole-cell pipette solution and physiological buffer bathing the cells were: E_{Na} = +65 mV, E_{Ca} > +120 mV, E_{Cl} = -2 mV, and E_K = -82 mV.

For excised-patch and whole-cell recordings with clamped internal Ca^{2+} concentrations the solution included (in mM): 126 KCl, 10 NaCl, 20 HEPES, 1 $MgCl_2$, 0.1 fura-2, pH 7.1. Free Ca^{2+} concentrations ranging from 0 to 1000 nM were achieved by adding appropriate amounts of H_2K_2EGTA and CaK_2EGTA as described (44, 47). The final EGTA concentration was always 10 mM in Ca^{2+}-clamped cells. The pipette solution for single-channel recordings consisted of a high Na^+ with 2 mM Ca^{2+} buffer solution already described. Ionic currents were amplified by a List EPC-7 patch-clamp amplifier with a headstage having switchable feedback resistors of 0.5 and 50 GΩ. Capacitance transients were measured for each cell during 10 ms pulses from a holding potential of -80 mV to a test potential of -70 mV. Capacity currents were filtered at a low-pass cutoff frequency of 8.4 KHz and digitized at 25 μs intervals. The whole- cell currents were filtered by an 8-pole low pass filter at a cutoff frequency of 400 Hz, digitized at 600 μs intervals. Excised and cell-attached patch-clamp recordings were filtered at a low-pass cutoff frequency of 1 KHz and digitized at 240 μs intervals. Data were stored and analyzed on computer using AxoBASIC software (Axon Instruments, Foster City, CA).

Much greater detail of the voltage-clamp protocol has been provided previously (38), consisting of a holding potential (HP) of -80 mV and a test potential (TP) every 5 sec. The test potential (+30 mV unless otherwise stated) had a duration of 323 ms; thus, the cells were depolarized for only 6.5% of the time. During the test potential whole-cell current amplitude, or single channel activity was measured.

RESULTS AND DISCUSSION
Ca^{2+}-activated K^+ channels

Single channel recordings have revealed that a variety of smooth muscle cells contain large-conductance Ca^{2+}-activated K^+ channels which are activated by membrane depolarization and subsarcolemmal Ca^{2+} concentrations (24, 27, 35, 36). The activity of these channels has been shown to be unaltered in either cell-attached mode or excised into patches (28). Although the channels are sensitive to Ca^{2+}, we use the term Ca^{2+}-activated because in many cell types the channels may open during depolarization even in the absence of Ca^{2+}. In secretory cells, such as the chromaffin cell, the K^+ channel is regulated solely by the membrane potential and subsarcolemmal Ca^{2+} concentrations (for review see (30). However, in cultured aortic smooth muscle intracellular agents such as cyclic AMP directly modulated the Ca^{2+}-activated K^+ channel (34). Due to the limitations of this review, we will discuss modulation of the channels by Ca^{2+} and voltage only.

Fig. 2 illustrates the voltage and Ca^{2+}-dependence of the whole-cell K^+ current from smooth muscle of porcine coronary artery. These cells were dialyzed with 10 mM Ca⁰ EGTA buffer solutions (described under Materials and Methods) in which free intracellular Ca^{2+} was clamped at 200, 400, 750, and 1000 nM Ca^{2+}. To ensure full buffering of intracellular Ca^{2+}, caffeine (5 mM) was applied externally to the cells. Caffeine induces a rapid release of Ca^{2+} from the SR which causes the largest increases in the fura-2 ratio (37, 38, 47, 48) and in the amplitude of the whole-cell Ca^{2+}-activated K^+ current in cells which were not Ca^{2+}-clamped (see Fig. 4). However, when Ca^{2+}-clamped solutions were dialyzed into cells, caffeine did not alter the whole-cell current or the fura-2 ratio, suggesting that Ca^{2+} released from the SR was rapidly buffered prior to binding any Ca^{2+}-sensitive ion channel. The amplitude of the whole-cell Ca^{2+}- activated K^+ current was measured at different test potentials from -50 to 50 mV and plotted in Fig. 2. In order to normalize the data for the different cell sizes, which obviously would affect the amplitude of the whole-cell current, the amplitude was divided by the capacitance of each cell (pA/pF). It is clear in Fig. 2 that as the free intracellular Ca^{2+} concentration increased from 0 to 1000 nM Ca^{2+} the current amplitude also increased at test potentials positive to -50 mV. Furthermore, at any given Ca^{2+} concentration, the current amplitude increased in response to greater depolarization. Thus, Fig. 2 summarizes the Ca^{2+}- and voltage-sensitivity of the Ca^{2+}- activated K^+ current in smooth muscle cells from porcine coronary artery. At +30 mV the sensitivity of the channels to subsarcolemmal Ca^{2+} concentrations is good, while the cells appear to tolerate a brief depolarization to that

68

potential quite well. Therefore, we have chosen +30 mV as our standard test potential for monitoring changes in the whole-cell Ca^{2+}-activated K^+ current. At a given membrane potential changes in the Ca^{2+}-activated K^+ current should reflect changes in the subsarcolemmal Ca^{2+} concentration, thus providing a way to monitor relative changes in subsarcolemmal Ca^{2+}. However, it was essential that we first rule out any direct action of pharmacological agents on channel activity; thus, the following set of experiments were conducted.

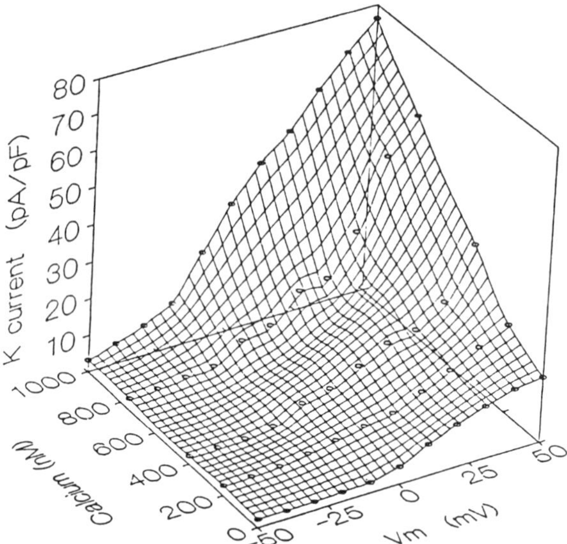

Figure 2. Ca^{2+}- and voltage- sensitivity of whole-cell Ca^{2+}-activated K^+ current. Free intracellular Ca^{2+} was clamped in porcine smooth muscle cells and the current amplitude (pA/pF) recorded at varying test potentials (Vm). Holding potential was -80 mV throughout recordings and test potential durations of 323 ms. Actual data points are shown as circles and the lines are a smooth extrapolation between the data points. The graph summarizes data collected from 15 cells (2-4 cells per data point).

Typical single channel recordings of Ca^{2+}-activated K^+ channels are shown in Fig. 3 from an excised patch (inside-out) from a bovine cell. This patch contains approximately 7 channels as determined by maximal stimulation of the channels in 2 mM Ca^{2+} at +50 mV. Single channel activity during test pulses to +50 mV are shown first under control conditions (control) with 120 mM K^+ and 400 nM free Ca^{2+} in the bath resulting in an equilibrium potential for K^+ of -33 mV. The mean unitary current at +50 mV was 9.8 pA, slope conductance was 200 pS with a reversal potential for K^+ of -33 mV. When K^+ was replaced by Cs^+ in the bath the channel activity was reversibly abolished (Fig. 3, Cs). The addition of 5 mM caffeine (CAF), 10^{-8} M ryanodine (ryan), and 10^{-6} M glyburide (glyb) to the 400 nM Ca^{2+}-clamped solution had no direct affect on channel activity as seen in Fig. 3. Therefore, pharmacological agents which we use to manipulate the SR Ca^{2+} store have no direct affect on the Ca^{2+}-activated K^+ channels other than by altering the Ca^{2+} concentration at the subsarcolemmal surface of the membrane.

Figs. 2 and 3 indicate the sensitivity of these channels to Ca^{2+} and membrane potential and show the lack of sensitivity to some pharmacological agents such as caffeine, ryanodine, and glyburide. Thus, the amplitude of the whole-cell Ca^{2+}-activated K^+ current may be used as an indicator of the subsarcolemmal Ca^{2+} concentrations, while fura-2 measures free myoplasmic Ca^{2+}. It is important to note that our photomultiplier measures of whole-cell fura-2 fluorescence give an estimate of the Ca^{2+} averaged over the entire cell (bulk myoplasm). Simultaneous recording of both whole-cell current and fura-2 allows monitoring of Ca^{2+} in the two compartments (subsarcolemmal and myoplasmic) as is shown in Fig. 4. The purpose of the protocol was to manipulate the SR Ca^{2+} store and monitor any affects on the subsarcolemmal or myoplasmic Ca^{2+} concentrations. The bovine cell shown in Fig. 4 was dialyzed with a mock intracellular solution containing fura-2 as described previously (Materials and Methods). Both the whole-cell Ca^{2+}-activated K^+ current amplitude at a test potential of +30 mV (circles) and the fura-2 ratio (triangles) are plotted in Fig. 4. At min 0 the cell was superfused with physiological buffer containing

2 mM Ca^{2+} with a holding potential of -80 mV and a test pulse to +30 mV every 5 sec (see Fig. 6 for voltage template). At min 1 the holding potential was depolarized to -40 mV while the test potential was maintained at +30 mV. The 3 min depolarization induced a slow increase in free myoplasmic Ca^{2+} as indicated by the rise in the fura-2 ratio (Fig. 4, triangles). This increase in Ca^{2+} during depolarization has been reported previously and fills the SR with Ca^{2+} (29, 37, 38, 42, 48). In contrast, the current amplitude (Fig. 4, circles) remained relatively stable during the depolarization. Following the depolarization, when the SR was Ca^{2+}-filled (37, 38, 42), the K^+ current amplitude increased significantly, indicating that subsarcolemmal Ca^{2+} concentrations increased when the SR was Ca^{2+}-filled. The current amplitude returned to resting levels gradually during min 5 through 9. In contrast, the fura-2 signal quickly returned to resting levels and remained steady during the same time period. At min 9 caffeine (5 mM) was added to the physiological buffer solution and the cell responded with increases in both the myoplasmic and subsarcolemmal Ca^{2+} concentrations. When caffeine was removed at min 10, the myoplasmic Ca^{2+} concentration (triangles) returned to resting levels while the current amplitude (circles) dropped below resting levels. We conclude that following exposure to caffeine the SR was emptied of a majority of its Ca^{2+}. Preferential release of Ca^{2+} toward the SR (noted during min 4 - 7) was abolished and subsequent subsarcolemmal Ca^{2+} concentration decreased while myoplasmic Ca^{2+} concentrations returned to resting. Therefore, a correlation was noted between the amount of Ca^{2+} in the SR and the subsarcolemmal Ca^{2+} concentration (when the SR was filled subsarcolemmal Ca^{2+} concentrations were high, when the SR was emptied subsarcolemmal Ca^{2+} concentrations decreased).

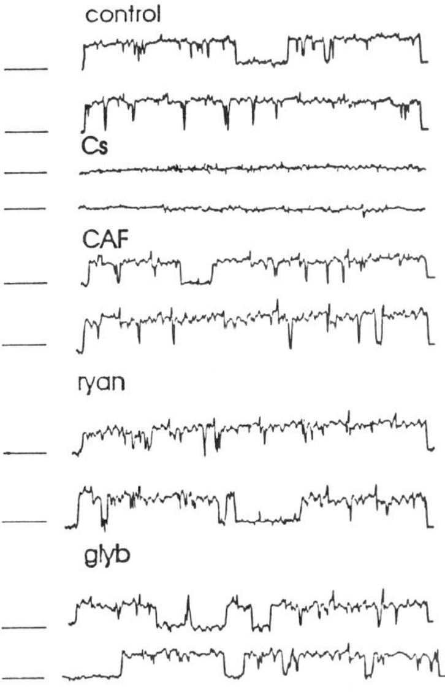

Figure 3. Single channel recordings from inside-out membrane patch of bovine cell. The holding potential was -80 mV and a test potential of +50 mV for a duration of 140 ms was applied every 2 s. Horizontal bars on the left indicate 0 mV current (closed channels). Drugs were added to the bath (cytoplasmic face of the membrane). Bath Ca^{2+} concentration was maintained at 400 nM. First, single channel activity is shown under control conditions (prior to application of drugs). Cesium (Cs) blocked the outward current while caffeine (CAF), ryanodine (ryan), and glyburide (glyb) had no direct affect on the channel activity.

70

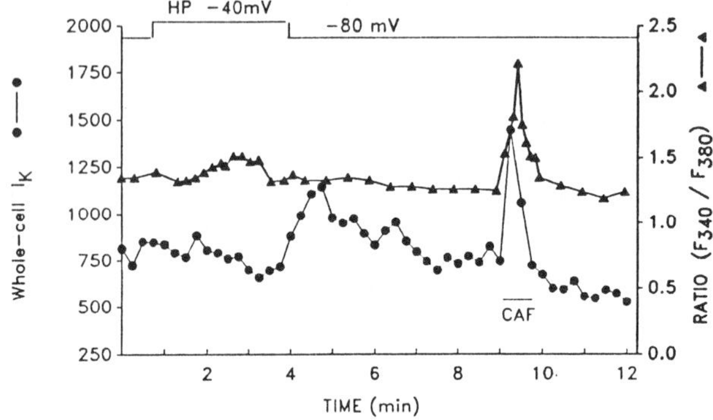

Figure 4. Simultaneous monitoring of whole-cell Ca^{2+}-activated K^+ current (in pA, circles) and fura-2 ratio fluorescence (triangles) illustrate changes in subsarcolemmal and myoplasmic Ca^{2+} concentrations, respectively. A single bovine cell was patch- clamped with a holding potential (HP) of -80 or -40 mV, but always with a test potential every 5 s to +30 mV.

If the increase in subsarcolemmal Ca^{2+} from min 4 to 5 in Fig. 4 (circles) was due to preferential release of Ca^{2+} from the SR then, we predicted, the addition of low concentrations of caffeine or ryanodine would increase the preferential release of Ca^{2+} from the SR, without increasing myoplasmic Ca^{2+} concentrations. Thus, the whole-cell Ca^{2+}-activated K^+ current would increase but fura-2 ratios would be unaltered. The experiments shown in Fig. 5 were designed to test that hypothesis. Low concentrations of ryanodine (10^{-8} M) (Fig. 5A) or caffeine (10^{-4} M) (Fig. 5B) were added resulting in sustained increases in the Ca^{2+}-activated K^+ current (circles) without changes in the fura-2 ratio (triangles). The increase in subsarcolemmal Ca^{2+} appears to be due to release (unloading) of Ca^{2+} from the SR as subsequent exposure to 5 mM caffeine induced only a small increase in the current and in the fura-2 ratio (compare with caffeine response shown in Fig. 4). Therefore, low concentrations of caffeine and ryanodine enhanced the release of Ca^{2+} from the SR, which increased subsarcolemmal Ca^{2+}, but not myoplasmic Ca^{2+} concentrations. Further, these drugs emptied the SR and/or desensitized the SR Ca^{2+} store to subsequent caffeine (high concentration) exposure.

Another link between SR Ca^{2+} release and sarcolemmal K^+ currents is the identification of spontaneous transient outward currents (STOCs). STOCs have been characterized previously in vascular smooth muscle (3, 9, 10, 13, 27, 38). They have been shown to be due to a bolus release of Ca^{2+} from the SR that activates many K^+ channels (see Fig. 1) thus causing a rapid, transient increase in the whole-cell K^+ current (10). Fig. 6 shows examples of STOCs from bovine and porcine coronary artery smooth muscle cells. We define a STOC as a 20% or greater increase in the whole-cell current amplitude following plateau during the test potential (again +30 mV). To determine whether STOCs in these cells were due to rapid changes in subsarcolemmal Ca^{2+} concentrations we again dialyzed Ca^{2+}-clamped solutions into the cell and monitored the occurrence of STOCs. In 23 bovine and 15 porcine cells monitored with intracellular Ca^{2+} clamped at 0 to 1500 nM Ca^{2+} STOCs were not seen. In contrast, in cells not clamped with Ca^{2+}_o EGTA buffers STOCs were noted in 9.5% of the bovine cells and 44% of the porcine cells monitored. The STOCs seen in both cell types were associated with SR Ca^{2+} release, as emptying the SR Ca^{2+} store with 5 mM caffeine abolished further STOC activity for an average of 5 min (shown in Fig. 8). Along with species variation in the frequency of STOCs, the magnitude of the STOCs also differed between porcine and bovine cells. As shown in Fig. 6 STOCs from porcine cells were typically larger (greater percent increase over the plateau amplitude). In Fig. 6 a STOC is shown from a bovine cell in which the whole-cell current amplitude increased 33% during the STOC and a porcine cell with a STOC whose current amplitude was 5 times greater than the plateau amplitude. The species differences in STOC frequency and amplitude may be due to species-related differences in the release of

Ca^{2+} from the SR of resting cells. We believe that Ca^{2+} is released slowly and uniformly from the SR of quiescent bovine smooth muscle cells preferentially toward the sarcolemma to be extruded from the cell (39). In contrast, resting porcine cells appear to release Ca^{2+} in bolus amounts that is resequestered (recycled) by the SR rather than being transported from the cell (38, 39). Results shown in Fig. 7 are consistent with that hypothesis.

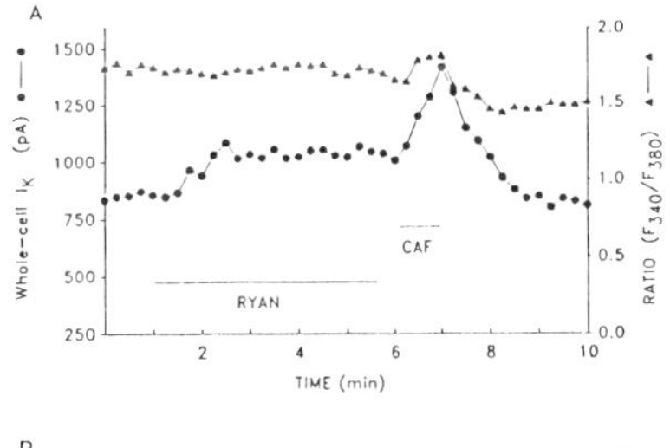

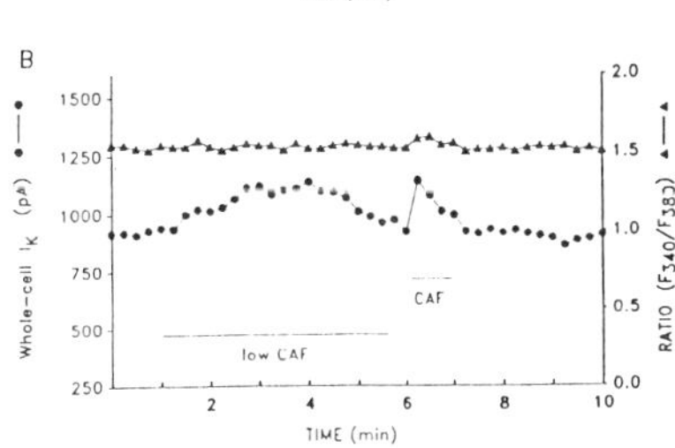

Figure 5. Low concentrations of ryanodine (A) or caffeine (B) mimic the affects seen during recovery from depolarization (Fig. 4) by increasing the whole- cell Ca^{2+}-activated K^+ current (circles) without changes in the fura-2 ratio (triangles). (A) The bovine cell was exposed to 10^{-8} M (RYAN) from min 1 to 6, resulting in an increase in the current. Subsequently, the cell was exposed to 5 mM caffeine (CAF) for 1 min. (B) The bovine cell was exposed to 10^{-4} M caffeine (low CAF) followed by application of 5 mM caffeine at min 6.

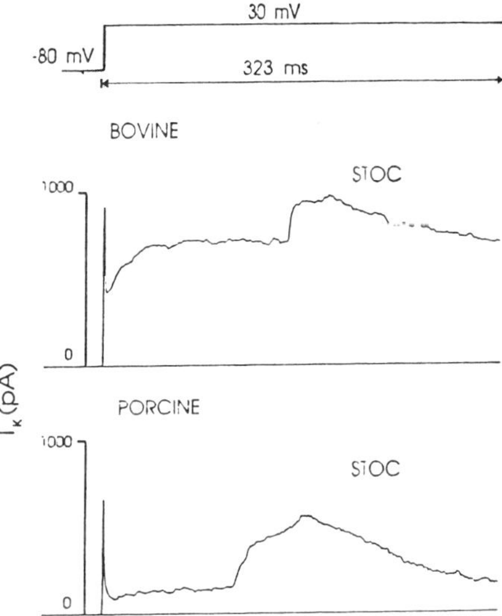

Figure 6. Typical STOCs from bovine and porcine cells during test potential of +30 mV from holding potential of -80 mV. The difference in the plateau amplitude of the whole- cell Ca^{2+}-activated K^+ current was due to different sizes in the cell as determined by the membrane capacitance (32 pF for bovine and 18 pF for porcine cells, respectively).

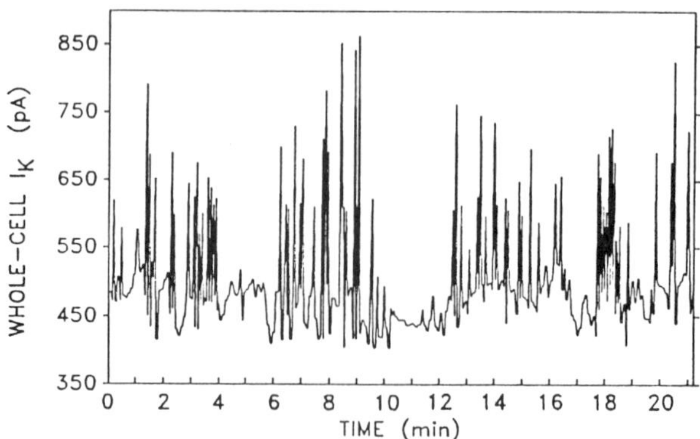

Figure 7. Frequency pattern of STOCs in a single porcine cell. Holding potential and test potential throughout the experiment were -80 and +30 mV, respectively with a test potential applied every 5 s. During each test potential the amplitude of the whole-cell Ca^{2+}-activated K^+ current was plotted. Upward spikes indicate the amplitude and time course of STOCs during the test potentials.

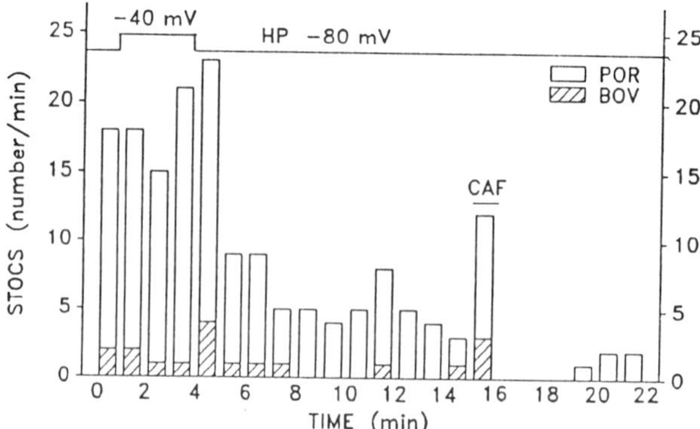

Figure 8. The number of STOCs per min illustrate the difference in STOC frequency in 2 typical porcine (open bars, POR) and 2 typical bovine (crosshatch, BOV) smooth muscle cells. The holding potential (HP) was -80 mV until min 1 when the cells were depolarized to a holding potential of -40 mV for 3 min. The test potential was +30 mV throughout the protocol. At min 15 caffeine (5 mM, CAF) was applied.

The notion that STOCs represent Ca^{2+} released from and resequestered by the SR was first introduced by Benham and Bolton (3) and Ohya et al. (27). Benham and Bolton (3) found that STOC activity in rabbit arterial smooth muscle was not dependent on extracellular Ca^{2+} and they noted a correlation in some cells in the latency between STOCs and the amplitude of the STOCs. The authors suggested that STOCs in arterial smooth muscle were due to release and resequestration of Ca^{2+} by the SR, which stimulated Ca^{2+}-activated K^+ channels. Fig. 7 illustrates the frequency pattern of STOCs over a 21 min period from a porcine coronary artery smooth muscle cell bathed in physiological buffer without pharmacological intervention. The current amplitude is plotted every 5 sec. STOC amplitudes and frequencies are shown by the upward spikes. At min 4.25, 10.5 and 16.5 periods void of STOC activity were noted and spontaneous return of STOCs occurred 1.5 to 2.5 min later. This oscillatory pattern in STOC activity was seen frequently when porcine cells were monitored without pharmacological intervention for long periods of time. However, the frequency pattern of the STOCs (time course of periods lacking STOCs) varied between cells. This oscillatory pattern in STOC frequency supports the hypothesis that in porcine cells STOCs represent Ca^{2+} released by the SR and subsequently resequestered. Bursts of STOC activity would release a large quantity of Ca^{2+} from the SR which would require time be resequestered by the SR thus explaining the periods of time lacking STOC activity. Dependence of the oscillatory STOC behavior on extracellular Ca^{2+} would be determined definitively in Ca^{2+}-free solutions and be voltage-gated Ca^{2+} channel blockers (11). STOCs were seen too infrequently in bovine smooth muscle cells to quantify any oscillatory pattern (see Fig. 8).

The histogram in Fig. 8 illustrates the frequency of STOCs shown as the number of STOCs per min for porcine (open bars) and bovine (crosshatch) cells. This protocol has been used previously to compare the frequency of STOCs in porcine smooth muscle from exercise trained and sedentary pigs (38). Cells were depolarized for 3 min with a holding potential (HP) of -40 mV from min 1 to 4. This depolarization fills the SR with Ca^{2+} (38). Following filling of the SR (HP returned to -80 mV) the STOC frequency increased slightly in porcine cells to 23 STOCs per min while STOC frequency in bovine cells remained relatively steady. At min 15 caffeine was applied and STOC activity increased transiently and then was abolished in both cell types for 3 min. The results indicate that the SR Ca^{2+} store is involved in STOC activity, as the frequency of STOCs was greatest when the SR was Ca^{2+}-filled (during and following depolarization) and STOCs were abolished when the SR was Ca^{2+}-emptied (following caffeine). Furthermore, STOCs were seen at a much higher frequency and with larger amplitudes in porcine than bovine cells. Endothelin, which releases Ca^{2+} from a subset of the caffeine- sensitive SR Ca^{2+} store also altered STOC activity in porcine cells. In 16 of 25 porcine cells tested, application of 5×10^{-8} M endothelin abolished subsequent STOC activity, in contrast with full abolition of STOCs in all cells when caffeine was applied.

Ca^{2+}-activated Cl^- and non-selective cation channels

Recent evidence indicates that there are both non-selective cation channels (4, 5, 14, 49) and Ca^{2+}-activated chloride (Cl^-) channels (2, 18, 22, 49) present in the sarcolemmal membrane of many types of smooth muscle (18, 49). This was shown using conventional whole-cell and single-channel recording techniques and stimulation with various agonists. Agonists such as caffeine (see Figs. 9-13), endothelin (18), acetylcholine (4), and norepinephrine (49) elicit an inward current of 10-500 pA. At negative membrane potentials the amplitude of the current elicited from the non-selective cation channel is typically much smaller (10-50 pA) than the inward current through the Cl^- selective channel (50-500 pA) (49).

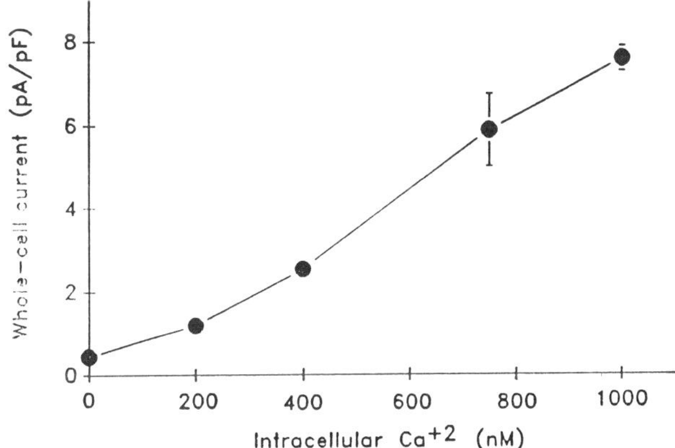

Figure 9. The Ca^{2+}- dependence of the whole-cell inward current from porcine cells (n=11) is shown. Cells were Ca^{2+}- clamped and the amplitude of the whole- cell inward current was measured at a constant membrane potential of -80 mV (E_K = -82 mV).

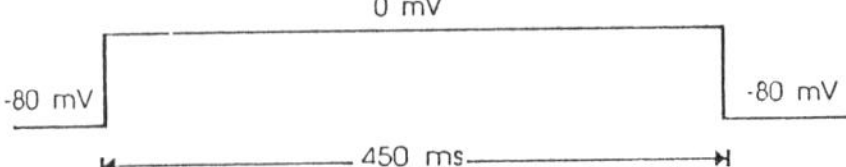

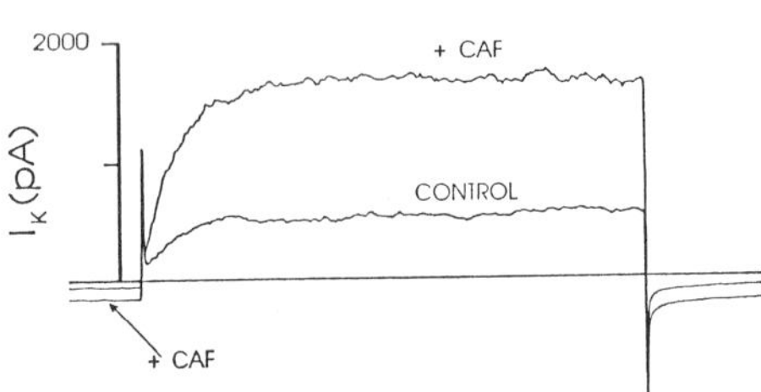

Figure 10. Whole-cell recording of inward and outward currents in a porcine cell before and during caffeine (CAF, 5×10^{-3} M) application with a holding potential of -80 mV and a test potential of 0 mV.

Cl⁻ current The Ca^{2+}-activated Cl⁻ channel can be found in many different types of smooth muscle, including porcine coronary artery (18), human mesenteric artery (18), rabbit portal vein (49), rabbit ear artery (2). The Ca^{2+}-activated Cl⁻ channel was shown to be Cl⁻- selective using norepinephrine to evoke current whose reversal closely followed E_{Cl}. When external Cl⁻ was replaced by nitrate or bromide, the reversal potential of the norepinephrine- evoked current shifted to more negative potentials (following E_{Cl}), indicating that an increase in Cl⁻ conductance contributes to the norepinephrine-activated current (2).

Non-selective cation current The existence of Ca^{2+}-activated inward cation currents has been well-documented in cardiac muscle cells (7, 8, 21, 25) and numerous other cell types (22). However, the inward currents activated by intracellular Ca^{2+} release in cardiac muscle are largely due to Na-Ca exchange (19, 21, 25) and only a minor component is explained by Ca^{2+}-activated non-selective cation channels (8, 21). These findings make it difficult to ascribe a physiological role to non-selective cation currents in cardiac muscle. There is some dispute as to whether Ca^{2+}-activated non-selective cation channels are also present in smooth muscle despite evidence for ligand-gated (receptor-operated) non-selective cation channels that are permeable to Ca^{2+}, but not activated by an increase in subsarcolemmal Ca^{2+} (4, 14). Data from this laboratory suggest that non-selective cation channels activated solely by an increase in subsarcolemmal Ca^{2+} are present in porcine coronary smooth muscle. Ion flux through the non-selective cation channel found in portal vein smooth muscle appears to be voltage dependent, since this channel displayed inward rectification and passed little outward current at positive potentials (49). Furthermore, this channel required receptor occupation by acetylcholine for activation, but its activation was enhanced by an increase in intracellular Ca^{2+}.

Clamping the intracellular Ca^{2+} concentration allows for identification of several types of Ca^{2+}-dependent ion channels in the sarcolemma. Free intracellular Ca^{2+} was clamped in porcine cells at 0, 200, 400, 750 and 1000 nM Ca^{2+}. The amplitude of the inward current at -80 mV is plotted in Fig. 9. Again the data have been normalized for differences in cell sizes by dividing the current amplitude by each cell's capacitance (pA/pF). The amplitude of the whole-cell inward current at -80 mV is clearly sensitive to the subsarcolemmal Ca^{2+} concentration. The current is not entirely the result of K^+ flux through K^+-selective pores because the membrane potential was set at E_K of -80 mV for these experiments. However, it is difficult to distinguish whether this current is primarily due to current through Ca^{2+}-activated Cl⁻ channels or Ca^{2+}-activated non-selective cation channels because Cl⁻, Na^+, and K^+ equilibrium potentials were more positive than -80 mV. These individual currents through these two types of channels can be differentiated by using ion replacement.

Evidence that the caffeine-induced inward current was carried partially by cations is shown in Fig. 11. Porcine smooth muscle cells were maintained at a holding potential of -80 mV with test pulses applied to varying negative potentials (0 to -110 mV). The ionic equilibrium potentials were: E_{Na} = +67 mV, E_{Cl} = -29 mV, E_{Ca} = 120 mV, E_K = + infinity, and E_{Cs} = - infinity due to replacement of intracellular K^+ with Cs^+. The whole-cell current amplitude at each test potential is plotted in Fig. 12 under control conditions (CON, physiological buffer) and during a caffeine response (CAF). The outward K^+ current was blocked by replacing K^+ in the pipette with Cs^+. Caffeine enhanced the inward current amplitude at each negative membrane potential tested. Since inward current persisted with the application of caffeine at a test potential -30 mV (near the equilibrium potential for Cl^-), the inward current cannot be attributed solely to outward Cl^- flux, but likely to inward cation movement through Ca^{2+}-activated non-selective cation channels.

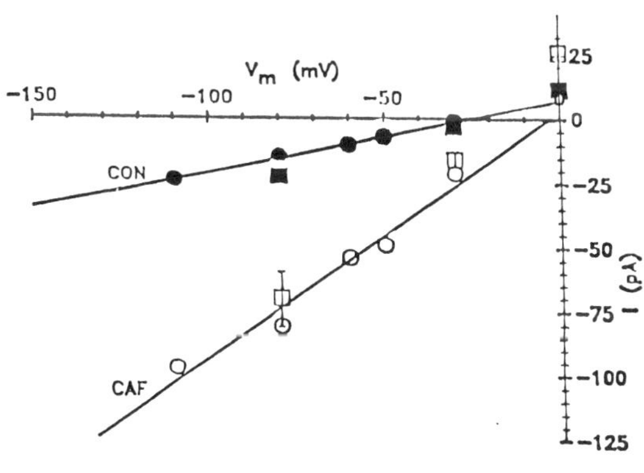

Figure 11. Current/voltage relationship of inward current at negative membrane potentials with under control conditions (CON) and during the application of caffeine (CAF). Current was measured at various voltages using the whole-cell recording configuration of patch clamp. The pipette solution contained Cs^+ to block outward current and allow inward current to be measured. Cl^- equilibrium potential was set at approximately -30 mV. Circles are results from single cells at each test potential with repeated caffeine exposures; squares indicate mean $\pm$ SE of n=21, 14, and 4 cells at membrane potentials of -80, -30, and 0 mV, respectively.

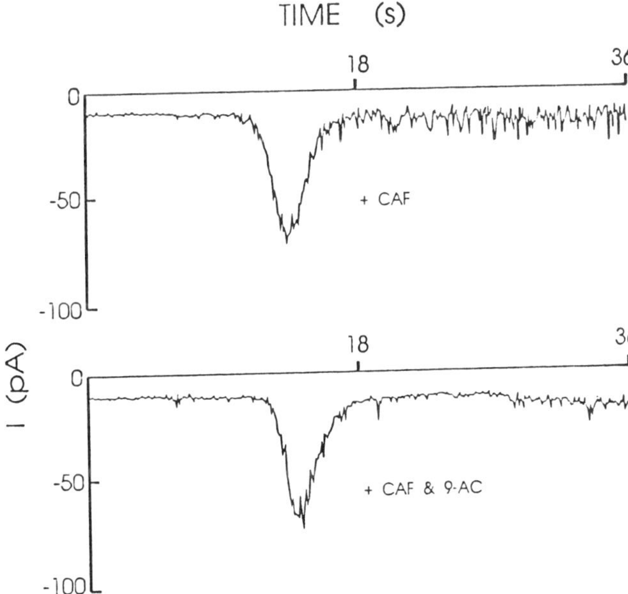

Figure 12. Effects of Cl^- channel blocker, 9-AC, on caffeine-activated inward current. Caffeine (CAF, 5×10^{-3} M) or caffeine in the presence of the Cl-channel blocker 9-AC (CAF & 9-AC) was applied at time 0. The holding potential was a constant -80 mV during the 72 s the current was monitored. Peak currents for caffeine and caffeine plus 9-AC were 71 pA and 72 pA, respectively.

If the caffeine-induced current were carried solely by the Ca^{2+}-activated non-selective cation current then one would expect a reversal potential of 0 mV. However, the mean current amplitude shown in Fig. 11 at 0 mV was approximately 25 pA (open square). Therefore, the ion carrying that current must have a reversal potential negative to 0 mV. As noted in the preceding paragraph the major ion with a normal negative reversal potential is Cl^-. A Cs^+-selective current would be outward at all potentials tested here. Therefore, we conclude that both the Ca^{2+}-activated Cl^- current and the Ca^{2+}-activated non-selective cation current comprise the caffeine-induced increase in current amplitude. The portion of total current carried by each has not been determined and deserves further investigation.

The relationship between subsarcolemmal Ca^{2+} concentrations and ion channel activation is better understood with the help of various agents which alter free intracellular Ca^{2+} concentrations without directly affecting sarcolemmal Cl^- and non-selective channels. For example, caffeine augments both Ca^{2+}-activated K^+ currents during a test potential of 0 mV and it augments the Ca^{2+}-sensitive inward current during the holding potential of -80 mV. (Fig. 10) This augmentation primarily is due to release of intracellular Ca^{2+} from the SR as caffeine application in Ca^{2+}-clamped cells fails to alter the whole-cell currents either at positive or negative membrane potentials (39). Therefore, both the inward and outward currents appear to be Ca^{2+}-sensitive.

We provided evidence in Fig. 11 using manipulation of equilibrium potentials that a portion of the caffeine-induced inward current shown in Fig. 10 during the -80 mV holding potential was due to opening of Ca^{2+}-activated non-selective channels. However, the current could have been carried by Ca^{2+}-activated Cl^- channels. As shown in Fig. 12, 4 mM 9- anthracenecarboxylic acid (9-AC) (50), a Cl^- channel blocker, did not decrease the magnitude of the caffeine-induced inward current generated by caffeine alone. The cell was superfused with physiological buffer and every 4 to 5 min repeatedly exposed to caffeine (5 mM). We have previously noted that subsequent to the first caffeine exposure the amplitude of the caffeine- induced changes in current and in the fura-2 ratio were constant for each consecutive caffeine application (P. Obye, unpublished observation, (42)). Therefore, we compared the caffeine- induced inward current amplitude during the second caffeine exposure with the amplitude during the third caffeine exposure plus 9-AC. As shown in Fig. 12 neither the time course nor the amplitude of the caffeine-induced response were altered by the application of 9-AC. From the results in Fig. 12 one might conclude that none of the caffeine-induced current is carried by Cl^-; however, in identical experiments where the caffeine-induced response was not as robust as that shown in Fig. 12, 9-AC did appear to diminish the amplitude of the current slightly. One possible explanation for the discrepancy is that 9-AC may require more time to act on the Cl^- channel. We monitored fluorescent emission during the protocol shown in Fig. 12. 9-AC is autofluorescent under 340 and 380 nm wavelength light used in our fura-2 excitation. The fluorescence emission rose dramatically approximately 10 s prior to the change in the current. Therefore, 9-AC was present in the bath prior to the caffeine response. However, 10 s may not be enough time for the drug to have its action. Furthermore, even under optimal conditions using epithelial cells, 8 mM 9-AC only blocked 50% of the inward current at -80 mV (50). The 9-AC concentration for these experiments was 4 mM, possibly explaining the lack of effect shown in Fig. 12.

Caffeine application does not appear to directly alter the inward current. In Ca^{2+}-clamped porcine cells (protocol described above), the application of caffeine failed to enhance the inward current (n=15). This is in contrast to the results obtained when caffeine was applied to cells without clamped internal Ca^{2+} (Fig. 10-13). We conclude that although caffeine elicited release of Ca^{2+} from the SR of Ca^{2+}-clamped cells, it was quickly buffered by 10 mM EGTA, thus not altering the inward current. This result along with those shown in Fig. 9 suggest that the caffeine-induced increase in inward current may be secondary to a rise in intracellular free Ca^{2+}. To test this hypothesis we simultaneously monitored the whole-cell inward current and fura-2 fluorescence. If the caffeine-induced increase in the inward current is due to a rise in Ca^{2+}, then the fura-2 signal (indicating the free intracellular Ca^{2+} concentration) should increase prior to or simultaneous with the current change.

Results from a typical porcine coronary artery smooth muscle are shown in Fig. 13. The cell was held at -80 mV for a long duration recording of 72 s, while caffeine was applied and inward current and fura-2 fluorescence were monitored simultaneously. We noted that the time course for activation of inward current was simultaneous with the increase in intracellular Ca^{2+} (fura-2). Furthermore, increases in both inward current and intracellular Ca^{2+} were blocked by 10 mM EGTA in the pipette solution. These data suggest that the caffeine-induced augmentation of the inward current is most likely modulated by intracellular Ca^{2+}.

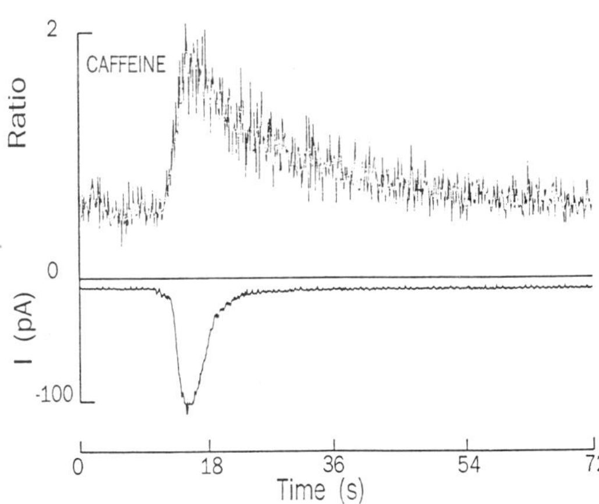

Figure 13. Simultaneous measurements of intracellular Ca^{2+} concentration (fura-2 ratio) and whole-cell current in order to observe the time course of the responses to caffeine. CAF was applied to porcine coronary artery smooth muscle cells during a pulse at -80 mV with a duration of 72 s. Caffeine exposure was maintained through the time shown. K^+-filled pipette solution and physiological bath solution, giving E_K = -82 mV as described in Materials and Methods.

SUMMARY

The main findings of this study are that the major Ca^{2+}-sensitive currents in bovine and porcine coronary artery smooth muscle cells are via the Ca^{2+}-activated K^+ channels at positive membrane potentials and through the Ca^{2+}-activated non-selective cation channels and Ca^{2+}- activated Cl^- channels at negative potentials. A component of the Ca^{2+}-activated inward current appears to be carried by Cl^- as a small outward current is noted at 0 mV when the reversal potential for the cations are positive while E_{Cl} was -29 mV (Fig. 11). Both the outward in inward whole-cell currents were found to be sensitive to free intracellular Ca^{2+} (Fig. 2 and 9). Both currents are augmented by caffeine (Fig. 10); however, this augmentation is dependent on changes in the subsarcolemmal Ca^{2+} concentration (Fig. 3 and 13) and does not occur when free intracellular Ca^{2+} is held constant with 10 mM EGTA.

While free intracellular Ca^{2+} has many roles in coronary artery smooth muscle cells, often overlooked is the ability of Ca^{2+} to activate sarcolemmal channels thus altering ionic currents. The importance of these channels and regulation by Ca^{2+} under physiological conditions has not been fully established. However, the K^+ channels in particular offer a unique method of monitoring subsarcolemmal Ca^{2+} concentrations within a 50 nM distance from the plasma membrane (39).

A major question still remaining is whether Ca^{2+}-activated non-selective cation channels have a physiological role in vascular smooth muscle cells. Despite the existence of these channels in many different cell types (see Ref. (22) for review), only in neutrophils has the Ca^{2+} flux through the channels been sufficient to contribute to the rise in free intracellular Ca^{2+} (46). We have preliminary evidence that Ca^{2+} flux through the channels may contribute to refilling of the SR Ca^{2+} store (40). Since vascular smooth muscle cells have a very large surface-to- volume ratio, perhaps minimal Ca^{2+} flux through Ca^{2+}-activated non-selective cation channels will also be sufficient to increase free Ca^{2+}. Indeed, ^{24}Na and ^{45}Ca influx studies in rabbit aorta indicate a predominant Ca^{2+}-dependent Na^+ and Ca^{2+} influx pathway distinct from voltage-gated channels (1).

Finally, the differences in STOC frequency and magnitude in porcine and bovine smooth muscle suggests that SR Ca^{2+} release is highly regulated occurring in large bolus amounts (as seen most frequently in porcine cells) or in a uniform constant release of SR

Ca^{2+} (as noted in bovine cells) (39). The results are consistent with a model of SR Ca^{2+} release that is species specific and modulates sarcolemmal currents.

ACKNOWLEDGEMENTS

This work was supported by grants from the American Heart Association-Missouri Affiliate, Institutional Biomedical Support Grant RR07053, Marion, Merrill Dow Inc., and NIH HL41033 to M.S., a grant from the American Physical Therapy Association to L.SB. and support from NIH Training Grant T-32 HL07094.

REFERENCES

1. Aaronson, P.I. and A.W. Jones, Am. J. Physiol. 254, C75-C83 (1988).
2. Amededee, T., W.A. Large, and Q. Wang, J. Physiol. (Lond) 428, 501-516 (1990).
3. Benham, C.D. and T.B. Bolton, J. Physiol. (Lond) 381, 385-406 (1986).
4. Benham, C.D., T.B. Bolton, and R.J. Lang, Nature 316, 345-347 (1985).
5. Benham, C.D. and R.W. Tsien, Nature 328, 275-278 (1987).
6. Casteels, R. and G. Droogmans, J. Physiol. (Lond) 317, 263-279 (1981).
7. Clusin, W.T., Nature 301, 248-250 (1983).
8. Colquhoun, D., E. Neher, H. Reuter, and C.F. Stevens, Nature 294, 752-754 (1981).
9. Desilets, M., S.P. Driska, and C.M. Baumgarten, Circ. Res. 65, 708-722 (1989).
10. Ganitkevich, V. and G. Isenberg, Circ. Res. 67, 525-528 (1990).
11. Ganitkevich, V.Y. and G. Isenberg, J. Physiol. (Lond) 426, 19-42 (1990).
12. Harder, D.R., Circ. Res. 46, 372-377 (1980).
13. Hume, J.R. and N. Leblanc, J. Physiol. (Lond) 413, 49-73 (1989).
14. Inoue, R. and G. Isenberg, J. Physiol. (Lond) 424, 73-92 (1990).
15. Ito, Y., K. Kitamura, and H. Kuriyama, J. Physiol. (Lond) 294, 595-611 (1979).
16. Itoh, T., Y. Kanmura, H. Kuriyama, and K. Sumimoto, J. Physiol. (Lond) 375, 515-534 (1986).
17. Kanaide, H., M. Hasegawa, S. Kobayashi, and M. Nakamura, Biochem. Biophys. Res. Commun. 143, 532-538 (1987).
18. Klockner, U. and G. Isenberg, Pflugers Arch. 418, 168-175 (1990).
19. Leblanc, N. and J.R. Hume, Science 248, 372-376 (1990).
20. Leijten, P.A.A. and C. Van Breemen, J. Physiol. (Lond) 357, 327-339 (1984).
21. Lipp, P., S. Mechmann, and L. Pott, Pflugers Arch. 410, 121-131 (1987).
22. Marty, A., Trends Neurosci. 12, 420-424 (1989).
23. Matsumoto, T., H. Kanaide, J. Nishimura, Y. Shogakiuchi, S. Kobayashi, and M. Nakamura, Biochem. Biophys. Res. Commun. 135, 172-177 (1986).
24. Mccann, J.D. and M.J. Welsh, J. Physiol. (Lond) 372, 113-127 (1986).
25. Miura, Y. and J. Kimura, J. Gen. Physiol. 93, 1129-1145 (1989).
26. Nishimura, J., R.A. Khalil, and C. VanBreemen, Hypertension 13, 835-844 (1989).
27. Ohya, Y., K. Kitamura, and H. Kuriyama, Am. J. Physiol. Cell Physiol. 252, C401-C410 (1987).
28. Pallotta, B.S., J.R. Helper, S.A. Oglesby, and T.K. Harden, J. Gen. Physiol. 89, 985-997 (1987).
29. Pankow, W., K. Neumann, J. Ruschoff, J. Heymanns, and P. Von Wichert, J. Immunol. Methods 129, 127-133 (1990).
30. Petersen, O.H. and Y. Maruyama, Nature 307, 693-696 (1984).
31. Ratz, P.H. and S.F. Flaim, Circ. Res. 54, 135-143 (1984).
32. Ratz, P.H. and S.F. Flaim, J. Pharmacol. Exp. Ther. 234, 641-647 (1985).
33. Rembold, C.M., J. Physiol. (Lond.) 429, 77-94 (1990).
34. Sadoshima, J.-I., N. Akaike, H. Kanaide, and M. Nakamura, Am. J. Physiol. 255, H754-H759 (1988).
35. Sadoshima, J.-I., N. Akaike, H. Tomoike, H. Kanaide, and M. Nakamura, Am. J. Physiol. 2550, H410-H418 (1988).
36. Singer, J.J. and J.V. Walsh Jr., Pflugers Arch. 408, 98-111 (1987).
37. Stehno-Bittel, L., M.H. Laughlin, and M. Sturek, Am. J. Physiol. Heart Circ. Physiol. 259, H643-H647 (1990).
38. Stehno-Bittel, L., M.H. Laughlin, and M. Sturek, J. Appl. Physiol. (In press) (1991).
39. Stehno-Bittel, L. and M. Sturek, (Submitted).
40. Sturek, M. and L. Bowman, FASEB J. 3, A1194 (Abstract) (1989).
41. Sturek, M., W.M. Caldwell, and C. Wagner-Mann in Electrophysiology and Ion Channels of Vascular Smooth Muscle Cells and Endothelial Cells, N. Sperelakis and H. Kuriyama, eds. (Elsevier, New York 1991) in press.

42. Sturek, M., K. Kunda, and Q. Hu, J. Physiol. (Lond.) (1991).

43. Sturek, M., L. Stehno-Bittel, C. Wagner-Mann, P.K. Obye, and Q. Hu, in Electrophysiology and Ion Channels of Vascular Smooth Muscle Cells and Endothelial Cells N. Sperelakis and H. Kuriyama, eds (Elsevier, New York 1991) in press.

44. Thayer, S.A., M. Sturek, and R.J. Miller, Pflugers Arch. 412, 216-223 (1988).

45. Van Breemen, C. and K. Saida, Annu. Rev. Physiol. 51, 315-329 (1989).

46. Von Tscharner, V., B. Prodhom, M. Baggiolini, and H. Reuter, Nature 324, 369-372 (1986).

47. Wagner-Mann, C.C., L. Bowman, and M. Sturek, Am. J. Physiol. Cell Physiol. 260, C763-C770 (1991).

48. Wagner-Mann, C.C. and M. Sturek, Am. J. Physiol. Cell Physiol. 260, C771-C777 (1991).

49. Wang, Q. and W. A. Large, J. Physiol. (Lond) 435, 21-39 (1990).

50. Welsh, M.J., Science 232, 1648-1650 (1986).

ION CHANNELS INVOLVED IN RESPONSES TO MUSCARINIC RECEPTOR ACTIVATION IN SMOOTH MUSCLE

RYUJI INOUE

Department of Pharmacology, Kyushu University, Fukuoka 812, JAPAN

INTRODUCTION

ACh is a ubiquitous neurotransmitter, which is found in both the central and peripheral nervous systems, and produces a wide variety of responses on various tissues. Recent progress in the methods of drug synthesis has allowed the manufacture of many muscarinic agonists and antagonists, and studies of their actions have led to the suggestion that there is considerable heterogeneity amongst muscarinic receptors (18). Subsequently, molecular biological techniques have been successfully applied to the investigation of the receptors, and combination of the above two lines of observations has established that several subtypes of muscarinic receptors, which are classified on their pharmacology, reflect a real difference in molecular structures and are coupled to several distinct effector systems including ion channels (15).

Muscarinic receptor activation produces contraction in some smooth muscles and relaxation in others (7, 29). The mechanisms responsible for this variability are now thought to involve several distinct ion channels which would be indirectly or directly affected upon muscarinic receptor activation and are found not only in the plasma membrane but also in the sarcoplasmic reticulum membrane. The mode of their activations are rather complicated and in some cases associated with biochemical events including G-proteins, the phosphatidylinositol breakdown cascade and other factors which would be released as a consequence of receptor activation.

This short article attempts to briefly overview what is known about such channels. Emphasis will be placed mainly on the excitatory action of ACh, particularly on non-selective cation channels (NS channels), which have been found in various types of tissues such as neurons, muscles, secretory glands, endothelium and blood vessels, and whose understanding is increasingly being promoted since the emergence of patch clamp techniques.

INHIBITORY ACTION OF ACH

It has long been known that in many large and small blood vessels ACh causes dilation by relaxing smooth muscles. An important breakthrough was made by Furchgott and his collaborators (16) who discovered using intact and de-endothelialized strips of rabbit thoratic aorta that the presence of the endothelium is a prerequisite for the relaxing effect of ACh. As a result of more recent work, it is now agreed that many vasoactive agents in addition to ACh stimulate the endothelium of blood vessels to release two inhibitory factors, EDRF (endothelium-dependent relaxing factor), and EDHF (endothelium-dependent hyperpolarizing factor) (51). These factors are believed to diffuse in some way to neighbouring smooth muscle cells and then cause hyperpolarization and relaxation. The most likely candiadate of EDRF is nitric oxide, which is thought to stimulate a soluble form of guanylate cyclase and in turn elevate the intracellular cGMP level, while the EDHF has not yet been biochemically identified.

Recently, several investigators have attempted to account for the molecular mechanism of such ACh actions using single channel recordings, and have proposed a number of different hypotheses:

(1)   In the rabbit and rat mesenteric arteries (48), it was demonstrated that a certain class of ATP-sensitive K channel is present in the preparation, and its blocker, glibenclamide, inhibits the muscle relaxation induced by various vasodilatory agents including ACh. This led the authors to suggest that openings of the ATP-sensitive K channels may be implicated in the relaxing action of ACh. Further support for this view has come from a recent microelectrode experiment, in which the ACh-induced

Published 1991 by Elsevier Science Publishing Company, Inc.
Ion Channels of Vascular Smooth Muscle Cells and Endothelial Cells
Sperelakis and Kuriyama, Editors

hyperpolarization is shown to be blocked by the ATP-sensitive K channel blockers, glibenclamide and Ba (8).

(2) In the cultured bovine aorta (54), several vasodilatory agents, eg. nitroprusside, adenosine and atrial natriuretic factor, which are known to increase the cGMP level, have been shown to increase the open probability of a large conductnace, Ca-dependent K channel (the so-called "maxi K" or "BK" channels), by causing a leftward shift of the $[Ca^{2+}]_i$-availability curve of the channel. The effect was mimicked specifically by a direct perfusion of cGMP (1-100 μM) to the cytosolic side of the channel. Provided that the effect occurrs due solely to cGMP, another potent cGMP-increasing agent, ACh, might also exert its hyperpolarizing (and thus relaxing) action partly through this mechanism.

(3) Olsen et al. (39) have reported that ACh can elicit a pertusis toxin-insensitive inward rectifying K conductance in endothelial cells isolated from the calf aorta. Since there is some evidence that endothelial cells and vascular smooth muscle cells communicate electrically via gap junctions, they have further postulated that the ACh-induced hyperpolarization initiated in the endothelial cell may spread into neighbouring smooth muscle cells, resulting in hyperpolarization of the muscle.

(4) Although it may correlate with muscle contraction rather than relaxation, Ganitkevich and Isenberg has reported an interesting action of ACh on the membrane potential in the guinea-pig coronary artery (17). In this muscle spike-like hyperpolarizations (SLHs), which probably result from activation of Ca-dependent K channels due to a periodic Ca relaese from the sarcoplasmic recticulum, occur spontaneously, and ACh increases the magnitude and frequency of the SLHs. They therefore hypothesized that in multicellular systems the SLHs sum up and comrpise the resting K conductance, which may be increased by ACh.

EXCITATORY ACTION OF ACH

The excitatory action of ACh in smooth muscle has been supposed to involve three major Ca mobilizing mechanisms, Ca release from an internal store site (sarcoplasmic reticulum; SR), depolarization and subsequent changes in electrical activity, and Ca-influx through a voltage-independent pathway opened by ACh (7). In recent years, some of these ideas have been substantiated in terms of whole-cell or single channel recordings.

Ion Channels and Intracellular Ca

Several subtypes of muscarinic receptor have been shown to activate via a G-protein a membrane bound phospholipase C, which hydrolyses the phosphatidylinositol and generates two important second messengers, inositol trisphosphate (IP_3) and diacylglycerol (5). It has been shown that in smooth muscles IP_3 causes Ca release from a certain type of SR (e.g. dog trachea, Hashimoto et al. ,1985). This is probably due to opening of IP_3-gated Ca channels in the SR membrane (14). The IP_3-gated Ca channel shows a sensitivity to $[Ca^{2+}]_i$ in the physiological range (6) but differs from the Ca-activated Ca channel that also resides on the SR and is affected by ATP, caffeine and ryanodine (47).

Ca release from the SR results in a rapid elevation in the intracellular Ca concentration ($[Ca^{2+}]_i$), and a number of Ca-dependent ionic conductances can consequently be induced (36, 37). It is known that smooth muscle membrane is abundant in charybdotoxin-sensitive large conductance Ca-dependent K channels (BK channels). The BK channel is very sensitive to the $[Ca^{2+}]_i$ in the range of 10 -1000 nM (46) and can be activated in response to Ca influx via action potentials (APs) or to agonist-induced Ca release via IP_3 production (3, 28, 30). On the other hand, little has been reported about other Ca-dependent conductances (e.g. Cl channels, NS channels) on muscarinic receptor activation in smooth muscle. Only recently in the rat portal vein, Loirand et al. (35) have described a large conductance, Ca-dependent NS channel that opens on application of ACh, norepinephrine or caffeine, or by depolarization. The physiological role of Ca-dependent channels in muscarinic receptor activation is unknown. In analogy to their role in the plateau or repolarization phase of APs, the channels might participate in an elaborate control of the membrane potential, and also in the regulation of pumps and carriers by changing intracellular ionic contents.

Ion Channels Responsible for Depolarization

It is well known that ACh depolarizes the membrane of gut muscle and modulates APs and slow waves in their frequency and pattern. Two different lines of evidence, increased Na permeability ($I_{ns,ACh}$) and decreased K permeability (M-current), have so far been provided to account for this depolarization. Although these mechanisms appear mutually exclusive, they show a surprisingly similar voltage-dependency, suggesting that their effects on the electrical activity will be similar. As will be described later, however, the $I_{ns,ACh}$ exhibits a number of unique properties.

M-current. It was first described in bullfrog sympathetic ganglion that muscarinic receptor activation brings about a decrease in a voltage-dependent K conductance, the properties of which can be distinguished from other voltage-dependent K conductances, eg. the delayed rectifying K current (10). This novel type of K current was designated 'M-current', and later identified in many types of tissues (9) including gastric smooth muscles (31, 43).

The M-current in smooth muscles has been best charaterized in amphibian stomach (43). Reversal potential measurements on changing $[K]_0$ revealed that the current is strictly selective for K ion. The current is not affected by Ca withdrawal from the extracellular space, but strongly suppressed by milimolar Ba as was also seen in sympatethic ganglion. Voltage jump experiments showed that the M-current seems to deactivate on hyperpolarization, its activation threshold being about -70 mV and its steady state activation curve being a sigmoid with a slope factor of 9 mV and a half maximum activation voltage of -49 mV. The maximum conductance (observed at -40 - -20 mV) amounts to 1nS per cell, the value being comparable to the input resistance of smooth muscle cells at rest. The physiological importance of such voltage-dependent gating parameters is obvious and will be discussed for the case of the $I_{ns,ACh}$ channel below.

It is known that the M-current receives interesting control from several bioactive substances, such as substance-P, lutenizing hormone releasing hormone (LHRH) and somatostatin (9). Also in smooth muscle, the possibility is demonstrated that some tachykinins, substance-P and substance-K, suppress the M-current (45). This action is not additive to that elicited by ACh and can occur in the presence of atropine, suggesting different types of receptors regulate the common effector channel. In contrast to such inhibitory actions, isoproterenol augments the M-current which seems to be endogenously active (44). This counter-regulating effect is mimicked by forskolin or membrane-permeable cAMP analogues, all being antagonized by ACh. A target of such dual regulation by isoproterenol and ACh could exist in the steps which follow the one controlling the level of cAMP, presumably cAMP-dependent kinase and/or its substrate proteins including the M channel itself. The physiological importance of this mechanism may be found in the observation that in this muscle β-adrenergic agonists cause hyperpolarization accompanied by an increase in K conductance and subsequent relaxation.

Table 1 Permeability ratios of various cations to Cs in the $I_{ns,ACh}$ channel.
In order to simplify the calculation of permeability ratios, biionic conditions in symmetrical concentrations were used. Reversal potentials were obtained from 2 to 4 experiments and taken from Inoue and Isenberg, 1990b. The activity of cations has not been corrected.

cations (mM)		reversal potential	permeability
bath	pipette	(mV)	(P_X/P_{Cs})
130Na	130Cs	+1.6±4.2	1.06
30Na	130Cs	-32.6±3.7	1.22
130Li	130Cs	+1.5±2.3	1.06
130K	130Cs	-1.8±2.4	0.93
110Ba	130Cs	+22.7±2.6	2.44
110Ca	130Cs	+19.9±9.5	2.03
110Mg	130Cs	-28.3±3.1	0.13

Nonselective cation channels of muscarinic receptor ($I_{ns,ACh}$). Most early electrophysiological and flux studies have supported the view that in gut smooth muscles the ACh-induced depolarization is due to increased cationic conductances such as Na, K and Ca (7). This view was confirmed by Benham et al. (4) applying a patch clamp method to disaggregated jejunal smooth muscle cells of the rabbit. They demonstrated that ACh elicits potential-dependent inward currents which are accompanied by a clear increase in the noise level. They also suggested that the currents discriminate poorly between Na and K, and possibly Ca. In subsequent years, more details of this nonselective cationic current have been elucidated in the longitudinal smooth muscle of guinea-pig ileum (22, 24-27) (Table 1).

To study the current $I_{ns,ACh}$, it needs to be separated from other concomittant muscarinic receptor activated conductances which obscure the measurement of the $I_{ns,ACh}$. This could be done by loading Cs-aspartate into the voltage-clamped cells via a patch electrode, as usually performed to isolate voltage-gated Ca currents (eg., 32; Fig. 1Aa). Under this condition, the ACh concentration - $I_{ns,ACh}$ current relationship was well described by a Michaelis-Menten equation (Hill coefficient : 1.0, half activation at 10 μM ACh). A nonspecific muscarinic antagonist, atropine (1 μM) completely abolished the $I_{ns,ACh}$, whilst a relatively M1 selective antagonist pirenzepine 1 μM failed to block the current, suggesting that the receptor coupling to the $I_{ns,ACh}$ channel is not M1, but M2 or M3 subtype. With a latency in the order of several hundred miliseconds, activation and deactivation of the $I_{ns,ACh}$ requires a few seconds (Fig. 1Aa), the values being of a similar order to those observed for ACh-activated K channel (I_{KACh}) in cardiac myocytes to which a G-protein is supposed to couple directly (Fig.1Ab). The involvement of a pertusis toxin sensitive G-protein in the muscarinic receptor - $I_{ns,ACh}$ channel pathway has been supported by an experiment using internal dialysis of various gunanine nucleotide analogues (25).

Voltage-Dependency

The $I_{ns,ACh}$ shows a voltage-dependent profile very similar to that of the M-current (26). When the steady state attained at a holding potential was abruptly perturbed by hyperpolraization or depolarization, the $I_{ns,ACh}$ quickly jumped to its instantaneous peak and then reached a new steady state level (Fig. 1B). This transient change is called current relaxation, which was first observed by Adams in nicotinic receptor-activated currents (1) and later for I_{KACh} channel current and M-current. The time course of relaxations can be approximated by a monoexponential term and their time constants appear to be independent of the preceding holding potential but a function of the applied potential, suggesting that a simple two state transition (open-closed) is involved in this voltage-dependent behaviour. The steady state activation curve is a sigmoid which can be fitted by the following equation:

$$G(V_m) = G_{max} / (1 + \exp(-(V_m - V_h) / k))$$

$$G_{max} = 1.5 \text{ nS}, k = 15 \text{ mV}, V_h = -50 \text{ mV}$$

where V_m, V_h and k denote the membrane potential, half maximum activation potential and a slope factor, respectively, and $G(V_m)$ and G_{max} denote the $I_{ns,ACh}$ conductance at V_m and the maximum $I_{ns,ACh}$ conductance per cell (that is usually observed at 0 mV), respectively. It is noteworthy that the $I_{ns,ACh}$ resembles the M-current described above, in all voltage-dependent gating properties such as steady state activation variables and relaxation time course. These observations could, to a first approxaimation, be interpreted in terms of a minimal reaction scheme:

$$\text{Step 1} \qquad\qquad\qquad \text{Step 2}$$

$$\text{ACh} + \text{R} \underset{K_{-1}}{\overset{k_{+1}}{\rightleftharpoons}} \text{ACh-R (Closed)} \underset{\alpha}{\overset{\beta}{\rightleftharpoons}} \text{ACh-R}^* \text{ (Open)}$$

If it is assumed that in the first step intervening biochemical processes might be equilibrated and unaffected during short changes in the membrane potential, the scheme abbreviates to a simple two state model in which only voltage-dependent open and closed transitions occur. Now at full receptor activation, the time constant of relaxation could be expressed as $1/(\alpha + \beta)$, and the fraction of the $I_{ns,ACh}$ channel remaining open, i.e. $G(V_m)/G_{max}$, as $\beta/(\alpha + \beta)$. It is then possible to calculate the rate constants of closing and opening, α and β, for various membrane potentials. Over a wide range of membrane potentials, the rate of opening was found to be relatively unchanged, while that of closing greatly reduced by depolarization, implying that the mean open life time of the $I_{ns,ACh}$ channel is prolonged on depolarization (Fig.8 in 26). This property was confirmed by single channel recording of the $I_{ns,ACh}$ channel (24), and seems to be shared by the M-current (42).

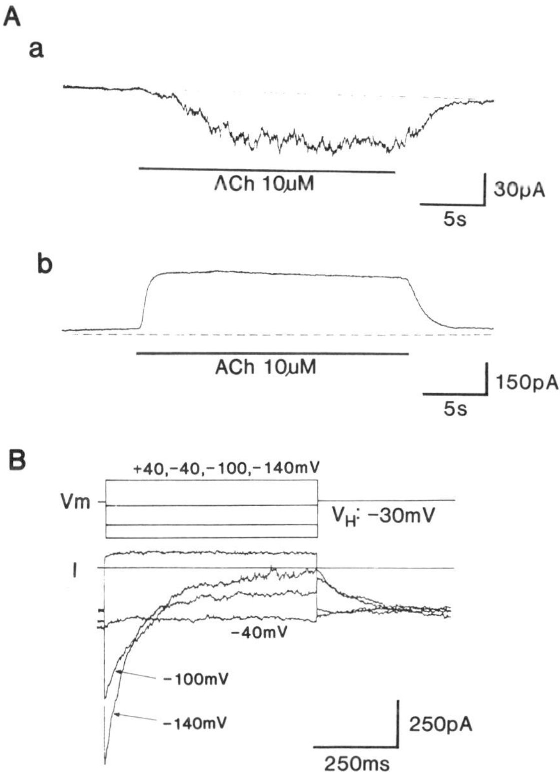

Figure 1. A: a; Time course of activation and deactivation of $I_{ns,ACh}$ in the guinea-pig ileum on rapid application of $10\,\mu M$ ACh using a concentration jump apparatus described previously (23). The pipette and bath contained Cs-aspartate solution and physiological saline solution (PSS), respectively. Holding potential was -60 mV. b; I_{kACh} current recorded from a guinea-pig atrial myocyte during a concentration jump (ACh $10\,\mu M$) using the same method as in a. The pipette and bath contained KCl internal solution and PSS, respectively. Holding potential was set to -60 mV. B: ACh-sensitive nonselective cation currents ($I_{ns,ACh}$) during voltage jumps to various membrane potentials (Vm; -140, -100, -40 and 40 mV) from a holding potential of -30 mV. The currents are defined as the difference of net currents between in the presence and absence of ACh (see also 26).

86

Physiological importance of this voltage-dependency may be found in the following. With the V_h value of -50 mV, even small oscillations around the resting membrane potential could greatly affect the ability of ACh to depolarize the membrane. At membrane potentials between -80 and -30 mV, the voltage change required to cause an e-fold increase in the open probability of the $I_{ns,ACh}$ channel is about 35 mV, which is of similar magnitude to that of slow waves. Indeed, it has been demonstrated that in current clamp experiments in which the membrane was electrically hyperpolarized to -70 or -80 mV subsequent application of ACh produced only slowly rising and strongly attenuated depolarizations (see Fig. 7, 25). It seems therefore likely that in physiological situations the depolarizing action of ACh would interact with other slow membrane oscillatory mechanisms, eg. slow waves, and produce complex pattern of modulation on action potentials and contractions.

Striking similarities in voltage dependency of the $I_{ns,ACh}$ and the M-current may make us question whether either of the currents might have inadequately been characterized. However, there are some essential differences between the $I_{ns,ACh}$ and the M-current in their biophysical nature, which would assure distinction of the two currents.

Ca Permeability

The $I_{ns,ACh}$ channels have been shown to be permeable to various cations including Ca but not to Cl (26). Table 1 summarizes relative permeabilities of several cations through the $I_{ns,ACh}$ channel, which have been determined by reversal potential measurement under biionic conditions. Applying a modified Goldmann-Hodgkin-Katz equation (34), Ca appears to be two times more permeable through the channel than other cations such as Na, if there were no interaction among cations on permeation. This gives an estimation that in physiological salt concentrations about 10% of the $I_{ns,ACh}$ current could be carried by Ca, the percentage being also demonstrated by Fig. 2A. In this respect, the $I_{ns,ACh}$ channel may be the most likely pathway which permits Ca entry independent of the membrane potential and corresponds to the notion proposed by Bolton of a receptor-operated Ca channel. However, the true ability of the $I_{ns,ACh}$ to permeate Ca may differ from this estimation and its proof needs direct measurement of $[Ca^{2+}]_i$, as performed for the ATP-activated non-selective cation channel (2). Nevertheless, it is worth noting that Ca inward currents with a very low magnitude close to the detection limit (1pA or less) have been shown to produce a significant elevation in $[Ca^{2+}]_i$ in non-excitable cells (40).

$[Ca^{2+}]_i$ Sensitivity

Although Ca is not a primary activator, the $I_{ns,ACh}$ has also been found to be sensitive to the changes in $[Ca^{2+}]_i$ caused by several different procedures, voltage-dependent Ca-influx (I_{Ca}), internal dialysis of various values of pCa using Ca-EGTA or agents releasing Ca from the SR (25, 27). Fig. 2B shows that in the presence of ACh I_{Ca} evoked by a step depolarization is accompanied by a large tail current, which is never observed in the absence of ACh. The nature of the tail current is identical to that of $I_{ns,ACh}$ in ionic selectivity and voltage-dependency, and the size of the tail current can be increased by prolonging depolarizing step, whereas in the presence of inorganic Ca antagonists the effect of depolarization is completely abolished. Ca influx needed to cause the half maximum facilitation of $I_{ns,ACh}$ is calculated to be 2-4pC by integrating I_{Ca} over the time. Assuming that a typical action potential in the smooth muscle could carry 3-4pC per cell (cell capacitance 50pF and the magnitude of action potential 60 - 80 mV), a single action potential seems obviously sufficient to cause a large increment in the activity of $I_{ns,ACh}$. In other types of experiments using internal dialysis of various value of $[Ca^{2+}]$ (10 - 1400 nM), the most efficient range of $[Ca^{2+}]_i$ for the facilitation is found between 100 nM and 1 μM. Considering that the physiological $[Ca^{2+}]_i$ changes dynamically within this range, the $[Ca^{2+}]_i$ sensitivity of the $I_{ns,ACh}$ channel might serve as a key mechanism to control the ACh-induced depolarizations and contractions in physiological situations. This mechanism would become particularly important when a subthreshold concentration of ACh for the $I_{ns,ACh}$ (< 100 nM) is present in the vicinity of the channel. The well-known prolongation of action potential observed in the presence of ACh might be ascribed mainly to this mechanism.

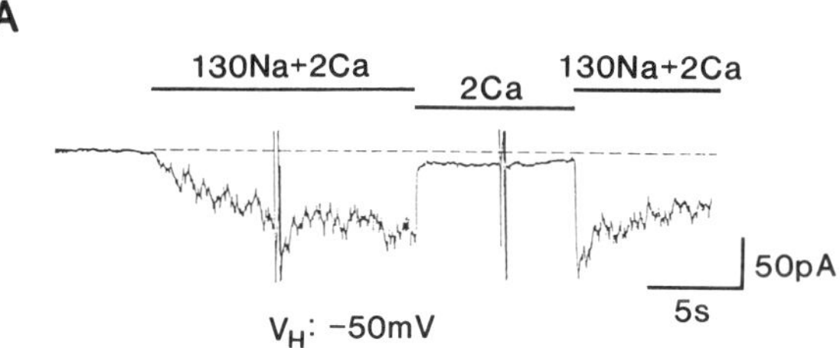

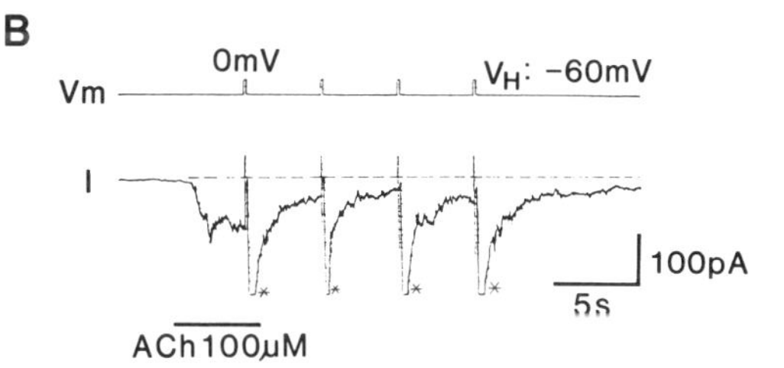

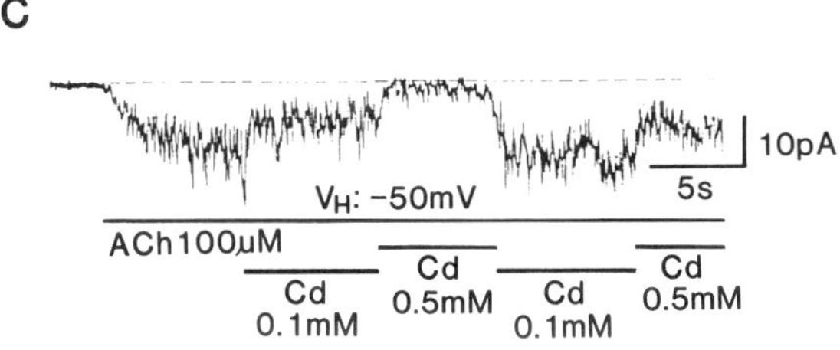

Figure 2. **A:** The $I_{ns,ACh}$ is carried predominantly by Na rather than Ca near the resting membrane potential. The apparent percentage of Ca influx through the $I_{ns,ACh}$ channel is estimated about 10% - 15% of the total current at this potential (-50 mV). At bars, the ionic content of the external solution was suddenly changed using another concentration jump method (the so-called sewer-pipe method; 22, 55). On removal (Tris replacement) or restoration of Na, the size of $I_{ns,ACh}$ was reduced or recovered in a very quick fashion. **B:** Immediately after Ca currents were evoked by a 200ms long depolarizing step to 0 mV, large tail currents appeared (their peaks are out of scale and marked by *). 100 μM ACh was superfused into the bath at the bar. **C:** In the continued presence of ACh (100 μM), two concentrations of Cd2+ were externally added (lower bars) using a concentration jump method. The $I_{ns,ACh}$ was blocked in a quick and reversible manner.

Modulation by External Polyvalent Cation of the $I_{ns,ACh}$

Another important mechanism affecting the activity of the $I_{ns,ACh}$ channel is a modulatory effect of external polyvalent cations (21, 22). As has been reported for other types of neurotransmitter-activated NS channels (eg. NMDA channels, 33), Ca related divalent cations, Cd, Ni, Mn or Co, which are known as inorganic Ca channels blockers, inhibit the $I_{ns,ACh}$ in a rapid and reversible fashion (Fig. 2C). The mode of the inhibition is voltage-independent and not associated with changes either in ionic selectivity or in the degree of receptor activation. Physiological divalent cations such as Mg and Ca have also been found to affect the $I_{ns,ACh}$, the former blocking while the latter augmenting it. The action of external Ca is unique among various divalent cations, since it has the ability to strongly affect the voltage-dependent properties of the $I_{ns,ACh}$ channel.

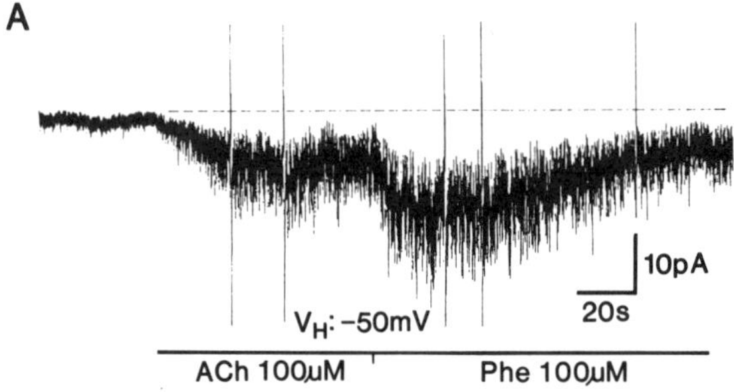

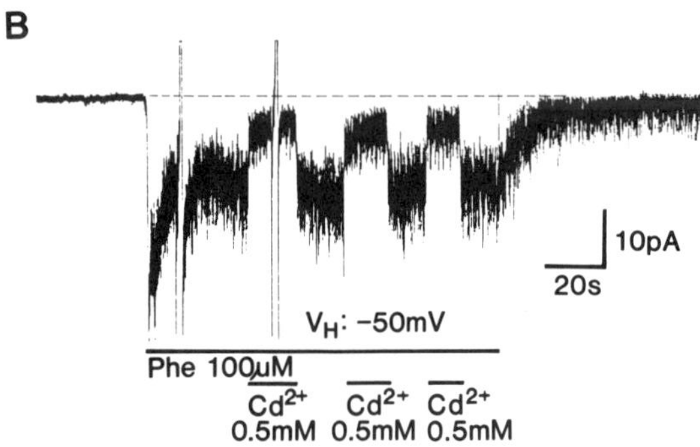

Figure 3. **A:** At a holding potential of -50 mV, ACh and Phe elicited inward currents (non-selective cationic) in the rabbit portal vein. Cs-aspartate was used as the pipette soultion. ACh and Phe were applied in a consecutive manner using a concentration jump apparatus as in Fig. 2. **B:** Cd2+ (0.5 mM) blocked an inward current induced by 100 μM Phe in a manner very similar to the case of $I_{ns,ACh}$ (compare with Fig.2C). Vertical reflections represent voltage steps or ramps.

Comparison With Other NS Channels:

It is well known that many of neurotransmitters and autacoids as well as depolarization of the membrane can activate non-selective cation channels (NS channels) in a wide variety of excitable and nonexcitable tissues (50). According to the mode of activation, the NS channels can be devided into three classes, Ca-activated NS channels, cyclic nucleotide-activated NS channels and neurotransmitter-activated NS channels. The main role of NS channels is thought, in general, to be production of a sustained depolarization and modulation of the frequency and duration of APs. In spite of a diversity in their biophysical propeties such as voltage-dependency and Ca-permeability, some of fundamental properties of the NS channels seem well conserved, eg. the single channel conductance of the majority of NS channels is around 20 - 30pS. The $I_{ns,ACh}$ channel is found to have a single channel conductance of about 20 - 25 pS (24) and monovalent cations, Na, K, Li and Cs, permeate almost equally. However, the $I_{ns,ACh}$ channel also shows additional properties that characterize it as a special type of NS channel; it is Ca-permeable, voltage-dependent and very sensitive to both intracellular Ca and external polvalent cations. Nevertheless, there are a number of NS channels reported in the literature, which resemble the $I_{ns,ACh}$ channel. For instance, in guinea-pig chromaffin cells, it has been reported that muscarinic agonists induce a NS conductance which allows Ca entry and has a negative slope in its steady state i-v relationship at negative potentials (20). The activation of this NS channel seems to involve a pertusis toxin-sensitive G-protein, all of the features entirely resembling those of the $I_{ns,ACh}$ channel. It is therefore tempting to address the question of whether a similar type of NS channel is broadly distributed in other types of smooth muscles. In the following, I tested the possibility in the rabbit portal vein, where ACh is known to contract the muscle (49).

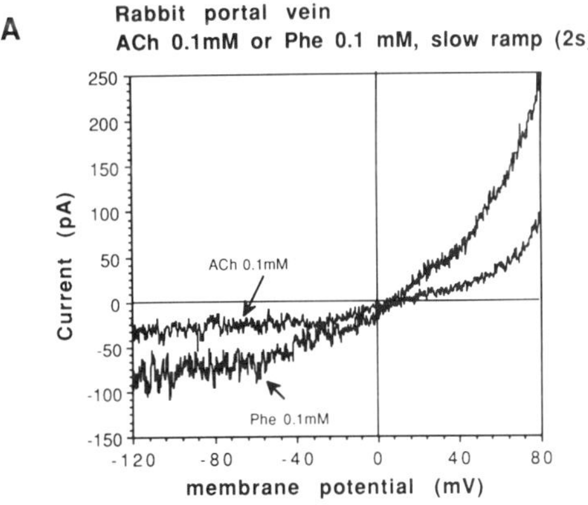

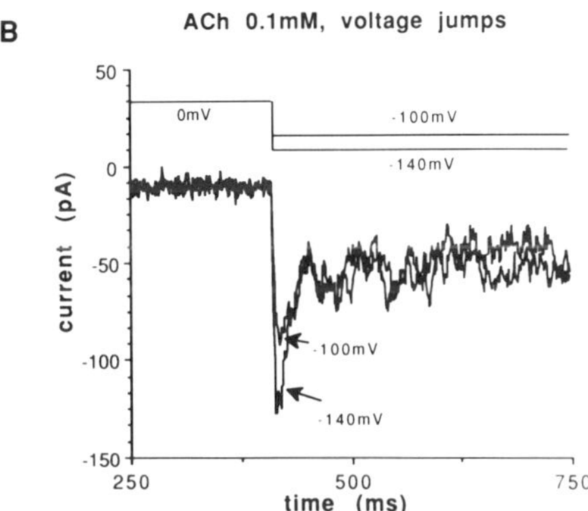

Figure 4. A: Voltage-dependency of the ACh and Phe-sensitive inward currents evaluated by a slow rising ramp (-120 - 80 mV, 2s). The ramp currents were measured as the difference of net currents between in the presence and absence of ACh or Phe. The potency of ACh was consistently weaker than that of Phe. **B:** ACh-sensitive inward currents show voltage-dependent relaxations on hyperpolarizing steps. In order to maximally activate the current, the membrane potential was first held at 0 mV and then set to -140 or -100 mV.

With loading Cs-aspartate into a cell to block Ca-dependent K and Cl conductances (11), a sustained inward current was recorded at membrane potentials negative to 0 mV on exposure to ACh or phenylephrine (Phe) (Fig.3A). The i-v relationship of the ACh- or Phe-induced current is outwardly rectifying, indicating that at more negative potentials the degree of current activation is reduced (Fig.4A). The current shows a relaxation on hyperpolarizing steps, the time course being accerelated by more negative jumps (Fig.4B).

The reversal potentials of the ACh- and Phe-induced currents are close to 0 - 10 mV in normal $[Na]_o$ (150 mM outside), but shift in a negative direction by halving $[Na]_o$, suggesting that the currents are of a nonselective type. Externally applied Cd, which quickly inhibits the $I_{ns,ACh}$, reduces the amplitude of both of the ACh- and Phe-induced currents in a quick and reversible fashion (Fig.4B). These results provide two important conclusions. First, muscarinic and α-adrenergic receptors may regulate a common NS channel. Second, the NS channel channel in the rabbit portal vein resembles the $I_{ns,ACh}$ channel in many aspects. Although it is too premature to generalize the above results, this opens a possibility that many different types of receptors observed in smooth muscle may couple to similar NS channels stemming from a single common family.

OTHER CHANNELS MODULATED BY MUSCARINIC RECEPTOR ACTIVATION

It has been widely accepted that many of voltage-dependent and Ca dependent channels receive a control from nerves, hormones and autacoids. One of the best known examples is the voltage-gated Ca channel (I_{Ca}) of ventricular myocytes, in which norepinephrine and ACh reciprocally regulate the availability of I_{Ca} channels by changing the cAMP level via two G-proteins linking to adenylate cyclase (52).

Dihydropyridine-sensitive L-type Ca channels (I_{Ca}) in some smooth muscles have been found to be regulated through muscarinic receptors. In the bullfrog's stomach (12) and in the rabbit coronary artery (38), ACh increases I_{Ca} with no discernable shift in the i-v relationship, and this effect is blocked by pretreatment of atropine. It has further been shown that a synthetic diacylglycerol analogue mimicks the effect of ACh on I_{Ca}, suggesting that protein kinase C might be involved in the ACh action (53). However, the converse effect of ACh on I_{Ca} has been found in rabbit jejunal longitudinal muscle (41). These authors have described that suppression of I_{Ca} occurs in the presence of ryanodine or high concentrations of EGTA in the patch pipette or using Ba as a charge carrier, therefore concluding that the effect would not be due to Ca-dependent inactivation of I_{Ca}, which could secondarily be caused by a rise in $[Ca^{2+}]_i$.

Muscarinic receptors also seems to modulate other channels than I_{Ca} in smooth muscle. Cole et al. (13) has reported that ACh reduces the Ca-dependent K conductance by shifting the activation vs. voltage curve of the BK channels in a positive direction without changing the $[Ca^{2+}]_i$. In their view, muscrainic suppression of BK channels is consistent with the effect of ACh on the multicellular tissue, and may be involved in prolongation by ACh of slow waves.

CONCLUSION

Muscarinic receptor-coupled channels in smooth muscle are diverse in their biophysical profile and mode of activation or modulation. They may not simply serve as an independent mechanism to transduce cellular signalling but seem to mutually interact via numerous feedback controls. A simple example has been described for the case of the $I_{ns,ACh}$ channel and other Ca-mobilizing mechanisms (i.e. $[Ca^{2+}]_i$ sensitivity). Future experiments are expected to disentangle the complicated feedback networks, leading to more comprehensive understanding of the receptor-channel relationship.

ACKNOWLEDGEMENT

I am grateful to Prof.Hirosi Kuriyama for persistent encouragement and help during the course of writing, and also to Dr.Alison F.Brading for improving the manuscript.

REFERENCES

1.   Adams, P.R., Br. J. Pharmacol. 53, 308-310 (1974). .
2.   Benham, C.D., J.Physiol. 419, 689-701 (1989).
3.   Benham, C.D. and Bolton, T.B., J. Physiol. 381, 385-406 (1986).
4.   Benham, C.D., Bolton, T.B. and Lang, R.J., cell. Nature 316, 345-347 (1985).

5. Berridge, M.J. and Irvine, R.F., Nature $\underline{312}$, 315-321 (1984).
6. Bezprozvanny, I., Watras, J., and Ehrlich, B.E., Nature $\underline{351}$, 751-754 (1991).
7. Bolton, T.B., Physiol. Rev. $\underline{59}$, 606-718 (1979).
8. Brayden, J.E., Am. J. Physiol. $\underline{259}$, H668-673 (1991).
9. Brown, D.A., Trends. Neurosci. $\underline{11}$, 294-299 (1988).
10. Brown, D.A. and Adams, P.R., Nature $\underline{283}$, 673-676 (1980).
11. Byrne, N.G. and Large, W.A., J. Physiol. $\underline{404}$, 557-573 (1988).
12. Clapp, L.H., Vivaudou, M.B., Walsh, J.V., and Singer, J.J., Proc.Natl.Acad.Sci. USA. $\underline{84}$,2092-2096 (1987).
13. Cole, W.C., Carl, A., and Sanders, K.M., Am. J. Physiol. $\underline{257}$, C481-C487 (1989).
14. Ehrlich, B.E. and Watras, J., Nature $\underline{336}$, 583-586 (1988).
15. Fukuda, K., Kubo, T., Maeda, A., Akiba, I., Bujo, H., Nakai, J., Mishima, M.,Higashida, H., Neher, E., Marty, A., and Numa, S., Trends in Pharmacol. Sci. suppl. 4-10 (1989).
16. Furchgott, R.F. and Zawadzki, J.V., Nature $\underline{288}$, 373-376 (1980).
17. Ganitkevich, V. and Isenberg, G., Cir. Res. $\underline{67}$, 525-528 (1990).
18. Hammer, R., Berrie, C.P., Birdsall, N.J.M., Burgen, A.S.V., and Hulme, E.C., Nature $\underline{283}$, 90-92 (1980).
19. Hashimoto, T., Hirata, M., and Ito, Y., Br. J. Pharmacol. $\underline{86}$, 191-199 (1985).
20. Inoue, M. and Kuriyama, H., J. Physiol. $\underline{436}$, 511-529 (1991).
21. Inoue, R., J. Physiol. $\underline{430}$, 119P (1990).
22. Inoue, R., J. Physiol, in press (1991).
23. Inoue, R. and Brading, A.F., Br. J. Pharmacol. $\underline{100}$, 619-625 (1990).
24. Inoue, R., Kitamura, K., and Kuriyama, H., Pflugers Archiv $\underline{410}$, 69-74 (1987).
25. Inoue, R. and Isenberg, G., Am J. Physiol. $\underline{258}$, C1173-C1178 (1990).
26. Inoue, R. and Isenberg, G., J. Physiol, $\underline{424}$, 57-71 (1990).
27. Inoue, R. and Isenberg, G., J.Physiol. $\underline{424}$, 73-92 (1990).
28. Kluckner, U. and Isenberg, G., Pflugers Archiv $\underline{405}$, R61 (1985).
29. Kuriyama, H., Ito, Y., Suzuki, H., Kitamura, K., and Itoh, T., Am. J. Physiol. $\underline{243}$, H641-H662 (1982).
30. Komori, S. and Bolton, T.B., J.Physiol. $\underline{433}$, 495-517 (1991).
31. Lammel, E., Deitmer, P. and Noack, T., J.Physiol. $\underline{432}$, 259-282 (1991).
32. Lee, K.S. and Tsien, R.W., Nature $\underline{297}$, 498-501 (1982).
33. Legendre, P. and Westbrook, G.L., J. Physiol. $\underline{429}$, 429-449 (1990).
34. Lewis, C., J. Physiol. $\underline{286}$, 417-445 (1979).
35. Loirand, G., Pacaud, P., Baron, A., Mironneau, C., and Mironneau, J., J.Physiol. $\underline{437}$, 461-475 (1991).
36. Marty, A., Trends in Neurosci. $\underline{12}$, 420-424 (1989).
37. Marty, A., Evans, M.G., Tan, Y.P., and Trautmann, A., J. Exp. Biol. $\underline{124}$, 15-32 (1986).
38. Matsuda, J.J., Volk, K.A., and Shibata, E.F., $\underline{427}$, 657-680 (1990).
39. Olsen, S.P., Davies, P.F., and Clapham, D.E., Circ. Res. $\underline{62}$, 1059-1064 (1988).
40. Penner, R. and Matthews, G., and Neher, E., Nature $\underline{334}$, 499-504 (1988).
41. Russel, S.N. and Aaronson, P.I., J.Physiol. $\underline{426}$, 23P (1990).
42. Sims, S.M., Clapp, L.H., Walsh, J.V., and Singer, J.J., Pflugers Archiv $\underline{417}$, 291-302 (1990).
43. Sims, S.M., Singer, J.J., and Walsh, J.V., J. Physiol. $\underline{367}$, 503-529 (1985).
44. Sims, S.M., Singer, J.J., and Walsh, J.V., Science $\underline{239}$, 190-193 (1988).
45. Sims, S.M., Walsh, J.V., and Singer, J.J., Am. J. Physiol. $\underline{251}$, C580-C587 (1986).
46. Singer, J.J. and Walsh, J.V., Pflugers Archiv $\underline{408}$, 98-111 (1987).
47. Smith, J.S., Coronado, R., and Meissner, G., J. Gen.Physiol. $\underline{25}$, 236-244 (1986).
48. Standen, N.B., Quayle, J.M., Davies, N.W., Brayden, J.E., Huang, Y., and Nelson, M.T., Science $\underline{245}$, 177-180 (1989).
49. Sutter, M.C., Pharmacol. Rev. $\underline{42}$, 288-325 (1990).
50. Swandulla, D. and Partridge, L.D., in Potassium Channels, N.S. Cook, ed., (Ellis Horwood Ltd., Chichester 1990).
51. Taylor, S.G. and Weston, A.H., Trends in Pharmacol. Sci. 9, 272-274 (1988).
52. Trautwein, W. and Heschler, J., Ann. Rev. Physiol. $\underline{52}$, 257-274 (1990).
53. Vivaudou, M., Clapp, L.H., Wlash, J.V. Jr., and Singer, J.J., FASEB J. 2, 2497-2504 (1988).
54. Williams, D.L.Jr., Katz, G.M., Roy-Constancin, L., and Reuben, J.P., Proc. Natl. Acad. Sci. USA. $\underline{85}$, 9360-9364 (1988).
55. Yellen, G., Nature $\underline{296}$, 357-359 (1982).

CYTOSOLIC CALCIUM TRANSIENTS IN VASCULAR SMOOTH MUSCLE

KATSUYA HIRANO[1], MAYUMI HIRANO[1], SHIMAKO ABE[2], and HIDEO KANAIDE[3]

Division of Molecular Cardiology, Research Institute of Angiocardiology, Faculty of Medicine, Kyushu University, 3-1-1 Maidashi, Higashi-ku, Fukuoka 812, Japan

INTRODUCTION

In vascular smooth muscle cells (VSMCs), changes in cytosolic calcium concentration ($[Ca2+]_i$) play a central role in regulating contractions and relaxations (6, 32) and also in cell growth (43). VSMCs control $[Ca^{2+}]_i$ rapidly and precisely by mobilizing and restoring Ca^{2+} of intra- and extra-cellular pools. Recent developments in optical techniques in biology and in fluorescent- indicator dyes have made it feasible to quantitatively and continuously monitor changes in $[Ca^{2+}]_i$ in living cells under physiological conditions (5, 42).

Recently, we developed a microfluorometric technique using quin2 and smooth muscle cells of rat aorta in primary culture, to determine changes in $[Ca^{2+}]_i$ in a small area ($<1\ \mu m^2$) of the cytosol (19, 25). We examined the characteristics if intracellular Ca^{2+} store sites and determined effects of some vasorelaxants on $[Ca^{2+}]_i$.

We also developed front-surface fluorometry of whole heart and vascular strips, using optic fibers, the objective being to minimize optical artifacts which can arise from contractile movements of tissue (8, 11, 18, 26). Using fura-2 and front-surface fluorometry, we simultaneously determined $[Ca^{2+}]_i$ and force during contractions of vascular strips of porcine coronary artery (1, 8, 26). We characterized the $[Ca^{2+}]_i$-tension relationship of agonist- induced contractions and examined effects of some vasorelaxants on $[Ca^{2+}]_i$ and contractions.

In this review, we summarized the results of our studies on the characteristics of intracellular Ca^{2+} storage sites of cultured smooth muscle cells of rat aorta, quantitative relationship between $[Ca^{2+}]_i$ and developed force during contractions and relaxations of porcine artery. Also, we discuss the mechanisms of actions of vasoactive substances such as Ca^{2+} antagonists or nitroglycerine, especially in terms of alteration of $[Ca^{2+}]_i$ and of the relationship between $[Ca^{2+}]_i$ and force.

QUIN2 MICROFLUOROMETRY OF CULTURED VASCULAR SMOOTH MUSCLE CELLS

For the quin2 microfluorometry, we used rat aortic smooth muscle cells in primary culture (45). Cells just before confluency (5-7 culture day) were loaded with quin2 in acetoxymethylester form by incubating with growth medium containing $50\ \mu M$ quin2/AM for 60 min at 37° C (Figure 1). Details of the quin2 microfluorometry are described elsewhere (19). We used a fluorescence microscope (model Standard 18, Zeiss) equipped with a photon-counting system (Zeiss), a water-immersion objective lens, and an appropriate combination of filters (Zeiss and Toshiba), in which the cells were excited at wavelengths between 350 nm and 360 nm and analyzed at wavelength between 470 nm and 560 nm (Figure 2). By using a pinhole diaphragm (Zeiss) in the light axis, the fluorescence in a small spot ($<1\ \mu m^2$) in the cytosol of VSMCs was measured. An input-output calculator (model 97S, Hewlett-Packard) was used to read the fluorescence intensities. The changes in $[Ca^{2+}]_i$ were expressed in arbitrary units of fluorescence intensity. Using the method of Tsien et al. (42), an estimation of $[Ca^{2+}]_i$ in relation to fluorescence signal was made.

---

[1] Present address; Muscle Biology Group, Department of Animal Science, University of Arizona, Tucson, Arizona 85721, U.S.A.

[2] Present address; Graduate School of Health and Nutrition Sciences, Nakamura-Gakuen, 5-7-1 Befu, Johnan-ku, Fukuoka 814, Japan.

[3] To whom correspondence should be sent.

Published 1991 by Elsevier Science Publishing Company, Inc.
Ion Channels of Vascular Smooth Muscle Cells and Endothelial Cells
Sperelakis and Kuriyama, Editors

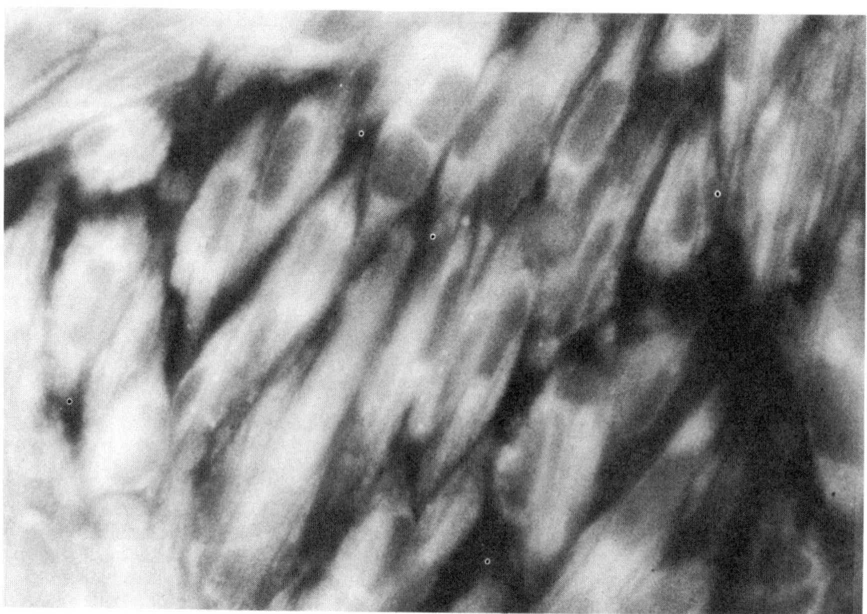

Figure 1. A fluorescence microphotograph of rat vascular smooth muscle cells in primary culture loaded with quin2.

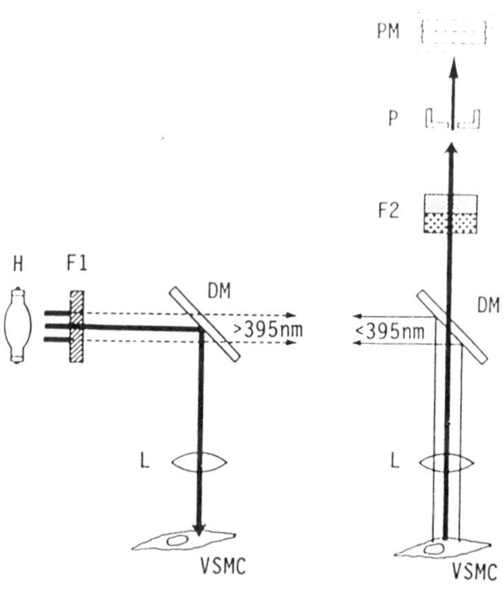

Figure 2. Scheme of quin2-microfluorometry. **H:** mercury light source, **F1:** filter to select excitation light, **DM:** dichroic mirror, reflecting light with wave length shorter than 395 nm and transmitting light with wave length longer than 395 nm, **L:** water- immersible objective lens, **F2:** filter to select emission light, P: pin hole (50 μm) diaphragm, **PM:** photomultiplier.

In optical measurement, each cells was exposed to the excitation light only once and for not longer than 2 sec to avoid the photobleaching effects on the dye. A cytosolic spot 3

μm apart from the non-fluorescent nucleus was chosen for measurements to avoid possible fluctuations caused by the uneven thickness of the cells. The VSMCs appeared as "hills and valleys" on the culture slide and cells in one layer in the "valleys" were selected for measurements to avoid possible fluctuations in the light signals produced by a pile-up of the cells in the "hills".

During measurements, neither contraction nor swelling of cells occurred in the presence and absence of quin2, as determined by phase contrast microscope (19, 25). However, cells were contractile, especially in early days of culture. On day 3, cells in primary culture and with a loose contact with the culture slide contracted to 10^{-5} M norepinephrine (NE) in the absence of extracellular Ca^{2+} (19).

Fura-2 is a most widely used fluorescence Ca^{2+} indicator, because of its improved property of fluorescence; higher quantum yield and extinction coefficient (5). Fura-2 also has a higher dissociation constant of Ca^{2+} binding, thus, less Ca^{2+}-buffering action. In fact, we found that tension development of fura-2-loaded strips was about equal to that obtained in non-loaded strips in high K^+-depolarization (ref. 10 and see below). However, there are several advantages with quin2 staining, when measuring fluorescence changes in a small spot of the cytosol. The dye is homogeneously distributes in the cytosol, and nucleus and stress fibers are not stained (Figure 1; ref. 19, 25). The fluorescence of quin2 fades faster in vitro than does that of fura-2. however, when in the cells, photobleaching of quin2 fluorescence is slower than that of fura-2 (17). Because of these advantages associated with quin2, we prefer quin2 for the microfluorometry of $[Ca^{2+}]_i$ of a cytosolic spot in a VSMC in primary culture.

SIMULTANEOUS MEASUREMENTS OF $[Ca^{2+}]_i$ AND FORCE IN FURA-2-LOADED VASCULAR STRIPS, USING FRONT-SURFACE FLUOROMETRY

We developed front-surface fura-2 fluorometry of vascular strips, using optic fibers, to simultaneously measure $[Ca^{2+}]_i$ and force during contractions and relaxations (1, 8, 26). A block diagram of fura-2 front-surface fluorometry is shown in Figure 3. Vascular strips (1 x 5 x 0.1 mm) were prepared from the coronary artery of pig, immediately after slaughter. To load with fura-2 in a physiological condition, strips were incubated in medium containing 25 μM fura- 2/AM (an acetoxymethylester form of fura-2) and 5% fetal bovine serum for 3-4 h at 37° C. The fura-2-loaded strips were washed with normal physiological saline to remove dye in the extracellular space and further incubated in physiological saline for 1 h before measurements. Strips were mounted vertically in a quartz organ bath. Resting tension of strips was adjusted to 200-250mg. Contractile force was monitored using a force transducer (TB-612T, Nihon Koden, Tokyo Japan). The fura-2 fluorescence of the strips was monitored with a front-surface fluorometer (CAM-OF-1,Japan Spectroscopic Inc., Tokyo, Japan), designed specifically for fura-2 fluorometry (see block diagram in Figure 3). Strips were illuminated by alternating excitation light (340 nm and 380 nm) through quartz optic fibers. Surface fluorescence was collected by glass optic fibers and introduced through a 500 nm band-pass filter into a photomultiplier. Fluorescence intensities at 340 nm and 380 nm excitation were measured at 500 nm and their ratio also recorded. Developed tension and fluorescence ratio were expressed as percent; the values in normal physiological saline (5.9 mM K, 1.25 mM Ca^{2+}) and during 118 mM K^+-depolarization were assumed to be 0% and 100%, respectively. The absolute values of $[Ca^{2+}]_i$ were estimated according to methods of Grynkiewcz et al (5).

Fura-2-loaded strips showed the fluorescence spectra characteristic of fura-2 (10), thereby suggesting that fura-2/AM was effectively incorporated into cells and converted intracellularly to fura-2 free acid (Figure 4-A). Ca^{2+} indicator dyes have a Ca^{2+} buffering action, hence may decrease the contractility. Formaldehyde released on acetoxymethylester hydrolysis may also cause tissue damage by possible acidification of cells. To examine the effects of dye-loading on contractility of pig coronary arterial strips, the responsiveness to 118 mM K^+- depolarization was determined before and after loading with fura-2 (Figure 4-B). Loading the strips with fura-2 did not alter the time course and the maximum levels of tension development during contractions (10). Only after fura-2-loading, were there fluorescence changes indicating increase in $[Ca^{2+}]_i$ during contraction. These observations suggested that one can measure tension development and fura-2-fluorescence simultaneously and carry out quantitative elevations of the $[Ca^{2+}]_i$-tension relationships in intact tissues.

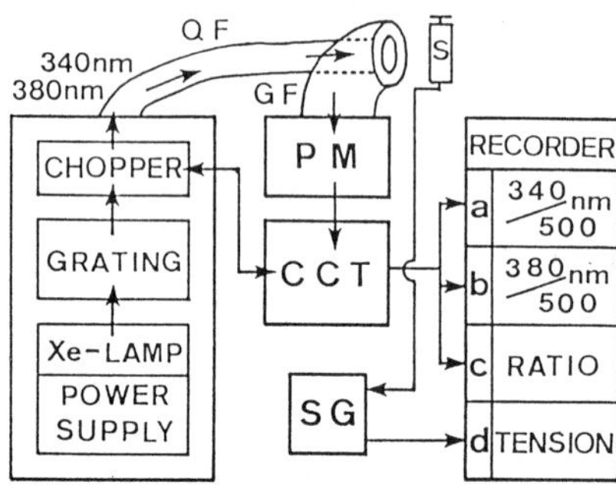

Figure 3. Block diagram of a front-surface fluorometer. Excitation light was obtained spectroscopically from a Xenon light source. The strip (1 x 5 x 0.1 mm) were illuminated by guiding the alternating dual wave length excitation lights (340 nm and 380 nm) through quartz optic fibers arranged in a concentric inner circle (diameter = 3mm). Surface fluorescence of strip was collected by glass optic fibers arranged in an outer circle (diameter = 7mm) and introduced through a 500 nm band-pass filter into photomultiplier. Fluorescence intensities at 340 nm and 380 nm excitation were measured at 500 nm emission and their ratio calculated by control circuit was also recorded. CCT: control circuit, GF: glass optic fibers, PM: photomultiplier, QF: quartz optic fibers, SG: strain gauge, S: vascular strip.

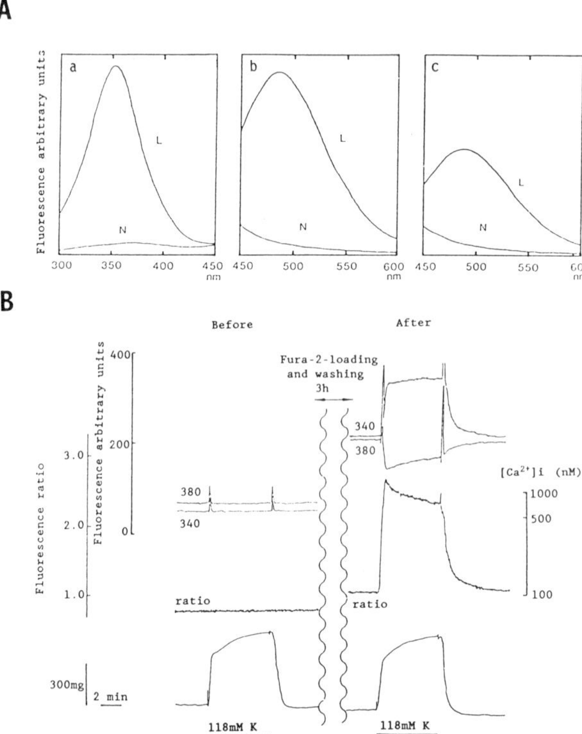

Figure 4. Characteristics of the fura-2-loaded vascular strips. **A:** Fluorescence spectra of fura-2-loaded vascular strip. (a) Excitation spectrum recorded at 500 nm emission. (b) Emission spectrum recorded at 340 nm excitation. (c) Emission spectrum recorded at 380 nm excitation. L: loaded strip. N: non-loaded strip (ref. 10). **B:** Effects of fura-2-loading on contractility of vascular strips. The response to 118 mM K^+-depolarization was recorded before and after fura-2-loading. The first and second traces from the top show changes in 500 nm-fluorescence at 340 nm and 380 nm excitations. The third trace shows changes in ratio of fluorescence intensities at 340 nm excitation to that at 380 nm excitation. The lowest trace shows tension development (ref. 10).

CALCIUM TRANSIENTS OF RAT VASCULAR SMOOTH MUSCLE CELLS IN PRIMARY CULTURE

VSMCs in primary culture responded to contractile stimuli and increased $[Ca^{2+}]_i$. Figure 5 shows the time course of fluorescence changes in cytosolic small spots, when VSMCs were exposed to histamine (HIS; 10^{-5} M), norepinephrine (NE; 10^{-5} M), caffeine(CF; 10 mM), and high K^+-depolarization (100 mM KCl). HIS and NE induced a rapid and transient elevation of $[Ca^{2+}]_i$ both in the presence and absence of extracellular Ca^{2+} (4, 29). In normal physiological saline, HIS and NE elevated $[Ca^{2+}]_i$ with a peak at 2 min and 1 min (the first component), respectively, followed by declines to a fairly steady level (the second component) within 8 min, despite the continuous applications of these agents. When cells were exposed to Ca^{2+}-free media, $[Ca^{2+}]_i$ decreased gradually and reached a steady state within 6 min. This level remained unchanged for at least 60 min. In Ca^{2+}-free media, HIS and NE induced only the first component of the $[Ca^{2+}]_i$ elevations. The similarity in time course of the first component between the presence and absence of extracellular Ca^{2+} and the lack of the second component in Ca^{2+}-free media suggest that the first component is due to Ca^{2+} release from intracellular storage sites and the second component depends on the extracellular component (possibly Ca^{2+}-influx). Specific receptor antagonists revealed that $[Ca^{2+}]_i$ transients induced by HIS and NE in cultured VSMCs are mainly mediated through H_1-receptor and alpha$_1$-receptor, respectively (4, 29).

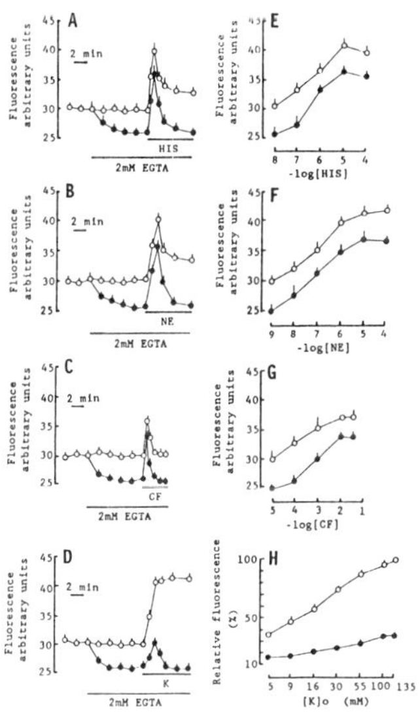

Figure 5. Ca^{2+} transients of cultured VSMCs in response to histamine (A,E), norepinephrine (B,F), caffeine (C,G) and K^+- depolarization (D,H), both in the presence (O) and absence (O) of extracellular Ca^{2+}. A-D: Time-profile curves. E-H: Concentration response curves (ref. 17, 20).

CF induced only rapid and transient elevations of $[Ca^{2+}]_i$, both in the presence and absence of extracellular Ca^{2+} (19, 20). Elevation of $[Ca^{2+}]_i$ by CF was rapid, reached the peak at 30 sec and declined rapidly to the level before stimulation. When applied during high K^+-depolarization in the presence of extracellular Ca^{2+}, CF induced a transient further elevation of $[Ca^{2+}]_i$ and subsequently decreased the level of $[Ca^{2+}]_i$ elevated by high K^+-depolarization (unpublished observation). These data suggest that the Ca^{2+} transient induced by CF is mainly due to Ca2+ release from intracellular storage sites and the $[Ca^{2+}]_i$-reducing action of CF may cause an early peak and rapid decline of [Ca2+]$_i$ transients.

When VSMCs were exposed to 100 mM K^+-depolarization in the presence of extracellular Ca^{2+}, $[Ca^{2+}]_i$ elevated rapidly to the peak at 2 min; and this peak level was sustained throughout exposure to high K^+ media (19, 25). In Ca^{2+}-free media, K^+-

depolarization induced a transient elevation of $[Ca^{2+}]_i$, which peaked at 2 min and then declined to a pre-stimulation level within 4 min (23). The level of $[Ca^{2+}]_i$ elevation obtained with high K^+-depolarization in Ca^{2+}-free media was much lower than that observed in the presence of extracellular Ca^{2+}. These results suggest that K^+-depolarization causes Ca^{2+} release from intracellular storage sites and that the marked and sustained elevation of $[Ca^{2+}]_i$ observed in the presence of extracellular Ca^{2+} is mainly due to an influx of extracellular Ca^{2+}. Using patch clamp method, we observed the voltage-dependent Ca^{2+}-inward current and characterized both L-type and T-type Ca^{2+} channels in cultured VSMCs. We also found that a population of cells showing predominantly T-type Ca^{2+} channel activity was high in the early days of culture and decreased with the duration of cell culture (2).

CHARACTERISTICS OF INTRACELLULAR CALCIUM STORAGE SITES OF CULTURED VASCULAR SMOOTH MUSCLE CELLS

After 10 min exposure to Ca^{2+}-free media, the longer the cells were exposed to Ca^{2+}-free media, the lower was the level of $[Ca^{2+}]_i$ elevations induced by the first application of HIS or NE (Figure 6- A; ref. 21, 30). On the other hand, the level of CF-sensitive $[Ca^{2+}]_i$ transient was little affected (21, 30). The half time of decay of the peak levels of NE- and CF-sensitive Ca^{2+} transients were 13 min and 190 min, respectively (21). The time course of decay of the levels of HIS-sensitive $[Ca^{2+}]_i$ transient was similar to that of NE- sensitive ones (30). Thus, HIS- and NE-sensitive stored Ca^{2+} is easily and rapidly depleted with exposure to Ca^{2+}-free media, while CF-sensitive stored Ca^{2+} has greater resistance to depletion. When cells were repeatedly stimulated by HIS in Ca^{2+}-free media, the peak level of $[Ca^{2+}]_i$ elevation at each stimulation progressively became smaller and the third application of HIS induced little or no $[Ca^{2+}]_i$ elevation (30). Thus, HIS-sensitive Ca^{2+} stores were considered to be practically depleted (Figure 6B-a). Similar phenomena were observed with NE, CF, or high K^+ stimulations (21, 24, 30). The number of repetitive stimulations required to deplete stored Ca^{2+} sensitive to NE, CF and high K^+-depolarization were three, five and five, respectively. Once extracellular Ca^{2+} was replenished (by incubating cells in 1 mM Ca^{2+}-containing media for 3 min) after the depletion of HIS-sensitive stores, HIS induced a transient $[Ca^{2+}]_i$ elevation after re-exposure to Ca^{2+}-free media; the extent and the time course was the same as those observed after the first exposure to Ca^{2+}-free media (30). Thus, HIS-sensitive Ca^{2+} stores were considered to be completely replenished by such treatment. This was also the case with NE-sensitive stored Ca^{2+} (21). In contrast, the CF-sensitive Ca^{2+} stores were not replenished (21, 30). These findings also suggest that HIS- and NE-sensitive stores were functionally different from CF-sensitive ones. The HIS- and NE-sensitive stores easily lose stored Ca^{2+} when exposed to Ca^{2+}-free media and are easily replenished by extracellular Ca^{2+} once depleted. On the other hand, the CF-sensitive stores are resistant to lose stored Ca^{2+} in Ca^{2+}-free media and once depleted these are little restored by extracellular Ca^{2+}.

After depletion of the HIS-sensitive stores by repeated applications of HIS, the subsequent application of CF induced a transient $[Ca^{2+}]_i$ elevation and the peak level of the $[Ca^{2+}]_i$ transient was almost the same as that induced by CF with the same incubation time in Ca^{2+}-free media and without pretreatment with HIS (Figure 6B-a; ref 30). Conversely, after CF-sensitive Ca^{2+} stores were depleted by repetitive stimulations of CF, the subsequent application of HIS induced a transient $[Ca^{2+}]_i$ elevation; the peak level being equal to that induced by HIS for the same duration of exposure to Ca^{2+}-free media and without pretreatment wit CF (Figure 6B-b; ref. 30). These findings indicate that the HIS-sensitive Ca^{2+} store differ from the CF-sensitive one. The relationship between NE-sensitive Ca^{2+} stores and CF-sensitive ones is definitely similar to that between HIS and CF (21).

Transient elevations of $[Ca^{2+}]_i$ repeatedly appeared in response to repetitive HIS-stimulation in Ca^{2+}-free media, with progressive reduction in peak levels. The level of $[Ca^{2+}]_i$ transient induced by NE after (n-1) time application of HIS was almost equal to that observed at n-th application of HIS (Figure 6B-c; ref. 30). Conversely, the level of $[Ca^{2+}]_i$ transient induced by HIS after (n- 1) time applications of NE was almost equal to that observed at n- th application of NE (data not shown). These results indicate that HIS- and NE-sensitive Ca^{2+} stores may be functionally identical in cultured VSMCs of rat aorta. The relationship between CF- and high K^+-sensitive stores was similar to that between HIS- and NE- sensitive ones (24). Thus, we propose that the location and mechanisms of release and uptake of HIS-sensitive Ca^{2+} stores are identical to those of NE-sensitive ones and that CF- and high K^+- depolarization-sensitive store are identical but might be functionally different from those sensitive to HIS and NE in rat aortic VSMCs in primary culture.

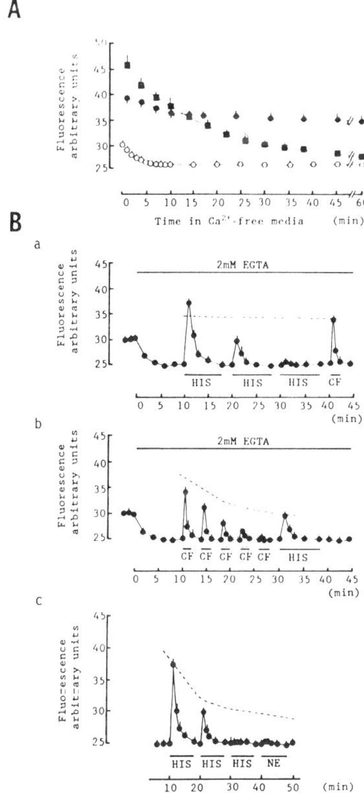

Figure 6. Characteristics of intracellular Ca^{2+} storage sites of cultured VSMCs. A: Effects of duration of exposure to Ca^{2+}-free media on peak levels of $[Ca^{2+}]_i$ induced by the first application of norepinephrine (10^{-5} M,) and caffeine (10 mM, O). When VSMCs were exposed to Ca^{2+}-free media with no stimulations, $[Ca^{2+}]_i$ decreased to a new steady state levels within 6 min and remained at the level for at least 60 min (O). (ref. 17, 21, 32) B: Effects of repetitive applications of histamine on the subsequent elevation of $[Ca^{2+}]_i$ induced by caffeine (a), and norepinephrine (c). Effects of repetitive applications of caffeine on the subsequent elevation of $[Ca^{2+}]_i$ induced by histamine (b). Broken line ( ) indicates the estimated peak level of $[Ca^{2+}]_i$ transient induced by the first application of caffeine (a), histamine (b), and norepinephrine (c) after exposure to Ca^{2+}-free media for the time on the abscissa (see Figure A). (ref. 30)

HIS and NE are shown to produce at least two intracellular second messengers, inositol 1,4,5-trisphosphate (IP3) and diacylglycerol with activation of H_1-receptors and $alpha_1$- adrenoceptors, respectively (4, 38). IP3 release intracellularly stored Ca^{2+} (44). It is suggested that the difference in Ca^{2+} release mechanisms between HIS/NE-sensitive stores and CF/high K^+- depolarization-sensitive stores is mainly linked to difference in messengers which induce Ca^{2+} release. However, we found that endothelin, which was also shown to produce IP3, induced Ca^{2+} release mainly from CF-sensitive stores in cultured VSMCs (14). The relation between IP3-sensitive Ca^{2+} stores and agonist-sensitive ones remain to be elucidated

CALCIUM TRANSIENTS AND FORCE DEVELOPMENT

In the vascular strips of pig coronary artery loaded with fura-2, we simultaneously determined $[Ca^{2+}]_i$ and developed tension and quantitatively evaluated the relationship between $[Ca^{2+}]_i$ and force (1, 8, 26), HIS induced, in a concentration-dependent manner, rapid and transient elevations of $[Ca^{2+}]_i$ and tension both in the presence and absence of extracellular Ca^{2+} (Figure 7A-b, 7B-b; ref. 9). In the presence of extracellular Ca^{2+} (1.25 mM), HIS induced abrupt elevation of $[Ca^{2+}]_i$, which reached the first peak within several seconds (the first component). After a slight dip at 30 s, $[Ca^{2+}]_i$ reached a second peak at 3 min and then gradually declined but remained higher than the pre-stimulation level (the second component). In Ca^{2+}-free media, HIS induced only transient increase of $[Ca^{2+}]_i$, which reached its peak within 1 min. As with cultured VSMCs, it was suggested that the first component is due to Ca^{2+} release from the intracellular storage sites and that the second component depends on extracellular Ca^{2+}, probably the influx of Ca^{2+}.

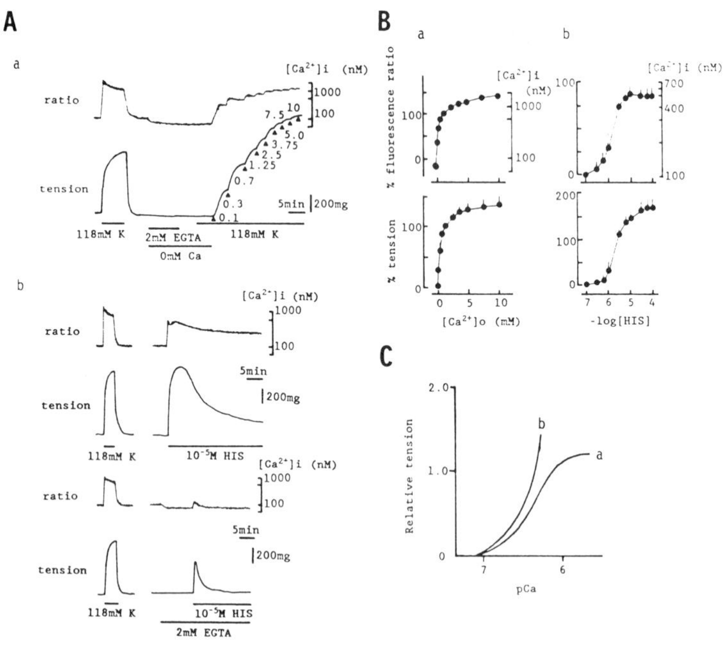

Figure 7. Changes in $[Ca^{2+}]_i$ and tension during contractions. **A:** Representative recordings of changes in ratio of fluorescence intensity at 340 nm excitation to that at 380 nm excitation and tension developments. **B:** Dose-response curves. **C:** $[Ca^{2+}]_i$-tension relation curves. (a): Contraction induced by cumulative applications (0 mM to 10 mM) of extracellular Ca^{2+} during 118 mM K^+- depolarization. (b): Contractions induced by histamine. Dose- response curve and $[Ca^{2+}]_i$-tension relation curve for histamine- induced contraction were obtained by cumulative applications of histamine in the presence of extracellular Ca^{2+} (ref. 9).

When vascular strips were exposed to high K^+-depolarization (118 mM), fluorescence ratio and tension reached maximum levels within 30 sec and 3 min, respectively. These levels were either sustained or slightly reduced during depolarization (Figure 7A-a). The level of $[Ca^{2+}]_i$ and the extent of tension development were concentration dependent. After incubation of strips in Ca^{2+}-free media containing 2 mM EGTA followed by 5 min exposure to Ca^{2+}-free media without EGTA, contraction could be initiated by cumulative applications of extracellular Ca^{2+} during 118 mM K^+-depolarization (9). Fluorescence ratio and tension increased stepwise with elevation of extracellular Ca^{2+} (Figure 7A-a). $[Ca^{2+}]_i$ increased from 75 nM to about 3000 nM, while tension developed to about 150% of that obtained by 118 mM K^+-depolarization in the 1.25 mM Ca^{2+}-containing media (Figure 7B-a).

The relationships between $[Ca^{2+}]_i$ and tension were examined in $[Ca^{2+}]_i$-tension curves (Figure 7C). The $[Ca^{2+}]_i$-tension curve of contractions induced by cumulative applications of extracellular Ca^{2+} during high K^+-depolarization was similar to the Ca^{2+}-tension relationship obtained in saponin-skinned fibers (9). Namely tension started to develop at about 100 nM of $[Ca^{2+}]_i$ and reached a maximum at 3000 nM of $[Ca^{2+}]_i$, $[Ca^{2+}]_i$ at 50% of maximum tension development was about 400 nM. The $[Ca^{2+}]_i$-tension relationship of contractions induced by cumulative applications of HIS (10^{-7} M to 10^{-4} M) shifted to the left of that obtained with K^+-depolarization. The extent of tension development for a given change in $[Ca^{2+}]_i$ in HIS-induced contractions was greater than that of K^+- depolarization-induced contractions, especially at $[Ca^{2+}]_i$ over 300 nM (Figure 7B-c).

The relation between $[Ca^{2+}]_i$ and tension varied with time after the initiation of contractions. The $[Ca^{2+}]_i$-tension relationship in the early, rising phase of HIS-induced contraction was similar to that obtained during depolarization (Figure 8A). At the time of maximum tension development , the relation was greater than that observed during depolarization, which persisted in the phase of declining tension. Thus, temporal change in $[Ca^{2+}]_i$-tension relationship during contraction induced by HIS in the presence of extracellular Ca^{2+} showed a counter-clockwise rotation with time (Figure 8-A). Treating strips with H-7, a relatively specific inhibitor of protein kinase C, shifted the $[Ca^{2+}]_i$-tension relation of the HIS-induced contraction, especially at the late phase of contraction, towards the level observed with high K^+- depolarization; the result being clockwise rotation of time course of $[Ca^{2+}]_i$-tension relationship (9). On the contrary, the $[Ca^{2+}]_i$-tension relationship in the early phase of contraction either in the presence and absence of extracellular Ca^{2+} was little affected by H-7. Thus, the contractile mechanisms in the late phase of contraction may differ from those at the early phase. Although H-7 also inhibits myosin light chain kinase with a 16 fold less potency than in the case of inhibition of protein kinase C, these findings suggest that HIS-induced contractions especially in the late phase, are maintained with increased Ca^{2+} sensitivity of contractile apparatus, which appear to be linked to the activation of H-7- sensitive mechanisms, probably protein kinase C. Thus, HIS induced a greater contraction for a given change in $[Ca^{2+}]_i$ during the steady state of contraction than did K^+-depolarization. This greater contraction seems to be mediated by activation of an H-7- sensitive mechanism.

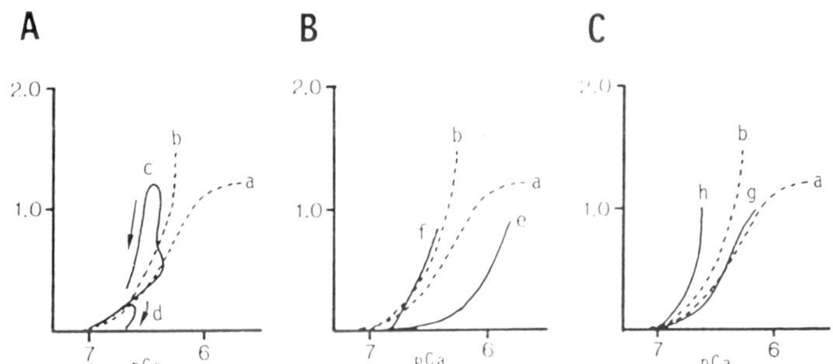

Figure 8. Alteration of $[Ca^{2+}]_i$-tension relationships by H-7 (A), nitroglycerine (B) and diltiazem (C). The $[Ca^{2+}]_i$-tension curves of contraction induced by cumulative applications of extracellular Ca^{2+} during K^+-depolarization in the absence (a; from Figure 7C-a) and presence of 10^{-5} M nitroglycerine (e) or 10^{-6} M diltiazem (g). The $[Ca^{2+}]_i$-tension curve induced by cumulative applications of histamine in the absence (b; from Figure 7C-b) and presence of 10^{-5} M nitroglycerine (f) or 10^{-6} M diltiazem (h). Temporal changes in $[Ca^{2+}]_i$-tension relationship during histamine-induced contraction in the absence (c) and presence (d) of 10^{-5} M H-7. Arrows indicate the direction of time (ref. 1, 9, 10).

There are several reports showing that receptor-mediated stimulation increased Ca^{2+}-sensitivity of contractile elements (7, 31, 39, 40). The difference between agonist-induced contractions and depolarization-induced contractions in intact muscle seems to be related to difference in the activation of membrane-bound proteins. The activation of G-proteins was reported to enhance the Ca^{2+}-sensitivity of myofilaments in alpha-toxin permeabilized smooth muscle (22, 36). In saponin or alpha-toxin permeabilized smooth muscle, phorbol ester increased the amplitude of contraction evoked by Ca^{2+} (12, 34). However, the role of protein kinase C in the regulation of smooth muscle contraction remains controversial (37).

EFFECTS OF VARIOUS AGENTS ON CALCIUM TRANSIENTS
Nitrovasodilators

In cultured VSMCs, nitroglycerine (NG) decreased $[Ca^{2+}]_i$ to the lower steady-state levels, regardless of whether VSMCs were at rest, in a steady-state of Ca^{2+}-depletion or at K^+-depolarization (15, 25). When VSMCs were exposed to Ca^{2+}-free media in the presence of NG, the rate of reduction of $[Ca^{2+}]_i$ was accelerated and the steady state level of $[Ca^{2+}]_i$ was lower than that observed in the absence of NG (25). In the absence of NG, 1.73 ± 0.31 min (n = 6) were required for $[Ca^{2+}]_i$ to fall to $1/e$ (time constant), and the steady- state level was 17.7 ± 0.8 fluorescence units (n = 6). In the presence of 10^{-5} M NG, time constant and the steady-state levels were 0.78 ± 0.02 min. (n = 6) and 13.7 ± 0.6 fluorescence units (n = 6), respectively . When VSMCs were treated with NG in Ca^{2+}-free media, the first application of CF induced a transient increase in $[Ca^{2+}]_i$ to the level observed in the absence of NG, suggesting that NG had no effect on the CF-sensitive Ca^{2+} release from the intracellular storage sites (15, 25). However , in the presence of NG, transient increase in $[Ca^{2+}]_i$ caused by the second application of CF was attenuated, and there was little or no increase in $[Ca^{2+}]_i$ during the third and subsequent applications, while in the absence of NG the third and fourth application of CF induced significant increases in $[Ca^{2+}]_i$ in Ca^{2+}-free media (Figure 6B-b; ref. 25). These findings suggested that NG might decrease $[Ca^{2+}]_i$ by activating extrusion of Ca^{2+} from cells and that the amount of Ca^{2+} restored to the intracellular storage sites for the subsequent release by CF is markedly and progressively reduced. On the other hand, a transient increases in $[Ca^{2+}]_i$ induced by the first application of NE in the Ca^{2+}-free media was markedly reduced. This finding is consistent with observations that NE-sensitive stored Ca^{2+} is easily and rapidly depleted in Ca^{2+}-free media. Nitrovasodilators such as NG increase intracellular cGMP levels, cGMP-dependent protein kinase (PKG) activity and phosphorylation of some protein substrates for PKG (11). We observed that 8-bromo cGMP, a membrane permeable cGMP analog, activity decreased $[Ca^{2+}]_i$ in VSMCs,irrespective of the level of $[Ca^{2+}]_i$, which was similar than observed with NG (16). We suggested that acceleration of Ca^{2+} extrusion through plasma membrane plays a major role in the reduction of $[Ca^{2+}]_i$ during cGMP-mediated vasorelaxation in cultured VSMCs.

We also determined the effects of NG on $[Ca^{2+}]_i$ and force development in the porcine coronary artery loaded with fura-2 (1). NG reduced in a concentration-dependent manner both $[Ca^{2+}]_i$ and tension, irrespective of whether the vascular strips were in a resting state or during exposure to high K^+-depolarization or to HIS stimulation. In contrast with the case of cultured VSMCs. NG did not enhance the decrease in $[Ca^{2+}]_i$ observed when vascular strips were exposed to Ca^{2+}-free media. NG had no effect on the $[Ca^{2+}]_i$ transient induced by repetitive applications of CF in the Ca^{2+}-free media. These findings indicate that NG does not affect Ca^{2+}-release by CF nor does it deplete CF-sensitive stored Ca^{2+} of vascular strips (1). However, tension development induced by the first application of CF was significantly inhibited by NG, Thereby suggesting that NG inhibited the CF-induced contraction without affecting $[Ca^{2+}]_i$. NG strongly inhibited the transient elevations of $[Ca^{2+}]_i$ and contraction induced by HIS (1). However, when NG was washed out in Ca^{2+}-free media, the subsequent application of HIS induced transient elevations of both $[Ca^{2+}]_i$ and tension, the extent of these elevations being greater than those observed during the first application of HIS in the presence of NG. In addition, the second application of HIS induced a marked contraction despite evidence peak $[Ca^{2+}]_i$ was significantly lower than that at the resting level. These findings are consistent with the observation that NG inhibited NE-sensitive Ca^{2+} release of cultured VSMCs in Ca^{2+}-free media. However, it is suggested that in vascular strips, NG attenuated transient elevations of $[Ca^{2+}]_i$ induced by HIS in Ca^{2+}- free media, not only by depleting HIS-sensitive stored Ca^{2+} but also by inhibiting the release of Ca^{2+} (1). An active reduction in $[Ca^{2+}]_i$ may result in depletion of stored Ca^{2+}.

In addition to the finding that $[Ca^{2+}]_i$ of vascular strips decreased during relaxation induced by NG, NG relaxed the porcine coronary artery to a greater extent than that expected from the reduction in $[Ca^{2+}]_i$, regardless of whether vascular strips were at last, during exposure to high K^+-depolarization or to HIS stimulation (1). The $[Ca^{2+}]_i$-tension relation curve of contractions induced by stepwise increments of external Ca^{2+} during high K^+-depolarization in the presence of NG shifted to the right of that obtained without NG (Figure 8B). The $[Ca^{2+}]_i$-tension relation curve of contraction induced by cumulative applications of HIS in the presence of NG shifted slightly to the right of that obtained without NG (Figure 8B). Thus, NG relaxes the porcine coronary artery by reducing $[Ca^{2+}]_i$ and also by directly controlling contractile elements mediated through second messengers not related to changes in $[Ca^{2+}]_i$.

Ca^{2+} Antagonists

In the presence of extracellular Ca^{2+} (1mM), Ca^{2+} antagonists, verapamil and diltiazem, concentration-dependently inhibited extracellular Ca^{2+}-dependent elevations of $[Ca^{2+}]_i$, as induced by high K^+-depolarization (19) and NE or HIS (28) stimulation of cultured VSMCs. When VSMCs were exposed to 10^{-5} M HIS, verapamil and diltiazem inhibited the second component of $[Ca^{2+}]_i$ elevation, the IC_{50} being 0.09 and 0.18 μM, respectively (28). When VSMCs were exposed to high K^+-depolarization, verapamil and diltiazem inhibited the second component with the IC_{50} being 0.47 μM and 0.31 μM, respectively (19). These values are consistent with those which were reported to indicate a specific binding of these drugs to plasma membrane or the IC_{50} for inhibition of voltage-dependent Ca^{2+} currents (13, 27). Thus, inhibition of the extracellular Ca^{2+}-dependent elevation of $[Ca^{2+}]_i$ by Ca^{2+} antagonists is due to a specific inhibition of the Ca^{2+} influx through Ca^{2+} channels in plasma membrane. Why the HIS-induced Ca^{2+} influx of extracellular Ca^{2+} is sensitive to Ca^{2+} antagonists remains to be elucidated. It was reported that HIS depolarized the membrane potential of some species of artery (3). Other studies showed that agonist stimulation enhanced the Ca^{2+} current through voltage-dependent Ca^{2+} channels (33). We also observed that HIS enhanced voltage-activated Ca^{2+} inward currents in cultured VSMCs (unpublished data).

At high concentrations, verapamil and diltiazem inhibited transients elevation of $[Ca^{2+}]_i$ induced by NE and HIS in Ca^{2+}-free media but had no effect on the Ca^{2+} release induced by CF or K^+-depolarization. The IC_{50} of verapamil and diltiazem for inhibition of the Ca^{2+} release by HIS was 8.7 μM and 95.7 μM, respectively; much higher than that required to inhibit second components. The order of potency for inhibition of the first component (diltiazem > verapamil), differed from that for inhibition of $[Ca^{2+}]_i$ elevations due to Ca^{2+} influx through voltage-dependent Ca^{2+} channels (verapamil > diltiazem), suggesting that Ca^{2+} antagonists inhibit the first component through mechanisms independent of or different from the inhibition of Ca^{2+} influx. We found that verapamil and diltiazem inhibited specific binding of $[^3H]$ mepyramine to membrane, the dissociation constants binding 7.1 μM and 114 μM, respectively (28). Thus, it is suggested that Ca^{2+} antagonists at high concentrations inhibit competitively the binding of HIS to the H_1-receptor. The similarity between IC_{50} for inhibition of Ca^{2+} release and Ki values for inhibition of mepyramine binding suggest that Ca^{2+} antagonists inhibit Ca^{2+} release mainly by competing with HIS for the H_1-receptor. This was also the case with NE stimulation (19, 35). Thus, we suggested that Ca^{2+} antagonists inhibit the extracellular Ca^{2+}-dependent elevation of $[Ca^{2+}]_i$ by inhibiting Ca^{2+} influx (specific effects of Ca^{2+} antagonist) and that at high concentrations they inhibit HIS- or NE-induced Ca^{2+} release from intracellular storage sites, mainly by interfering with the binding of HIS or NE to H_1-receptors or alpha -adrenoceptors, respectively.

The effects of diltiazem on $[Ca^{2+}]_i$ in the porcine coronary artery are similar to those observed in cultured VSMCs (10). Namely, diltiazem, within the range of 18^{-8} M to 10^{-5} M, inhibited the second component of $[Ca^{2+}]_i$ elevation with no effect on the first component, as induced by 10^{-5} M HIS. Only at a high concentration (over 10^{-5} M), did diltiazem inhibit the first component of $[Ca^{2+}]_i$ elevation induced by HIS, in the presence and absence of extracellular Ca^{2+}. As in the case of $[Ca^{2+}]_i$, diltiazem at a lower concentration inhibited only the second component of tension development and shortened the time to reach the peak. Only a high concentration of diltiazem (over 10^{-5} M) inhibited the first component, in addition to inhibiting the second component. We quantitatively evaluated the role of $[Ca^{2+}]_i$ reduction in relaxing smooth muscle by examining the effects of diltiazem on the $[Ca^{2+}]_i$-tension relationship of concentrations, as induced by cumulative applications of extracellular Ca^{2+} during high K^+-depolarization (13) or by cumulative applications of HIS in the presence of extracellular Ca^{2+} (unpublished observation). The $[Ca^{2+}]_i$-tension relationship of the Ca^{2+}-induced contraction obtained in the presence of diltiazem practically overlapped that obtained without diltiazem. This suggests that diltiazem had no direct effects on the contractile element. However, in case of the HIS-induced contraction, diltiazem displaced the $[Ca^{2+}]_i$-tension relationship of the HIS-induced contraction to the left of the curve obtained in the absence of diltiazem. This observation suggests that diltiazem caused a reduction of tension proportional to the reduction of $[Ca^{2+}]_i$, without directly affecting contractile elements or mechanisms by which HIS produces a greater tension for given changes in $[Ca^{2+}]_i$.

In summary, Ca^{2+} antagonists, at low concentrations, specifically inhibit extracellular Ca^{2+}-dependent elevation of $[Ca^{2+}]_i$, possibly by inhibiting Ca^{2+} influx through Ca^{2+} channels, and secondarily cause a relaxation proportional to a reduction of $[Ca^{2+}]_i$, with no direct effect on contractile mechanisms. Within this range, a reduction of $[Ca^{2+}]_i$ plays a major role in the relaxation induced by Ca^{2+} antagonists. At a high concentration, Ca^{2+} antagonists inhibit the Ca^{2+} release induced by receptor agonists such as HIS or NE but not those induced by CF or high K^+- depolarization. The inhibition of agonist-induced Ca^{2+} releases may partly result from the inhibition of binding of agonists at receptor sites.

SUMMARY AND CONCLUSION

We have presented herein an overview on Ca^{2+} transients in cultured VSMCs and vascular strips and characterized intracellular Ca^{2+} stores of cultured rat aortic VSMCs. Our findings suggest that there are two functionally different store sites with different responses to Ca^{2+}-depletion and different mechanisms of Ca^{2+} release or uptake. By determining $[Ca^{2+}]_i$ and tension simultaneously, we evaluated the quantitative relationship between $[Ca^{2+}]_i$ and tension development during contractions and relaxations. Changes in $[Ca^{2+}]_i$ play an important and central role in regulating contractions and relaxations of vascular strips. However, there are additional regulatory mechanisms which directly control the contractile apparatus through second messengers not directly related to changes in $[Ca^{2+}]_i$.

ACKNOWLEDGMENTS

This work was supported in part by Grant-in-aid for Scientific Research on Priority Area (No.03253208), for General Scientific Research (No.01480250) and for Developmental Scientific Research (03557043) from the Ministry of Education, Science and Culture, Japan and Grants from the "Research Program on Cell Calcium Signals in the Cardiovascular System", from Uehara Memorial Foundation, from the CIBA-GEIGY Foundation (Japan) for the Promotion of Science, from the Mitsukoshi Prize of Medicine 1990, and from the Japan Research Foundation for Clinical Pharmacology.

REFERENCES

1. Abe, S., H. Kanaide, and M. Nakamura, Br. J. Pharmacol. 101, 545-552 (1990).
2. Akaike, N., H. Kanaide, T. Kuga, M. Nakamura, J. Sadoshima, and H. Tomoike, J. Physiol. 416, 141-160 (1989).
3. Casteels, R. and H. Suzuki, Pflugers Arch. 387, 17-25 (1980).
4. Daum, P.R., C.P. Downes, and J.M. Toung, J. Neurochem. 43, 25-32 (1984).
5. Grynkiewicz, G., M. Poenie, and R.Y. Tsien, J. Biol. Chem. 260, 3440-3450 (1985).
6. Hartshorne, D.J., in Physiology of the Gastrointestinal Tract 2nd ed, L.R. Johnson, ed. (Raven Press, New York 1987) pp. 423-482.
7. Himpens, B. and A.P. Somlyo, J. Physiol. 395, 507-530 (1988).
8. Hirano, K., H. Kanaide, and M. Nakamura, Br. J. Pharmacol. 98, 1261-1266 (1989).
9. Hirano, K., H. Kanaide, S. Abe, and M. Nakamura, Br. J. Pharmacol. 102, 27-34 (1991).
10. Hirano, K., S. Abe, H. Kanaide, and M. Nakamura, Br. J. Pharmacol. 101, 273-280 (1990).
11. Ignarro, L.J. and P.J. Kadowitz, Annu. Rev. Pharmacol. Toxicol. 25, 171-191 (1987).
12. Ito, T., Y. Kubota, and H. Kuriyama, J. Physiol. 397, 401-419 (1988).

13. Janis R.A. and A. Sriabine, Biochem. Pharmacol. 32, 3499-3507 (1983).
14. Kai, H., H. Kanaide, and M. Nakamura, Biochem. Biophys. Res. Commun. 158, 235-243 (1989).
15. Kai, H., H. Kanaide, and M. Nakamura, J. Pharmacol. Exp. Ther. 251, 1174-1180 (1989).
16. Kai, H., H. Kanaide, T. Matsumoto, and M. Nakamura, FEBS Lett. 221, 284-288 (1987).
17. Kanaide, H., Asia Pacific J. Pharmacol. 5, 177-183 (1990).
18. Kanaide, H., R. Yoshimura, R. Makino, and M. Nakamura, Am. J. Physiol. 242 (Heart Circ.Physiol. 11), H980-H989 (1982).
19. Kanaide, H., S. Kobayashi, J. Nishimura, M. Hasegawa, Y. Shogakiuchi, T. Matsumoto, and M. Nakamura, Circ. Res. 63, 16-26 (1988).
20. Kanaide, H., T. Matsumoto, and M. Nakamura, Biochem. Biophys. Res. Commun. 140, 195-203 (1986).
21. Kanaide, H., Y. Shogakiuchi, and M. Nakamura, FEBS Lett. 214, 130-134 (1987).
22. Kitazawa, T., S. Kobayashi, K. Horiuti, A.P. Somlyo, and A.V. Somlyo, J. Biol. Chem. 264, 5339-5342 (1989).
23. Kobayashi, S., H. Kanaide, and M. Nakamura, Biochem. Biophys. Res. Commun. 129, 877-884 (1985).
24. Kobayashi, S., H. Kanaide, and M. Nakamura, J. Biol. Chem. 261, 15709-15713 (1986).
25. Kobayashi, S., H. Kanaide, and M. Nakamura, Science 229, 553-556 (1985).
26. Kodama, M., H. Kanaide, S. Abe, K. Hirano, H. Kai, and M. Nakamura, Biochem. Biophys. Res. Commun. 160, 1302-1308 (1989).
27. Kuga, T., J. Sadoshima, H. Tomoike, H. Kanaide, N. Akaike, and M. Nakamura, Circ. Res. 67, 368-375 (1990).
28. Matsumoto, T., H. Kanaide, J. Nishimura, T. Kuga, S. Kobayashi and M. Nakamura, Am. J. Physiol. 257 (Heart Circ.Physiol. 26), H563-H570 (1989).
29. Matsumoto, T., H. Kanaide, J. Nishimura, Y. Shogakiuchi, S. Kobayashi, and M. Nakamura, Biochem. Biophys. Res. Commun. 135, 172-177 (1986).
30. Matsumoto, T., H. Kanaide, Y. Shogakiuchi, and M. Nakamura, J. Biol. Chem. 265, 5610-5616 (1990).
31. Morgan, J.P. and K.G. Morgan, J. Physiol. 351, 155-167 (1984).
32. Murphy, R.A., Annu. Rev. Physiol. 51, 275-349 (1989).
33. Nelson, M.T., N.B. Standen, J.E. Brayden, and J.F. Worley III, Nature 336, 382-385 (1988).
34. Nishimura, J. and C. van Breemen, Biochem. Biophys. Res. Commun. 163, 929-935 (1988).
35. Nishimura, J., H. Kanaide, and M. Nakamura, J. Pharmacol. Exp. Ther. 236, 789-793 (1986).
36. Nishimura, J., M. Kolber, and C. van Breemen, Biochem. Biophys. Res. Commun. 157, 677-683 (1988).
37. Nishizuka, Y. Science 233, 305-312 (1986).
38. Raymond, J.R., M. Hnatowich, R.J. Lefkowitz, and M.G. Caron, Hypertension 15, 119-131 (1990).
39. Rembold, C.M. and R.A. Murphy, Circ. Res. 63, 593-603 (1988).
40. Sato, K., H. Ozaki, and H. Karaki, J. Pharmacol. Exp. Ther. 247, 466-472 (1988).
41. Shogakiuchi, Y., H. Kanaide, S. Kobayashi, and M. Nakamura, Biochem. Biophys. Res. Commun. 135, 9-15 (1986).
42. Tsien, R.Y., T. Pozzan, and T.J. Rink, J. Cell Biol. 94, 325-334 (1982).
43. Whitfield, J.F., in Calcium, Cell Cycles, and Cancer (CRC Press, Boca Raton 1990) pp. .
44. Yamamoto, H. and C. van Breemen, Biochem. Biophys. Res. Commun. 130, 270-274 (1985).
45. Yamamoto, H., H. Kanaide, and M. Nakamura, Br. J. Exp. Pathol. 64, 156-165 (1983).

POTASSIUM CHANNEL CURRENTS OF VASCULAR MUSCLE

KENT HERMSMEYER

Cardiovascular Research Laboratory, Earle A. Chiles Research Institute, Providence Medical Center and Departments of Medicine and Cell Biology and Anatomy, Oregon Health Sciences University Portland, Oregon 97213

ABSTRACT

Vascular muscle contains multiple K^+ channels that can be identified by voltage, kinetics, and blocking agents. Several different identifiable K^+ channels can be recognized in a single contracting vascular muscle cell. In the electrical spike, the first K^+ current to occur is the transient pacemaker outward current ($I_{K,TP}$), followed by the large conductance (big) current ($I_{K,B}$), which is both Ca^{2+} and ATP-dependent. When activated under depolarizing conditions, $I_{K,B}$ dominates membrane currents because the conductance is large (200 pS). K^+ channel opener vasodilators increase a large current that may be a subset of $I_{K,B}$ or a related Ca^{2+} and ATP-dependent I_K. Pinacidil (10 μM), cromakalim (10 μM), and EMD 52692 (3 μM) increase an $I_{K,B}$ that enhances relaxation by causing a 5-25 mV hyperpolarization, both at rest and during norepinephrine (300 nM) exposure. $I_{K,B}$ induced by pinacidil, cromakalim, or EMD is graded by ATP/ADP in the pipette solution and can be partially inhibited by 10 μM glyburide; $I_{K,B}$ is also sensitive to Ca^{2+} and pH of the myoplasm. Smaller I_K which are also Ca^{2+} sensitive and with voltage domains that include resting E_m are also increased by pinacidil, cromakalim, and EMD.

INTRODUCTION

Vasodilators considered to be potassium (K^+) channel openers are effective anti-hypertensives, which suggests the importance of K^+ channels for determination of membrane potential and thus regulation of contraction (7). Such agents include cromakalim, pinacidil, and EMD 52692, which were studied in these experiments. The differences in responses to K^+ channel openers, K^+ channel blocking agents, and the dependence of the increase in K^+ current on intracellular Ca^{2+}, pH, and ATP/ADP ratios suggest the existence of several different types of K^+ channel in a single vascular muscle cell.

MATERIALS AND METHODS

Experiments were carried out on isolated single cells from azygos veins of neonatal rats cultured according to procedures we have previously published (5, 8, 9). The cells were cultured in CV3M consisting of 4 mM L-glutamine, 20 μg/mL gentamicin, 20 mM HEPES buffer (pH = 7.4), and 15% horse serum added to 85% minimal essential medium (MEM) Earle's salts. Cells were dispersed by collagenase and trypsin, separated by differential sedimentation (6), and studied from 2-21 days after dispersion, with 5 μM 5-bromo-2'-deoxyuridine (BrdU) always in the culture medium. Vascular muscle cells kept in a 5% CO_2 incubator at 95% humidity and 37^0 C showed spontaneous contractions, sensitive responses to norepinephrine, and maintenance of K^+ and Ca^{2+} channels.

K^+ movements through vascular muscle cell membranes were quantitated by recordings under voltage clamp conditions using tight seal or patch-clamp pipettes containing (mM): 120 K^+ glutamate, 20 KCl, 10 HEPES, 5 K_2ATP, 1 EGTA, 0.5 $MgCl_2$, and 0.1 GTP. Extracellular perfusing solution contained 150 mM KCl and 10 mM HEPES buffer. Data were recorded on a Hewlett-Packard Vectra computer using a Labmaster TL-1 interface and an Axon Instruments Axopatch 1B amplifier, with filtering at 2 kHz and digitizing at 5 kHz.

Published 1991 by Elsevier Science Publishing Company, Inc.
Ion Channels of Vascular Smooth Muscle Cells and Endothelial Cells
Sperelakis and Kuriyama, Editors

RESULTS

On exposure to the K^+ channel openers pinacidil, cromakalim, or EMD 52692, there were increases in K^+ current, especially during the sustained phase of outward current. These currents were blocked by 1 mM TEA and were also partially blocked (Fig. 1) by charybdotoxin (ChTX) (1 nM), but not by apamin (100 nM). K^+ currents were increased by 10 μM pinacidil (Fig. 2) and by 10 μM cromakalim or 3 μM EMD 52692. The K^+ current increases appeared to be due primarily to the large conductance K^+ channel, which was 200 pS in a symmetrical K^+ solution. However, the TEA or ChTX did not completely block these currents and in some experiments, pinacidil could be shown to increase currents even in the presence of TEA or ChTX.

ATP sensitive K^+ channels were also found in these vascular muscle cells. These currents had a conductance of 50 pS in the symmetrical K^+ solutions and could be blocked by 10 μM glyburide. This ATP-dependent channel was not blocked by high concentrations of ATP (3-5 mM), as were those in cardiac muscle, and is thus more likely of the pancreatic type (1).

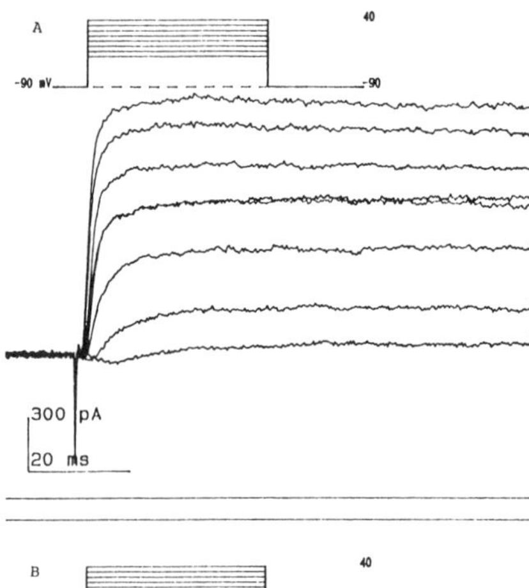

Figure 1. Control currents in an azygos venous muscle cell recorded in a whole cell voltage clamp were large under control conditions (A) and reduced by 60% after 10 min exposure to 1 nM Charybdotoxin (B).

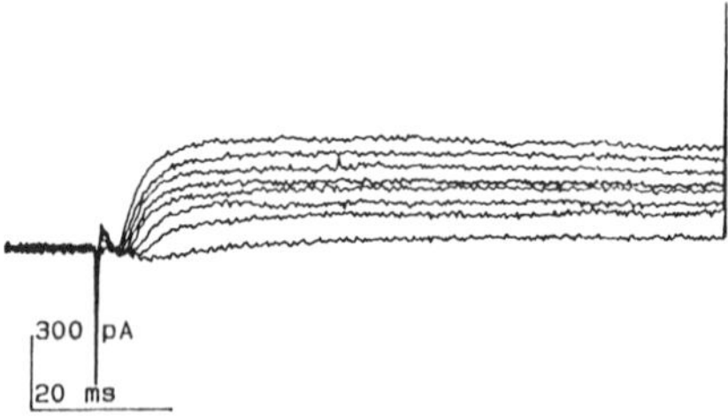

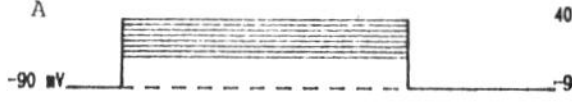

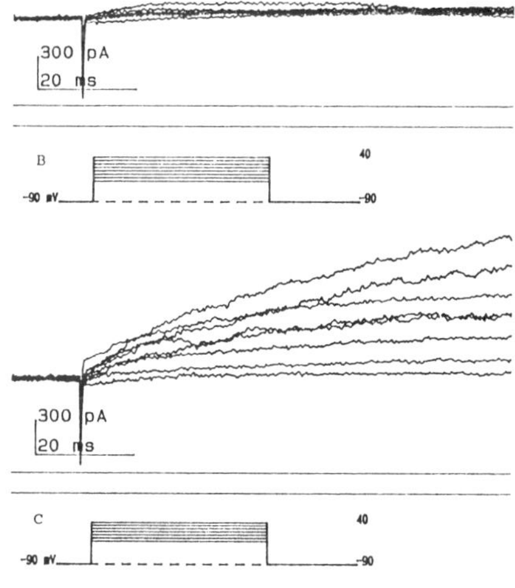

Figure 2. Whole cell voltage clamp currents in a rat azygos vein vascular muscle cell were small under control conditions (A), but increased by more than 10X after 3 min exposure to 10 μM pinacidil (B), which was partially blocked by 10 nM Charybdotoxin for 10 min (C).

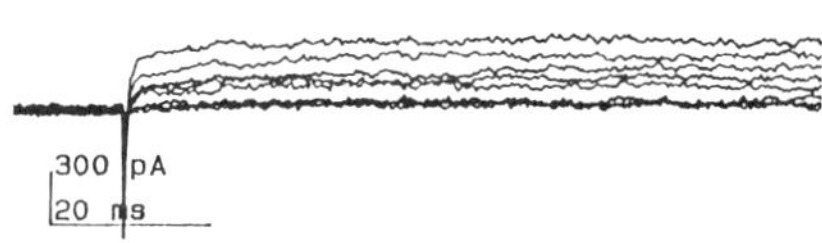

Another channel was found that showed a transient outward current, similar to that which has been noted as a pacemaker current. This transient outward K^+ current was prominent in cells showing pronounced pacemaker activity, and seemed to correlate with a rapid inward Na^+ channel that is described in a separate publication (10).

Another K^+ channel that was kinetically slow and was open at -60 to -40 mV had characteristics of a voltage pacemaker, being relatively insensitive to any of these K^+ channel blocking agents or K^+ channel openers, but was highly sensitive to Ca^{2+}. This pacemaker channel, which appears to have a conductance of 50 pS in symmetrical high K^+ solutions, was resistant to all of the blockers tried, even TEA. These K_{2}^+ channels may represent the slow pacemaker oscillatory channels that, together with Ca^{2+} channels and because of their Ca^{2+} sensitivity probably confer pacemaker properties (4).

DISCUSSION

The significance of finding several K^+ channels in a single kind of vascular muscle cell is suggested by the possibility for control of E_M and thus contraction by a number of mechanisms that involve intracellular messengers and metabolic processes. These experiments show that Ca^{2+}, pH, Mg^{++} concentration, ATP, and GTP are all important parameters for modulation of K^+ channels.

Thus agents which directly affect non-membrane mechanisms would also have an E_M component of action. With these different K^+ channel identities and the possibility for varying distribution of K^+ channels in different blood vessels, it is highly likely that

110

manipulations of K^+ channels should allow specificity of drug treatment especially in pathophysiologic conditions involving abnormal reactivity.

The group of K^+ channel openers available at present offers the possibility for effective vasodilation, but has not fully developed the potential for selectivity which would appear to exist for this new class of drugs. Further exploration will be necessary to determine how effectively different blood vessels can be targeted by different K^+ channels and possibly different distributions and mixes with Ca^{2+} channels. These considerations will be important for future analysis and design of vasodilators. One important commonality of Ca^{2+} channels with K^+ channels should be noted here. The subsarcolemmal accumulation of intracellular Ca^{2+} that occurs during activation and entry of trigger Ca^{2+} has a dual relaxant action, causing not only inhibition of Ca^{2+} channels, (11) but also activation of Ca^{2+}-dependent K^+ channels, which must exist in at least 3 major Ca^{2+}-dependent K^+ channel types, as shown in these experiments. The existence of different K^+ channel types has made the identification of the action of K^+ channel openers more complicated than was at first expected. Multiple classes of K^+ channels that can be affected by these 3 K^+ channel openers and the likelihood that the primary actions of pinacidil, cromakalim, and EMD 52692 may be on the intermediate conductance (50 pS) K^+ channel would explain divergent interpretations about the type of K^+ channel being affected (2).

Since it is known that other cells (e.g., endothelium) are also regulated by a Ca^{2+}-dependent K^+ channel, there may be multiple cell effects within the blood vessel wall, all of which would tend to cause relaxation, perhaps accounting for the effectiveness of K^+ channel blockers against activation by a number of agonists. Further investigation of the actions of these K^+ channel openers and consideration of their possible role on intracellular Ca^{2+} (3), will be necessary to better appreciate the role of K^+ channels in concert with other mechanisms in the function of these vascular muscle cells as the final common path in the blood vessel wall.

ACKNOWLEDGEMENTS

This study was supported by National Institutes of Health grants HL 38537 and HL 38645, and by Eli Lilly and E Merck Darmstadt.

REFERENCES

1.   Ashcroft, F.M., Ann. Rev. Neurosci. 11, 97-118 (1988).
2.   Brown, D., Annu. Rev. Physiol. 52, 215-42 (1990).
3.   Erne, P. and K. Hermsmeyer, Arch. Pharmacol., in press (1991).
4.   Hermsmeyer, K. and H. Akbarali, Prog. Appl. Microcirc. 15, 32-40 (1989).
5.   Hermsmeyer, K. and K. Bian, J. Vasc. Med. Biol. in press (1991).
6.   Hermsmeyer, K. and R. Mason, Circ. Res. 50, 627-632 .
7.   Hermsmeyer, K., J. Cardiovasc. Pharmacol. 12(Suppl 2), S17-S22 (1988).
8.   Marvin, W., R. Robinson, and K. Hermsmeyer, Circ. Res. 45, 528-540 (1979).
9.   Rusch, N.J. and K. Hermsmeyer, Mol. Cell. Biochem. 80, 87-93 (1988).
10.   Self, D.A. and K. Hermsmeyer, Submitted for publication (1991).
11.   Sturek, M. and K. Science 233, 475-478 (1986).

MODULATION OF Ca-ACTIVATED K CHANNELS FROM CORONARY SMOOTH MUSCLE

L. TORO and F. SCORNIK

Dept. Molecular Physiology and Biophysics. Baylor College of Medicine, Houston, TX 77030

Abstract

Coronary smooth muscle has predominant Ca-activated K channels. These channels can be regulated by various vasoactive substances such as: thromboxane A2, arachidonic acid, angiotensin II, endothelin, acethylcholine, nitroglycerin and cGMP. This large spectrum of regulatory elements (lipids, vasoactive peptides, neurotransmitters and nucleotides) may be an index of the importance that these channels have in the regulation of coronary function. In addition, the regulatory character of Ca-activated K channels seems to be a general property of this class of channels as has been demonstrated in many other tissues (Table 1). Both, direct and indirect mechanisms of action have been proposed. Final steps in the indirect mechanisms are: variations in intracellular Ca, channel regulation by G proteins (62), and changes in the phosphorylation state of the channel. However, the action of the regulatory agents is not a simple process and more than one mechanism can be involved in the modulatory response. This may explain the fact that in the same type of channel the same agonist can exert both up- and down-regulation. Finally, recent evidence supports the view that Ca-activated K channels form stable regulatory complexes with other membrane proteins (e.g. G proteins, receptors, protein kinases and protein phosphatases).

INTRODUCTION

The physiological role for Ca-activated K [$K_{(Ca)}$] channels in smooth muscle has not yet been well established. Their voltage and Ca sensitivities have suggested a predominant role in the repolarization phase of action potentials and slow waves (69, 70). In addition, due to their large conductance (200-300 pS), their high density per cell (>15,000/cell), and the high input resistance of smooth muscle cells (>1 GΩ), $K_{(Ca)}$ channels have also been suggested to contribute to the resting membrane potential of smooth muscle (66).

Nevertheless, voltage and calcium are not the only modulatory factors for these channels. Early experiments in neurons demonstrated that calcium-activated K conductances could be down-regulated by noradrenaline and histamine (22, 36). Since then a number of hormones, neurotransmitters, lipids and nucleotides have been shown to have a role in the regulation of $K_{(Ca)}$ channels (*for review see*, ref. 62). These investigations have given light to the concept that $K_{(Ca)}$ channel modulation may be an important cell strategy for controlling smooth muscle function. Thus, the understanding of the mechanism(s) of action by which different metabolites regulate $K_{(Ca)}$ channel activity is essential to understand the physiological role of these channels in each tissue.

The purpose of this review is to discuss some of the recent evidence about the regulation of $K_{(Ca)}$ channels with special emphasis in coronary artery smooth muscle (Table 1).

Published 1991 by Elsevier Science Publishing Company, Inc.
Ion Channels of Vascular Smooth Muscle Cells and Endothelial Cells
Sperelakis and Kuriyama, Editors

Table I. Regulatory properties of Ca-activated K channels

Cell type or tissue	Regulatory agent	Effect	Type of Ca-activated K channel	Ref.
CSM	A.A.	(+)	maxi K(Ca)	Fig. 4
	ACh	(+)	?	37
	AGII	(-)	maxi K(Ca)	27
	cGMP, Nitroglycerin	(+)	maxi K(Ca)	41
	Endothelin	(+,-)	maxi K(Ca), ?	14, 35
	TXA2	(-)	maxi K(Ca)	19
SM	A.A., fatty acids	(+)	maxi K(Ca)	62,63
	ACh	(-)	maxi K(Ca)	64
	Adrenergic agents	(+)	maxi K(Ca)	54,65,66
	Adenosine, ANF	(+)	maxi K(Ca)	67
	Cholesterol	(+)	maxi K(Ca)	68
	GMP, cGMP, GMP, GTP	(+)	maxi K(Ca)	67
	$[H^+]$	(-)	maxi K(Ca)	58
	Histamine	(+,-)	?	43,69
	Histamine	(+)	?, V.I.	44
	Substance P	(+,-)	maxi K(Ca)	70,71
Brain	ATP	(+)	maxi K(Ca), Type 2	53
Chromaffin	ACh	(-,+)	maxi K(Ca), ?	46,72
Epithelia	$[H^+]$	(-)	maxi K(Ca)	56,57
Fibroblasts	Bn, GRP, NmB, Bk	(+)	?	49
Gonadotrophs	GRH	(+)	small K(Ca)	73,74
HeLa Cells	Histamine	(+)	small K(Ca), V.I.	75
Kidney	ADH	(+)	maxi K(Ca)	76
Lacrimal cells	ACh	(+)	?	77
Neuroblastoma-glioma hybrid cells	Bk	(+)	?	78
Neurons	$[H^+]$	(-)	?	60
	Histamine	(+)	?	79
	Adrenergic agents	(-)	?	79,80
Oocytes	ADP	(+)	small K(Ca)	81
Pancreatic B-cells	$[H^+]$	(-)	maxi K(Ca)	55
Pituitary tumor cells	Somatostatin	(+)	maxi K(Ca)	45
Skeletal muscle	$[H^+]$	(-)	maxi K(Ca)	59

CSM, coronary smooth muscle; SM, smooth muscle; A.A., arachidonic acid; ACh, acethylcholine; AGII, angiotensin II; TXA2, thromboxane A2; ANF, atrial natriuretic factor; Bn, bombesin; GRP, gastrin releasing peptide; NmB neuromedin B; Bk, bradykinin; GRH, gonadotrophin releasing hormone; ADH, antidiuretic hormone; (+), activation; (-), inhibition; maxi K(Ca), channels with conductances >100 pS; ?, not known; small K(Ca), channels with conductances <100 pS; V.I., voltage independent.

CORONARY SMOOTH MUSCLE

Coronary smooth muscle is regulated by chemical stimuli, such as, hormones, neurotransmitters, lipids and nucleotides. Most of this knowledge has been obtained from studies *in vivo* with perfused hearts and with strips or rings from the vessel. While these types of approaches are useful to study an overall response that emerges from the interactions between multiple variables, the underlying molecular mechanisms are difficult to define. In the last few years the study of the interaction of these vasoactive substances at the single channel level, in single cell preparations or in reconstituted channels, has led to the fundamental conclusion that $K_{(Ca)}$ channels from the coronary artery are common targets of vasoconstrictors and vasorelaxants.

Single cells from coronary smooth muscle (Fig. 1) possess robust $K_{(Ca)}$ currents. Wilde and Lee (72) described two types of $K_{(Ca)}$ currents sensitive to internal Ca ($[Ca^{2+}]_i$) in smooth muscle of dog coronary arteries, and noticed that these type of K currents were the predominant currents in these cells. We have made similar observations using isolated cells from porcine coronary artery (61), and recently Volk et al. (68) have shown that $K_{(Ca)}$ currents represent an important component of the outward current in rabbit coronary smooth muscle. Consistent with these results is our finding that "maxi" $K_{(Ca)}$ channels (245, 295 pS in 50/250 mM KCl gradient) are abundant and have functional "isoforms" in isolated plasmalemmal membranes from porcine coronary smooth muscle (64), and the finding of other investigators that patches of the corresponding cells contain primarily $K_{(Ca)}$ channels (148 pS in 5.4/140 mM K gradient; 216 pS and 300 pS in 140/140 mM K) (16, 24, 58).

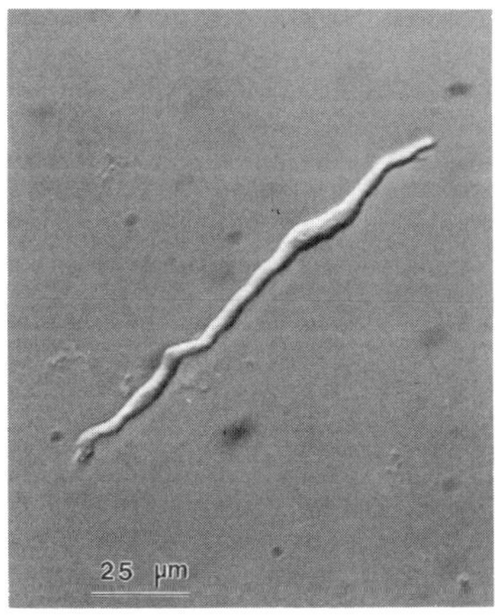

Figure 1. Isolated cell from pig coronary smooth muscle. Single cells are obtained by enzymatic dispersion.

$K_{(Ca)}$ channels from coronary smooth muscle are voltage and Ca-sensitive, similarly to the ones described in other tissues (*for review see* ref. 33). An interesting feature, however, is that in isolated cells from coronary artery $K_{(Ca)}$ channels may be active at $[Ca^{2+}]_i$ as low as $10^{-8}M$ (24). Thus, the abundance of $K_{(Ca)}$ channels, their diversity, their apparent high Ca-affinity, and their regulatory characteristics may warrant their importance in coronary function.

Thromboxane A2 (TXA2)

One of the events during platelet aggregation at the site of coronary artery damage (where smooth muscle is exposed) is the activation of the arachidonic acid metabolic pathway. Arachidonic acid from the plasma membrane of platelets is metabolized by the enzyme cyclooxygenase to produce cyclic endoperoxides. These compounds are subsequently metabolized to TXA2 (primarily) and prostaglandin I2 (Fig. 2). Thromboxane A2 reaches the smooth muscle at the coronary lesion causing vasoconstriction (54). The mechanism of action of TXA2 on the coronary smooth muscle is not well understood. However, in astrocytoma cells it has been shown that it may activate phospholipase C via a pertussis toxin-insensitive G protein (43).

We have recently found that the TXA2 mimetic U-46619 (Fig. 2) inhibits $K_{(Ca)}$ channels from coronary smooth muscle incorporated into lipid bilayers (Fig. 3). We used the stable analog U46619, since TXA2 is chemically unstable (half life = 30 s, at 37^0 C) (49). U46619 (50-150 nM) reduced channel activity between 15% and 60% of the control open probability. The inhibitory effect of the TXA2 agonist modified both the open and closed states of the channel, and the inhibition could be reversed by internal Ca^{2+}. This inhibition is specific since it takes place only when U46619 is added to the external side of the channel. Moreover, the inactive hydrolysis derivative TXB2 (1 μM) had no effect on channel activity and the TXA2 receptor antagonist SQ29,548 (up to 750 nM) prevented the action of U46619 (55). These findings are in agreement with the vasoconstrictor action of TXA2, since inhibition of K channels would lead to depolarization and contraction of coronary smooth muscle. Since our experimental protocol did not include GTP, Mg or ATP in the solution facing the internal side of the channel, a mechanism involving the activation of an endogenous G protein or an endogenous protein kinase seems to be unlikely. Therefore, it is possible that the channel itself has a site for the TXA2 action or

114

that a TXA2 receptor protein is closely associated with the channel. This "direct" mechanism of inhibition of $K_{(Ca)}$ channels by TXA2 may be parallel to the activation of phospholipase C in the intact cell. It would be interesting to test in single cells if TXA2 inhibits $K_{(Ca)}$ channels through the activation of phospholipase C according to the following metabolic cascade: **(1)** interaction of TXA2 to its receptor, **(2)** activation of a G protein after TXA2 receptor occupancy, **(3)** activation of phospholipase C or in addition inhibition of $K_{(Ca)}$ channels, by the activated G protein, **(4)** phospholipase C mediated formation of inositol-1,4,5-trisphosphate, IP3, and diacylglycerol, **(5)** activation of protein kinase C by diacylglycerol, and **(6)** protein kinase C mediated phosphorylation of $K_{(Ca)}$ channels or of a closely related molecule that inhibits channel activity. Also, the possibility that IP3 or diacylglycerol may by themselves regulate channel activity has to be explored.

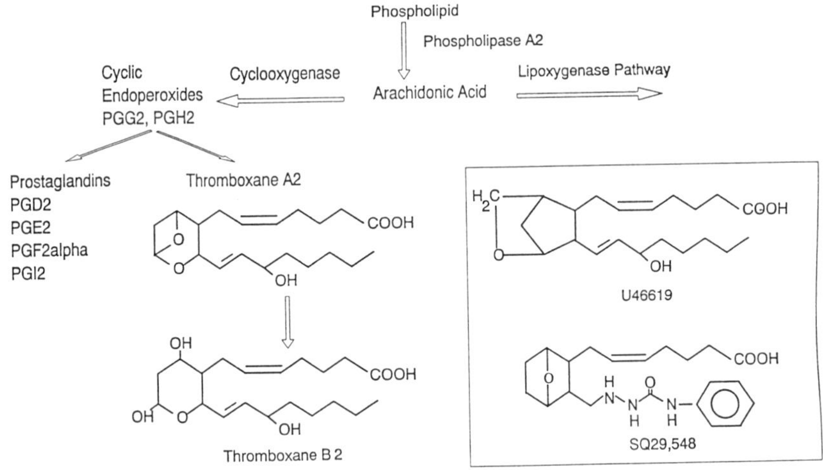

Figure 2. Cyclooxygenase pathway of arachidonic acid metabolism, and chemical structure of Thromboxane A2 and related compounds. Thromboxane A2 is a natural metabolite of the arachidonic acid cascade. It has a half life of 30 s, and its hydrolysis produces the inactive form Thromboxane B2. U46619 is a stable analog and SQ29,548 is a competitive antagonist of Thromboxane A2. PG, prostaglandin.

Arachidonic acid

This lipid relaxes coronary vessels (54). Accordingly, we have observed that arachidonic acid (10 - 50 μM) increases coronary $K_{(Ca)}$ channel activity in a dose-dependent manner (Fig. 4). This increase in $K_{(Ca)}$ channel activity may induce hyperpolarization and relaxation of the coronary artery. However, the specificity of this effect has to be determined in order to correlate this effect with a physiological role of this lipid. Besides the role of arachidonic acid and its metabolites as external vasoactive substances, the possibility also exists that these compounds are synthesized in the smooth muscle itself and that they act as intracellular modulators of channel activity. This type of action has been demonstrated in cardiac cells where lipoxygenase metabolites of arachidonic acid (derived after phospholipase A2 cleavage of membrane phospholipids) modulate the muscarinic K channel (31). A role for arachidonic acid in the regulation of $K_{(Ca)}$ channels from pulmonary artery (47) and aorta (26) has also been suggested. The question remains open whether this lipid itself modulates channel activity or if one or more of its metabolites exert a regulatory action on $K_{(Ca)}$ channels *in vivo*.

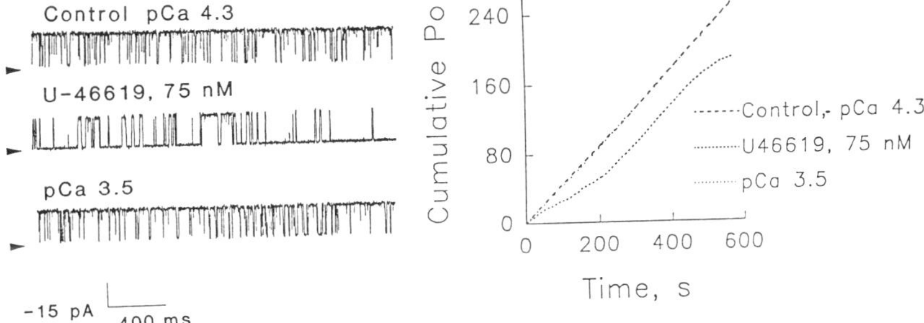

Figure 3. Inhibition of coronary $K_{(Ca)}$ channel activity by the thromboxane A2 agonist U46619, and its reversion by intracellular calcium. **A:** Records of channel activity before (control), after addition of external U46619 (75 nM), and after raising intracellular calcium. **B:** Corresponding cumulative P_O vs. time plots in the absence (P_O = 0.92), in the presence of 75 nM U46619 (P_O = 0.69), and after increasing internal calcium concentration (P_O = 0.92). Holding potential, V_H = -40 mV. Channel conductance = 350 pS. The experiment was performed using symmetric KCl (250/250) in the bilayer system (cis/trans). The plots represent the cumulative open probability calculated every 2 seconds during 10 minutes of continuous recording. Arrows mark the closed state of the channel.

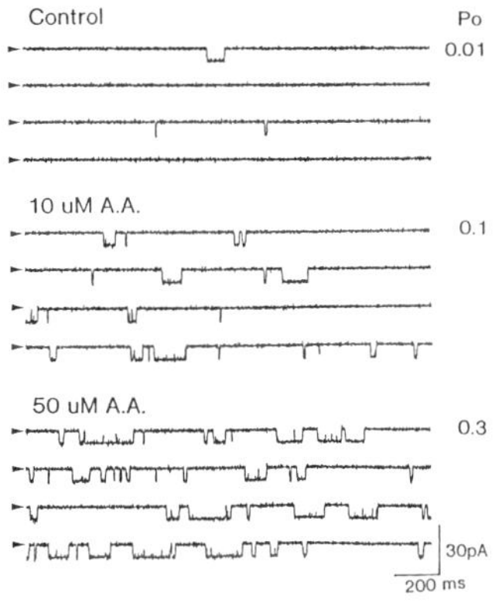

Figure 4. Arachidonic acid activates $K_{(Ca)}$ channels from coronary smooth muscle. Records of channel activity in the absence (control), and in the presence of arachidonic acid (10 and 50 μM). V_H = -20 mV. pCa = 5.8. Channel conductance = 300 pS. The experiment was performed using symmetric KCl (250/250) in the bilayer system. Arrows mark the closed state of the channel.

Angiotensin II (AGII)

This peptide is one of the most potent vasoconstrictors of the coronary vessels. Plasma membrane receptors for AGII have been detected in a number of vascular smooth muscles, and its action on cell metabolism has been widely studied (20). Binding of AGII to its receptor activates the formation of IP_3 and diacylglycerol through a G protein-mediated activation of phospholipase C. Consequently, there is an increase in intracellular Ca that is caused by IP_3-induced Ca release from internal stores. Besides these actions, AGII may also produce an increase of Ca currents, blockade of K currents (4, 46), and depolarization of the plasma membrane (25).

We have recently reported that AGII may inhibit coronary $K_{(Ca)}$ channels incorporated into lipid bilayers (65). The inhibition produced by AGII was dose dependent with a $K_{1/2}$ of 58 nM. AGII affected both the open and closed states of $K_{(Ca)}$ channels. Since the experiments were performed in the absence of GTP, Mg^{2+} or ATP, it was difficult to postulate an inhibition mediated by an endogenous G protein or an endogenous protein kinase. A more suitable explanation was a direct effect of AGII on $K_{(Ca)}$ channels. This does not mean, however, that in the intact tissue parallel pathways of regulation do not take place leading to the same final response: depolarization and contraction of coronary arteries. Therefore, it is of interest to determine the mechanism(s) by which AGII is regulating $K_{(Ca)}$ channels. Is the channel itself being modified by AGII? Is there a receptor coupled to the channel? The involvement of a specific receptor may be tested with the peptide antagonist [Sar^1, Ala^8]-angiotensin II (saralasin), or non-peptide antagonists like DuP753 (to AGII-1 type receptor) and PD123177 (to AGII-2 type receptor). Testing DuP753 would be specially interesting since it has antihypertensive properties (74).

Endothelin

This vasoactive peptide was isolated from the supernatant of cultured endothelial cells (75). Endothelin has been shown to produce either contraction or relaxation of the coronary vessels (15, 28). Relaxation has been explained as mediated by the release of endothelium-derived relaxing factor (EDRF) and prostacyclin from endothelial cells (13). On the other hand, contraction is thought to occur by several mechanisms that raise $[Ca^{2+}]_i$ including the activation of Ca^{2+} channels and induction of phosphoinositide breakdown after stimulation of phospholipase C (19, 67).

Recently Hu et al. (24) have demonstrated that endothelin may modulate $K_{(Ca)}$ channel activity in a dual manner (inhibition and activation), depending on the peptide concentration. This evidence provides an alternate explanation for endothelin opposing effects on coronary smooth muscle. Endothelin at concentrations between 0.1 nM to 1 nM increased channel activity in cell attached patches of isolated coronary smooth muscle cells. On the contrary, when endothelin concentration was raised to 10-100 nM, $K_{(Ca)}$ channel activity diminished. Klockner and Isenberg (27), however, found that 100 nM endothelin increased the frequency of spontaneous transient outward currents (STOCs), thought to be the result of simultaneous activation of $K_{(Ca)}$ channels (2). This discrepancy seems not to be related to differences in species since both investigations were carried out using porcine coronary smooth muscle cells. The main reason for these differences using high endothelin concentrations could be that the inhibitory effect was observed at the single channel level in cell attached experiments while the activation was seen using the whole cell configuration. It is possible that in the whole cell experiments an important intermediate of the endothelin induced inhibition of $K_{(Ca)}$ channel activity is lost due to diffusion into the patch pipette. Another explanation could be that since enzymatic treatment may modify the structure of membrane proteins (receptors and channels), different cell dissociation procedures give rise to the discrepancies in the results. Nevertheless, both groups propose that endothelin regulates $K_{(Ca)}$ channel activity through changes in the internal Ca concentration.

Many questions remain open: Can endothelin regulate $K_{(Ca)}$ channels directly? Is modulation by endothelin through a metabolite of the phosphoinositide cascade? Are phosphorylation and/or dephosphorylation involved?

Acetylcholine

The role of acetylcholine on the modulation of $K_{(Ca)}$ channels from coronary smooth muscle has recently been addressed at the single cell level (17). $K_{(Ca)}$ channel activity was studied at the macroscopic level by measuring STOCs and "spikelike hyperpolarizations" (SLHs), while single channel activity was studied in cell attached patches. Perfusion with 10 μM ACh caused an increase of SLHs leading to hyperpolarization of the membrane, an increase of the STOCs, and an enhancement of single $K_{(Ca)}$ channel activity. In agreement with a role of $K_{(Ca)}$ channels on ACh induced SLHs and STOCs, these events disappeared after dialysis of the internal solution with 10 mM EGTA. The fact that activation of $K_{(Ca)}$ channels was observed in cell attached patches when ACh was present in the bath, suggested to the authors that activation occurred via a cytosolic messenger. This messenger seems to be calcium released from intracellular stores, since heparin which inhibits IP_3-induced calcium release suppressed the ACh increased STOCs but not the basal STOCs. In addition, the ACh induced STOCs were reversibly blocked by atropine, suggesting a muscarinic receptor associated mechanism.

It is interesting to note that in colonic smooth muscle ACh (10 μM) produces inhibition of $K_{(Ca)}$ channels. This inhibition is probably due to a direct effect of a pertussis toxin sensitive G protein on $K_{(Ca)}$ channels (for review see ref. 62). Therefore, it is of interest to determine if $K_{(Ca)}$ channels from coronary smooth muscle can be modulated by ACh directly or via a closely associated molecular complex.

Nitroglycerin and cGMP

The chemical identity of endothelial relaxing factor (EDRF) is thought to be nitric oxide (NO) (48). Although, more recent data suggest that EDRF is not solely or mainly composed by NO (42). NO containing compounds such as nitroglycerin are important as pharmacological and therapeutical agents due to their vasorelaxing properties. These agents activate the soluble form of guanylate cyclase (GC), leading to increased levels of cGMP followed by relaxation. In addition, NO causes hyperpolarization of coronary smooth muscle at rest (60), suggesting that it may activate K conductances. In accordance, Fujino et al. (16) have recently reported that 4-aminopyridine-sensitive $K_{(Ca)}$ channels from coronary smooth muscle cells, grown in explants, are stimulated by nitroglycerin (10 μM). These authors also tested the effect of cGMP on $K_{(Ca)}$ channel activity. The permeant analog Br-cGMP (300 μM) was used in cell attached patches, while cGMP (300 μM) was added to the intracellular side of inside-out patches. As expected for a NO-related relaxing effect, cGMP or its permeant analog induced the activation of $K_{(Ca)}$ channels. Thus, activation of $K_{(Ca)}$ channels by NO-compounds via cGMP may account in part for the relaxation of coronary smooth muscle. On the other hand, the role of cGMP in smooth muscle relaxation has been attributed to the activation of cGMP-dependent protein kinase (PKG), which leads to a decrease in internal $[Ca^{2+}]$ (35). Thus, it would be interesting to establish if activation of PKG is also involved in the increase of $K_{(Ca)}$ channel activity (via phosphorylation), and/or if cGMP may by itself enhance channel activity. Since Fujino et al., (16) reported that cGMP could increase channel activity in inside-out patches where soluble PKG may be lost, it is plausible that this cyclic nucleotide may act directly on the gating of $K_{(Ca)}$ channels (Fig. 5).

OTHER SMOOTH MUSCLES AND TISSUES

The regulatory properties of $K_{(Ca)}$ channels seem to be a common characteristic of this type of membrane proteins. We have recently reviewed their regulation by neurotransmitters, hormones, nucleotides and lipids (61). Therefore, we shall only discuss the more recent contributions to this field.

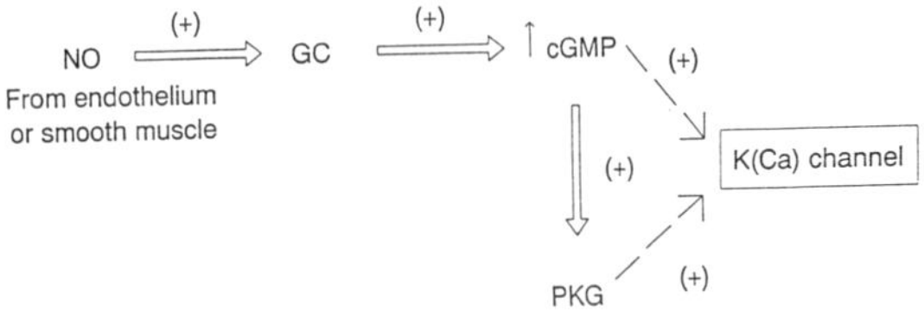

118

NO (From endothelium or smooth muscle) ⟹ (+) GC ⟹ (+) ↑cGMP → (+) K(Ca) channel ↑ (+) ; cGMP → (+) PKG → (+) K(Ca) channel

Figure 5. Schematic representation of a hypothetical pathway for the activation of coronary $K_{(Ca)}$ channels by NO-compounds. In a simplistic view NO (nitric oxide) activates guanylate cyclase (GC) which increases cGMP levels. cGMP could activate $K_{(Ca)}$ channel directly or via the activation of a cGMP-dependent protein kinase (PKG). Solid arrows mark known events while dashed arrows show hypothetical steps of the mechanism of $K_{(Ca)}$ channel activation by NO. (+), activation; [cGMP, increase in cGMP concentration.

Histamine

The binding of histamine to their receptors may cause inhibition of STOCs or activation of $K_{(Ca)}$ channels (for review see ref. 62). Recently, dual effects have been shown in single cells from carotid artery (14) and activation in a vas deferens derived cell line, DDT_1 MF-2 (40).

Desilets et al. (14), described a major inhibitory effect of histamine after a transient small activation on single cells from carotid artery. Under voltage clamp conditions, histamine (10 μM) reversibly inhibited spontaneous spikes of outward current (I_{to}) and a related transient hyperpolarization. I_{to} similarly to STOCs is most likely due to a cyclic $K_{(Ca)}$ channel activation by Ca from internal stores. This assumption is supported by the fact that I_{to} was abolished by internal EGTA or agents that impair sarcoplasmic reticulum function like caffeine 10 mM and ryanodine 1μM. Moreover, I_{to} was voltage dependent and sensitive to TEA. However, the mechanism of the inhibition of I_{to} by histamine is not well understood. It is presumed that histamine depletes the Ca stores and Ca is no longer available to activate I_{to}. In addition to the histamine induced inhibition of I_{to} already described, this current was preceded by a slow hyperpolarization and an outward current hump. This phenomenon was explained by the authors as an early activation of $K_{(Ca)}$ channels due to the initial release of Ca from the internal stores.

On the other hand, histamine in DDT_1 MF-2 cells activated an outward current that is Ca-sensitive and voltage independent (40). H1 receptor stimulation seems to be involved in this process since the histamine-induced K current was not affected by the H2-receptor antagonist cimetidine (10 μM), but it was completely blocked by the H1 antagonist mepyramine (10 μM). This histamine-induced current was not affected by apamin (300 nM), 3-4, diaminopyridine (100 μM) nor by glipizide (100 μM) an ATP-sensitive K channel blocker. Associated events to the histamine-induced outward current were an increase in internal Ca, IP_3 and inositol (1,3,4,5)-tetrakisphosphate (IP_4) concentrations. However, IP_3 formation was fast and transient and did not correlate well with the time course of the current. On the other hand, the sustained rise in IP_4 formation was related to the rise in internal Ca and the histamine-induced outward current. Moreover, addition of IP_4 (10 μM) and Ca (10 μM) to the cytoplasm revealed that these agents were necessary to activate the outward current. Based on these findings, the authors propose that Ca and IP_4 are important regulators of K channels responsible for the histamine-evoked K current. Schematically, the binding of histamine to its receptor would activate a G-protein followed by phosphoinositide breakdown. Formation of IP_4 (after a fast phosphorylation of IP_3) in conjunction with Ca released from internal stores by IP_3 would activate K channels.

It may be concluded that the Ca-activated K channels underlying the inhibition (most likely maxi $K_{(Ca)}$ channels) and the activation of the outward current by histamine are different types of channels. This is supported by the fact that the former are voltage dependent and the latter are not. Therefore, it would be of great importance to determine at the single channel level the identity of these histamine regulated K channels.

Finally, even though histamine has opposite effects in the two systems described above (carotid artery and vas deferens), both groups relate the observed modulation to changes in internal Ca. However, the mechanistic linkage between the initial increase in I_{to} current by histamine and its subsequent inhibition is not clear. Is histamine depleting internal stores from Ca or in addition to its effect on the release of calcium, is it directly inhibiting $K_{(Ca)}$ channels? Are other metabolites involved in the response?

Somatostatin

White et al. (71) demonstrated that the neuropeptide somatostatin increased K currents through the activation of maxi $K_{(Ca)}$ channels (120 pS in 140/140 mM KCl) in rat pituitary tumor cells (GH$_4$C$_1$). GH$_4$C$_1$ cells were studied through nystatin-permeabilized membrane patches to avoid the loss of intracellular components during measurements of macroscopic currents, and with cell free patches to study single channel events. Somatostatin (100 nM) increased a $K_{(Ca)}$ current at all voltages studied (0 to +40 mV). Addition of 2 mM CoCl$_2$ to the bath decreased the outward current and blocked the effect of somatostatin. This suggested to the authors the involvement of calcium-activated K channels in the somatostatin response. $K_{(Ca)}$ channels would be activated by the entrance of external Ca through voltage sensitive Ca channels. In agreement, both charybdotoxin (30 nM) and tetraethylammonium (1 mM) blocked the outward current and the effect of somatostatin as well. The involvement of maxi $K_{(Ca)}$ channels was corroborated at the single channel level. Experiments in cell attached patches demonstrated that 100 nM somatostatin added to the bath increased the open probability of $K_{(Ca)}$ channels from 0.05 to ~ 0.5. The facts that somatostatin could exert its regulatory property without being in direct contact with the channel (was absent in the pipette solution), and that the maximum response was reached at 5-10 minutes were a strong indication that an intracellular metabolic pathway was involved in the process. Furthermore, the diffusible second messenger did not seem to be calcium since somatostatin reduced calcium entry through voltage activated channels.

Several lines of evidence supported the view of a second messenger: a) somatostatin effect was abolished if the cells were pretreated overnight with 100 ng/ml of pertussis toxin, suggesting a process coupled to a G-protein; b) the phosphatase inhibitor, okadaic acid prevented the stimulation of $K_{(Ca)}$ channels by somatostatin, indicating that dephosphorylation of the channel could be involved; and c) somatostatin could stimulate $K_{(Ca)}$ channel activity previously inhibited by a diffusible analog of cAMP (0.1 mM). These results taken together indicated to the authors that phosphorylation may inhibit $K_{(Ca)}$ channel activity, while dephosphorylation triggered by somatostatin activate $K_{(Ca)}$ channels (Fig. 6).

At present, the role of the G protein in the somatostatin activation of $K_{(Ca)}$ channels is not clear. The authors discard the possibility that a G protein directly activates $K_{(Ca)}$ channels. The evidence given to support this conclusion was: 1) somatostatin applied to the bath could stimulate channel activity in cell attached patches suggesting a second messenger pathway, and 2) the somatostatin effect was not reproduced in cell free patches by addition of 100 μM GTP to the cytoplasmic side of the channel. At present two hypothesis exist, that somatostatin activates a phosphatase responsible for dephosphorylation of $K_{(Ca)}$ channels increasing their activity, or that somatostatin inhibits the PKA responsible for maintaining the channel in a less active phosphorylated state. In both cases, the role of a putative G protein in the inhibition of the somatostatin effect by pertussis toxin remains to be explored.

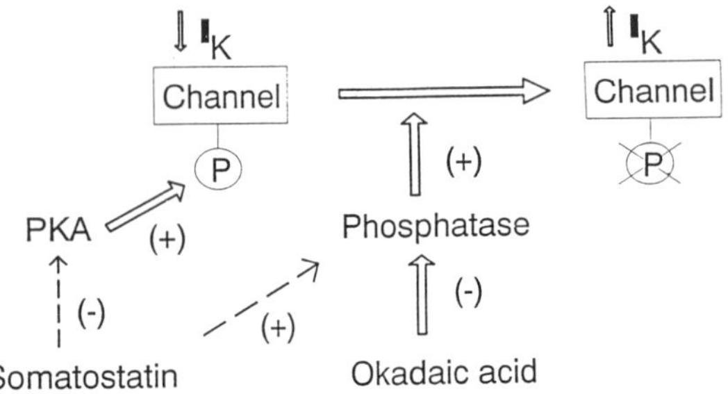

Figure 6. Putative metabolic diagram for the activation of $K_{(Ca)}$ channels from GH_4C_1 cells by somatostatin. The proposed scheme shows that channel activity is increased by somatostatin via dephosphorylation. Somatostatin may inhibit the cAMP-dependent protein kinase (PKA) or may increase phosphatase activity. In both cases dephosphorylation will render the channel more active. Solid arrows show known events while dashed arrows are showing hypothetical steps of $K_{(Ca)}$ channel activation by somatostatin. (-), inhibition; (+), activation; ↑I_K, increase in channel activity; ↓I_K, decrease in channel activity; P, phosphorylated state.

Acetylcholine

Acetylcholine can produce both activation of Ca-activated K currents and inhibition of maxi $K_{(Ca)}$ channels when applied to the extracellular space (see ref. 62, Table 1). Recently, the role of ACh applied to the internal side of the channel was studied in inside-out membrane patches of chromaffin cells (18). The author found that ACh applied to the intracellular side of maxi $K_{(Ca)}$ channels (130-220 pS) reduced both, channel amplitude and their open probability (P_0). These ACh effects were dose-dependent and produced at high neurotransmitter concentrations (> 10 mM). The $K_{(Ca)}$ channel present in chromaffin cells had a high open probability (~0.8) when the internal Ca concentration was 250 nM at potentials from -40 to +40 mV. The reduction of P_0 after ACh addition was essentially voltage independent. ACh shortens the channel open time while prolongs the closed time intervals. On the other hand, the "fast" blockade of ACh which produces the reduction in channel amplitude was slightly voltage dependent with a change in affinity of e-fold per 104 mV. The fraction of the voltage sensed at the ACh blocking site was 0.2, and the K_d at 0 mV was 65 mM. These results suggested to the authors that ACh may act as a fast (accounting for the reduction in channel amplitude) and slow blocker (explaining the reduction in P_0). However, the possibility also exists that the changes in P_0 produced by ACh are due to a change in the gating mechanism of the channel. Since ACh is a charged amine similar to tetraethylammonium, it would be interesting to investigate if they compete for a common binding site.

Bombesin (Bn), Gastrin Releasing Peptide, Neuromedin B, and Bradykinin.

Bombesin, gastrin releasing peptide and neuromedin B belong to a large gene family of mitogenic peptides (59). Bradykinin is a circulating vasoactive peptide which has several functions depending on the target tissue. Its functions include neurotransmitter release, muscle contraction, secretion, and cell growth (6). All these peptides (1 nM to 1 μM) caused an increase in a transient outward K current in fibroblasts from Swiss 3T3 cell lines (32). The K current involved seemed to be Ca-dependent, since inclusion of 5-10 mM EGTA diminished the outward current. In addition these cells did not respond to bombesin in the presence of 30 mM of TEA, or 15 mM Ba^{2+} in the external solution.

Since it is known that bradykinin receptors may be coupled to various G proteins inducing the activation of different membrane proteins like phospholipase C or phospholipase A_2 (3); it would be interesting to test the effects of the metabolites of these cascades on the activity of $K_{(Ca)}$ channels. Similarly, the role of the second messengers derived from the activation of phospholipase C after bombesin stimulation of the complex receptor-Gplc protein has to be addressed. Many questions remain unanswered: What is the single channel nature of the $K_{(Ca)}$ currents stimulated by these peptides? Which is the mechanism of action? Are G-proteins involved? Can any of these peptides directly regulate channel activity? Is phosphorylation or dephosphorylation involved?

Phosphorylation

Neurotransmitters and hormones acting through distinct second-messenger pathways have as common end the alteration of the phosphorylation status of proteins (56). Recent evidence has demonstrated that $K_{(Ca)}$ channel proteins are also targets of this chemical modification, and that phosphorylation/dephosphorylation processes may be responsible for the physiological response induced by neurotransmitters. Examples are adrenergic agents (for review see ref. 62) and somatostatin (this review).

The role of phosphorylation/dephosphorylation on $K_{(Ca)}$ channel activity has been studied by Reinhart et. al. (51) using the bilayer technique. This system has the advantage that the experimental milieu surrounding the channel is better controlled than in patch clamp experiments. The authors study two main classes of maxi $K_{(Ca)}$ channels (Type 1 and 2) from membrane vesicles of rat brain. Besides their differences in gating kinetics they have different charybdotoxin (CTX) sensitivities. Type 1 are CTX sensitive, while Type 2 are CTX insensitive.

Type 1 channels could be subclassified in terms of their phosphorylation properties. The predominant class was activated by phosphorylation, while the other was inhibited. Activation by phosphorylation was achieved after addition of 400 nM of the cAMP-dependent protein kinase catalytic subunit (PK-A), resulting in the channels being open most of the time. This increase in channel activity was reversed by addition of 30 nM of the catalytic subunit of the protein phosphatase 2A (PP-2A). Dephosphorylation by PP-2A was specific since protein phosphatase 1 (PP-1) was ineffective in reversing the effect of PK-A. Similarly, inhibition by phosphorylation in the less frequent class of type 1 channels was observed when PK-A was added to the intracellular side, and this decrease in channel activity could be reversed by PP-2A.

It was also demonstrated that the predominant class of Type 1 channels could be in a phosphorylated state previous to their incorporation into the bilayer, since addition of phosphatase (PP-2A, 30 nM) to a "freshly" incorporated channel could induce inhibition of channel activity.

Type 2 channels (CTX-insensitive), on the other hand, responded to treatment with PK-A with a decrease in single channel open probability due to a large increase in closed times and a decrease in the open times. As expected for a phosphorylation process, channel activity could be restored by dephosphorylating the channel with phosphatase 2A.

It was also observed that activation due to phosphorylation caused an increase in the calcium affinity of the channel, without modifying the Hill coefficient of the process. Since the modification of $K_{(Ca)}$ channel gating by phosphorylation is comparable to the one observed by Ca, the possibility exists that phosphorylation could affect the Ca sensor of $K_{(Ca)}$ channels. Thus, different states of phosphorylation may explain the differences in Ca affinity between channels of the same class.

An interesting observation made by the authors is that phosphorylation of Type 1 channels by PK-A produced large variations of channel activity which in average could be related to an increase in channel activity. The authors postulate that these variations may be due to the existence of a membrane bound phosphatase closely associated with the channel. On the other hand, Type 2 channels (inhibited by exogenous PK-A) may be activated by ATP or ATP-γ-S, suggesting the presence of an endogenous protein kinase different from PKA. Moreover, the increase in the channel open probability produced by ATP is also oscillatory, suggesting again the presence of an endogenous phosphatase in close association with the channel (9). These findings give light to a new form of channel modulation via cycles of phosphorylation/ dephosphorylation and supports the idea that $K_{(Ca)}$ channels may form tight complexes with other molecules in the membrane (9, 63).

Intracellular pH

Several investigations have lead to the conclusion that $K_{(Ca)}$ channels can also be modulated by intracellular pH. This has been observed in several tissues like pancreatic B-cells (11), epithelia (8, 12), smooth muscle (30), skeletal muscle (34), and neurons (50). The majority of these studies have been performed looking at single channel activity of maxi $K_{(Ca)}$ channels using the patch clamp technique. In general, it has been observed that acidification reduces channel activity. This reduction in channel activity resulted in a parallel shift of the voltage activation curves without any change in the slope factor. On the other hand, the calcium activation curves indicate that lowering pH makes the binding affinity for Ca to diminish (curves shift towards higher Ca concentrations) and the apparent cooperativity of the Ca binding to decrease (Hill coefficient diminishes).

The physiological role of the pH-induced changes on $K_{(Ca)}$ channel gating has to be determined. However, it is important to stress the possibility that not only $K_{(Ca)}$ channel proteins may be directly modified by the $[H^+]$ concentration of the intracellular milieu, but $[H^+]$ may also modify $K_{(Ca)}$ channel interaction with other molecule(s) involved in the physiological response to external (neurotransmitters, hormones, etc.), and internal (second messengers) signals.

CONCLUSIONS

There is general consensus that $K_{(Ca)}$ channels from coronary smooth muscle and other tissues are heavily regulated not only by variations in internal Ca, but by many other agents. In most cases the mechanisms of action are not yet clarified due to the complexity of the cellular machinery. Moreover, the apparent close association of other membrane proteins to the channel makes difficult to precisely determine a mechanism of action in simpler experimental systems (reconstitution in bilayers, excised patches). However, in all studied cases including phosphorylation the up- or down-regulation is related to a change in the channel affinity for Ca. How the Ca-binding site of the channel is modified to change its affinity for Ca, may be answered only with structure-function relationship studies. In this respect, recent molecular biology studies indicate that $K_{(Ca)}$ channels may possess several regulatory regions including a Ca-binding domain and phosphorylation sites (1). Unfortunately, the functional expression of the cloned channel has not been successful preventing the answer to basic questions such as if the channel has independent phosphorylation and Ca domains.

What is clear is that K(Ca) channels are ubiquitously distributed, are abundant and diverse in coronary smooth muscle and other tissues, and are targets of many regulatory factors. These properties assure their contribution to the regulation of primary functions such as contraction and relaxation.

ACKNOWLEDGMENTS

Supported by Grant-in-Aid 900963 from the American Heart Association, National Center (L. Toro) and grants HD-25616 and H137044 from NIH (E. Stefani).

REFERENCES

1. Atkinson, N.S., G.A. Robertson, and B. Ganetzky, Science 253, 551-555 (1991).

2. Benham, C.D. and T.B. Bolton, J. Physiol. 381, 385-406 (1986).

3. Birnbaumer, L., J. Abramowitz, A. Yatani, K. Okabe, R. Mattera, R. Graf, J. Sanford, J. Codina, and A.M. Brown, Critical Rev. Biochem. Molec. Biol. 25, 225-244 (1990).

4. Bkaily, G.,M. Peyrow, A. Sculptoreanu, D. Jacques, M. Chahine, D. Regoli and N. Sperelakis, Pflugers Arch. 412, 448-450 (1988).

5. Bolotina, V., V. Omelyanenko, V. Heyes, U. Ryan, and P. Bregestovski, Pflugers Arch. 415, 262-268 (1989).

6. Burch, R.M., S.G. Farmer, and L.R. Steranka, Med. Res. Rev. 10, 237-269 (1990).

7. Carl, A., N. G. McHale, N.G. Publicover, and K.M. Sanders, J. Physiol. 429, 205-221 (1990).

8. Christensen, O. and T. Zeuthen, Pflugers Arch. 408, 249-259 (1987).

9. Chung, S., P.H. Reinhart, B.L. Martin, D.Brautigan, and I.B. Levitan, Science 253, 260-262 (1991).

10. Cole, W.C., A. Carl, and K.M. Sanders, Am. J. Physiol, 257, C481-C487 (1989).

11. Cook, D.L., M. Ikeuchi, and W.I., Fujimoto, Nature 311, 269-271 (1984).

12. Copello, J., Y. Segal, and L. Reuss, J. Physiol., 434, 577-590 (1991).

13. De Nucci, G., R. Thomas, P. D'Orleans, E. Antunes, C. Walder, T.D. Warner, and J.R. Vane, Proc. Natl. Acad. Sci. 85, 9797-9800 (1988).

14. Desilets, M., S.P. Driska, and C.M. Baumgarten, Clr. Res. 65, 708-722 (1989).

15. Folta, A., I.G. Joshua, and R.C. Webb, Life Sci. 45, 2627-2635 (1989).

16. Fujino, K., S. Nakaya, T. Wakatsuki, Y. Miyoshi, Y. Nakaya, H. Morin and I. Inoue, J. Pharmacol. Exper. Ther. 256, 371-377 (1991).

17. Ganittrevich V. and G. Isenberg, Circ. Res. 67, 525-528 (1990).

18. Glavinovic, M.I., Neuroscience 39, 815-822 (1990).

19. Goto, K., Y. Kasuya, N. Matsuki, Y. Takuwa, H. Kurihara, T. Ishikawa, S. Kimura, M. Yanagisawa, and T. Masaki, Proc. Natl. Acad. Sci. USA, 86, 3915-3918 (1989).

20. Griendling, K.K., T. Tsuda, B.C. Berk, and R.W. Alexander, J. Cardiovasc. Pharmacol. 14(Suppl.6), S27-S33 (1990).

21. Guggino, E.S., B.A. Suarez-Isla, W.B. Guggino, and B. Sacktor, Am. J. Physiol. 249, F448-F455 (1985).

22. Haas, H.L. and A. Konnerth, Nature 302, 432-434 (1983).

23. Higashida, H. and D.A. Brown, Nature, 323, 333-335 (1986).

24. Hu, S., H. S. Kim, and A.Y. Jeng, Eur. J. Pharmacol. 194, 31-36 (1991).

25. Johns D.W. and N. Sperelakis, Eur. J. Pharmacol. 187, 183-191 (1990).

26. Katz, G., L. Roy-Contancin, T. Bale, and J.P. Reuben, Biophys. J. 57, 506a (1990).

27. Klockner, U. and G. Isenberg, Pflugers Arch. 418, 168-175 (1991).

28. Kodama, M., H. Kanaide, S. Abe, K. Hirano, H. Kai, and M. Nakamura, Biochem. Biophys. Res. Commun. 160, 1302-1308 (1989).

29. Kume, H., A. Takai, H. Tokumo, and T. Tomita, Nature 341, 152-154 (1989).

30. Kume, H., K. Takagi, T. Satake, H. Tokuno, and T. Tomita, J. Physiol. 424, 445-457 (1990).

31. Kurachi, Y., H. Ito, T. Sugimoto, T. Shimizu, I. Miki, and M. Ui, Nature 337, 555-557 (1989).

32. Kusano, K. and H. Gainer, Am. J. Physiol. 260, C701-C707 (1991).

33. Latorre, R., A. Oberhauser, P. Labarca, and O. Alvarez, Annu. Rev. Physiol. 51, 385-399 (1989).

34. Laurido, C., D. Wolff, and R. Latorre, Biophys. J. 57, 507a, (1990).

35. Lincoln, T.M. and T.L. Cornwell, Blood Vessels 28, 129-137 (1991).

36. Madison, D.V. and R.A. Nicoll, Nature 299, 636-638 (1982).

37. Mason, W.T. and D.W. Waring, Neuroendocrinology 43, 205-219 (1986).

38. Mayer, E.A., D.D.F. Loo, A. Kodner, and S. Narashima Reddy, Am. J. Physiol. 257, G887-G897 (1989).

39. Mayer, E.A., D.D.F. Loo, W.J. Snape Jr., and G. Sachs, J. Physiol., 420, 47-71 (1990).

40. Molleman, A., B. Hoiting, M. Duin, J. van den Akker, A. Nelemans, and A. Den Hertog, J. Biol. Chem. 266, 5658-5663 (1991).

41. Morris, A.P., D.V. Gallacher, R.F. Irvine, and O.H. Petersen, Nature 330, 653-655 (1987).

42. Myers, P.R., R. Guerra Jr., and D.G. Harrison, Am. J. Physiol., 256, H1030-H1037 (1989).

43. Nakahata N., I. Matsuoka, T. Suzuki, and H. Nakanishi, Eur. J. Pharmacol. 162, 407-417 (1989).

44. Neely, A. and C.J. Lingle, Biophys. J. 57, 110a (1990).

45. Neliat, G., F. Masson, and Y.M. Gargouil, Pflugers Arch. 414, S186-S187 (1989).

46. Ohya, Y. and N. Sperelakis, Circ. Res., 68, 763-771 (1991).

47. Ordway, R.W., L.H. Clapp, A.M Gurney, J.J Singer, and J.V. Walsh Jr., J. Gen. Physiol. 94, 37a (1989).

48. Palmer, R.M.J., A.G. Ferrige, and S. Moncada, Nature 327, 524-526 (1987).

49. Patscheke, H., Blut. 60, 261- 268 (1990)

50. Peers, C. and F.K. Green, J. Physiol. 437, 589-602 (1991).

51. Reinhart, P.H., S. Chung, B.L. Martin, D.L. Brautigan, and I. Levitan, J. Neurosc. 11, 1627-1635 (1991).

52. Sadoshima, J.I., N. Akaike, H. Kanaide and M. Nakamura, Am. J. Physiol., 255, H754-H759 (1988).

53. Sauve, R., C. Simoneau, L. Parent, R. Monette, and G. Roy, J. Membr. Biol. 96, 119-208 (1987).

54. Schrader, B.J. and S.I. Berk, Clinical Pharmacy 9, 118-124 (1990).

55. Scornik, F.S. and L. Toro, Am. J. Physiol. (submitted) (1991).

56. Shenolikar, S., FASEB J. 2, 2753-2764 (1988).

57. Sikdar, S.K., R.P. McIntosh, and W.T. Mason. Brain Res. 496, 113-123 (1989).

58. Silberberg, S. and C. van Breemen, BBRC 172, 517-522 (1990).

59. Spindel, E., TINS 9, 130-333 (1986).

60. Tare., M., H.C. Parkington, H.A. Coleman, T.O. Neild, and G.J. Dusting, Nature 346, 69-71 (1990).

61. Toro L., S.D. Silberberg, A.M. Brown, and E. Stefani, Biophys. J. 55, 543a (1989).

62. Toro, L. and E. Stefani, J., Bioenerg. Biomemb. 23, 561-576 (1991).

63. Toro, L., J. Ramos-Franco, and E. Stefani, J. Physiol. 96, 373-394 (1990).

64. Toro, L., L. Vaca, and E. Stefani, Am. J. Physiol. 260, H1779-H1789 (1991).

65. Toro, L., M. Amador, and E. Stefani, Am. J. Physiol. 258, H912-H915 (1990).

66. Trieschmann, U. and G. Isenberg, Pflugers Arch. 414, S183-S184 (1989).

67. Van Renterghem, C.,P. Vigne, J. Barhaini, A. Schmid-Alliana, C. Frelin, and M. Lazdunski, Biochem. Biohys. Res. Commun. 157, 977-985 (1988).

68. Volk, K.A., J.J. Matsuda, and E.F. Shibata, J. Physiol. 439, 751-768 (1991).

69. Walsh, J.R. and J.J. Singer, Cell Calcium 4, 321-330 (1983).

70. Weigel, R.J., J.A. Connor, and C.L. Prosser, Am. J. Physiol. 237, C247-C256 (1979).

71. White, R.E., A. Schonbrunn, and D.L. Armstrong, Nature 351, 570-573 (1991).

72. Wilde, D.W. and K.S. Lee, Circ. Res. 65, 1718-1734 (1989).

73. Williams, Jr., D.L., G.M. Katz, L. Roy-Contancin, and J.P. Reuben, Proc. Natl. Acad. Sci. 85, 9360-9364 (1988).

74. Wong, P.C., S.D. Hart, A.M. Zaspel, A.T. Chiu, R.J. Ardecky, R.D. Smith, and P.B. Timmermans, J. Pharmacol. Exper. Ther. 255, 584- 592 (1990).

75. Yanagisawa, M., H. Kurihara, S. Kimura, Y. Tomobe, M. Kobayashi, Y. Mitsui, Y. Yazaki, K. Goto, and T. Masaki, Nature 3323, 411-415 (1988).

76. Yoshida, S., S. Plant, A.I. McNiven, and C.R. House, Pflugers Arch. 415, 516-518 (1990).

EFFECT OF CROMAKALIM AND LEMAKALIM ON WHOLE-CELL AND SINGLE-CHANNEL K^+ CURRENTS IN CANINE COLONIC, RENAL AND CORONARY SMOOTH MUSCLE CELLS

CRAIG H. GELBAND, ANDREAS CARL, JOSEPH M. POST, SUSAN M. BOWEN, TOMOHISA ISHIKAWA, KATHLEEN D. KEEF, KENTON M. SANDERS, and JOSEPH R. HUME

Department of Physiology, University of Nevada School of Medicine, Reno, NV 89557

ABSTRACT

Although cromakalim has been demonstrated to activate an ATP-sensitive K^+ channel in pancreatic β-cells, cardiac and skeletal muscle, it's mechanism of action in smooth muscle is controversial. ATP-sensitive K^+ channels, large conductance Ca^{2+}-activated K^+ channels, delayed rectifier K^+ channels, and small conductance Ca^{2+}-activated K^+ channels have all been demonstrated to be modulated by potassium channel openers in smooth muscle. In colonic smooth muscle under whole-cell voltage clamp cromakalim (20 μM), but not lemakalim (the (-) optical isomer of cromakalim, 20 or 100 μM) inhibited L-type Ca^{2+} current. In colonic, coronary and renal artery isolated smooth muscle cells, cromakalim (20 μM) and lemakalim (10-2O μM) increased the amplitude of a time-independent (quasi-instantaneous) component of whole cell K^+ current. In single channel experiments on colonic smooth muscle cells, lemakalim (2O μM) increased the open probability of a 265 pS Ca^{2+}-activated K^+ channel by a factor of 4.3 in inside-out patches and by 3.2 in cell-attached patches. In renal artery smooth muscle cells, lemakalim (1 μm) also increased NP_O of a 217 pS voltage dependent K^+ channel. When TEA (200 μM) was added to the cell-attached patch pipette solution, the actions of lemakalim were prevented. We conclude that cromakalim and lemakalim activate large conductance, voltage-dependent $Ca^{2+}-$activated K^+ channels in colonic and renal artery smooth muscle cells. We cannot conclude that large conductance $Ca^{2+}-$activated K^+ channels are solely responsible for the hyperpolarization and dilation of smooth muscle cells. It may be that there are other types of K^+ conductances with a sparser distribution that may also be modulated by potassium channel openers.

INTRODUCTION

Recently, a new class of vasodilators (potassium channel openers, PCO'S) which act by opening membrane K^+ channels have been described (14, 24). Cromakalim and its active isomer, lemakalim, have been reported to cause membrane hyperpolarization, lower arterial blood pressure, enhance the efflux of tracer $^{86}Rb^+$- or $^{42}K^+$ and relax precontracted vascular smooth muscle (13, 15, 23, 25, 38, 48). Hypoglycemic sulfonylureas, like tolbutamide and glibenclamide, that block ATP-sensitive K^+ channels (K_{ATP}) in pancreatic β-cells (38) and cardiac (43) and skeletal (41) muscle, have been demonstrated to antagonize the above effects of cromakalim in smooth muscle with a K_i of 0.3-0.5 μM suggesting the existence of K_{ATP} channels in smooth muscle (7, 11, 16, 40, 49). However, the pharmacological target site for the K^+ channel openers remains controversial. Gelband *et al* (20) and Kluckner *et al* (29) have demonstrated that cromakalim activates large conductance Ca^{2+}-activated K^+ (BK) channels in vascular smooth muscle. Standen *et al.* (45) have isolated a K_{ATP} channel from rabbit mesenteric artery that is activated by cromakalim and blocked by glibenclamide. Beech and Bolton (1) have demonstrated that cromakalim stimulates a voltage-dependent, but calcium insensitive K^+ current (I_{CKM}) in rabbit portal vein that resembles the delayed rectifier K^+ current. Finally, Okabe *et al* (36) and Kitamura and Kuriyama (28) have demonstrated that cromakalim increased the probability of opening of a small conductance, Ca^{2+} and ATPsensitive K^+ channel.

The type of potassium channel most often observed in membranes of a number of cell types including smooth muscle cells is the BK channel (5, 37). In smooth muscle, for example, this channel has been studied in rabbit aorta (20), ear artery (3) mesenteric artery (4), jejunum (4) and portal vein (27, 36); canine trachea (33) and colon (8) and in toad stomach (44). The probability of opening, P(open), of these K^+ channels is increased by membrane depolarization or by an increase in $[Ca^{2+}]i$. It has a single channel conductance of 200-300 pS in symmetrical K^+ solutions, but in spite of this high conductance, it possesses a high selectivity for K^+ over other monovalent cations. Tetraethylammonium (27) and the scorpion toxin, charybdotoxin (21), characteristically block BK channels from the extracellular surface of the channels with high affinity, while Ba^{2+} blocks from the intracellular surface of the cell (3, 35). BK channels have been proposed to play a role in the repolarization of smooth muscle membranes during an action potential or other depolarizing stimuli and in slow wave termination of visceral smooth muscle. Due to the abundance of BK channels in smooth muscle and its potential importance in regulating both membrane potential and vasorelaxation under physiological and pathophysiological conditions, we performed experiments to further elucidate the actions of cromakalim and lemakalim on whole-cell currents and single K^+ channels in isolated canine colon, renal, and coronary smooth muscle cells.

METHODS
Electrophysiological Measurements

Single smooth muscle cells from the circular layer of the canine proximal colon, renal and coronary artery were prepared using an enzymatic procedure previously described (30). Ionic currents were recorded using the whole-cell, cell-attached or inside-out configuration of the standard patch-clamp technique (22). High resistance seals (>5 GΩ) were formed using patch pipettes of borosilicate glass (3-10 MΩ). Whole-cell currents were elicited by voltage-step (10 mV) or ramp depolarizations (-80 to 80 mV over 4 seconds) were recorded with standard amplifiers (AXOPATCH-lD, Axon Instruments (renal and coronary cells) or a Dagan 8900, Dagan Instruments (colon cells). Data were digitized on-line and stored on an IBM compatible AT computer. Whole-cell currents were filtered at 1 (colon) or 5 (renal) kHz and digitized at a sampling rate of 0.5 kHz. Data analysis for the whole-cell measurements was performed using PCLAMP software (Version 5.5, Axon Instruments). Experiments were performed at 22-25° C (renal) or 35° C (colon and coronary).

Single channel currents from cell-attached patches were recorded with a EPC-7 amplifier (List Electronics). Data were filtered and digitized at 1 kHz and stored on an IBM-AT computer. Data analysis was performed using pCLAMP 5.5 software and a program generously provided by Dr. Mark T. Nelson. Single channel currents measured from inside-out patches were recorded using an Axopatch-lD amplifier and stored on video tape with a PCM-1 digital VCR-instrumentation recorder adaptor (Medical Systems). The data were replayed through an 8-pole Bessel filter (Ithako) at 4 kHz and digitized with pClamp Software (Version 5.5, Axon Instruments) at a sampling rate of 10 kHz. Unitary amplitudes were measured with an amplitude-window detector (IPROC software, Axon Instruments). The current relative to baseline current (leakage) was averaged over time using IPROC and divided by the single channel amplitude to determine NP. (N: Number of channels per patch, P_O: open probability). In some experiments voltage ramp protocols were applied using an IBM compatible AT clone, pClamp software and a digital analog interface (Tecmar). The patch potential was changed over 4 sec in decrements of 0.24 mV starting at positive potentials. During each ramp 2000 data points were sampled at a frequency of 500 Hz. Each ramp was repeated 5-15 times. All experiments were performed at 22-25° C.

Solutions

Whole-cell configuration. Pipette solution for K^+ currents in colon cells (in mM): K gluconate, 110; KCl, 20; $MgCl_2$, 0.5; K_2ATP, 5; phosphocreatine, 5; EGTA, 0.1; and HEPES, 5 (pH 7.2, KOH). Bath solution for colon cells contained (in mM): NaCl, 125; KCl, 5.4; $NaHCO_3$, 4.2; Na_2HPO_4, 0.4; KH_2PO_4, 0.4; $CaCl_2$ 1.8; $MgCl_2$ 0.5; $MgSO_4$, 0.4; D-glucose, 10; sucrose, 2.9; HEPES, 10 (pH 7.4, NaOH). For Ca^{2+} current measurements, the pipette solution was (in mM): Cs-aspartate, 110; $CsCl_2$, 20; Na_2ATP, 4; NaGTP, 1; phosphocreatine, 2; $MgCl_2$ 1; EGTA, 1; HEPES, 5. **The** bath solution used for studying Ca^{2+} currents contained (in mM): TEA-CL, 40; NaCl, 90; $CaCl_2$, 2.5; KCl, 5.4; $MgCl_2$ 0.5; $NaHCO_3$, 4.2; Na_2HPO_4, 0.25; $MgSO_4$, 0.4; D-glucose, 10; sucrose, 2.9; HEPES, 10 (pH

7.4, NaOH). Pipette solution for renal and coronary cells (in mM): K gluconate, 120; KCl, 20; $MgCl_2$ 0.5; EGTA, 5 (renal), 0.1 (coronary); Hepes, 5; (pH 7.2, KOH). Bath solution for renal and coronary cells (in mM): NaCl, 130; $NaHCO_3$, 10; KCl, 4.2; KH_2PO_4, 1.2; $MgCl_2$, 0.5; $CaCl_2$, 1.5; D-glucose, 5.5; Hepes, 10; (pH 7.4, NaOH).

Cell-attached configuration. Pipette and bath solutions for colon cells each contained (in mM): KCl, 145; $MgCl_2$, 2.3; D-glucose, 5.5; EGTA, 0.1; and HEPES, 5.0 (pH 7.4, KOH). For renal artery cells the pipette and bath solutions contained (in mM): K gluconate, 140; KCl, 10; $MgCl_2$, 0.5; $CaCl_2$ 1.5; D-glucose, 5; Hepes, 5; (pH 7.2, KOH).

Inside-out configuration. Pipette solution (in mM): KCl, 140; Ca, 2.0; D-glucose, 10; EGTA, 1; HEPES, 10 (pH 7.4, KOH). Bath solution (in mM) : KCl, 140; Ca^{2+}, 0.61 to 0.941; Dglucose, 10; EGTA, 1; HEPES, 10 (pH 7.4, KOH). The concentration of Ca^{2+} in bath and pipette solutions was buffered by EGTA to obtain activities between 10^{-7} and 10^{-6} M. Association constants for H^+, K^+, and Ca^{2+} binding to EGTA were taken from Fabiato and Fabiato (18) and Fabiato (17) and calculated with a program developed by Chi-Ming Hai (U. of VA., Charlottesville). Ca^{2+} concentrations of 0.61 and 0.941 mM were added to solutions containing 1 mM EGTA to obtain Ca^{2+} activities of 10^{-7} and 10^{-6} M, respectively.

Drugs

Cromakalim (BRL 34915) and lemakalim (BRL 38227) were generously provided by SmithKline Beecham Laboratories and glibenclamide was obtained from Sigma. 20 mM stock solutions were made in dimethyl sulfoxide (DMSO) and stored at 40° C. 0.2% DMSO in 140 mM KCl solution had no effect on channel activity in inside-out patches (n = 5). The final concentration of DMSO in the bath solution was 0.2% or below.

RESULTS

In intact canine colonic smooth muscle, Post *et al* (39) demonstrated that cromakalim (10 μM) caused a pronounced hyperpolarization of the resting membrane potential from an average of -82 $\pm$ 1.8 mV to -89 $\pm$ 3 mV (n = 6) which was accompanied by a reduction in both the amplitude (51 $\pm$ 4 to 47 $\pm$ 7 mV) and in frequency (4.2 $\pm$ 0.3 to 2.9 $\pm$ 0.06 cycles per min) of slow waves. Cromakalim reduced the upstroke velocity by 50 $\pm$ 6% (n = 6) and abolished the plateau phase of the slow waves. Lemakalim (1 μM) produced similar effects on the slow wave as cromakalim. The effects of both drugs were reversible in intact tissue; however, washout required 10-15 min. to return to control levels. Pretreatment of the tissue with glibenclamide (10 μM) prevented the cromakalim (10 μM) induced hyperpolarization and decrease in duration and upstroke velocity of the slow waves. In addition, glibenclamide added after cromakalim, reversed the effects of cromakalim on slow wave activity. The hyperpolarization caused by cromakalim and lemakalim suggest that these drugs may increase K^+ conductance. However, inhibition of the plateau phase could result from either a decrease in L-type Ca^{2+} current or an increase in the K^+ conductance regulating repolarization (42). In order to determine the mechanism(s) responsible for the effects of cromakalim and lemakalim, whole-cell patch clamp experiments were performed to characterize the effects of these drugs on Ca^{2+} and K^+ currents in canine colon.

Inward currents in canine colon are greatly reduced by the addition of nisoldipine (1 μM) to the external solution. In addition, since the threshold for activation is positive to -40 mV, and its inactivation kinetics are slow, these Ca^{2+} currents appear to result primarily from activation of L-type Ca^{2+} channels. Cromakalim (20 μM) inhibited the peak whole-cell inward Ca^{2+} current by an average of 60.75 $\pm$ 11.9% (Fig. *1A*, n = 8). At all membrane potentials investigated, the inhibitory effect was not associated with a shift in the voltage dependent activation of Ca^{2+} current as compared to control (Fig. 1A bottom panel). In cells previously incubated in glibenclamide (10 μM), cromakalim inhibited the peak whole-cell Ca^{2+} current by 66.4 $\pm$ 7.1% (n = 3) (39).

128

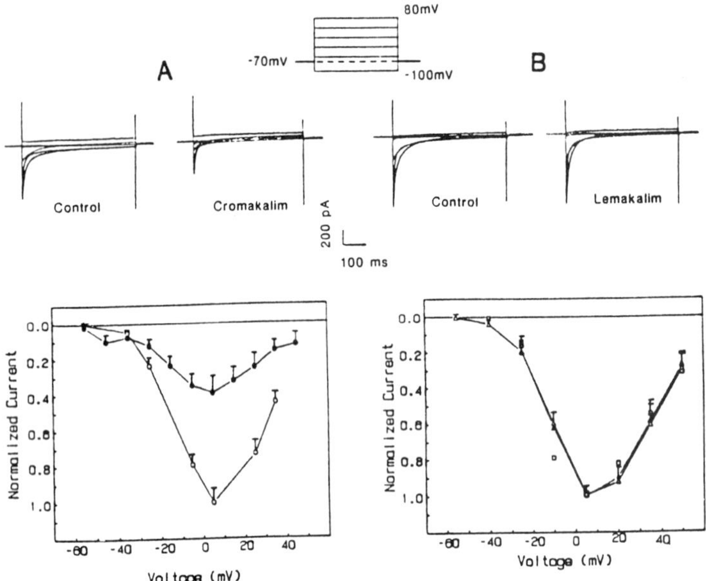

Figure 1. Effect of cromakalim (20 μM; Panel A) and lemakalim (20 μM; Panel B) on Ca^{2+}, currents in colon cells. **A:** Ca^{2+} currents were normalized with respect to the peak inward current in control conditions [absence (0) and presence (●) of cromakalim (n = 8)]. **B:** Normalized current-voltage relationships in the absence (▲, ■) and presence of 20 (Δ; n = 4) and 100 μM (□; n = 1) lemakalim. (Reproduced with permission from ref. 39).

Lemakalim (20 or 100 μM), in contrast to cromakalim, did not affect inward Ca^{2+} current (Fig. 1B, n = 4). Inhibition of Ca^{2+} channels by cromakalim is consistent with the decrease in slow-wave duration observed in the whole tissue (42); however, Ca^{2+} channel inhibition does not account for the cromakalim induced hyperpolarization. Also, the absence of an effect on the Ca^{2+} current by lemakalim suggests that the lemakalim induced hyperpolarization, decreases in slow-wave duration and upstroke velocity occurs via different mechanisms. Therefore, studies were conducted to evaluate the effect of cromakalim and lemakalim on the whole-cell K^+ current in canine colon, renal, and coronary smooth muscle cells.

Numerous studies have described the properties of macroscopic K^+ currents in a variety of smooth muscle cells (34, 46). Using the whole-cell voltage clamp technique, the outward current most often described is a Ca^{2+}-, time-, and voltage-dependent K^+ current of large conductance. Step depolarization activates this current, while, extracellular TEA, EGTA or Mn^{2+} inhibit it. A majority of the total outward K^+ current of smooth muscle cells is believed to be carried by Ca^{2+}-activated BK channels. Another type of macroscopic K^+ current is a time-independent (quasi-instantaneous) current, which can be resolved after settling of capacitative transients in smooth muscle cells (26). This current has been suggested to regulate the resting membrane potential and has been shown to be modulated by the K^+ channel agonists, minoxidil sulfate (32) and cromakalim (1) in rabbit portal vein cells. The type of single channel underlying this current is still unknown. Finally, in some vascular smooth muscle cells a delayed rectifier K^+ current has been described. This current has been best characterized in the rabbit portal vein (1, 32). Activation of this K^+ current occurs at potentials positive to -40 mV, rectifies in the outward direction, is insensitive to Ca^{2+}, and exhibits slow inactivation. This current has not been well characterized in other types of smooth muscle. A 50 pS channel described in rabbit jejunum (2) may underlie the delayed rectifier current. This channel shows voltage dependent activation, is insensitive to Ca^{2+} and does not inactivate.

We investigated the effects of cromakalim and lemakalim on macroscopic K^+ currents in smooth muscle cells isolated from canine colon, renal and coronary artery. In colon cells, cromakalim (20 μM) was found to increase both the time-independent (quasi-instantaneous) and sustained components of outward current in 3 of 4 cells (39). Lemakalim (20 μM) increased the former component of K^+ current in 6 of 7 cells by an average of -43.0 $\pm$ 14.3 pA at -100 mV (Fig. 2, see arrows). When the holding potential was set near the resting membrane potential of the colon cells, both cromakalim and lemakalim induced a small outward shift in holding current, consistent with hyperpolarization of the cell's resting membrane potential. Pretreatment of cells with glibenclamide (10 lM) reduced the effect of lemakalim such that the lemakalim-stimulated increase in the time-independent component of current was only -6.3 $\pm$ 6.9 pA at -100 mV.

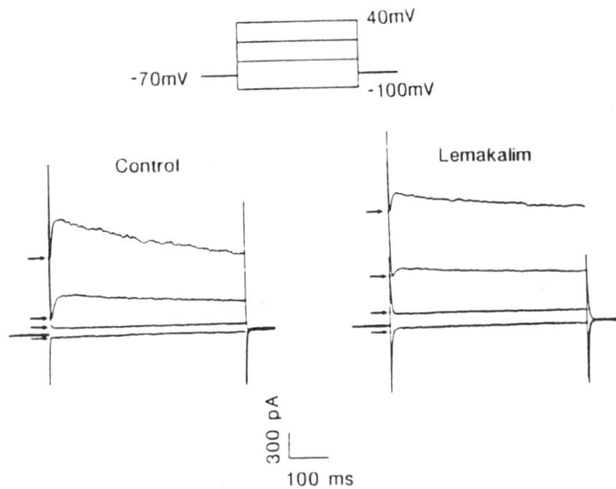

Figure 2. Effect of lemakalim on outward K^+ current in colon cells. Currents were evoked by test pulses from a holding potential of -70 mV to the levels indicated. Time-independent component of K^+ current was evident following settling of capacitative transients (arrows) during control and 5 minutes after addition of lemakalim (20 μM; reproduced with permission from ref. 39).

In 6 of 6 renal artery cells, lemakalim (10 μM) also increased both the time-independent (quasi-instantaneous) and sustained components of outward current (Fig. 3, see arrows at +20 and +40 mV). When voltage ramps were applied to renal artery cells (Fig. 4), two components of current were usually evident. One activated between -40 and -30 mV and was small in amplitude, while the second component was larger in amplitude and activated positive to +10 mV. Lemakalim caused a shift of the current-voltage relationship in the hyperpolarizing direction and increased the amount of current recorded at each membrane potential in 6 of 7 cells. This is consistent with lemakalim shifting the activation range of outward current or activating a third component of outward current. If the former was true then lemakalim increased the amount of current generated by the cell at each membrane potential by increasing the probability of K^+ channel opening or shifting its activation properties. Thus during voltage-step or ramp depolarizations of colon and renal artery cells, lemakalim enhanced mainly a time-independent outward current while currents with a slow (> 100 msec.), time-dependent onset, like the delayed rectifier current, were not affected. Qualitatively similar effects of lemakalim on whole-cell currents were also observed in coronary arterial cells (Fig. 5).

130

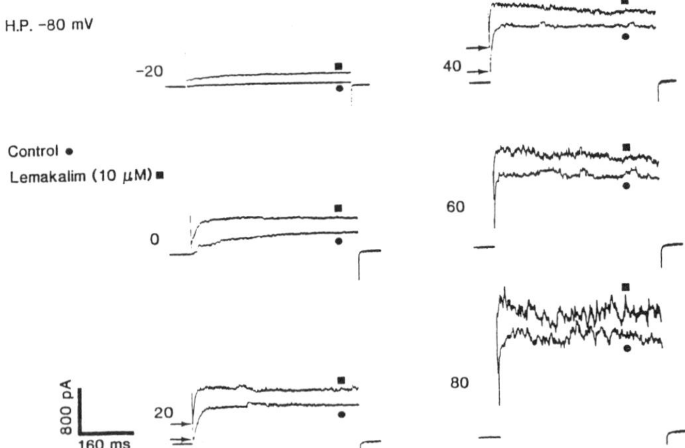

Figure 3. Effect of lemakalim on whole-cell K^+ currents elicited by voltage step depolarizations in renal artery cells. Lemakalim (10 μM, ■) increased outward K^+ current at potentials greater than -20 mV. An increase was observed in both the time-independent (quasi-instantaneous, see arrows at +20 an +40 mV) and sustained components of current at each test potential.

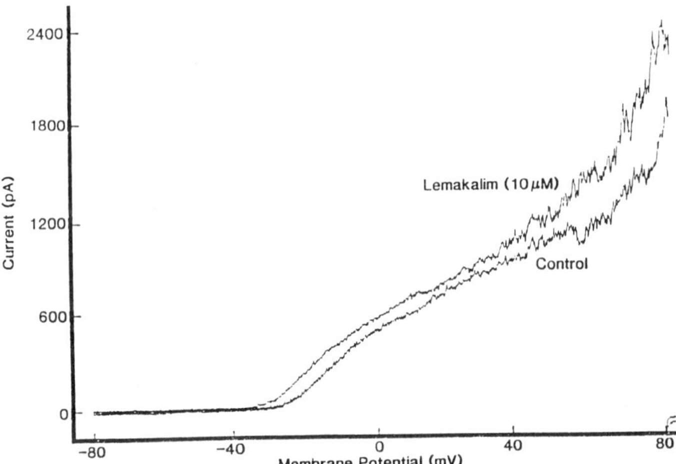

Figure 4. Effect of lemakalim on whole-cell K^+ currents elicited by ramps of voltage in renal artery cells. Lemakalim (10 μM) shifted the current-voltage relationship in the hyperpolarizing direction. Both the small and large components of K^+ current are potentiated.

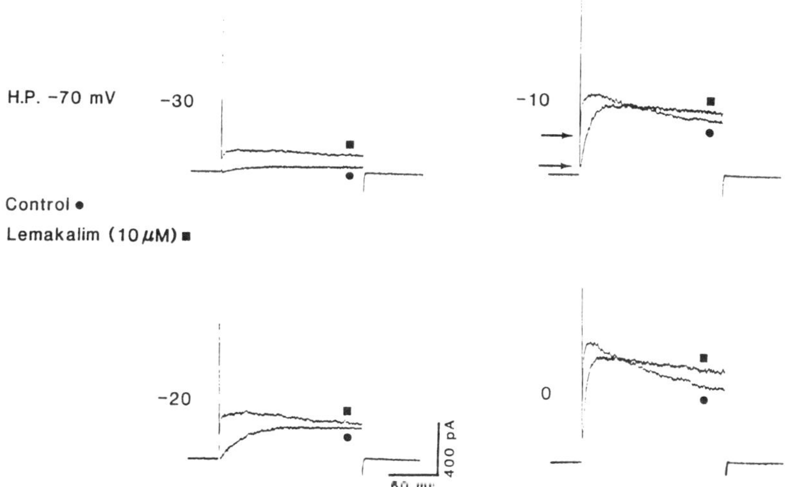

Figure 5. Effect of lemakalim on macroscopic K^+ currents elicited by voltage step depolarizations in a cell isolated from coronary artery. Lemakalim (10 μM, ■) increased outward K^+ current at potentials greater than -40 mV. An increase was observed in the time-independent (quasi-instantaneous, see arrows at -10 mV) current.

In both colon and renal artery cells, the predominant type of single channel recorded in cell-attached or excised inside-out patches is the BK channel. We tested, therefore, whether drugs such as cromakalim and lemakalim might modulate the activity of BK channels. In inside-out patches of colon cells, cromakalim increased probability of channel opening. Figure 6A illustrates that under control conditions (10^{-7} M Ca^{2+}, +50 mV), simultaneous openings of up to two channels were mainly evident (NP_O = 0.17). Upon application of cromakalim (10 μM) to the bath solution, single channel activity increased (NP_O = 0.45). It is clear that three channels are active and a fourth level of single channel opening is evident. In other experiments, the effects of lemakalim on BK channels in insideout patches of colonic cells were reversed by glibenclamide. At a holding potential of -50 mV and a $[Ca^{2+}]$ of 10^{-6} M, lemakalim (10 μM) increased probability of channel opening by 175%. Application of glibenclamide (10 μM) reversed the increase in probability of channel opening to below control levels. Removal of the glibenclamide caused a rapid increase in probability of channel opening by 162% of control. Finally washout of the lemakalim reduced probability of channel opening to control levels. Lemakalim had similar results and increased NP_O by a factor of 4.3 ± 1.3 (n = 6) (10).

In a similar set of experiments, lemakalim (20 μM) increased BK channel activity in cell-attached patches of colonic smooth muscle at a patch potential of +50 mV. Probability of channel opening increased from 0.080 ± 0.016 to 0.26 ± 0.063 (p < 0.05, n = 6) in the presence of lemakalim (10). When the cells were bathed with both lemakalim (20 μM) and glibenclamide (20 μM), there was no increase in probability of channel opening from control (control, 0.067 ± 0.021; lemakalim + glibenclamide, 0.017 ± 0.032, n = 3) (10). The time course of this effect was very slow taking several minutes for the onset to occur suggesting that lemakalim must first cross the cell membrane and then bind to an internal site near the channel pore. The average increase in channel activity by lemakalim was 3.2 ± 1.2 fold (n = 9) (10). Glibenclamide by itself did not significantly alter NP. in cell-attached patches (control, 0. 13 ± 0. 1; glibenclamide, 0.20 ± 0.08, n = 3) (45). Since BK channels are notorious for spontaneous short term and long term changes in open probability, control experiments were performed to test the time course of NP. in cell-attached patches over 60 min. In standard bath solution channel activity remained fairly stable over 60 min. and, if at all, slightly declined (n = 2) (10).

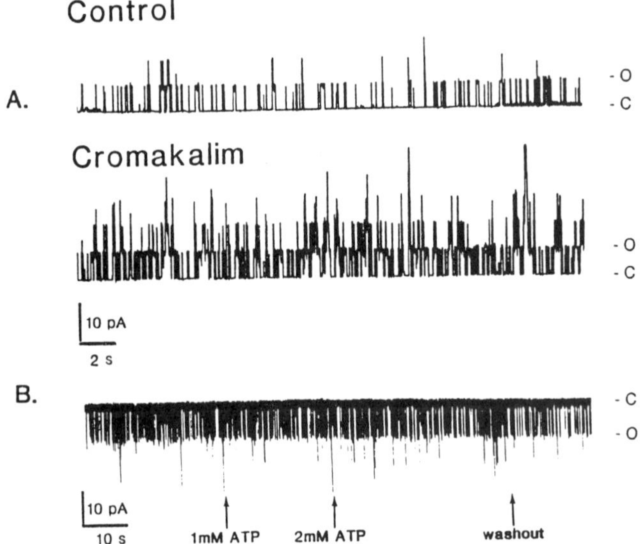

Figure 6. Effect of cromakalim and ATP on colonic BK channels. Panel A illustrates application of cromakalim (10 μM) to the bath solution resulted in activation of BK channels in an inside-out patch (V_H = +50 mV). C and 0 represent the closed and open states of the channel. Panel B illustrates a continuous recording of BK channel activity from an inside-out patch held at -50 mV with a bath Ca^{2+} of 10^{-7} M. Application of 1 and 2 mM ATP to the bath solution did not affect open probability. All single channel openings in Panel B are downward deflections of current.

Since Standen *et al* (45) reported that a large conductance K^+ channel was sensitive to both intracellular ATP and cromakalim, experiments were performed to test whether BK channels from canine colon are modulated by intracellular ATP. Inside-out patches of membrane were perfused with standard bath solution containing 10^{-7} M Ca^{2+}. Channel activity was recorded at a holding potential of -50 mV and either 1 or 2 mM ATP was perfused in the bath solution. ATP (1 or 2 mM) did not change NP. or single channel amplitude of colonic BK channels (n = 10) (Fig. 6B).

Experiments were performed to examine the possibility that ensemble average current responses from cell-attached patches of colon cells could mimic the results obtained in whole-cell experiments. Cell-attached patches were depolarized from 0 to +70 mV and current responses from 128 such voltage steps were averaged. Results from one such experiment are shown in Figure 7. Capacitive and leak currents were compensated for by subtracting a single sweep without channel openings from each record. During control voltage steps very few single channel openings were evident in each sweep (Fig. 7A). The opening of two channels simultaneously was rare and at no time did we observe the opening of three channels simultaneously. When lemakalim (20 μM) was added to the bath solution for approximately 10 minutes, the resultant single channel activity increased. At this time it was apparent that three channels were active in the presence lemakalim (Fig. 7D). When the single channel currents were ensemble averaged, the total current after lemakalim was increased three fold (Fig 7B and 7E). The activation of the lemakalim induced current was very rapid, but the time constant for activation was not significantly different from control (control 28 msec, lemakalim 42 msec). The rapid activation of this current increased the likelihood that under whole-cell conditions this current would appear as a time-independent (quasi-instantaneous) component of outward current. Without subtraction of leak and capacitive current, the activation of this current was hidden in the capacitive transient (Fig. 7C and 7F).

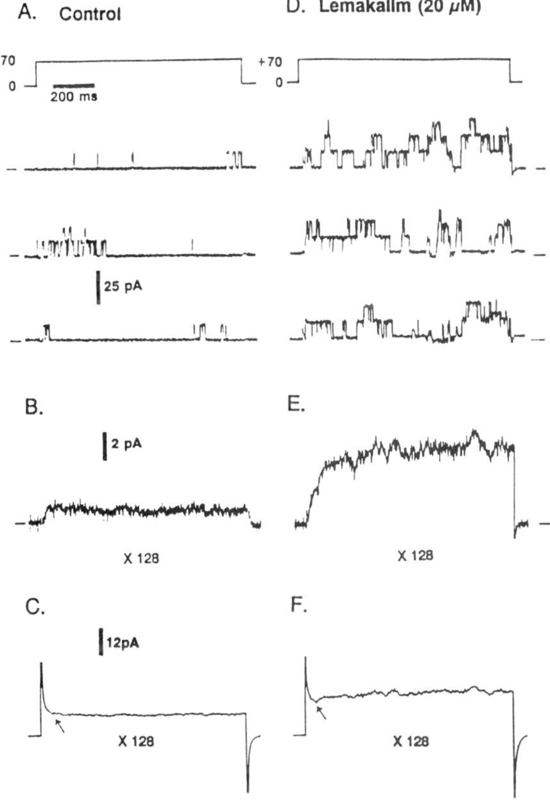

Figure 7. Effect of lemakalim on single colonic BK channels and ensemble average current. Channel openings in response to a step depolarization from 0 to +70 mV membrane potential were recorded before (Panel A) and after application of lemakalim (20 μM) to the bath solution (Panel D). Currents are corrected for leak and capacitive transients. Panel B and E show the ensemble average current from 128 such sweeps. Panels C and F show the average current from the same cell, however leak and capacitive current are not subtracted.

In renal artery cells, the most predominant type of single K^+ channel recorded from cell-attached patches was a large conductance, voltage-dependent channel. Figure 8A illustrates that as a patch of membrane was depolarized from 0 to +70 mV, large conductance channels were activated. Channel activity increased with membrane depolarization and the slope conductance of these channels was 217 ± 5 pS (n = 6). When lemakalim (1 μM) was added to the bath solution a dramatic increase in probability of channel opening was observed (Fig. 8B). In this experiment in control conditions (H.P. = +30 mV), NP. was 0.077. Upon application of lemakalim, NP. increased to 0.168 and refumed to 0.049 after washout of the drug. There was no apparent change in single channel amplitude after lemakalim application. Lemakalim increased NP_0 in a similar fashion in 6 of 6 cells tested.

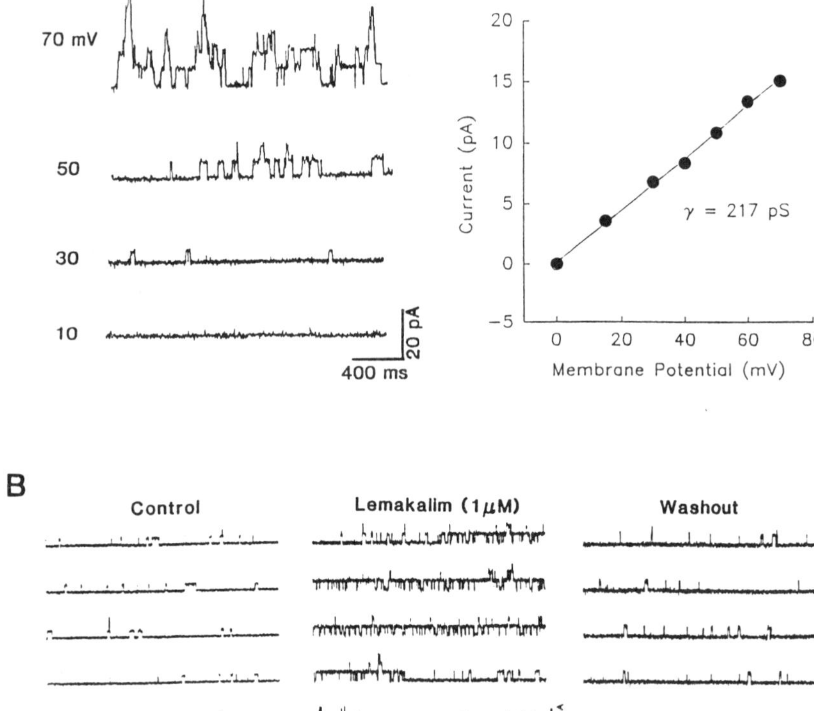

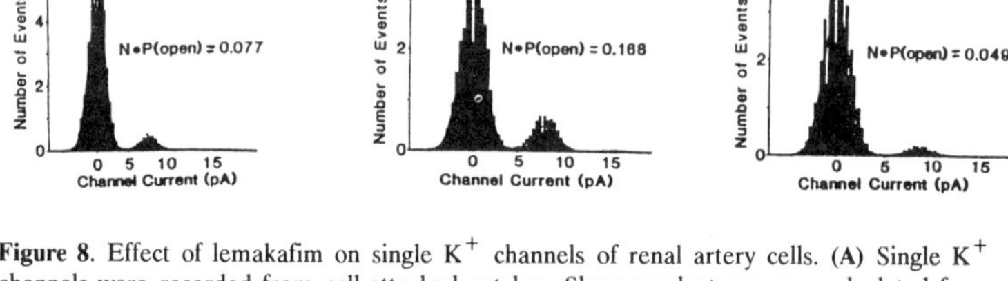

Figure 8. Effect of lemakafim on single K^+ channels of renal artery cells. **(A)** Single K^+ channels were recorded from cell-attached patches. Slope conductance was calculated from linear regression analysis. **(B)** When a constant potential of $+30$ mV was applied to the patch of membrane under control conditions, NP. was 0.077. Upon bath addition of lemakalim (1 μM), NP_O increased to 0.168. Washout decreased NP_O to control levels. AH single channel openings are upward deflections of current.

Experiments were performed in order to examine the possibility that cell-attached patches of renal artery cells contained another large conductance channel activated by lemakalim. Tetraethylammonium (TEA) characteristically blocks BK channel activity from the extracellular solution with a K_j of approximately 200 μM while having no effect on K_{ATP} channels at that concentration (31). When TEA (200 μM, $V_H = +50$ mV) was included in the patch pipette, single channel current amplitude was apparently reduced about 50% due to the characteristic flicker block produced by TEA (Fig. 9). When lemakalim (1 μM) was added to the bath solution, NP_0 was not affected (n = 3). Similar results were also obtained in canine colon cells (n = 3) (10).

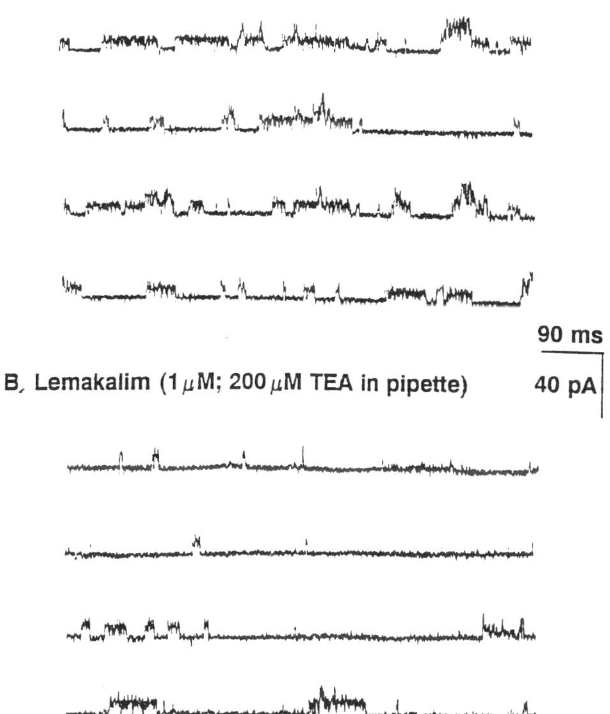

Figure 9. TEA prevents the action of lemakalim on single K^+ channels in renal artery cells. Panel (A) illustrates that when TEA (200 μM) was included in the patch pipette (V_H = +50 mV), single channel current amplitude was reduced. Panel (B) illustrates bath application of lemakalim (1 μM) under these conditions caused no increase in NP. (n = 3).

DISCUSSION

It is well accepted that potassium channel openers (PCO's) increase the activity of ATP-sensitive K^+ channels (K_{ATP}) in pancreatic β-cells, cardiac and skeletal muscle (41, 43, 50). However, little is known about the type of ion channel responsible for the hyperpolarization achieved when PCO's are applied to smooth muscle tissue or isolated smooth muscle cells. A number of K^+ channels have been implicated in the effect of PCO's in smooth muscle including large conductance Ca^{2+}-activated K^+ (BK) channels (10, 20, 29), K_{ATP} channels (45), delayed rectifier K^+ channels (1) and small conductance Ca^{2+} and ATT-sensitive K^+ channels (28-36). This study examined the effects of cromakalim and lemakalim on whole-cell and single channel ionic currents in colonic, renal and coronary artery smooth muscle cells.

Our data suggest that the membrane hyperpolarization by cromakalim and lemakalim is related to an increase in K^+ conductance in each of these preparations. Both cromakalim and lemakalim appear to significantly increase the magnitude of a time-independent (quasiinstantaneous) component of K^+ current in colonic, renal and coronary artery cells. In addition, lemakalim increases a sustained component of current in renal artery cells. The BK channel is thought to be the major contributor to the sustained component of outward K^+ current at potentials positive to $+10$ mV; however, the type of channel that underlies the time-independent background current is currently unknown. It has been reported that this time-independent current is blocked by TEA and Ba^{2+} in rabbit portal vein (26) and the results in Figure 7 and 9 suggest that activation of BK channels may contribute to this macroscopic K^+ current. Lemakalim increased the magnitude of ensemble average current to step depolarizations in cell-attached patches. The time constant for activation of this ensemble average current was approximately 30 msec and is in the same range for BK channels from colonic myocytes in inside-out patches (8). This fast activating current may appear under whole-cell conditions as a time-independent (quasi-instantaneous) current. The patch clamp experiments were performed at room temperature, while the whole-cell voltage clamp experiments were performed at 37° C. Increasing bath temperature from 25° C to 37° C causes a further decrease in the time constant for activation of BK ensemble average current (Carl, unpublished observation).

In inside-out patches of colonic smooth muscle cells and cell-attached patches of renal artery and colon cells, lemakalim and cromakalim increased the open probability of BK channels to a similar extent. The time course of activation was much slower in cell-attached patches than inside-out patches, suggesting that the lemakalim and cromakalim may have to diffuse through the lipid bilayer in order to exert their effect. Glibenclamide inhibited the cromakalim and lemakalim-stimulated increase in the time-independent component of current. Similarly, glibenclamide reversed the lemakalim induced increase in NPO in inside-out patches, but only partially, in cell-attached patches. These data suggest, that K^+ channel openers like lemakalim and cromakalim and the sulfonylureas may not be specific for K_{ATP} channels at the concentrations studied. Our findings resemble those of Gelband et al, (19) who reported that cromakalim and pinacidil, another PCO, increased the open probability of BK channels in smooth muscle cells from rabbit aorta and was inhibited by glibenclamide.

The K_{ATP} channel in vascular smooth muscle was reported to be $[Ca^{2+}]_i$, TEA, and, voltage insensitive, inhibited by glibenclamide and millimolar concentrations of ATP and activated by cromakalim (45). Since this channel had a slope conductance of 135 pS at 0 mV in 60/120 mM KCl, it would be difficult to distinguish a K_{ATP} channel from a BK channel in our experiments. Thus, we examined the possibility that the increase in channel activity by lemakalim may have resulted from a different type of large conductance channel rather than an increase in open probability of BK channels. In inside-out patches, bath application of ATP had no effect on open probability of the large conductance channels (Fig. 6). Also under conditions that would reveal a TEA insensitive, large conductance channel where the BK channel was half maximally blocked by TEA, (Fig. 9), lemakalim activated no other single channel conductance. Finally, under conditions where BK channels were maximally activated with 10^{-5} M Ca^{2+}, lemakalim failed to further increase ensemble average current (10). It might be that such a channel does not exist in membranes of colonic or renal artery myocytes, or that the density or conductance of this channel may be very low. Occasionally we found a 80-100 pS channel in our patches that was also activated by lemakalim. This channel had a similar voltage and Ca^{2+} dependence as the BK channel and was never observed under conditions, where the BK channel is inactive. It is not clear, whether this channel is actually a separate entity, or a subconductance state of the BK channel. This suggests that the increase in large conductance channel activity observed at steady-state membrane potentials is due to an increase in open probability of BK channels.

In canine colon, BK channels have been implicated to participate in the repolarization phase of slow waves (8). Under resting conditions, open probability of these channels is very low and it is not clear whether they also contribute to the resting membrane potential. It has been suggested that BK channels may contribute to resting membrane potential in porcine vascular smooth muscle (47). The open probability of BK channels near the resting membrane potential can be extrapolated from measurements at high Ca^{2+} and positive membrane potentials by fitting the activation curves with a

Boltzmann function $Po = 1/(1 + \exp[-K(V-V_{1/2})])$. Using data acquired from colon cells (8) ($V_{1/2} = +107$ mV at 10^{-7} M Ca^{2+}, $K^{-1} = 16.9$ mV) estimates a P(open) of 1.67 x 10^{-4} at -40 mV and 2.83 x 10^{-5} at -70 mV. Using a BK channel density of 23000/cell and a single channel current of 2.32 pA at -40 mV and 0.58 pA at -70 mV, the current carried by BK channels would be 8.9 pA at -40 mV and 0.38 pA at -70 mV. With an input resistance of 0.84 GΩ (30) such current would generate transmembrane potentials of 7.5 mV and 0.3 mV, respectively. These calculations from measurements on excised patches have to be interpreted with caution however, since the activity of BK channels may be further modulated by changes in $[Ca^{2+}]_i$, intracellular second messengers and mechanisms like phosphorylation (9). These observations suggest that BK channels may contribute to the cell's resting membrane potential in canine colon, but are not solely responsible for setting the cells' membrane potential. Thus due to the cell's high input resistance, even a small increase, in open probability of BK or other K^+ channels induced by PCO'S, like lemakalim and cromakalim, may cause hyperpolarization of the intact tissue.

ACKNOWLEDGEMENTS

The authors would like to thank Dr. Mark T. Nelson for the use of single channel analysis program, Nancy Horowitz for preparing the colon cells and Mylah Lizares and Lan Xue for their excellent technical assistance. We thank SmithKline Beecham for the gifts of the cromakalim and lemakalim. This work was supported by National Institute of Health (NIH) Grants DK-41315 (to K.M.S. and J.R.H.) and HL-40399 (K.D.K), and a Grant-in-Aid from the American Heart Association (to J.R.H.). J.R.H. is an Established Investigator of the American Heart Association, C.H.G. and J.M.P. are NIH Postdoctoral fellows, and T.I. is a Postdoctoral fellow of the Nevada Affiliate of the American Heart Association. Address reprint requests to J.R. Hume.

REFERENCES

1. Beech, D.J. and T.B. Bolton, Br. J. Pharmacol. 98, 851-864 (1989).
2. Benham, C.D., and T.B. Bolton, J. Physiol. (Lond) 340, 469-486 (1983).
3. Benham, C.D., T.B. Bolton, R.J. Lang, and T. Takewaki, Pflugers Archiv. 403, 120-127 (1985).
4. Benham, C.D., T.B. Bolton, R.J. Lang, and T. Takewaki, J. Physiol. (Lond) 371, 45-67 (1986).
5. Blatz, A.L., and K.L. Magleby, Trends in Neurosc. 10, 463-467 (1987).
6. Buckingham, R.E., J.C. Clapham, T.C. Hamilton, S.D. Longman, J. Norton, and R.H. Poyser, J. Cardiovasc. Pharmacol. 8, 798 804 (1986).
7. Buckingham, R.E., T.C. Hamilton, D.R. Howlett, S. Mootoo, and C. Wilson, Br. J. Pharmacol. 97, 57-64 (1988).
8. Carl, A., and K.M. Sanders, Am. J. Physiol. 257, C470-C480 (1989).
9. Carl, A., J.L. Kenyon, D. Uemura, N. Fusetani, and K.M. Sanders, Am. J Physiol. (In press).
10. Carl, A., S. Bowen, C.H. Gelband, K.M. Sanders, and J.R. Hume, (submitted 1991).
11. Cavero, I., S. Mondot, and M. Mestre, J. Pharmacol. Exp. Ther. 248, 1261-1268 (1989).
12. Clapham, J.C., and S.D. Longman, Eur. J Pharmacol. 171, 109-117 (1989).
13. Cook, N.S., S.W. Weir, and M.C. Danzeisen, Br. J. Pharmacol. 95, 741-752 (1988).
14. Cook, N.S., Trends in Pharmol Sci. 9, 21-28 (1988).
15. Cook, N.S., U. Quast, R.P. Hof, Y. Baumlin, and C. Pally, J. Cardiovasc. Pharmacol. 11, 90-99 (1987).
16. Eltze, M., Eur. J. Pharmacol. 165, 231-239 (1989).
17. Fabiato, A, J. Gen. Physiol. 78, 457-497 (1981).
18. Fabiato, A., and F. Fabiato, J. Physiol. (Paris). 75, 463-505 (1979).
19. Gelband, C.H., J.R. McCullough, and C. van Breemen, Circ. Res. (submitted (1991).
20. Gelband, C.H., N.J. Lodge, and C. van Breemen, Eur. J. Pharmacol. 167, 201-210 (1989).
21. Gimenez-Gallego, G., M.A. Navia, J.P. Reuben, G.M. Katz, G.J. Kaczorowski, and M.L. Garcia, Proc. Natl. Acad Sci. USA, 85, 3329-3333 (1988).
22. Hamill, O.P., A. Marty, E. Neher, B. Sakmann, and F.J. Sigworth, Pflugers Archiv. 391, 85-100 (1981).
23. Hamilton T.C., S.W. Weir, and A.H. Weston, Br. J. Pharmacol. 88, 103-111 (1986).
24. Hamilton, T.C. and A.H. Weston, Gen. Pharmacol. 20, 1-9 (1989).
25. Hof, R.P., U. Quast, N.S. Cook and S. Blarer, Circ. Res. 62, 679-686 (1988).
26. Hume, J.R. and N. Leblanc., J. Physiol. (Lond) 413, 49-73 (1989).
27. Inoue, R., K. Kitamura, and H. Kuriyama, Pflugers Archiv. 405, 173-179 (1985).

28. Kitamura, K. and H. Kuriyama, J. Molec. Cell. Cardiol. 23, (suppl. Ill) S.98 (1991).

29. Klockner, U., U. Trieschmann, and G. Isenberg, Drug Research 39, 120-126 (1989).

30. Langton, P.D., and K.M. Sanders, Am. J. Physiol. 257, C451-460 (1989).

31. Langton, P.D., M.T. Nelson, Y. Huang, and N. Standen., Am. J. Physiol. 260, H927-H934 (1991).

32. Leblanc, N., D.W. Wilde, K-D. Keef, and J.R. Hume, Circ. Res. 65, 1102-1111 (1989).

33. McCann, J.D., and M.J. Welsh, J. Physiol. (Lond) 372, 113-127 (1986).

34. Nelson, M.T., J.B. Patlak, J. F. Worley, and N.B. Standen, Am. J. Physiol. 259, C3-Cl8 (1990).

35. Neyton, J., and C. Miller, J. Gen. Physiol. 92, 569-586 (1988).

36. Okabe, K., S. Kajioka, K. Nakao, K. Kitamura, H. Kuriyama and A.H. Weston, J. Pharmacol. Exp. Ther. 252, 832-839 (1990).

37. Petersen, O.H., and Y. Maruyama, Nature 307, 693-696 (1984).

38. Post, J.M., J. M. Smith, and A.W. Jones, J. Pharmacol. Exp. Ther. 250, 591-597.

39. Post, J.M., R.J. Stevens, Y-M. Sanders, and J.R. Hume, Am. J. Physiol. 260, C375-C382 (1991).

40. Quast, U. and N.S. Cook, J. Pharmacol. Exp. Ther. 250, 261-271 (1989).

41. Quasthoff, S.A., A. Spuler, F. Horn-Lehamn, and P. Grafe, Pflugers Archiv. 414, S179-S180 (1989).

42. Sanders, K.M., E.P. Burke, A. Carl, W.C. Cole, P. Langton, and S. Ward, in Frontiers in Smooth Muscle Research, N. Sperelakis and J. Wood, eds. (Alan R. Liss, Inc. 1990) pp. 307-321.

43. Sanguinetti, M.C., A.L. Scott, G.J. Zingaro, and P.K.S. Siegl, Proc. Natl. Acad. Sci. 85, 8360-8364 (1988).

44. Singer, J.J., and J.V. Walsh Jr., Pflugers Archiv. 408, 98-111 (1987).

45. Standen, N.B., J.M. Quayle, N.W. Davies, J.E. Brayden, Y. Huang, and M.T. Nelson, Science 245, 177-180 (1989).

46. Tomita, T., Jap. J. Physiol. 38, 1-18 (1988).

47. Trieschmann, U., and G. Isenberg, Pflugers Archiv. 414(Suppl. 1), S183-184 (1989).

48. Weir S.W. and A.H. Weston, Br. J. Pharmacol. 88, 121-128 (1986).

49. Winquist, R.J., L.A. Heaney, A.A. Wallace, E.P. Baskin, R.B. Stein, M.L. Garcia, and G.J. Kaczorowski, J. Pharmacol. Exp. Ther. 248, 149-155 (1989).

50. Zunkler, B.J., S. Lenzen, K. Manner, U. Panten, and G. Trube, Naunyn-Schmiedeberg's Arch. Pharmacol. 336, 225-230 (1988).

MECHANISMS OF HYPERPOLARIZATION INDUCED BY K⁺-CHANNEL OPENERS IN VASCULAR SMOOTH MUSCLE CELLS.

ZHILING XIONG, HIROSI KURIYAMA and KENJI KITAMURA

Department of Pharmacology, Faculty of Medicine, Kyushu University, Fukuoka 812, Japan

INTRODUCTION

The tone of vascular smooth muscle, which is one of the main factors in the control of blood pressure, is closely related to its intracellular Ca^{2+} concentration. Therefore, most the vasodilating agents have been targeted at sites which enclosed intracellular Ca^{2+} concentration to be directly modified. For example, α-blockers and angiotensin converting-enzyme inhibitors interfere with excitatory pathways by reducing the number of excitatory agents which bind to their receptors and nitro compounds stimulate guanylate cyclase, and lead to the pumping out of Ca^{2+} ions from the cytosol. Furthermore, one type of potent vasodilators, the Ca^{2+}-channel blockers, reduce intracellular Ca^{2+} concentration by direct inhibition of Ca^{2+} influx at the voltage-dependent Ca^{2+} channel (mostly L-type channel). These vasodilators all directly modulate the mechanisms regulating intracellular Ca^{2+} concentrations.

On the other hand, the K^+-channel openers are thought to inhibit the voltage-dependent Ca^{2+} channel indirectly via membrane hyperpolarization. This is because the K^+-channel openers especially cromakalim and its isomer, lemakalim, strongly inhibit agonist-induced contractions but not contraction induced by an excess K^+ solution, in which no hyperpolarization of the membrane is produced by these K^+-channel openers. Pharmacological profiles of the hyperpolarization induced by K^+-channel openers were first investigated more than 10 years ago; however, due to lack of knowledge of the ion channels in the smooth muscle cell membrane, the ionic mechanisms underlying the hyperpolarization induced by the K^+-channel openers remained obscure for a long time.

The recent development and application of the patch-clamp technique to smooth muscle cells, especially to various vascular smooth muscle cells has revealed the biophysical and pharmacological properties of several K^+ channels distributed in smooth muscle cells. The first report on K^+ channels in smooth muscle cells published in the Journal of Physiology by Benham et al. (5) showed-and later investigations by several researchers confirmed- that the large conductance Ca^{2+}-dependent K^+ channel was widespread in smooth muscle cell membranes, including those of vascular smooth muscle cells. Since then, we have accumulated knowledge of the properties of the ion channels and on the mechanisms of channel activation and inactivation. In this article, we review the properties of the ion channels that are activated by K^+-channel openers in smooth muscle cells.

GENERAL FEATURES OF THE ACTIONS INDUCED BY THE K⁺-CHANNEL OPENERS

Nicorandil, an anti-anginal agent, was introduced as a nitro compound. This drug had a potent vasodilating action comparable to that of papaverin, but no cardio-depressant activity (56, 58). In left atrial muscle fibers of the dog, Yanagisawa and Taira (65) found that nicorandil ($\geq 10 \, \mu M$) hyperpolarized the membrane and reduced the duration of the action potentials, suggesting that nicorandil increased the K^+ conductance in this cell. Similar hyperpolarization, but with a larger amplitude, was reported in the coronary and mesenteric arterial cells of the pig and guinea-pig on application of lower concentrations of nicorandil than those applied to cardiac cells (18, 31). Hyperpolarization was not induced by nicorandil in these smooth muscle cells in excess K^+ solution, containing more than $30mM \, K^+$, because in such solutions, the membrane potential was coincident with the K^+ equilibrium potential. The same authors also reported that, even in the presence of $10mM$ tetraethylammonium (TEA), a common K^+ channel blocker known to be effective in smooth muscle cells, nicorandil produced a hyperpolarization of the membrane.

Published 1991 by Elsevier Science Publishing Company, Inc.
Ion Channels of Vascular Smooth Muscle Cells and Endothelial Cells
Sperelakis and Kuriyama, Editors

One of the interesting observations in this report (18) was that the hyperpolarization induced by nicorandil did not persist even in the continued presence of nicorandil when the membrane was depolarized by TEA. Later work confirmed that the hyperpolarization induced by cromakalim or nicorandil in the guinea-pig mesenteric artery did not persist for longer than 15min (38). The action potentials evoked by a depolarizing pulse in the presence of 10mM TEA were abolished during the membrane hyperpolarization but reappeared during depolarizing phase in the presence of nicorandil. Therefore, Furukawa et al. (18) concluded that nicorandil did not actually block the action potential, but abolished it because of the membrane hyperpolarization. As nicorandil has nitroglycerin-like actions, hyperpolarization of the membrane was thought to play a minor role in causing vasodilation. However, KRN2391, which has a chemical structure similar to that of nicorandil, has been recently introduced as a more selective and potent K^+-channel opener. KRN2391 was reported to be more potent than nicorandil without possessing any nitroglycerin-like action (35), suggesting that membrane hyperpolarization might be quite an important mechanism for nicorandil-induced vasodilation too.

Cromakalim is a key drug for the understanding of K^+-channel openers, because the vasodilating mechanism of this drug is thought to be entirely due to the marked membrane hyperpolarization it induces (1, 22,47, 59, 60). Indeed, in guinea-pig trachealis and rat portal vein, both the relaxation of the tissue and the hyperpolarization of the membrane occur within the same concentration range of the drug (1, 22). However, several pieces of evidence against such a conclusion have been reported in rat uterus, rat portal vein and guinea-pig mesenteric artery (17, 21, 22, 38, 43). For example, in rat myometrium, the induced relaxation was associated with only a small membrane hyperpolarization and without a detectable change in $^{86}Rb^+$ efflux (21). The same authors also reported that cromakalim abolished the spontaneously generated action potentials in the rat myometrium without any detectable change in the amplitude of the voltage-dependent Ca^{2+} current. They therefore concluded that cromakalim affected the pacemaker activity of this smooth muscle cell. In rat portal vein and guinea-pig urinary bladder, spontaneous electrical discharges were inhibited by low concentrations of cromakalim without any detectable change in the membrane potential or $^{86}Rb^+$ efflux (17, 22, 52). Nakao et al. (38) also reported a lack of association between the membrane hyperpolarization and muscle relaxation induced by cromakalim in the guinea-pig mesenteric artery. In this tissue, cromakalim inhibited the generation of the action potentials evoked by a membrane depolarization, even when the hyperpolarization was diminished by a long application of the cromakalim (>15min). After removal of cromakalim, depolarization of the membrane by an electrical stimulation evoked an action potential without a change in the membrane potential. Furthermore, Okabe et al. (43) demonstrated that cromakalim directly inhibited the voltage-dependent L-type Ca^{2+} current in the rat portal vein (Fig. 1). Post et al. (44) also observed that in dog colonic smooth muscle cells cromakalim, but not lemakalim, inhibited the L-type voltage-dependent Ca^{2+} current. Similarly, minoxidil has also been reported to inhibit the voltage-dependent Ca^{2+} current (37). These results indicate that another mechanism, in addition to the membrane hyperpolarization, may be involved in the vasodilating actions of cromakalim and minoxidil. However, the main mechanism underlying the vasodilating actions of cromakalim is still believed to be the membrane hyperpolarization, because in excess K^+ solution cromakalim hardly produced relaxation.

Pinacidil is an antihypertensive agent which was synthesized in the early 1970's and whose mechanism of action was at first unknown. It was re-examined after K^+ channel-opening by cromakalim was established as a vasodilating mechanism in smooth muscle cells and after pinacidil itself was found to produce membrane hyperpolarization and to increase $^{86}Rb^+$ efflux (10, 53, 61). Pinacidil synthesized neither cAMP nor cGMP, indicating that this drug had no nitroglycerin-like action (61). The voltage-dependent Ca^{2+} channel was not inhibited by pinacidil in rat and rabbit portal veins (Kajioka & Kitamura; unpublished observations). However, this compound did inhibit the spontaneous activation of the Ca^{2+}-dependent K^+ channels (spontaneous transient outward current , STOC; 6; oscillatory outward current , I_{OO}; 42, which was thought to be closely related to Ca^{2+} release from the intracellular Ca^{2+} store sites; 7, 48) in the rabbit portal vein (63). Xiong et al. (63) speculated that pinacidil interferes with a step somewhere on the pathway for the release of Ca^{2+} from its store sites and may inhibit Ca^{2+} release from, or deplete Ca^{2+} the stored in the store sites. This was because the single-channel current activity of the large conductance Ca^{2+}-dependent K^+ channel, simultaneous activation of which is thought to form I_{OO}, was not affected by pinacidil.

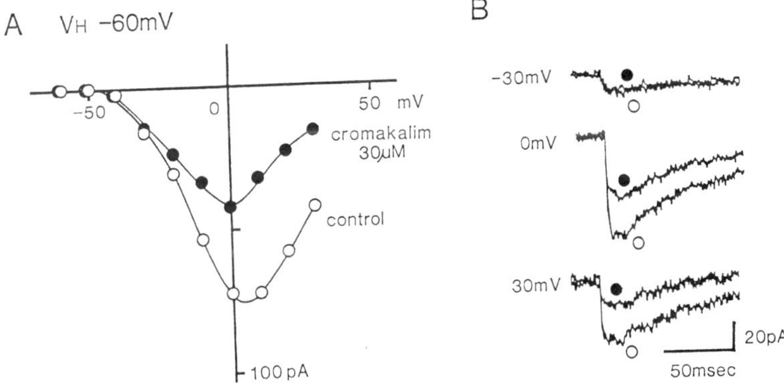

Figure 1. Effects of cromakalim (30 μM) on the voltage-dependent Ca^{2+} current recorded from the single smooth muscle cells of the rat portal vein at a holding potential of -60mV. **A:** current-voltage relationships in the presence and absence of cromakalim. **B:** traces observed at three different levels of membrane depolarization in the presence (closed circles) and absence (open circles) of cromakalim. From Okabe et al. (43) with permission.

Other agents, such as minoxidil, RP49356 and EMD52692, also have K^+ channel opening properties; however, more detailed studies will be necessary for a full understanding the mechanisms underlying the vasodilating actions of these drugs.

The pharmacological properties of the membrane hyperpolarization induced by several K^+-channel openers have been repeatedly investigated using various smooth muscle cells, including vascular smooth muscle cells. In guinea-pig mesenteric vein, portal vein and small intestine, nicorandil-induced hyperpolarization persisted in a solution containing no Ca^{2+}, suggesting that the K^+ channel responsible for the nicorandil-induced hyperpolarization was insensitive to extracellular Ca^{2+} concentration (34, 64). In dog trachea, procaine, a nonselective blocker of ion channels in high concentrations (·1mM), inhibited completely, but TEA (10mM) only partly inhibited the nicorandil-induced hyperpolarization, while in the dog mesenteric artery, a higher concentration of procaine was required to block the nicorandil-induced hyperpolarization (29). In guinea-pig trachealis, a similar concentration of TEA (8mM) significantly, but not completely inhibited the nicorandil- and cromakalim-induced hyperpolarizations (1). In the rat portal vein, the inhibitory actions of pinacidil on the integrated spontaneous mechanical activity was attenuated by TEA in concentrations higher than 1mM (61). Although nicorandil produced membrane hyperpolarization even in the presence of TEA (5mM in the dog mesenteric artery; 10mM in the pig coronary artery), the concentrations of nicorandil required to produce membrane hyperpolarization (1mM in the dog mesenteric artery; 0.1mM in the pig coronary artery) were much higher than those needed in the absence of TEA (18, 30). These results suggested that high concentratons of TEA (·5mM) inhibited the amplitude of the hyperpolarization induced by nicorandil. Inhibition of the nicorandil-induced hyperpolarization by procaine and other local anesthetics was also observed in the guinea-pig small intestine (64).

Apamin, a bee venom known to block certain Ca^{2+}-dependent K^+ channels, did not modify the hyperpolarization induced by nicorandil in the guinea-pig small intestine (64), the actions of cromakalim in guinea-pig taenia caeci (59), trachealis (1), rabbit aorta (10) and rat portal vein (46), or the reduction in the spontaneous mechanical activity of the rat portal vein by pinacidil (61). Cook and Hof (9) also observed the ineffectiveness of apamin on the cromakalim-induced hypotension in rat and rabbit. On the other hand, the crude venom of the Israeli scorpion, *Leiurus Quinquestriatus Hebraeus*, (LQV) has been reported to inhibit the ^{86}Rb efflux induced by cromakalim in the rat portal vein (46). As

LQV is known to inhibit the Ca^{2+}-dependent K^+ channels in the rat thymocytes, human erythrocytes and guinea-pig hepatocytes, the authors speculated that cromakalim may open a certain Ca^{2+}-dependent K^+ channel in the rat portal vein which had a sensitivity to LQV similar to that observed in the erythrocytes and hepatocytes. LQV contains, amongst other substances, charybdotoxin and leiurotoxin, which are respectively known to block the maxi-K^+ and the apamin-sensitive low conductance K^+ channels even in submicromolar concentrations (20). Since apamin had no inhibitory action on the responses induced by K^+-channel openers, this would suggest that the maxi-K^+ channel is a strong candidate for the K^+ channel activated by K^+-channel openers. Indeed, Hermesmeyer (25), Gelband et al. (19) and Hume (27) have clearly showned that the maxi-K^+ channel is indeed activated by K^+-channel openers. However, quite high concentrations of LQV were required to inhibit the cromakalim-induced ^{86}Rb efflux (46); therefore, it is uncertain whether LQV blocks the maxi-K^+-channel selectively or the cromakalim-sensitive K^+-channel nonselectively. Okabe et al. (43) and Kajioka et al. (33), working on the rat portal vein using the patch-clamp technique, reported that K^+-channel openers activated a very small conductance Ca^{2+}-dependent K^+ channel, but not the maxi-K^+ channel, and that these were inhibited by neither charybdotoxin nor apamin. Therefore, the findings that cromakalim and other K^+-channel openers open a Ca^{2+}-dependent K^+ channel in several vascular smooth muscle cells are inconsistent with observations in the guinea-pig mesenteric vein and small intestine, where the K^+-channel openers produce Ca^{2+}-insensitive hyperpolarization (19, 25, 27, 33, 34, 43, 64). These might suggest that K^+-channel openers open several different types of K^+ channel in the various smooth muscle cells. Using the guinea-pig mesenteric vein and artery, Nakao et al. (38) reported that the hyperpolarization induced by nicorandil and cromakalim was independent of the extracellular Ca^{2+} concentrations in the mesenteric artery, but that the slow component of the hyperpolarization was dependent on it in the mesenteric vein, suggesting at least two types of K^+ channel are participate in the membrane hyperpolarization induced by the K^+-channel openers in the mesenteric vein (Fig. 2). Foster et al. (17) also reported that the K^+ channel present in the smooth muscle cells of the guinea-pig urinary bladder, might be different from those distributed in the rat and guinea-pig portal veins, because cromakalim did not increase $^{86}Rb^+$ efflux in the urinary bladder but did increase it in the portal veins (47, 60). Recent studies with patch clamp techniques have revealed that several different K^+ channels are responsible for the actions of K^+-channel openers in various smooth muscle cells (see Table 1).

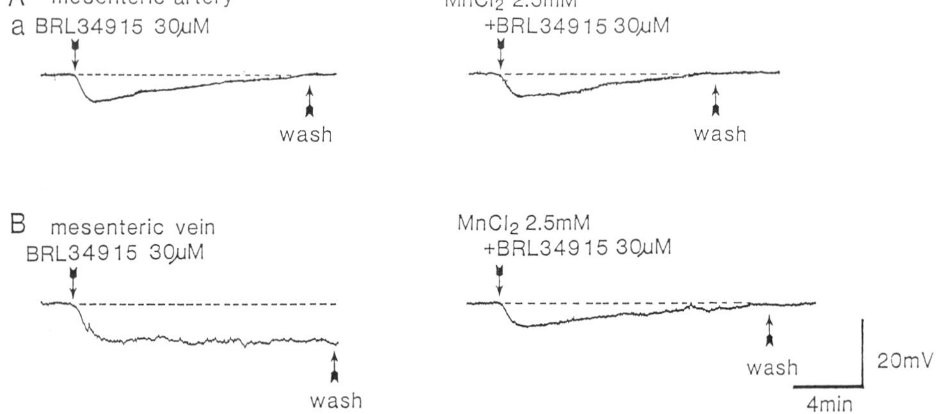

Figure 2. Effects of cromakalim (BRL 34915) on the membrane potential recorded from smooth muscle cells of the mesenteric artery and vein of the guinea-pig. Cromakalim ($30\,\mu M$) was applied in solutions containing 2.5mM Ca^{2+} (left side traces) or 2.5mM Mn^{2+} (right side traces). From Nakao et al. (38) with permission.

Table 1. TARGET CHANNELS FOR K-CHANNEL OPENERS IN SMOOTH MUSCLE CELLS

Species & Tissues (Reference)	Conductance	Channel Properties	Ionic Condition		Drug
			Pipette(Cis)	Bath(Trans)	
Rabbit & Bovine Aorta (Kusano et al. 1987)	200pS	Ca-Dependent TEA-Resistant (<1mM) ScTX-Sensitive	150mM KCl	150mM KCl	Out / Cromakalim
Rat Azygos Vein (Primary culture cell) (Hermesmeyer 1988)	200pS	Ca-Dependent	4.7mM KCl 1mM EGTA 1.8mM CaCl$_2$	140mM KCl 0.77mM EGTA 0.7mM CaCl$_2$	Out / Pinacidil
Rabbit Thoracic Aorta (Microsome) (Gelband et al. 1989)	337pS	Ca-Dependent	(250mM KCl) 1mM CaCl$_2$	(250mM KCl) 2mM EGTA	(Cis) (Trans) / In Croma.
Rabbit Mesenteric Artery (Standen et al. 1989)	135pS	Ca-Insensitive ATP-Sensitive TEA-Resistant	60mM KCl 60mM NaCl 1.8mM CaCl$_2$	104mM KCl 16mM KOH 5mM EGTA	Out / Cromakalim
Rat Portal Vein (Kajioka et al. 1990)	10pS (20pS)	Ca-(In)Sensitive ATP-Sensitive TEA-Resistant (<1mM) 4AP-Sensitive (>1mM)	6mM KCl 2.5mM CaCl$_2$ (140mM KCl)	140mM KCl 4mM EGTA (140mM KCl)	In / Nicorandil Pinacidil
Rabbit Portal Vein (Kajioka et al. 1991)	15pS (25pS)	Ca-Insensitive GDP & ATP-Sensitive TEA-Resistant 4AP-Sensitive	6mM KCl 2.5mM CaCl$_2$ (140mM KCl)	140mM KCl 4mM EGTA (140mM KCl)	Both / Cromakalim Pinacidil
Pig Coronary Artery (Inoue & Nakaya, 1989)	30pS	[Ca]o-Dependent ATP-Sensitive 4AP-Sensitive (5mM)	140mM KCl 10-100μM CaCl$_2$	140mM KCl 1.4mM CaCl$_2$	Both / Nicorandil Cromakalim
Guinea-Pig Portal Vein (Beech & Bolton, 1988)	5pS	Delayed K-Current Ca-Insensitive ATP-Insensitive TEA-Sensitive (>1mM)	140mM KCl 4mM EGTA	140mM NaCl EGTA+Cd	Whole-Cell

Although the hyperpolarization, $^{86}Rb^+$ efflux, $^{42}K^+$ efflux and other responses induced by various K^+-channel openers were inhibited by high concentrations of TEA and procaine, these K^+ channel blockers do not seem to block those responses, selectively. Glibenclamide and tolbutamide, both sulfonylureas, are known to increase insulin release from the pancreatic islet cells and to induce hypoglycemia (3, 24). The stimulating actions of tolbutamide on insulin release are associated with membrane depolarization and a reduction of $^{86}Rb^+$ efflux (Henquin & Meissner, 1982). On the other hand, diazoxide, a hypotensive and hyperglycemic agent with no diuretic action, increased $^{86}Rb^+$ efflux and hyperpolarized the membrane of pancreatic β-cells (23). These drugs were later found to act in pacreatic β-cells on ATP-sensitive K^+ channels which were first discovered in cardiac cells (41, 50, 51, 55-57), and diazoxide has been now categorized among the K^+-channel openers. Findings that glibenclamide inhibited the cromakalim-induced increase in the $^{86}Rb^+$ efflux in the rat portal vein and aorta led to the conclusion that in smooth muscle cells as in cardiac and pancreatic cells, the ATP-sensitive K^+ channel is the target channel for K^+-channel openers (8, 45, 62).

HETEROGENEITY OF THE K^+ CHANNELS ACTIVATED BY VARIOUS K^+-CHANNEL OPENERS IN VASCULAR SMOOTH MUSCLE

In 1988, Escande et al. (14) and Sanguinetti et al. (49) reported that cromakalim could activate the ATP-sensitive K^+ channel which was blocked by glibenclamide in cardiac muscle cells. Subsequently, several reports confirmed this important action of K^+-channel openers (2, 15, 16, 26, 39). However, as briefly mentioned above, the pharmacological profiles of the K^+ channels activated by K^+-channel openers are not uniform in over the range of smooth muscle cells and a wide variety of K^+ channels are thought to be opened. For example, the hyperpolarizations induced by cromakalim and nicorandil were Ca^{2+}-independent in the guinea-pig mesenteric artery but partly Ca^{2+}-dependent (slow component) in the mesenteric vein, and cromakalim opened a K^+

channel with a different permeability to Rb^+ in the guinea-pig urinary bladder than that observed in the rat portal vein (Fig. 1; 7,15).

So far, single-channel current recording using the patch-clamp technique has shown the several K^+ channels are activated by K^+-channel openers. Kusano et al. (36) reported that, in bovine and rabbit cultured aortic smooth muscle cells, cromakalim activated the large conductance Ca^{2+}-dependent K^+ channel (200pS with 150mM K^+ solution on both sides of the membrane). This K^+ channel had similar properties to the so-called maxi-K^+ channel, except in its TEA-sensitivity. Kusano et al. (36) reported that this K^+ channel was resistant to 1mM TEA, whereas the ordinary maxi-K^+ channel was markedly inhibited. Hermsmeyer (25) reported that pinacidil activated the large conductance Ca^{2+}-dependent K^+ channel in the cultivated rat azygos vein (200pS; 4.7mM K^+ and 140mM K^+ solutions on the outside and inside of the membrane, respectively). Furthermore, Gelband et al. (19) reported that cromakalim activated the large-conductance Ca^{2+}-dependent K^+ channel recorded from the microsome of the rabbit thoracic aorta using a planar membrane by increasing the mean open probability due to a reduction of the shut time (slow component) without a change in the mean open time. They concluded that the maxi-K^+ channel was a target for K^+-channel openers, as was the ATP-sensitive K^+ channel in smooth muscle cells. Recently, Hume (27) clearly demonstrated that cromakalim activated, and glibenclamide partly inhibited the activity of the maxi-K^+ channel in rabbit colonic smooth muscle cells. On the other hand, in the rat and rabbit portal veins, neither nicorandil nor pinacidil modified the activity of the large conductance Ca^{2+}-dependent K^+ channel (135 pS: 6 mM K^+ and 140mM K^+ solutions on either of the side membrane), which was sensitive to low concentrations of TEA (μ1 mM) and charybdotoxin but insensitive to 4AP ($\cdot$10 mM) and glibenclamide. Therefore, the target channel for K^+-channel openers is not the maxi-K^+ channel in rat and rabbit portal veins, but probably is this channel in other smooth muscle cells, such as bovine aorta, rabbit colon, renal artery and aorta and rat azygos vein. The single channel conductance and Ca^{2+}-sensitivity of the cromakalim activated K^+ channel in rabbit and bovine aorta (36) was quite similar to those reported for the maxi-K^+ channel in various smooth muscle cells, while the former type of channel was resistant and the latter, maxi-K^+ channel, was very sensitive to TEA (1 mM). Therefore, it is still possible that the large conductance Ca^{2+}-dependent K^+ channel activated by the K^+-channel openers is not identical with the maxi-K^+ channel. At present, we are unsure how to explain the different actions of K^+-channel openers on the activity of the maxi-K^+ channel in various smooth muscle cells. However, the pharmacological features of the hyperpolarization induced by various K^+-channel openers were not identical with those for the maxi-K^+ channel reported in a variety of smooth muscle cells (e.g. Ca^{2+}-dependency, TEA and 4AP-sensitivities etc.). Therefore, the maxi-K^+ channel may not be a major target for activation by K^+-channel openers in the smooth muscle cells, even if it is activated by these drugs.

In addition, Standen et al. (54) found an ATP-sensitive K^+ channel in the smooth muscle cells of the rabbit mesenteric artery and reported that cromakalim activated, but glibenclamide inhibited the open probability of the channel. The single-channel conductance of this ATP-sensitive K^+ channel in the mesenteric artery was larger than those observed in cardiac cells and pancreatic β-cells (135pS; 60mM K^+ and 104mM K^+ solution on either side of the membrane). As in cardiac cells and pancreatic β-cells, activation of the ATP-sensitive K^+ channel by cromakalim in the rabbit mesenteric artery could be seen in Ca^{2+}-free solution (5mM EGTA containing solution), suggesting that it is a Ca^{2+}-insensitive K^+ channel. As simultaneous application of Mg^{2+} with ATP to the inside of the membrane did not modify the inhibitory actions of ATP on the K^+ channel, the ATP-sensitive K^+ channel in the rabbit mesenteric artery possibly should be classified into the cardiac cell type rather than the pancreatic cell type. On the other hand, Kajioka et al. (32) have recently found a different ATP-sensitive K^+ channel in the rabbit portal vein which has a small unitary conductance (15pS in 6mM and 140mM K^+ solutions; 25pS in symmetrical 140mM K^+ solution). Simultaneous application of Mg^{2+} with ATP prevented the ATP-induced inhibition of the channel activity, and this Mg^{2+} effect was different from those observed in the rabbit mesenteric artery and cardiac cells (54). Kajioka et al. (32) also reported that the ATP-sensitive K^+ channel in the rabbit portal vein was easily inactivated by permeabilizing the cells or by excision of the membrane, suggesting A possible involvement of the intracellular regulators for channel activation. In cardiac cells and pancreatic β-cells, channel phosphorylation by ATP was thought to be important for channel reactivation. However, neither ATP nor other adenosine derivatives had any action on the reactivation of the 15pS channel in the rabbit portal vein, suggesting the involvement of different mechanisms for channel reactivation of the ATP-sensitive K^+ channels in cardiac cells as opposed to smooth muscle cells of the rabbit portal vein. Kajioka et al. (32) thought that GDP was the intracellular regulator for reactivation of the ATP-sensitive K^+ channel of the rabbit portal vein, because GDP, but not GDP-β-S, had the ability to reactivate the channel. Although GTP also had a weak channel-reactivating action (about one-tenth of the GDP action), Kajioka et al. (32) speculated that this apparent action of GTP was caused by contamination of the drug with GDP, because GTP is known to be easily hydrolyzed to GDP or GMP during delivery and storage (20% or less of the contents were estimated to be decomposed). GTP-binding protein is not involved in the reactivating mechanism of the ATP-sensitive K^+ channel of the rabbit portal vein, as GTP-γ-S had no action on the activity of this channel. Therefore, these authors speculated that the ATP-sensitive K^+ channel in the rabbit portal vein was in a pre-open state (operative state), when GDP bound at its site of action.

In the rat portal vein, cromakalim activated the outward K^+ current which was blocked by glibenclamide. However, the amplitude of the cromakalim-activated K^+ current in the rat portal vein was dependent on the extracellular Ca^{2+} concetration (43). The single-channel conductance of the K^+ channel was very small (10pS; 6mM K^+ and 140mM K^+ solutions on either side of the membrane) by comparison with that reported in the rabbit mesenteric artery. This K^+ channel was also activated by other K^+-channel openers, such as nicorandil and pinacidil, sensitive to intracellular Ca^{2+} and ATP concentrations, and inhibited by glibenclamide and millimolar concentrations of both TEA and 4AP (Figs. 3 & 4; 38). As ATP could inhibit the nicorandil-activated K^+ channel in the rat portal vein, but ATP in the simultaneous presence of Mg^{2+} had a lesser potency for inhibition, the ATP-sensitive K^+ channel in the rat portal vein was of the pancreatic β-cell type, the same as that in the rabbit portal vein and not like that in the cardiac cell type. They concluded that the K^+ channel activated by K^+-channel openers in the rat portal vein was a kind of ATP-sensitive K^+ channel; however, the properties of the K^+ channel differed from those observed in cardiac and pancreatic cells as well as from those in the rabbit mesenteric artery. Furthermore, in pig coronary artery, Inoue et al. (28) reported that nicorandil opened a novel K^+ channel with a single-channel conductance of 30pS (with 140mM K^+ solution on sides of the both membrane), which was activated by extracellular Ca^{2+} and inhibited by 5mM 4AP.

Using whole-cell voltage-clamp techniques, Beech and Bolton (4) concluded that cromakalim activated the delayed outward current (voltage-dependent K^+ current) in the rabbit portal vein. This was because the sequence of inhibitory actions on the cromakalim-induced K^+ current of various K^+ channel blockers (phencyclidine, quinidine, 4-AP, TEA and procaine) was the same as that observed on the delayed K^+ current, but different from that observed on the maxi-K^+ channel. In the rat portal vein, Okabe et al. (43) reached the opposite conclusion: that cromakalim inhibited the voltage-dependent K^+ current recorded in Mn^{2+} containing solution. Although Beech and Bolton (4) excluded the possible activation of the ATP-sensitive K^+ channel in the rabbit portal vein - because intracellularly applied ATP (1 mM) did not affect the current-voltage relationship of the cromakalim-induced K^+ current - they also found that glibenclamide (50 μM) completely inhibited the cromakalim-induced K^+ current, but only partly inhibited the delayed K^+ current. Using the same preparation (rabbit portal vein), Kajioka et al. (32) found that pinacidil (and lemakalim) activated an ATP-sensitive and intracellular Ca^{2+}-insensitive K^+ channel. Low concentrations of ATP applied intracellularly (100 μM) halved the open probability of the ATP-sensitive K^+ channel, but the simultaneous presence of Mg^{2+} with ATP markedly reduced the inhibitory action of ATP (32). This means that ATP^{4-}, but not Mg-ATP, has the ability to block the ATP-sensitive K^+ channel in the rabbit portal vein. As Beech and Bolton (4) applied ATP (1 mM) intracellularly with 1.2 mM Mg in the pipette solution, the concentration of ATP^{4-} would have been very low and this might be the reason why they failed to see the inhibitory action of ATP on the cromakalim-induced K current. Post et al. (44) also reported that delayed K^+ currents in the smooth muscle cells of the rabbit colon were not activated by K^+-channel openers, as the current activated by these drugs was generated with no delay.

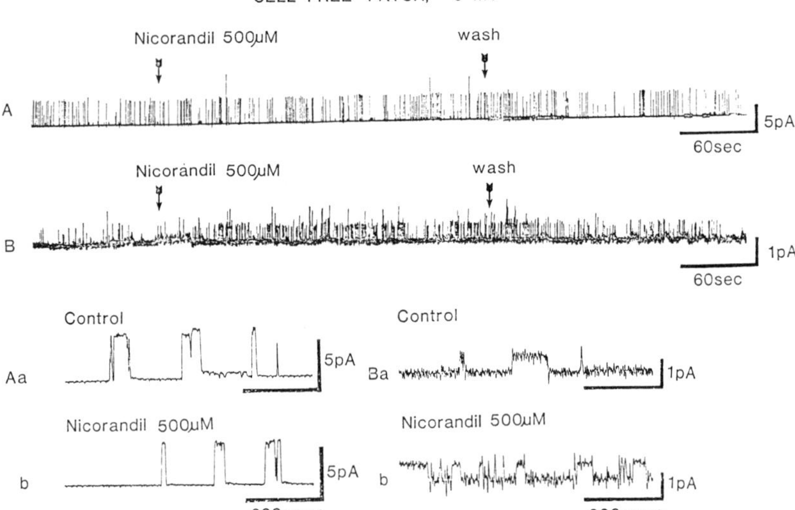

Figure 3. Effects of nicorandil on the large (A) and small (B) conductance K^+ channels recorded from smooth muscle cells of the rat portal vein (outside-out patch configuration). (Kajioka & Kitamura, unpublished observations).

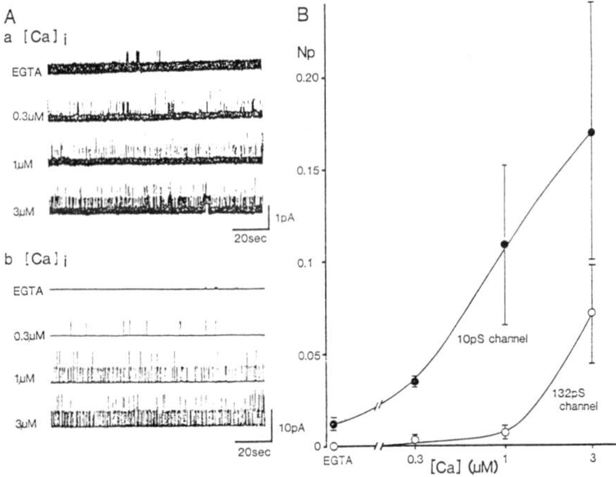

Figure 4. Effects of intracellular Ca^{2+} concentration on activities of the single channel currents of the large and small conductance K^+ channels (inside-out patch configurations). Aa and closed circles in B: small conductance K^+ channels. Ab and open circles in B: large conductance K^+ channels. From Kajioka et al. (33) with permission.

CONCLUSION

In the light of all the reported findings, it is clear that various K^+-channel openers open the K^+ channels sensitive to glibenclamide and intracellular ATP. However, the properties of the glibenclamide- (ATP-) sensitive K^+ channels are not uniform across the range of smooth muscle cells. Although heterogeneity of the K^+ channel activated by K^+-channel openers had been predicted from several results obtained from flux and microelectrode experiments, single-channel current recording using the patch clamp technique has confirmed that such heterogeneity is present in smooth muscle cells.

The physiological roles of the ATP-sensitive K^+ channel in the rat portal vein, pig coronary artery and rabbit mesenteric artery may also differ, because of their intracellular Ca^{2+}-sensitivity. All ATP-sensitive K^+ channels were spontaneously open under resting conditions (without K^+-channel openers). However, the channel activity of the K^+ channel in the rat portal vein should fluctuate during the excitation-relaxation cycle. As this vessel shows spontaneous burst discharge (and, as a consequence, the intracellular Ca^{2+} concentration also fluctuates during the burst cycles) the ATP-sensitive K^+ channel in the rat portal vein may at least participate in the regulation of the duration of the burst discharge of the action potentials. As the nicorandil-activated K^+ channel in the pig coronary artery is activated by extracellular Ca^{2+} ions, this channel plays an important role in producing the resting membrane potential in this cell. By contrast, the ATP-sensitive K^+ channel in the rabbit portal vein is closed in the absence of a K^+-channel opener, and glibenclamide did not modify the resting membrane current and potential, suggesting that there was no contribution to the resting membrane potential in the rabbit portal vein. From this evidence arises another important property of these K^+ channels, that is that endogenous or humoral activating factors should be present to activate the ATP-sensitive K^+ channels. Several intrinsic activators, such as calcitonin-gene related peptide, somatostatin and galanin have already been identified which open the ATP-sensitive K^+ channels in pancreatic b-cells and smooth muscle cells (11-13, 40). Further detailed studies on the ATP-sensitive K^+ and Ca^{2+}-dependent K^+ channels activated by various K^+-channel openers, including the intrinsic activators, are required to clarify the physiological nature of these K^+ channels.

ACKNOWLEDGEMENTS

This work was supported by the Grant-in-Aid for Scientific Research from the Ministry of Science, Education and Culture, Japan.

REFERENCES

1. Allen, S.L., J. Boyle, R.W. Foster, C.P. Morgan, & R.C. Small, Br. J. Pharmacol. 89, 395-405 (1986a).

2. Arena, J.P. & R.S. Kass, Am. J. Physiol. 257, H2092-2096 (1989).

3. Ashcroft, M.J., Ann. Rev. Neurosci. 11, 97-118 (1988).

4. Beech, D.J. & T.B. Bolton, Br. J. Pharmacol. 98, 582-588 (1989).

5. Benham, C.D. & T.B. Bolton, J. Physiol. 340, 469-486 (1983).

6. Benham, C.D. & T.B. Bolton, J. Physiol. 381, 385-406 (1986).

7. Bolton, T.B. & S.P. Lim, J. Physiol. 409, 385-401 (1989).

8. Buckingham, R.E., T.C. Hamilton, D.R. Howlett, S. Mootoo, & C. Wilson, Br. J. Pharmacol. 97, 57-64 (1989).

9. Cook, N.S. & R.P. Hof, Br. J. Pharmacol. 93, 121-131 (1988).

10. Cook, N.S., U. Quast, R.P. Hof, Y. Baumlin, & C. Pally, J. Cardiovasc. Pharmacol. 11, 90-99 (1988).

11. De Weille J.R., H. Schmid-Antomarchi, M. Fosset, & M. Lazdunski, Proc. Natl. Acad. Sci. USA, 85, 1312-1316 (1988).

12. De Weille J.R., H. Schmid-Antomarchi, M. Fosset, & M. Lazdunski, Proc. Natl. Acad. Sci. USA, 86, 2971-2975 (1989).

13. De Weille J.R., M. Fosset, C. Mourre, H. Schmid-Antomarchi, H. Bernardi, & M. Lazdunski, Pflugers Arch. Eur. J. Physiol. 414, (Suppl. 1) S80-S87 (1989).

14. Escande, D., D. Thuringer, S. Le Guern, & I. Cavero, Biochem. Biophys. Res. Commun. 154, 620-625 (1988).

15. Escande, D., D. Thuringer, S. Le Guern, J. Courteix, M. Laville, & I. Cavero, Pflugers Arch. Eur. J. Physiol. 414, 669-675 (1989).

16. Fan, Z., K. Nakayama, & M. Hiraoka, Pflugers Arch. Eur. J. Physiol. 415, 387-394 (1990).

17. Foster, C.D., K. Fujii, J. Kingdon, & A.F. Brading, Br. J. Pharmacol. 97, 281-291 (1989).

18. Furukawa, K., T. Itoh, M. Kajiwara, K. Kitamura, H. Suzuki, Y. Ito, & H. Kuriyama, J. Pharmacol. Exp. Ther. 218, 248-259 (1981).

19. Gelband, C.H., N.J. Lodge, & C. Van Breemen, Eur. J. Pharmacol. 167, 201-210 (1989).

20. Giacchi, G.G., G. Gimenez-Gallego, E. Ber, M.L. Garcia, R. Winquist, & M.A. Cascieri, J. Biol. Chem. 21, 10192-10197 (1988).

21. Hallingworth, M., T. Amedee, D. Edwards, J. Mironneau, J.P. Savineau, R.C Small, & A.H. Weston, Br. J. Pharmacol. 91, 803-813 (1987).

22. Hamilton, T.C., S.W. Weir, & A.H. Weston, Br. J. Pharmacol. 88, 103-111 (1986).

23. Henquin, J.C. & H.P. Meissner, Biochem. Pharmacol. 31, 1407-1415 (1982).

24. Henquin, J.C., Diabetologia 18, 151-160 (1980).

25. Hermsmeyer, K., Drugs 36, (Suppl. 7) 29-32 (1988).

26. Hiraoka, M. & Z. Fan, J. Pharmacol. Exp. Ther. 250, 278-285 (1989).

27. Hume, J., J. Mol. Cell. Cardiol. (1991).

28. Inoue, I., Y. Nakaya, S. Nakaya, & H. Mori, FEBS Lett. 255, 281-284 (1989).

29. Inoue, T., Y. Ito, & K. Takeda, Br. J. Pharmacol. 80, 459-470 (1983).

30. Inoue, T., Y. Kanmura, T. Fujisawa, T. Itoh, & H. Kuriyama, J. Pharmacol. Exp. Ther. 229, 793-802 (1984).

31. Itoh, T., K. Furukawa, M. Kajiwara, K. Kitamura, H. Suzuki, Y. Ito, & H. Kuriyama, J. Pharmacol. Exp. Ther. 218, 260-270 (1981).

32. Kajioka, S., K. Kitamura, & H. Kuriyama, J. Physiol. (1991) (in press).

33. Kajioka, S., M. Oike, & K. Kitamura, J. Pharmacol. Exp. Ther. 254, 905-913 (1990).

34. Karashima, T., T. Itoh, & H. Kuriyama, J. Pharmacol. Exp. Ther. 221, 472-480 (1982).

35. Kashiwabara, T., H. Odai, S. Kaneta, Y. Tanaka, H. Fukushima, & K. Nishikori, Eur. J. Pharmacol. 183, 1266 (1990).

36. Kusano, K., F. Barros, G.M. Katz, L. Roy-Contancin, & J.P. Reuben, Biophys. J. 51, 55a (1987).

37. Leblanc, N., D.W. Wilde, K.D. Keef, & J.R. Hume, Circ. Res. 65, 1102-1111 (1989).

38. Nakao, K., K. Okabe, K. Kitamura, H. Kuriyama, & A.H. Weston, Br. J. Pharmacol. 95, 795-804 (1988).

39. Nakayama, K., Z. Fan, F. Marumo, & M. Hiraoka, Circ. Res. 67, 1124-1133 (1990).

40. Nelson, M.T., Y. Huang, J.E. Brayden, J. Hesceler, & N.B. Standen, Nature 344, 770-773 (1990).

41. Noma, A., Nature 305, 147-148 (1983).

42. Ohya, Y., K. Kitamura, & H. Kuriyama, Am. J. Physiol. 251, C335-C346 (1987).

43. Okabe, K., S. Kajioka, K. Nakao, K. Kitamura, H. Kuriyama, & A.H. Weston, J. Pharmacol. Exp. Ther. 250, 832-839 (1990).

44. Post, J.M., R.J. Stevens, K.M. Sanders, & J.R. Hume, Am. J. Physiol. 260, C375-C382 (1991).

45. Quast, U. & N.S. Cook, Br. J. Pharmacol. 93, 204p (1989).

46. Quast, U. & N.S. Cook, Life Sci. 42, 805-810 (1988).

47. Quast, U. Br. J. Pharmacol. 91, 569-578 (1987).

48. Sakai, T., K. Terada, K. Kitamura, & H. Kuriyama, Br. J. Pharmacol. 95, 1089-1100 (1988).

49. Sanquinetti, M.C., A.L. Scott, G.J. Zingaro, & P.K. Siegel, Proc. Natl. Acad. Sci. USA, 85, 8360-8364 (1988).

50. Schmidt-Antomarch, H., J.R. De Weille, M. Fosset, & M. Lazdunski, J. Biol. Chem. 262, 15840-15844 (1987a).

51. Schmidt-Antomarch, H., J.R. De Weille, M. Fosset, & M. Lazdunski, Biochem. Biophys. Res. Commun. 146, 21-25 (1987b).

52. Shetty, S.S. & G.B. Weiss, Eur. J. Pharmacol. 141, 485-488 (1987).

53. Southerton, J.S., A.H. Weston, K.M. Bray, D.T. Newgreen, & S.G. Tailor, Naunyn- Schmiedeberg's Arch. Pharmacol. 338, 310-318 (1988).

54. Standen, N.B., J.M. Quayle, N.W. Davies, J.E. Brayden, Y. Huang, & M.T. Nelson, Science 245, 177-180 (1989).

55. Sturgess, N.C., R.Z. Kozolowski, C.A. Carrington, C.N. Hales, & M.L.J. Ashford, Br. J. Pharmacol. 95, 83-94 (1988).

56. Taira, N., K. Satoh, T. Yanagisawa, Y. Imai, & M. Hiwatari, Clin. Exp. Pharmacol. Physiol. 6, 301-316 (1979).

57. Trube, G., P. Rorsman, & T. Ohno-Shosaku, Pflugers Arch. Eur. J. Physiol. 407, 493-499 (1986).

58. Uchida, Y., N. Yoshimoto, & S. Murao, Jpn. Heart J. 19, 112-124 (1978).

59. Weir, S.W. & A.H. Weston, Br. J. Pharmacol. 88, 113-120 (1986a).

60. Weir, S.W. & A.H. Weston, Br. J. Pharmacol. 88, 121-128 (1986b).

61. Weston, A.H., K.M. Bray, S. Duty, A.D. McHarg, D.T. Newgreen, & J.S. Southerton, Drugs 36, (Suppl. 7) 10-28 (1988).

62. Wilson, C., J. Autonom. Pharmacol. 9, 9-16 (1988).

63. Xiong, Z., S. Kajioka, T. Sakai, K. Kitamura, & H. Kuriyama, Br. J. Pharmacol. <u>102</u>, 788-790 (1991).
64. Yamanaka, K., K. Furukawa, & K. Kitamura, Naunyn-Schmiedeberg's Arch. Pharmacol. <u>331</u>, 96-103 (1985).
65. Yanagisawa, T. & N. Taira, Naunyn-Schmiedeberg's Arch. Pharmacol. <u>312</u>, 69-76 (1980).
66. Zunkler, B.J., S. Lenzen, K. Manner, U. Panten, & G. Trube, Naunyn-Shmiedeberg's Arch. Pharmacol. <u>337</u>, 225-230 (1988).

CORONARY VASODILATION BY K$^+$ CHANNEL OPENERS

TERUYUKI YANAGISAWA and NORIO TAIRA

Department of Pharmacology, Tohoku University School of Medicine, Sendai 980, Japan

INTRODUCTION

Increasing attention has been paid to vasodilator drugs which activate or open K$^+$ channels in vascular smooth muscle and which eventually increase coronary blood flow or lower blood pressure. These drugs have been called "K$^+$ channel activators or openers" and include nicorandil (41, 55), pinacidil (1, 2) and cromakalim (19, 43, 44), although nicorandil also has a nitrate-like action (39, 40). The mechanism of action of nicorandil to open K$^+$ channels was at first found in cardiac muscles by us (51, 52, 55). In canine atrial muscle, these three K$^+$ channel openers reduce the force of contraction exclusively by opening K$^+$ channels, because their negative inotropic effects are totally blocked by the quaternary ammonium K$^+$ channel blockers, tetraethylammonium (TEA) or tetrabutylammonium (TBA) (53, 54). In vascular smooth muscle cells the resting membrane potential is less negative than the K$^+$ equilibrium potential (15). When the membrane is hyperpolarized towards the K$^+$ equilibrium potential by K$^+$ channel openers, excitation-induced Ca^{2+} influx *via* voltage-dependent Ca^{2+} channels is thought to decrease (8, 46). However, no direct measurement of the effects of K$^+$ channel openers on intracellular Ca^{2+} concentration ([Ca^{2+}]$_i$) has yet been reported. Furthermore, as K$^+$ channel openers are structurally different, it is possible that some possess other kinds of vasodilator mechanisms (9).

Although [Ca^{2+}]$_i$ is one of the main determinants of mechanical activity of vascular smooth muscle, various reports have shown that it is not the only determinant of this activity (23, 24). We have recently developed a microfluorimetric method, using the Ca^{2+} indicator, fura-2 (18), to measure simultaneously [Ca^{2+}]$_i$ with force in isolated vascular rings (56). In this article we will show the effects of K$^+$ channel openers on [Ca^{2+}]$_i$ and tension of isolated canine coronary arterial rings in comparison with nitroglycerin and verapamil. We further will show how their effects are modified by the nonselective K$^+$ channel blocker, TBA (32, 38) and the supposed adenosine triphosphate (ATP)-sensitive K$^+$ channel (K$_{ATP}$) blocker, glibenclamide (4, 7, 8), to identify what kind of K$^+$ channels are involved.

Since the isolated coronary arterial rings are essentially conduit arteries, we would like further to show the effects of K$^+$ channel openers on coronary arterial blood flow which indicates the mechanical activity of resistance arterioles by using the isolated-blood perfused papillary muscle preparations in comparison with nitroglycerin.

SPECIFIC K$^+$ CHANNEL OPENING ACTION AND [Ca^{2+}]$_i$

In canine coronary artery loaded with fura-2, depolarization by 30 - 90 mM KCl-phyiological salt solution (PSS) produced a concentration-dependent increase in [Ca^{2+}]$_i$ and force. A reduction of [Ca^{2+}]$_i$ below the basal level did not affect the tone while the threshold increase in [Ca^{2+}]$_i$ necessary to generate tension was about 10% of basal level. The observed increase in [Ca^{2+}]$_i$ and tension was verapamil-sensitive and thus almost certainly associated with an increased Ca^{2+} influx through voltage-dependent L-type Ca^{2+} channels, (16, 56).

In 5 mM KCl-PSS, cromakalim, pinacidil and nicorandil reduced [Ca^{2+}]$_i$ from the basal level, whereas the resting tone was not affected. When extracellular Ca^{2+} was removed by perfusion with Ca^{2+}-free-PSS containing 1 mM EGTA, [Ca^{2+}]$_i$ was reduced further. In the presence of 30 mM KCl-PSS, cromakalim, pinacidil and nicorandil reduced both [Ca^{2+}]$_i$ and contracture in a concentration-dependent manner, however, contraction was nearly abolished by pinacidil or nicorandil at their maximum concentrations, where a significantly increased [Ca^{2+}]$_i$ levels were remaining (Fig.1a, b; Table 1). There was no difference between the respective pD$_2$ values for cromakalim- or pinacidil-induced changes in both [Ca^{2+}]$_i$ and force observed in the presence of 5 mM and 30 mM KCl, although the

Published 1991 by Elsevier Science Publishing Company, Inc.
Ion Channels of Vascular Smooth Muscle Cells and Endothelial Cells
Sperelakis and Kuriyama, Editors

membrane potentials should have been different (Table 1). When the arterial rings were perfused with 45 mM or higher concentrations of KCl-PSS, cromakalim had no effect on either the force of contraction or the increased $[Ca^{2+}]_i$, whereas pinacidil or nicorandil inhibited the force of contraction in a concentration-dependent manner. However, the increased $[Ca^{2+}]_i$ was not changed by pinacidil or nicorandil at all (Table 1). Even during the washout of pinacidil or nicorandil the increased $[Ca^{2+}]_i$ did not change at all (Fig. 1c). In 45 and 90 mM KCl-PSS, cromakalim failed to change the increased $[Ca^{2+}]_i$ and force as has been shown in rat blood vessels (45). Thus, the effects of K^+ channel openers in reducing the increased $[Ca^{2+}]_i$ may only be achieved when the resultant membrane potential associated with the presence of a raised KCl concentration is less negative than the equilibrium potential for K^+ (15, 52) and when the K^+ channel openers can hyperpolarize the membrane potential to reduce the opening probability of voltage-dependent Ca^{2+} channels (45).

EFFECT OF VERAPAMIL ON THE INCREASED $[Ca^{2+}]_i$ INDUCED BY HIGH KCl-PSS IN THE PRESENCE OF PINACIDIL OR NICORANDIL

The observation that pinacidil or nicorandil relaxed coronary artery contracted by 45 or 90 mM KCl-PSS without changing the increased $[Ca^{2+}]_i$ promoted a series of experiments to determine whether the change in $[Ca^{2+}]_i$ was verapamil-sensitive. Fig. 1d shows the results of such experiments for pinacidil in arterial rings exposed to 45 mM KCl-PSS. Without changing the increased $[Ca^{2+}]_i$ pinacidil (10^{-4} M) reduced the force of contraction induced by 45 mM and 90 mM KCl-PSS to 22.4 $\pm$ 6.3% (n = 7) and 47.9 $\pm$ 11.6% (n = 6) (force produced by 45 mM or 90 mM KCl-PSS at 10 min = 100%), respectively. In the presence of 45 mM KCl-PSS containing 10^{-4} M pinacidil, addition of 10^{-5} M verapamil abolished the remaining force and the increased $[Ca^{2+}]_i$ (Fig. 1d). 10^{-3} M nicorandil reduced the force of contracture induced by 45 mM or 90 mM KCl-PSS to 15.9 $\pm$ 5.8% (n = 6) and 24.9 $\pm$ 2.2% (n = 8), respectively, with no effect on the increased $[Ca^{2+}]_i$. Exposure to 10^{-5} M verapamil in the presence of 10^{-3} M nicorandil abolished the remaining force and the increased $[Ca^{2+}]_i$. Since this increased $[Ca^{2+}]_i$ was reduced by 10^{-5} M verapamil to basal levels, it was due to an influx of extracellular Ca^{2+} through voltage-dependent Ca^{2+} channels.

Table 1. Effects of K+ channel openers, nitroglycerin and verapamil on the the force of contraction and $[Ca^{2+}]_i$

KCl (mM)	(n)	Reduction of force			Reduction of $[Ca^{2+}]_i$		
		Max. (%)	pD_2	P	Max. (%)	pD_2	P
Cromakalim							
5	(7)	no effect ($<10^{-5}$ M)			60.1±10.5	6.66±0.17	1.07±0.04
30	(7)	45.1± 6.4	6.04±0.20	1.11±0.13	40.1± 5.5	6.55±0.11	0.94±0.18
45	(6)	no effect ($<10^{-5}$ M)			no effect ($<10^{-5}$ M)		
Pinacidil							
5	(7)	no effect ($<3X10^{-4}$ M)			79.5± 8.6	5.40±0.11	1.05±0.10
30	(6)	100[a]	5.11±0.08	0.79±0.06	44.8± 6.7	4.94±0.16	1[a]
45	(7)	100[a]	4.47±0.04	1.07±0.08	no effect ($<3X10^{-4}$ M)		
90	(5)	100[a]	4.16±0.33	1.21±0.26	no effect ($<10^{-4}$ M)		
Nicorandil							
30	(7)	100[a]	4.33±0.18	1.04±0.17	35.9± 4.0	4.57±0.19	1[a]
45	(6)	100[a]	3.72±0.04	1.29±0.19	no effect ($<10^{-3}$ M)		
90	(8)	100[a]	3.64±0.12	1.02±0.13	no effect ($<10^{-3}$ M)		
Nitroglycerin							
30	(6)	100[a]	6.25±0.19	0.65±0.06	no effect ($<10^{-5}$ M)		
60	(6)	100[a]	5.54±0.10	0.62±0.09	no effect ($<10^{-5}$ M)		
90	(5)	100[a]	5.35±0.17	0.54±0.06	no effect ($<10^{-5}$ M)		
Verapamil							
45	(5)	100[a]	6.26±0.06	1.45±0.15	100[a]	6.34±0.24	0.87±0.10
90	(5)	100[a]	6.15±0.08	1.47±0.18	100[a]	5.99±0.16	0.72±0.28

The concentration-effect curves were analyzed by computer fitting to a logistic equation:
$$E=100-M \times A^P/(A^P+K^P)$$
where E is normalized response (basal value=100%), M is maximum response (Max.), A is drug concentration, K is EC_{50} value of each drug and P is slope parameter (29). EC_{50} values were presented as pD_2 (pD_2=-log EC_{50}). [a]: M and P were fixed 100 and 1, respectively (adopted from Refs. 56, 57).

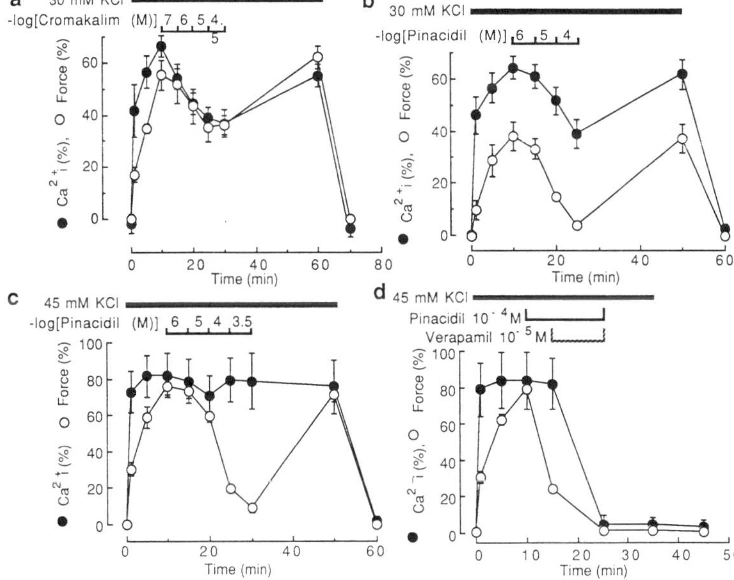

Figure 1. a. Effects of cromakalim (10^{-7}-3 x 10^{-5} M) on the force of contraction and increased $[Ca^{2+}]_i$ produced by 30 mM KCl-PSS in 7 canine coronary arterial rings. The force of contracture and the increased $[Ca2+]i$ are expressed as percentage of 10-min responses to 90 mM KCl-PSS. **b.** The summarized data of the effects of pinacidil (10^{-6}-10^{-4} M) on force of contraction and the increased $[Ca^{2+}]_i$ in 6 canine coronary arteries depolarized by 30 mM KCl-PSS. **c.** Summarized data of the effects of pinacidil (10^{-6}-3 x 10^{-4} M) obtained from 8 muscles perfused with 45 mM KCl. **d.** Effects of 10^{-4} M pinacidil and 10^{-5} M verapamil on changes in the increased $[Ca^{2+}]_i$ and force of contracture produced by 45 mM KCl-PSS (n = 7). (Adopted and modified from Ref. 57).

MODIFICATION BY GLIBENCLAMIDE OR TBA OF THE EFFECTS OF K$^+$ CHANNEL OPENERS

Fig. 2a shows modification by glibenclamide of the effects of 10^{-5} M cromakalim on the force of contracture and the increased $[Ca^{2+}]_i$ in canine coronary artery perfused with 30 mM KCl-PSS. Cromakalim (10^{-5} M) reduced both $[Ca^{2+}]_i$ and the force of contraction. When glibenclamide (10^{-6} and 10^{-5} M) was applied 10 min after perfusion with 30 mM KCl-PSS, no change in the increased $[Ca^{2+}]_i$ or force was observed. In contrast, the ability of cromakalim to reduce the increased $[Ca^{2+}]_i$ and force were greatly attenuated by exposure to glibenclamide at 10^{-6} M and abolished at 10^{-5} M (Fig. 2a). Although TEA is a very useful agent to block K$^+$ channels in various tissues, in canine coronary arterial rings, TEA is far less potent, *i.e.*, more than 10^{-2} M is necessary to block the effects of cromakalim. Thus, we chose TBA to block K$^+$ channels. The effects of cromakalim were also blocked by 10^{-3} M TBA. Although TBA per se had little effect on the force of contracture and the increased $[Ca^{2+}]_i$ in 30 mM KCl-PSS, its effects persisted after washout, since $[Ca^{2+}]_i$ after washout in 5 mM KCl-PSS was higher than the basal level. Thus, the blocking action of TBA may be due to the intracellular TBA molecules. Very recently, the putative intracellular and extracellular TEA binding sites have been elucidated by site-directed mutagenesis in *Drosophila Shaker* gene (20, 58). TBA (10^{-3} M) or glibenclamide (10^{-5} M) antagonized the ability of nicorandil (10^{-4} M and 10^{-3} M) or pinacidil (10^{-4} M and 3 x 10^{-4} M) to reduce $[Ca^{2+}]_i$ following exposure to 30 mM KCl-PSS, but had a small effect on nicorandil- or pinacidil-induced inhibition of contraction (Fig. 2b). In 5 mM KCl-PSS glibenclamide affected neither $[Ca^{2+}]_i$ nor the resting tension whereas TBA slightly increased $[Ca^{2+}]$ by about 10% without changing the resting tension

154

(data not shown). In normal (5 mM KCl) PSS, cromakalim, pinacidil and nicorandil reduced $[Ca^{2+}]_i$ below the basal level and such effects induced by pinacidil and nicorandil were blocked by TBA. This suggests that certain Ca^{2+} channels are open in normal PSS and that a presumed hyperpolarization of the plasma membrane by a K^+ channel opener is capable of closing this Ca^{2+}-entry pathway.

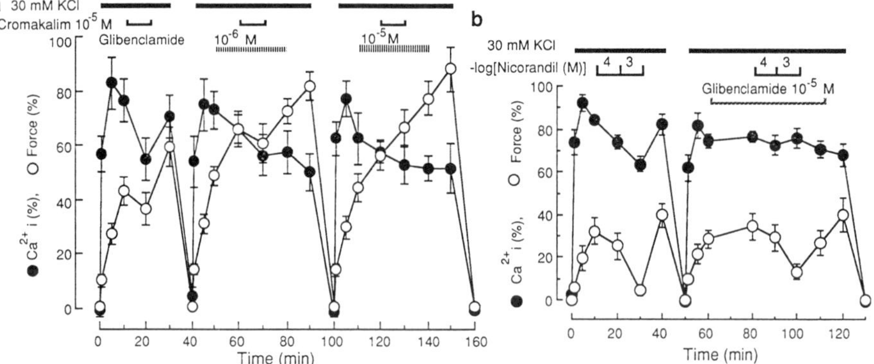

Figure 2. a. Modification by glibenclamide (10^{-6} and 10^{-5} M) of the effects of cromakalim (10^{-5} M) on the force of contraction and the increased $[Ca^{2+}]_i$ in canine coronary artery depolarized by 30 mM KCl-PSS (n = 5-9). When glibenclamide (10^{-6} and 10^{-5} M) was applied 10 min after perfusion with 30 mM KCl-PSS, no change in the increased $[Ca^{2+}]_i$ or force was observed. In contrast, the ability of cromakalim to reduce the increased $[Ca^{2+}]_i$ and force were greatly attenuated by exposure to glibenclamide at 10^{-6} M and abolished at 10^{-5} M. **b.** Modification by glibenclamide (10^{-5} M) of the effects of nicorandil (10^{-4}-10^{-3} M) in 6 canine coronary arterial muscles perfused with 30 mM KCl-PSS. Glibenclamide abolished the ability of nicorandil to reduce $[Ca^{2+}]_i$, but had a small effect on nicorandil-induced inhibition of contraction. (Adopted and modified from Ref. 57).

Although the vasodilator effect of K^+ channel openers is blocked by several nonselective K^+ channel blockers such as procaine, 4-aminopyridine and TEA (47), the identity of the K^+ channels involved has been uncertain. Recently, sulphonylureas like glibenclamide have been reported to block cromakalim-induced changes in cardiac (12, 14, 33, 34) and vascular smooth muscle (3, 10, 31, 49) suggesting a role for ATP-sensitive K^+ channels (37). In the present study glibenclamide blocked the effects of the three K^+ channel openers in reducing $[Ca^{2+}]_i$, although rather high concentrations ($>10^{-6}$ M) were needed. The finding that glibenclamide affected neither $[Ca^{2+}]_i$ nor force in both normal and high KCl-PSS indicates that glibenclamide-sensitive K^+ channels are closed under these conditions. And the finding that TBA increased $[Ca^{2+}]_i$ in normal KCl-PSS indicates that internal TBA blocks not only glibenclamide-sensitive K^+ channels (probably K_{ATP}) but also other K^+ channels opened under resting condition.

EFFECTS OF VERAPAMIL, NITROGLYCERIN, PINACIDIL AND NICORANDIL ON THE INCREASED $[Ca^{2+}]_i$ INDUCED BY HIGH KCl-PSS

To characterize the vasodilating effects of K^+ channel openers, it will be informative to examine and compare the effects of verapamil as a Ca^{2+} channel blocker or the effects of nitroglycerin as a classical nitrate in canine coronary arterial smooth muscles.

$[Ca^{2+}]_0$ and Force or $[Ca^{2+}]_i$ Relationship

Removal of extracellular Ca^{2+} produced a reduction of $[Ca^{2+}]_i$ below the basal level (-27 ± 2%; expressed as a % of the increase following 10-min exposure to 90 mM KCl-PSS), whereas the force was not changed. After removal of extracellular Ca^{2+} by perfusion with Ca^{2+}-free PSS, exposure of the muscle to 90 mM KCl failed to produce any increase in $[Ca^{2+}]_i$ (-31 $\pm$ 2%) or any change in force. After the perfusion of Ca^{2+}-free 90 mM KCl-PSS, the extracellular Ca^{2+} concentration ($[Ca^{2+}]_o$) was increased from 0.03 to 15 mM. As $[Ca^{2+}]_o$ was increased, $[Ca^{2+}]_i$ was increased and force was generated (Fig. 3a,b; Control). The results of computer fitting of the curve to the logistic equation are summarized in Table 2. The maximum is the difference between values at Ca^{2+}-free and 15 mM $[Ca^{2+}]_o$. The pD2 values for $[Ca^{2+}]_o$ of 3.05 $\pm$ 0.23 indicate the apparent K_D values of $[Ca^{2+}]_o$ for the binding constant at voltage-dependent Ca^{2+} channels opened by depolarization. The relationship between $[Ca^{2+}]_o$ and force of contracture was also analyzed and is summarized in Table 2. The maximum force produced by 90 mM KCl was 239 $\pm$ 41%, pD2 values for $[Ca^{2+}]_o$ were 2.62 $\pm$ 0.14, and the slope parameter was 1.22 $\pm$ 0.12. The difference between pD2 values for $[Ca^{2+}]_o$ of the increase in $[Ca^{2+}]_i$ and the increase in force are a consequence of the steepness of $[Ca^{2+}]_i$-force relationship (Fig. 4a-c; Control).

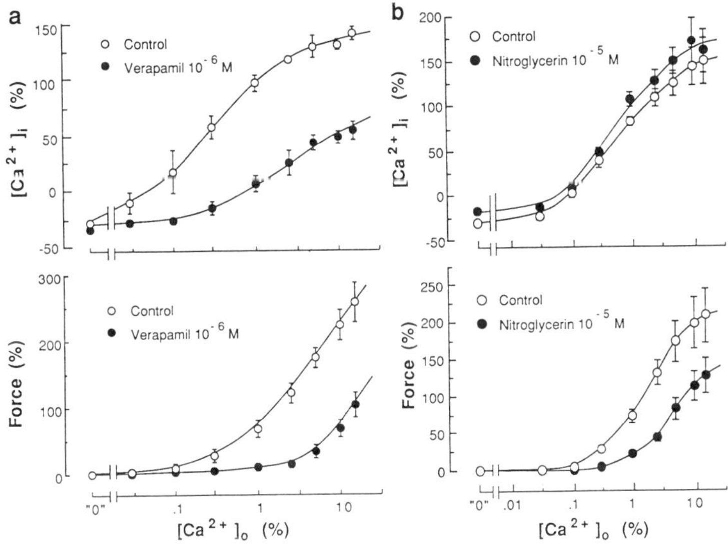

Figure 3. Changes in $[Ca^{2+}]_i$ and force produced by increasing $[Ca^{2+}]_o$ in canine coronary arterial muscles depolarized 90 mM KCl-PSS and the effects of (a) verapamil or (b) nitroglycerin on these changes. (Adopted and modified from Ref. 56).

Table 2. Effects of verapamil (10^{-6} M), nitroglycerin (10^{-5} M), pinacidil (10^{-4} M), and nicorandil (3×10^{-4} M) on the $[Ca^{2+}]_o$-force and $[Ca^{2+}]_o$-$[Ca^{2+}]_i$ relationship in 90 mM KCl-depolarized canine coronary artery

	n	Increase in force			Increase in $[Ca^{2+}]_i$		
		Max. (%)	pD2	p	Max. (%)	pD2	p
Control	5	376±78	2.44±0.27	1.04±0.06	158±35	3.23±0.29	1.11±0.24
Verapamil	5	223±45*	1.86±0.09*	1.67±0.23	64± 9*	2.80±0.14*	1.15±0.14
Control	6	225±44	2.73±0.05	1.17±0.04	190±37	3.19±0.13	0.83±0.19
Nitroglycerin	6	155±32*	2.32±0.04*	1.19±0.04	189±29	3.23±0.09	1.00±0.16
Washout	6	251±61	2.83±0.01	1.12±0.01	174±30	3.17±0.10	0.81±0.22
Control	6	239±41	2.62±0.14	1.22±0.12	160±23	3.05±0.23	0.96±0.14
Pinacidil	6	82±29*	2.05±0.15*	1.39±0.17	146±24	2.86±0.18	0.93±0.08
Washout	6	217±41	2.71±0.07	1.30±0.09	150±20	3.17±0.28	0.96±0.15
Control	5	309±85	2.53±0.06	1.39±0.07	174±25	2.99±0.07	0.88±0.11
Nicorandil	5	138±24*	1.97±0.13*	1.29±0.07	163±19	2.93±0.14	0.90±0.11
Washout	5	285±63	2.59±0.06	1.37±0.23	160±30	3.11±0.10	0.93±0.09

Increase in $[Ca^{2+}]_i$ was calculated from the difference between the value obtained in Ca^{2+}-free 90 mM KCl-PSS and the maximum value of $[Ca^{2+}]_i$ in each experiment (usually at 15 mM $[Ca^{2+}]_o$). * $P<0.05$ compared with control values (adopted from Refs. 56, 57).

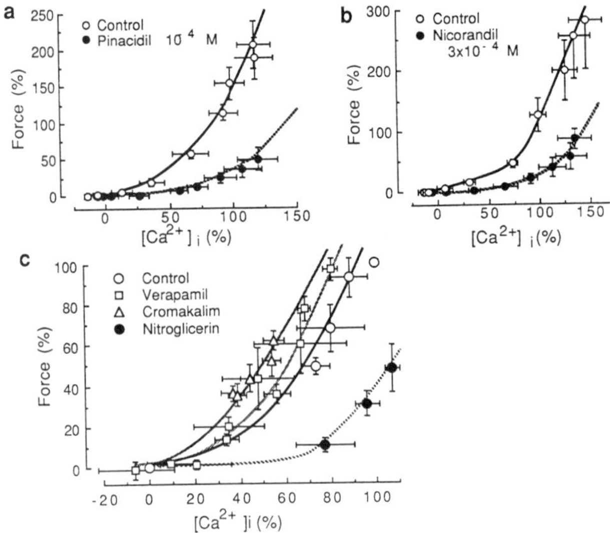

Figure 4. Relationship between $[Ca^{2+}]_i$ and force in the absence (control), and the presence of (a) 10^{-4} M pinacidil, (b) 3×10^{-4} M nicorandil and (c) cromakalim (10^{-7}-3×10^{-5} M), verapamil (10^{-7}-10^{-5} M) or nitroglycerin (10^{-5} M) in depolarized canine coronary artery. Vertical and horizontal bars are SEM. The curves were fitted by eye. (Adopted and modified from Refs. 56, 57).

Effects of Verapamil

When verapamil 10^{-6} M was applied to coronary arterial muscle perfused with Ca^{2+}-free 90 mM KCl-PSS, $[Ca^{2+}]_i$ was not changed. Thus, in the absence of extracellular Ca^{2+}, verapamil does not change $[Ca^{2+}]_i$ even the Ca^{2+} channels were opened by depolarization. In the presence of verapamil, the increases in $[Ca^{2+}]_o$ in 90 mM KCl-PSS induced smaller increases in $[Ca^{2+}]_i$, and the force of contraction was reduced. The inhibitory manner of verapamil on the increase in $[Ca^{2+}]_i$ seems to be a noncompetitive manner with the reduction of pD_2 values for $[Ca^{2+}]_o$ (Table 2). When the concentration of verapamil was increased to 10^{-5} M, the increases in $[Ca^{2+}]_i$ and force of contraction were abolished (data not shown). Thus, 10^{-5} M of verapamil is the suitable concentration for verapamil to eliminate the influx of Ca^{2+} through voltage-dependent Ca^{2+} channels.

Effects of Nitroglycerin

Nicorandil contains a nitroxy moiety within in its structure and is known to increase intracellular levels of cyclic GMP by activating guanylate cyclase in various vascular smooth muscle tissues (11, 21) and in cardiac muscle (53). The resultant activation of cyclic GMP dependent protein kinase is thought to be responsible for the vasodilator effect of nitrates (42). Thus, it is interesting to examine the effects of nitroglycerin as a classical nitrate on KCl-induced increased $[Ca^{2+}]_i$ and force of contraction (Fig. 3b) to compare with those of nicorandil. After the perfusion of Ca^{2+}-free 90 mM KCl-PSS, the extracellular Ca^{2+} concentration ($[Ca^{2+}]_o$) was increased from 0.03 to 15 mM. As $[Ca^{2+}]_o$ was increased, $[Ca^{2+}]_i$ was increased and force was generated. When the nitroglycerin 10^{-5} M was applied to coronary arterial muscle perfused with Ca^{2+}-free 90 mM KCl-PSS, $[Ca^{2+}]_i$ was not changed. Although increases in $[Ca^{2+}]_o$ produced the same increases in $[Ca^{2+}]_i$, the force produced by the increase in $[Ca^{2+}]_i$ was much smaller than that of control. Since the relationship between $[Ca^{2+}]_o$ and $[Ca^{2+}]_i$ was not changed by nitroglycerin, membrane potential and the influx of Ca^{2+} may have not been changed by nitroglycerin. Fig. 3b also shows the relaxant effect of nitroglycerin was larger in low concentration of $[Ca^{2+}]_i$ than in high concentrations. Thus, these results clearly show that nitroglycerin does not affect Ca^{2+} influx through Ca^{2+} channels, that its relaxant effect is operative without reducing $[Ca^{2+}]_i$ and that the relaxant potency of nitroglycerin is dependent on $[Ca^{2+}]_i$ but not on the degree of depolarization.

Effects of Pinacidil and Nicorandil

The relaxant potency of both pinacidil and nicorandil was dependent on the KCl concentration (Table 1). To clarify whether the extent of depolarization or the increase in $[Ca^{2+}]_i$ has the major influence on the relaxant effects of pinacidil and nicorandil, coronary muscle was at first perfused with Ca^{2+}-free PSS for 10 min and then with Ca^{2+}-free 90 mM KCl-PSS. During exposure to 90 mM KCl-PSS, $[Ca^{2+}]_o$ was increased from 0.03 to 15 mM with a consequent increase in both $[Ca^{2+}]_i$ and force. When pinacidil (10^{-4} M) or nicorandil (3×10^{-4} M) was applied to coronary arterial muscle perfused with Ca^{2+}-free 90 mM KCl-PSS, $[Ca^{2+}]_i$ was not changed. Although increases in $[Ca^{2+}]_o$ produced the same increases in $[Ca^{2+}]_i$ in the presence of pinacidil or nicorandil as in control, the force produced by the increased $[Ca^{2+}]_i$ was much smaller than that of control and the threshold values for $[Ca^{2+}]_o$ and $[Ca^{2+}]_i$ required to generate force were increased. Furthermore, the maximum force and the pD_2 values for $[Ca^{2+}]_o$ were significantly reduced in the presence of pinacidil or nicorandil. The inhibitory effects of pinacidil or nicorandil were reversible (Table 2).

EFFECTS OF K^+ CHANNEL OPENERS, VERAPAMIL AND NITROGLYCERIN ON THE RELATIONSHIP BETWEEN $[Ca^{2+}]_i$ AND FORCE IN KCl-DEPOLARIZED CORONARY ARTERY

Fig. 4 shows the relationship between $[Ca^{2+}]_i$ and force in the absence and the presence of pinacidil (10^{-4} M; Fig. 4a), nicorandil (3×10^{-4} M; Fig. 4b), cromakalim (10^{-7}-3 $\times 10^{-5}$ M; Fig. 4c), verapamil (10^{-7}-10^{-5} M; Fig. 4c) or nitroglycerin (10^{-5} M; Fig. 4c) in depolarized canine coronary artery. The curve in the presence of pinacidil, nicorandil or nitroglycerin was different from that of control and was shifted to the right and downward. These results obtained from the analysis of $[Ca^{2+}]_i$-tension relationship in the presence of pinacidil, nicorandil or nitroglycerin suggest that these vasodilators reduce the Ca^{2+} sensitivity of the contractile proteins. It is conceivable that nicorandil partly relaxes smooth muscle cells without reducing the increased $[Ca^{2+}]_i$ following Ca^{2+} influx by depolarization *via* mechanisms like those of nitroglycerin. Unlike nicorandil, pinacidil does not change cyclic GMP levels in either vascular smooth muscle (42) or in cardiac muscle (53). Thus, pinacidil exerts some of its relaxing effects independent of the involvement of cyclic nucleotides by a mechanism which remains to be determined.

The relaxant potencies of pinacidil and nicorandil in 90 mM KCl-PSS were reduced at the higher concentrations of the $[Ca^{2+}]_o$. This phenomenon was due to the changed relationship between $[Ca^{2+}]_i$ and force observed in the presence of pinacidil or nicorandil. A similar rightward shift of the $[Ca^{2+}]_i$-tension relationship by cyclic GMP-dependent mechanism has previously been reported (26, 30). The mechanism involved is uncertain but in the case of nicorandil, a cyclic GMP-mediated inhibition of myosin light chain phosphorylation may have been counteracted by the increase in $[Ca^{2+}]_i$.

N-K HYBRIDS AS VASODILATOR DRUGS

Thus, it is likely that nicorandil partly relaxes smooth muscle cells without reducing the increased $[Ca^{2+}]_i$ following Ca^{2+} influx by depolarization *via* mechanisms like those of nitroglycerin (21). We would like to call such a vasodilating mechanisms as a "N-K hybrid" indicating dual *i.e.*, nitrate (N) and potassium (K^+) channel activating, mechanisms in one molecule (39). Very recently another "N-K hybrid", KRN2391 (N-cyano-N'-(2-nitroxyethyl)-3-pyridinecarboximidamide methansulfonate), which has vasodilating and antihypertensive effects, has appeared (25, 28). The structure-activity relationships of pyridine derivatives as K^+ channel openers indicates that the nitroxy substituent is very important for this activity (28). Such a structure-activity relationship has been already shown for nicorandil (for review 9).

NICORANDIL INCREASES CORONARY BLOOD FLOW PREDOMINANTLY BY K^+ CHANNEL OPENING MECHANISM

As previously described, nicorandil is an antianginal drug with a hybrid property between nitrates and potassium (K^+) channel openers (39, 40). Nicorandil is highly vasoselective like parent compounds and in only very high doses which are clinically unavailable it exerts cardiac effects such as a negative inotropic effect, and abbreviations of cardiac action potentials and the refractory period of myocardium (41, 51, 52, 55). These, cardiac effects of nicorandil are explainable solely in terms of, K^+ channel opening (51-54). The mechanisms of action of nicorandil as a nitrate and as a K^+ channel opener both lead to vascular relaxation. However, it has been proposed that an increase in coronary blood flow produced by nicorandil be ascribable to the action as a K^+ channel opener, based on close similarity of the effect to that of cromakalim (39). Cromakalim has been shown to be more specific as a K^+ channel opener than nicorandil (6, 44). Nevertheless, it has not been fully settled whether an increase in coronary blood flow produced by nicorandil is really due to its action as a K^+ channel opener. Thus, we have wanted to address this question by using the blood-perfused papillary muscle preparation (59). Glibenclamide, an oral antidiabetic drug, whose mechanism of insulin releasing action involves blocking of ATP-dependent K^+ channels in pancreatic B-cells (36), has recently been shown to behave as a pharmacological antagonist of K^+ channel openers (3, 5, 10, 12, 14, 31, 33-35, 37, 49). Thus, in the present study the question raised was addressed by use of glibenclamide as a pharmacological tool, and cromakalim and nitroglycerin, a prototype of classic nitrates, as reference drugs.

Effects of Nicorandil, Cromakalim, and Nitroglycerin

As in previous experiments (17, 41, 55), under control conditions single injections of nicorandil (0.01 to 10 mmol) or cromakalim (0.001 to 0.3 mmol) into the anterior septal artery produced a dose-dependent increase in blood flow through this artery (simply called coronary blood flow hereafter). With lower doses of nicorandil or cromakalim, developed tension was not changed. With increasing doses of nicorandil or cromakalim, however, developed tension was decreased in a dose-dependent manner. With 10 mmol of nicorandil, in three of the seven preparations, ventricular fibrillation ensued after a profound decrease in developed tension but subsided spontaneously in about 3 min. At 0.3 mmol of cromakalim, in eight of the nine preparations, ventricular fibrillation occurred after a marked negative inotropic effect and ceased spontaneously in about 3 min. Although glibenclamide *per se* produced no changes in basal coronary blood flow nor force of contraction, after administration of glibenclamide 3 mmol/kg i.v. to support dogs, nicorandil or cromakalim became less effective in increasing blood flow (Fig. 5 and Table 3) and consequently the dose-response curve for increase in blood flow was shifted to the right (Fig. 5). Under these conditions, the negative inotropic effect of nicorandil was also reduced, and developed tension was increased slightly at medium doses of nicorandil. Ventricular fibrillation was no more elicited by nicorandil or cromakalim. After the cumulative dose of 10 mmol/kg i.v. of glibenclamide, the effects of nicorandil or cromakalim in increasing blood flow and decreasing developed tension were further reduced (Fig. 5 and Table 2).

Under control conditions nitroglycerin (0.03 to 100 nmol) injected intra-arterially produced only a dose-dependent increase in blood flow (Fig. 5). After 10 mmol/kg i.v. of glibenclamide, nitroglycerin gave that dose-response curve for increase in blood flow which was not significantly different from the control (Fig. 5); ED_{50} values of nitroglycerin were not significantly different (Table 3).

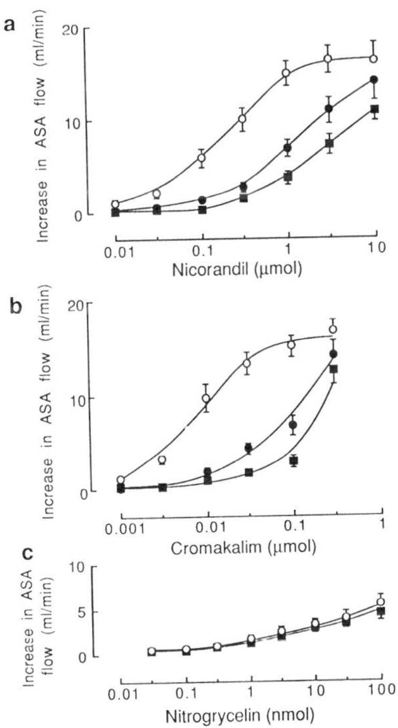

Figure 5. Dose-response curves for increase in coronary blood flow through the anterior septal artery (ASA) produced by intra-arterial (a) nicorandil, (b) cromakalim and (c) nitroglycerin in isolated, blood-perfused papillary muscle preparations of dogs, before (control, O) and after cumulative doses of 3 mmol/kg, i.v. (●) and 10 μ mol/kg, i.v. (■) of glibenclamide given to support dogs. Data points represent means $\pm$ SEM. (Adopted and modified from Ref. 59).

Table 3. ED$_{50}$ values and slope parameters of dose-response curves for increase in coronary blood flow produced by nicorandil, cromakalim and nitroglycerin before and after cumulative doses of glibenclamide given i.v. to support dogs

	Nicorandil	Cromakalim	Nitroglycerin
Control	0.31 $\pm$ 0.10	0.01 $\pm$ 0.01	0.02 $\pm$ 0.01
	(2.26 $\pm$ 0.77)	(1.47 $\pm$ 0.21)	(0.50 $\pm$ 0.06)
Glibenclamide 3μmol/kg	1.63 $\pm$ 0.52*	0.11 $\pm$ 0.02*	
	(1.30 $\pm$ 0.18)	(1.63 $\pm$ 0.35)	
Glibenclamide 10 μmol/kg	2.72 $\pm$ 0.63*	0.18 $\pm$ 0.01*	0.01 $\pm$ 0.00
	(1.29 $\pm$ 0.12)	(2.54 $\pm$ 0.40)	(0.55 $\pm$ 0.06)

ED$_{50}$ values (μmol) and slope parameters (in parentheses) of dose-response curves for increase in coronary blood flow through the anterior septal artery of the isolated blood-perfused papillary muscle preparations. Values are expressed in terms of mean $\pm$ SEM. Effects of nitroglycerin were examined after only a single dose of glibenclamide. No significant differences in corresponding values among two or three conditions. * $P<0.05$ compared with control values. (adopted and modified from Ref. 59).

These results were very similar to the findings obtained of rat or dog blood pressure with cromakalim and glibenclamide (5, 50). Thus, as in rat experiments (5), in the present experiments, too, glibenclamide behaved as a pharmacological antagonist of K$^+$ channel openers. The effect of nicorandil in increasing coronary blood flow was likewise antagonized by glibenclamide. Furthermore, the magnitude of parallel shifts of dose-response curves of nicorandil and cromakalim for increase in coronary blood flow was similar at corresponding doses (3 and 10 mmol/kg, i.v.) of glibenclamide. In contrast, the effect of nitroglycerin in increasing coronary blood flow did not significantly differ between before and after glibenclamide. Thus, although nicorandil has the mechanism of action as a nitrate (6, 21, 42), this action appears scarcely to contribute to its effect in increasing coronary blood flow. In other words, the effect of nicorandil in increasing coronary blood flow as a resistance vessel dilator appears predominantly due to the mechanism of action as a K^+ channel opener. This conclusion supports the previous supposition based on close similarity of the cardio-coronary vascular profile between nicorandil and cromakalim (39). The present conclusion also appears rational, as the predominant action of nitrates in coronary circulation is understood as a dilator action on large conductance arteries but not on resistance vessels (13, 27, 48).

Analysis of Dose-Response Curves

Dose-response curves for increase in blood flow produced by nicorandil or cromakalim before and after 3 and 10 mmol/kg i.v. of glibenclamide were parallel, because the slope parameters (p) were not changed, and ED_{50} values of nicorandil or cromakalim were increased by glibenclamide (Table 2). There were no significant differences in increase in ED_{50} values between nicorandil and cromakalim produced by two doses of glibenclamide. Schild analysis gave pA_2 values of 6.08 and 6.34 for glibenclamide against nicorandil and cromakalim, respectively. Dose-response curves for the negative inotropic effect of cromakalim were parallel under three conditions; before and after 3 and 10 mmol/kg of glibenclamide, and ED_{30} values were increased by glibenclamide. Schild analysis yielded a pA_2 of 5.64 for glibenclamide against cromakalim. The negative inotropic effect of nicorandil was also antagonized by glibenclamide, but development of positive inotropic effect made unable to subject the antagonism to Schild analysis. The almost same results have been obtained in isolated canine atrial muscles. The pA_2 values for glibenclamide against the negative inotropic effects of cromakalim, pinacidil and nicorandil were uniformly 6.06-6.35 (34).

Schild analysis of the antagonism yielded pA_2 values of 6.34 and 6.08 for glibenclamide against cromakalim and nicorandil, respectively, on the assumption that the slopes of Schild regressions were unity for both drugs. These pA_2 values are smaller by about one than those determined for glibenclamide against cromakalim, pinacidil and RP 49356 in guinea-pig pulmonary arteries constricted by 20 mM KCl (10). Nevertheless, uniform pA_2 values for glibenclamide obtained in the present experiments strongly suggest that glibenclamide competes with cromakalim and nicorandil for the same binding site at those K^+ channels of smooth muscle cells in resistance vessels of which stimulation hyperpolarizes membrane potentials and eventually causes relaxation. Glibenclamide was originally found to block K_{ATP} channels in pancreatic B-cells (36, 60). As a result cell membranes of B-cells depolarize and eventually insulin secretion occurs (36, 60). Thus, in pancreatic B-cells glibenclamide is an agonist which blocks K_{ATP} channels. Its ED_{50} values of 4×10^{-9} M (60) obtained in mouse pancreatic B-cells are close to its K_D values 3×10^{-10} M determined by a radioligand binding study in RINm5F insulinoma cells (36). Thus, K_B (dissociation constants calculated from pA_2 values) values determined for glibenclamide in the present experiments, $5-8 \times 10^{-7}$ M, are smaller by 2-3 orders of magnitude than ED_{50} or K_D values determined in pancreatic B-cells. Indeed, K_{ATP} channels have been demonstrated in vascular smooth muscle and glibenclamide has been shown to block the opening action of cromakalim there (37). Nevertheless, it appears still premature to conclude that those receptors associated with K^+ channels which were involved in the antagonism between glibenclamide and cromakalim or nicorandil in the present experiments are of the same types that are thought to be associated with K_{ATP} channels in pancreatic B-cells.

CONCLUSION

Cromakalim is a more specific K^+ channel opener than pinacidil and nicorandil and that vasodilation produced by cromakalim in canine coronary arterial rings is predominantly a result of a reduction of $[Ca^{2+}]_i$ due to the closure of voltage-dependent Ca^{2+} channels by hyperpolarization. In contrast, additional mechanisms are involved in the vasodilator actions of pinacidil and nicorandil. One of these is related to a reduction in the sensitivity of contractile proteins to Ca^{2+}. The latter mechanism of nicorandil is akin to that of nitroglycerin. K^+ channels opened by these K^+ channel openers may be ATP-sensitive ones which are blocked by glibenclamide. Nicorandil has a hybrid property between nitrates (N) and potassium (K^+) channel openers. We propose the term, "N-K hybrid". The effect of nicorandil in increasing coronary blood flow, like that of cromakalim, is predominantly due to the mechanism of action as a K^+ channel opener. The negative inotropy and ventricular fibrillation seen with high doses of nicorandil and cromakalim were also antagonized by glibenclamide, indicating that these effects are also due to K^+ channel opening.

ACKNOWLEDGEMENTS

These data in this article were obtained from studies supported by Grant-in-Aids from the Ministry of Education, Science and Culture, Japan. We are also grateful to Dr. T.C. Hamilton, Beecham Pharmaceuticals, The Pinnacles, Harlow, Essex, U.K. for the supply of cromakalim, to Shionogi & Co., Ltd., Osaka, Japan for pinacidil, to Chugai Pharmaceutical Co., Ltd., Tokyo, Japan for nicorandil and to Hoechst Japan Ltd., Tokyo, Japan and Yamanouchi Pharmaceutical Co., Tokyo, Japan for glibenclamide. The authors are grateful to Dr. M.P. Kingsbury for his helpful improvement of this manuscript.

REFERENCES

1. Arrigoni-Martelli, E., C.K. Nielsen, U.B. Olsen, and H.J. Petersen, Experientia 36, 445-447 (1980).
2. Bray, K.M., D.T. Newgreen, R.C. Small, J.S. Southerton, S.G. Taylor, S.W. Weir, and A.H. Weston, Br. J. Pharmacol. 91, 421-429 (1987).
3. Buckingham, R.E., T.C. Hamilton, D.R. Howlett, S. Mootoo, and C. Wilson, Br. J. Pharmacol. 97, 57-64 (1989).
4. Castle, N.A., D.G. Haylett, and D.H. Jenkinson, Trends Neurosci. 12, 59-65 (1989).
5. Cavero, I., S. Mondot, and M.J. Mestre, Pharmacol. Exp. Ther. 248, 1261-1268 (1989).
6. Coldwell, M.C. and D.R. Howlett, Biochem. Pharmacol. 36, 3663-3669 (1987).
7. Cook, N.S. and U. Quast in Potassium channels: Structure, classification, function and therapeutic potential, N.S. Cook, eds, (Ellis Horwood Ltd., Chester 1990) pp. 181-255.
8. Cook, N.S., Trend. Pharmacol. Sci. 9, 21-28 (1988).
9. Edwards, G. and A.H. Weston, Trends Pharmacol. Sci. 11, 417-422 (1990).
10. Eltze, M., Europ. J. Pharmacol. 165, 231-239 (1989).
11. Endoh, M. and N. Taira, Naunyn-Schmiedebergs Arch. Pharmacol. 322, 319-321 (1983).
12. Escande, D., D. Thuringer, S. Leguern, and I. Cavero, Bioch. Biophys. Res. Comm. 154, 620-625 (1988).
13. Fam, W.M. and M. McGregor, Circ. Res. 22, 649-659 (1968).
14. Fosset, M., J.R. De Weille, R.D. Green, H. Schmid-Antomarchi, and M. Lazdunski, J. Biol. Chem. 263, 7933-7936 (1988).
15. Furukawa, K., T. Itoh, M. Kajiwara, K. Kitamura, H. Suzuki, Y. Ito, and H. Kuriyama, J. Pharmacol. Exp. Ther. 218, 248-259 (1981).
16. Godfraind, T., R. Miller, and M. Wibo, Pharmacol. Rev. 38, 321-416 (1986).
17. Gotanda, K., K. Satoh, and N.J. Taira, J. Cardiovasc. Pharmacol. 12, 239-246 (1988).
18. Grynkiewicz, G., M. Poenie, and R.Y. Tsien, J. Biol. Chem. 260, 3440-3450 (1985).
19. Hamilton, T.C., S.W. Weir. and A.H. Weston, Br. J. Pharmacol. 88, 103-111 (1986).
20. Hartmann, H.A., G.E. Kirsch, J.A. Drewe, M. Taglialatela, R.H. Joho, and A.M. Brown, Science 251, 942-944, 1991.
21. Holzmann, S., J. Cardiovasc. Pharmacol. 5, 364-370 (1983).
22. Itoh, T., Y. Kanmura, H. Kuriyama, and T. Sasaguri, Br. J. Pharmacol. 84, 393-406 (1985).
23. Kamm, K.E. and J.T. Stull, Annu. Rev. Physiol. 51, 299-313 (1989).
24. Karaki, H., Trend. Pharmacol. Sci. 10, 320-325 (1989).
25. Kashiwabara, T., S. Nakajima, T. Izawa, H. Fukushima, and K. Nishikori, Europ. J. Pharmacol. 196, 1-7 (1991).
26. Kauffman, R.F., K.W. Schenck, B.G. Conery, and M.L. Cohen, J. Cardiovasc. Pharmacol. 8, 1195-1200 (1986).
27. Kurz, M.A., K.G. Lamping, J.N. Bates, C.L. Eastham, M.L. Marcus, and D.G. Harrison, Circ. Res. 68, 847-855 (1991).
28. Okada, Y., T. Yanagisawa, and N. Taira, Br. J. Pharmacol. (in press) (1991).
29. Parker, R.B. and D.R. Waud, J. Pharmacol. Exp. Ther. 177, 1-12 (1971).
30. Pfitzer, G., F. Hofmann, J. Disalvo, and J.C. Ruegg, Pflugers Arch. 401, 277-280 (1984).
31. Quast, U. and N.S. Cook, J. Pharmacol. Exp. Ther. 250, 261-271 (1989).
32. Rudy, B., Neurosci. 25, 729-749 (1988).
33. Sanguinetti, M.C., A.L. Scott, G.J. Zingaro, and P.K.S. Siegl, Proc. Natl. Acad. Sci. USA 85, 8360-8364 (1988).
34. Satoh, E., T. Yanagisawa, and N. Taira, Japan. J. Pharmacol. 54, 133-141 (1990).
35. Satoh, K., H. Yamada, and N. Taira, Naunyn-Schmiedebergs Arch. Pharmacol. 343, 76-82 (1991).

36. Schmid-Antomarchi, H., J. De Weille, M. Fosset, and M. Lazdunski, J. Biol. Chem. $\underline{262}$, 15840-15844 (1987).
37. Standen, N.B., J.M. Quayle, N.W. Davies, J.E. Brayden, Y. Huang, and M.T. Nelson, Science $\underline{245}$, 177-180 (1989).
38. Stanfield, P.R., Rev. Physiol. Biochem. Pharmacol. $\underline{97}$, 1-67 (1983).
39. Taira, N., Am. J. Cardiol. $\underline{63}$, 18J-24J (1989).
40. Taira, N., J. Cardiovasc. Pharmacol. $\underline{10}$ (Suppl. 8), S1-S9 (1987).
41. Taira, N., K. Satoh, T. Yanagisawa, Y. Imai, and M. Hiwatari, Clin. Exp. Pharmacol. Physiol. $\underline{6}$, 301- 316 (1979).
42. Waldman, S.A. and F. Murad, Pharmacol. Rev. $\underline{39}$, 163-196 (1987).
43. Weir, S.W. and A.H. Weston, Br. J. Pharmacol. $\underline{88}$, 113-120 (1986).
44. Weir, S.W. and A.H. Weston, Br. J. Pharmacol. $\underline{88}$, 121-128 (1986).
45. Weston, A.H., J. Cardiovasc. Pharmacol. $\underline{10}$ (Suppl. 8), S56-S61 (1987).
46. Weston, A.H., Pflugers Arch. $\underline{414}$ (Suppl. 1), S99-S105 (1989).
47. Wilson, C., M.C. Coldwell, D.R. Howlett, S.M. Cooper, and T.C. Hamilton, Europ. J. Pharmacol. $\underline{152}$, 331-339 (1988).
48. Winbury, M.M., B.B. Howe, and M.A. Hefner, J. Pharmacol. Exp. Ther. $\underline{168}$, 70-95 (1969).
49. Winquist, R.J., L.A. Heaney, A.A. Wallace, E.P. Baskin, R.B. Stein, M.L. Garcia, and G.J. Kaczorowski, J. Pharmacol. Exp. Ther. $\underline{248}$, 149-156 (1989).
50. Yamada, H., F. Yoneyama, K. Satoh, and N. Taira, Br. J. Pharmacol. $\underline{100}$, 413-416 (1990).
51. Yanagisawa, T. and N. Taira, Japan. J. Pharmacol. $\underline{31}$, 409-417 (1981).
52. Yanagisawa, T. and N. Taira, Naunyn-Schmiedebergs Arch. Pharmacol. $\underline{312}$, 69-76 (1980).
53. Yanagisawa, T., H. Hashimoto, and N. Taira, Br. J. Pharmacol. $\underline{95}$, 393-398 (1988).
54. Yanagisawa, T., H. Hashimoto, and N. Taira, Br. J. Pharmacol. $\underline{97}$, 753-762 (1989).
55. Yanagisawa, T., K. Satoh, and N. Taira, Japan. J. Pharmacol. $\underline{29}$, 687-694 (1979).
56. Yanagisawa, T., M. Kawada, and N. Taira, Br. J. Pharmacol. $\underline{98}$, 469-482 (1989).
57. Yanagisawa, T., T. Teshigawara, and N. Taira, Br. J. Pharmacol. $\underline{101}$, 157-169 (1990).
58. Yellen, G., M.E. Jurman, T. Abramson, and R.. MacKinnon, Science $\underline{251}$, 939-942 (1991).
59. Yoneyama, F., K. Satoh, and N. Taira, Cardiovasc. Drugs Ther. 4, 1119-1126 (1990).
60. Zunkler, B.J., S. Lenzen, K. Manner, U. Panten, and G. Trube, Naunyn-Schmiedebergs Arch. Pharmacol. $\underline{337}$, 225-230 (1988).

ELECTROPHYSIOLOGICAL CORRELATES OF VASODILATION: A PRIMARY ROLE FOR ATP-SENSITIVE POTASSIUM CHANNELS.

JOSEPH E. BRAYDEN

Department of Pharmacology and Vermont Center for Vascular Research, College of Medicine, The University of Vermont, Burlington, VT 05405

INTRODUCTION

The mechanisms by which activation and inhibition of vascular tone occur are diverse and complex. Interactions among the various pathways underlying vasoconstrictor and vasodilator responses probably exist. However, a central determinant of the level of vascular tone is the intracellular concentration of calcium ions ($[Ca^{+2}]_i$ and our understanding of how this process is regulated has greatly expanded in recent years. One important mechanism involved in regulation of $[Ca^{+2}]_i$ and the balance between excitatory and inhibitory influences on vascular smooth muscle cells is the level of the membrane potential (V_m) (14, 48, 58). Membrane depolarization leads to increased calcium influx and contraction; membrane hyperpolarization has the opposite effect. The following discussion will be concerned primarily with vasodilator mechanisms and will focus on the electrophysiological correlates of this response. Evidence will be presented that strongly supports the hypothesis that membrane hyperpolarization is an important mechanism involved in the vascular response to a variety of endogenous and pharmacological vasodilators.

CALCIUM CHANNELS AND VASCULAR TONE

Numerous mechanisms have been proposed by which a decrease in vascular tone can occur [see 80 for reviews]. As implied above, most of these mechanisms involve regulation of calcium ion concentration or other calcium sensitive cellular processes.

A fundamental mechanism by which vascular tone can be altered involves the regulation of pathways for movement of calcium into vascular smooth muscle cells. Intracellular concentrations of calcium and, consequently, vascular tone are maintained in large part by calcium influx across the cell membrane from the extracellular compartment (33). Therefore, factors that decrease calcium influx will result in inhibition of maintained tone.

Calcium influx may occur via several pathways. Two types of voltage-gated calcium channels have been identified in many excitable tissues, including vascular smooth muscle (3). They have been classified as transient (T-type), dihydropyridine-insensitive channels and as long-lasting (L-type), dihydropyridine-sensitive channels. Evidence for the existence of voltage-independent, receptor-operated calcium channels has been presented by numerous investigators (see 78 for review) and it is clear that receptor activation does stimulate calcium influx in vascular smooth muscle. Direct evidence regarding the specific calcium influx pathway involved in this response, i.e. from studies at the single channel level, is extremely limited. Receptor operated calcium channels activated by ATP have been identified in vascular smooth muscle (7). However, evidence has been presented indicating that calcium channels activated by other receptor ligands in vascular smooth muscle may in fact be voltage-gated calcium channels (59). Calcium influx by sodium-calcium exchange has been observed in vascular smooth muscle (12, 57), although usually only under rather non-physiological conditions, making the importance of this calcium influx pathway a matter of some debate. Although all these may make some contribution to total calcium influx in vascular smooth muscle, dihydropyridine-sensitive calcium influx appears to be the major determinant of steady-state tone in vascular beds such as the cerebral and coronary circulations and in particular in smaller "resistance" arteries in most vascular beds (23, 45, 62, 77). For this reason and in the context of the present topic, that is electrophysiological aspects of vascular control, voltage-gated, dihydropyridine-sensitive calcium channels and their role in mechanisms of vasodilation will be the focus of the following discussion.

Voltage-dependent calcium channels have been identified in vascular smooth muscle using microelectrode and patch-clamp techniques. In intact cerebral arteries of the rat, the presence of both T- and L-type calcium channels was demonstrated by Hirst and co-workers (48) using a single electrode voltage-clamp technique. Whole cell and single channel patch clamp recordings have revealed such channels in rabbit ear arteries (8) and in rat mesenteric arteries (4). Detailed analysis of the voltage-dependence of activation of dihydropyridine-sensitive calcium channels in rabbit mesenteric and cerebral arteries has recently been undertaken by Nelson and colleagues (59, 68, 83). The relationship between V_m and open-state probability (P_{open}) of this channel is exponential and can be described by a Boltzman function. In all cases, the relationship is quite steep; changes in V_m of 5-8 mV are associated with a 2.7 fold (e fold) change in open state probability of the calcium channel. Data from these studies suggest that even at low levels of activation, sufficient amounts of calcium could enter through voltage-gated calcium channels to cause a rise in $[Ca^{+2}]_i$ far in excess of that required to initiate and sustain contraction. A similar conclusion was reached by Hirst et al. in studies of intact cerebral arteries (48). An implication of this work is that relatively small changes in Vm should be associated with substantial changes in force and that the relationship between changes in Vm and force should be quantitatively similar to the relationship between Vm and open state probability of the calcium channel. This prediction has in fact been born out in studies of the rabbit mesenteric artery (59) in which the effects of excitatory stimuli and Vm on calcium channel activity and force were considered. Whether force is activated by raising $[K]_o$, by application of norepinephrine in the presence of various concentrations of $[K]_o$, or by a combination of norepinephrine, calcium channel agonist (Bay R 5417) and elevated $[K]_o$, the relationship between Vm and force is very similar to that described for Vm and P_{open} for the voltage-dependent calcium channel.

Numerous other investigators have examined the relationship between Vm and force in vascular smooth muscle (15, 40, 78). These studies have employed various means of inducing tone and have made simultaneous measurements of Vm and force or Vm and diameter. Elevation of extracellular potassium is a common means by which this relationship has been examined. Although there is some variation between tissues regarding the level of Vm at which changes in force are first detected (-50 to -40 mV), changes in Vm and force in these experiments are well-correlated. In several arteries, the relationship between changes in Vm and force is quite steep over the physiological range of Vm's, i.e. in the range of -55 to -30 mV. KCl-induced force development is greatly reduced or abolished by inhibitors of voltage-dependent calcium channels or by removal of extracellular calcium and thus apparently is due to influx of calcium through voltage-dependent calcium channels. Receptor agonists also depolarize vascular smooth muscle, but in this case the relationship between Vm and force can be different from that for potassium-induced depolarization and contraction. Several investigators have demonstrated sizable norepinephrine-induced contractions of vascular smooth muscle with little or no change in Vm (22, 28, 74). This and other similar data underlie the proposal that receptor agonists may activate a distinct class of voltage-insensitive calcium channels (receptor-operated calcium channels). However, as alluded to above, recent evidence indicates that norepinephrine may actually induce a shift in the relationship between Vm and open state probability of the voltage-dependent calcium channel such that at a given Vm, more calcium channels are open in the presence of norepinephrine than in its absence (59). In this instance, NE-induced contraction can occur with no change in Vm and yet such contractions may be due in large part to calcium influx through voltage-dependent calcium channels.

Other agonists may also activate contraction through effects on voltage-dependent calcium channels. For instance, serotonin depolarizes vascular smooth muscle (36, 43), and recent studies have demonstrated a direct effect of serotonin on voltage-dependent calcium channels (83). Open state probability of these channels in patches of membrane from rabbit cerebral arteries increased dramatically upon exposure to serotonin under conditions where Vm was held constant.

Elevation of transmural pressure is a physiological stimulus that results in induction of vascular tone in vitro (16, 44, 72). This form of myogenic tone is also due to activation of calcium influx through voltage-dependent calcium channels. Increased transmural pressure causes depolarization of vascular smooth muscle cells and contraction. The resulting tone is abolished by calcium channel blockers or by removal of extracellular calcium, indicating the role of voltage-dependent calcium channels in this response.

Similar observations have been made using pressurized coronary and mesenteric resistance arteries (J.E.B., unpublished).

MEMBRANE HYPERPOLARIZATION AND VASODILATION

An important prediction based on the aforementioned observations and the known properties of voltage-dependent calcium channels is that interventions resulting in increases in V_m, i.e. application of hyperpolarizing agents, should <u>decrease</u> the open state probability of voltage-dependent calcium channels. This will in turn result in a reduced level of calcium influx which should lead to a reduction in vascular tone. This potential mechanism of vasodilation will be examined in detail in the following paragraphs.

A number of possible mechanisms for hyperpolarization of vascular smooth muscle have been described. Numerous investigators have demonstrated that the sodium pump is electrogenic in vascular smooth muscle and have provided evidence that conditions can be established under which activation of the sodium pump causes membrane hyperpolarization and vasodilation (39, 46). Inhibition of the sodium pump activity also reduces hyperpolarizations caused by administration of some hyperpolarizing agents (16, 32). However, inhibition of sodium pump activity may also alter potassium channel conductances and this complicates interpretation of these results (47).

The V_m of vascular smooth muscle is determined by the relative membrane permeabilities to sodium, chloride and potassium (14). In general, the conductance to potassium predominates, but other factors must play a role since V_m is displaced in the positive direction from the equilibrium potential for potassium even in tissues that have not been activated by any means. However, potassium is the only ion having an equilibrium potential negative to the V_m of vascular smooth muscle under physiological conditions. Thus, it is reasonable to propose that a primary mechanism of hyperpolarization of vascular smooth muscle is via enhanced potassium conductance.

K Channels in Vascular Smooth Muscle

A substantial number of potassium channels that can be distinguished on the basis of their pharmacological and biophysical properties have been identified (25). Several of these channels are found in vascular smooth muscle and could conceivably play a role in a mechanism of vasodilation involving activation of a potassium conductance and membrane hyperpolarization.

One of the first identified and most common potassium channels in vascular smooth muscle is the calcium-activated potassium channel (K_{Ca}). These channels are activated by membrane depolarization and by elevation of $[Ca^{+2}]_i$ (6). Rapidly activating, large-conductance K_{Ca} channels have been identified in vascular smooth muscle from several species (9, 52). These channels are blocked with high affinity by TEA and by charybdotoxin. A calcium and voltage-sensitive, slowly activating potassium channel that is relatively insensitive to TEA has been observed in isolated cerebral arterioles (48). These channels may be activated by contractile agonists as a result of the elevation of $[Ca^{+2}]_i$ associated with stimulation and this may explain the failure of many vascular smooth muscle cells to exhibit regenerative action potential activity upon stimulation. A few studies indicate that vasodilators activate this channel (37), but this has not been confirmed by others (5).

An outwardly rectifying potassium current that is voltage-dependent has been observed in vascular smooth muscle cells (5, 30, 48, 61). This current is activated upon depolarization following a short delay and rectifies in the outward direction and is therefore called the delayed rectifier potassium current (K_{DR}). This current also shows time- and voltage-dependent inactivation. K_{DR} channels are blocked by high concentrations of TEA and by 4-aminopyridine. Again, as for the K_{Ca} channels, the physiological significance of K_{DR} channels in vascular smooth muscle is unclear and evidence of a role for these channels as mediators of the hyperpolarizing response to vasodilators is limited (5).

An inwardly rectifying potassium channel (K_{IR}) has been observed in many tissues (25) including vascular smooth muscle (31, 48). These channels are activated by hyperpolarization and are inhibited by barium ions. The maximum conductance of K_{IR} currents in arteriolar smooth muscle is increased when extracellular potassium ion concentration is increased. Unlike other tissues, in arteriolar smooth muscle K_{IR} channels are activated at potentials positive of E_K (48) and may make a substantial contribution to resting V_m in this tissue. Elevation of extracellular potassium concentration hyperpolarizes cerebral arteriolar smooth muscle and this has been proposed to be a mechanism of potassium-induced vasodilation long known to occur in cerebral arteries (51). However, any role of this channel in the response to receptor activation remains to be established.

Hyperpolarizing Vasodilators and K_{ATP} Channels

Recent experimental observations support the hypothesis that another distinct potassium channel, the ATP-sensitive (K_{ATP}) channel, is centrally involved in the mechanism of action of many vasodilator substances. This channel has been observed in a variety of tissues (1) and these findings have recently been extended to include vascular smooth muscle (73). A key characteristic of this channel is that it is inhibited by intracellular ATP [K_i: 0.1-3 mM depending on the tissue,(66)]. It is also specifically inhibited by antidiabetic sulphonylurea compounds such as glibenclamide and tolbutamide and by low concentrations of barium ions (30-100 μM). Other potassium channel blockers such as charybdotoxin and apamin have no effect on this channel and TEA is a very low affinity blocker (K_d: ~7mM). K_{ATP} channels are only weakly voltage-dependent and are not activated by changes in [Ca^{+2}]$_i$. This distinct profile of identifying characteristics has provided a clear means of testing the role of this channel as a mediator of the action of hyperpolarizing vasodilators.

Pharmacological vasodilators. Agents such as pinacidil, cromakalim, minoxidil and nicorandil are representative members of a new class of pharmacological vasodilators that have been classified as potassium channel openers (25, 60). These compounds relax vascular smooth muscle (65), increase efflux of labeled potassium or rubidium (67) and they hyperpolarize smooth muscle (42). Recent experiments have provided convincing evidence that potassium channel agonists activate K_{ATP} channels and that this is the primary mechanism of action of these agents. The hypotensive action and relaxant effects of these agents on vascular smooth muscle are inhibited by glibenclamide (73, 82). Cromakalim-induced potassium efflux in isolated vascular smooth muscle is blocked by glibenclamide (64) as are the hyperpolarizations induced by potassium channel openers (21) (Figure 1).

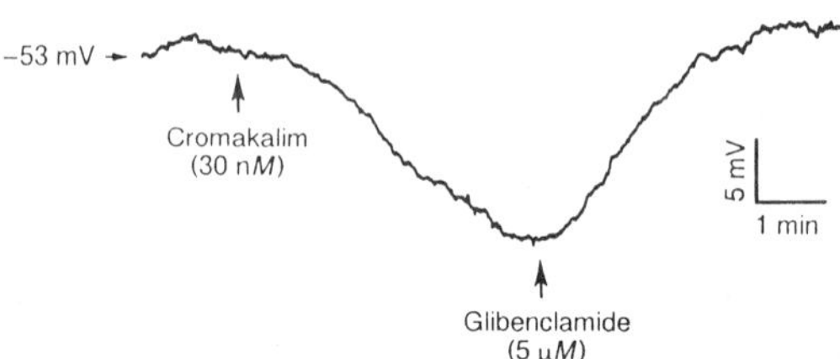

Figure 1. Hyperpolarization induced by cromakalim in an isolated, rabbit middle cerebral artery. The artery was mounted in a resistance artery myograph and a vascular smooth muscle cell was impaled using a glass microelectrode. (From reference 21, with permission).

Studies at the single channel level have confirmed the role of K_{ATP} channels in the response to potassium channel openers (73). Single, K_{ATP} channels in patches of membrane from smooth muscle cells isolated from the rabbit mesenteric artery are strongly activated by cromakalim (Figure 2).

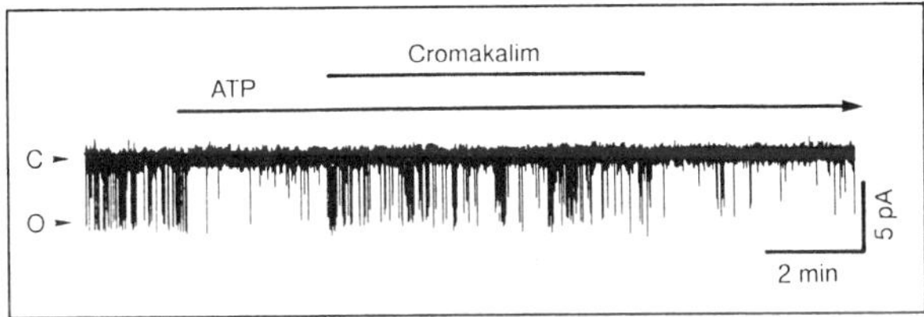

Figure 2. Activation of single K_{ATP} channels by cromakalim in excised, inside out patches of smooth muscle cell isolated from a rabbit mesenteric artery. Holding potential: -60mV. The pipette solution contained 60 mM K^+ and the bath solution contained 120 mM K^+ and 5 mM EGTA. (From reference 21, with permission).

The open state probability (P_{open}) of these channels can be increased from very low levels (nearly zero) to very high levels ($P_{open} > 0.75$) by cromakalim. This effect of cromakalim on K_{ATP} channels is inhibited by glibenclamide. Cromakalim, pinacidil and diazoxide relax norepinephrine-contracted mesenteric arteries and this effect can be reversed be glibenclamide and by barium chloride. These responses are not inhibited by blockers of calcium-activated potassium channels such as charybdotoxin or low concentrations of TEA (1 mM). Taken together, these observations provide convincing evidence that K_{ATP} channels mediate the inhibitory effects of K channel openers on vascular smooth muscle.

Neuropeptides. Other experiments provide evidence that activation of K_{ATP} channels is an important mechanism of action of several endogenous vasodilator substances, including neurotransmitters, endothelial factors and cellular metabolites.

Calcitonin gene-related peptide (CGRP) is a potent endogenous vasodilator found in perivascular nerves in many vascular beds. It is formed as a result of tissue-specific, alternative processing of messenger RNA from the calcitonin gene (70). A recent detailed study indicates that CGRP activates K_{ATP} channels and this accounts for part of the vasodilator response induced by CGRP (60). In these experiments, glibenclamide partially reversed CGRP-induced relaxations of rabbit mesenteric arteries. CGRP hyperpolarized the vascular smooth muscle cells in this artery by up to 22 mV (Figure 3).

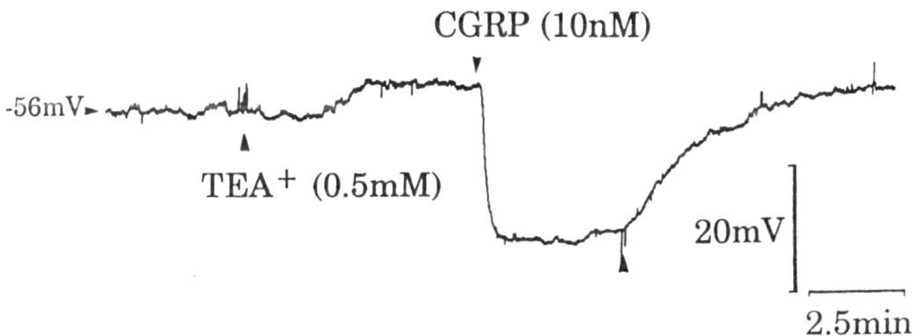

Figure 3. Effects of CGRP on membrane potential of a smooth muscle cell in an isolated rabbit mesenteric artery. The artery was mounted in a resistance artery myograph. Note large hyperpolarization in the presence of TEA. Glibenclamide (3 μM) was added at the time indicated by the third arrow head.

This hyperpolarization was abolished by glibenclamide, but was unaffected by TEA (1 mM) or charybdotoxin (30 nM). In addition, CGRP opened potassium channels on single smooth muscle cells of rabbit mesenteric artery in the presence of charybdotoxin or low concentrations of TEA. Glibenclamide greatly reduced the activity of the CGRP-activated potassium channels. Therefore, relaxations induced by this peptide are due in part to activation of K_{ATP} channels.

The failure of glibenclamide to reverse the entire CGRP-induced relaxation implies the presence of other pathways of dilation to CGRP that do not involve K_{ATP} channels. One such pathway may involve the plasma membrane associated Ca^{+2}-ATPase that is responsible for extrusion of calcium from intracellular sites (53, 69). Increased activity of this calcium pump will, in the face of steady or decreased rates of calcium influx or release, lead to decreased $[Ca^{+2}]_i$ and to inhibition of vascular tone. CGRP does stimulate formation of cyclic AMP within vascular smooth muscle cells (29) and increased intracellular concentrations of cyclic AMP or cyclic GMP result in activation of the Ca^{+2}-ATPase; this may account in large part for the role of cyclic nucleotides as vasodilator second messengers.

Cyclic nucleotides, by activation of specific kinases, exert their effects via phosphorylation of target proteins which in turn alters the functional activity of such proteins. Thus, increased cyclic nucleotide levels may cause membrane hyperpolarization-independent relaxation by a number of mechanisms. For instance, phosphorylation of contractile proteins such as myosin can alter enzyme activity (e.g. myosin ATPase) and decrease contractile function and vascular tone (41). Phosphorylation of contractile proteins may also directly alter their calcium-sensitivity which could lead to vasodilation even when $[Ca^{+2}]_i$ remains constant. Phosphorylation of MLCK decreases its activity, leading to reduced myosin phosphorylation and to changes in smooth muscle contractility (41). Finally, calcium channel proteins could be phosphorylated resulting in decreases in calcium influx via this pathway. Any of these mechanisms could occur concomitantly with the hyperpolarizing effects of substances such as CGRP and account for part of the vasodilator response.

VIP is also a potent vasodilator found in perivascular nerves and widely distributed in the cardiovascular system (38). The pharmacological profile of activity of this peptide is similar to that for CGRP. Relaxations of isolated rabbit cerebral arteries to VIP are partly reversed by glibenclamide and barium ions (J.E.B., unpublished) and VIP-induced hyperpolarizations of cerebral arteries are abolished by glibenclamide and barium (73). Although the effects of VIP on single potassium channels have not been reported, the evidence mentioned above would strongly support the suggestion that VIP activates K_{ATP} channels in cerebrovascular smooth muscle.

Endothelial factors. The vasodilator action of endothelium-derived substances is well-established (35). At least one of the inhibitory substances (EDHF) that is released from endothelial cells upon stimulation hyperpolarizes vascular smooth muscle (13, 17, 24, 50). This factor is probably distinct from nitric oxide (EDRF) (10, 17) although the observations of one group would suggest otherwise (75). All of the hyperpolarization and part of the vasodilator response induced by the endothelium-dependent vasodilator acetylcholine in cerebral arteries is inhibited by barium or glibenclamide, indicating a causal relationship between the hyperpolarization (mediated by K_{ATP} channels) and part of the dilator response (17). In this case, the hyperpolarization-independent response is probably mediated by nitric oxide and, secondarily by cyclic GMP, the formation of which is stimulated by nitric oxide. Application of other endothelium-dependent vasodilators (e.g. ADP) in other vascular beds (skeletal muscle and mesenteric resistance arteries) also leads to hyperpolarization of vascular smooth muscle cells, apparently via activation of K_{ATP} channels (18) (Figure 4). Blockade of the hyperpolarization reduces the vasodilation. Interestingly, a role for EDHF and K_{ATP} channels, although clearly present in large cerebral arteries (17), is reduced or absent in very small, distal pial arteries in the rabbit (18) and rat (56).

Recently, these studies have been extended to include analysis of the EDHF system and endothelium-dependent vasodilation of blood vessels isolated from hypertensive animals. Several investigators have reported that endothelium-dependent vasodilation is depressed in hypertensive animals (55, 76, 79) and humans (54). Indomethacin has been reported to normalize the responses of arteries from hypertensives in some cases (27). This raises the possibility that endothelium dependent vasodilators release a cyclooxygenase derived contractile factor from endothelial cells which blunts the endothelium-dependent dilator responses in arteries from hypertensives. A third possible explanation for the reduced endothelium-dependent response in hypertensive arteries is that the EDHF system is somehow compromised in these arteries and recent evidence to support this possibility is available. The magnitude of endothelium-dependent hyperpolarizations induced by acetylcholine was reduced in large and small arteries isolated from SHR and SHRSP rats compared to arteries from WKY control rats (20, 34, Figure 5).

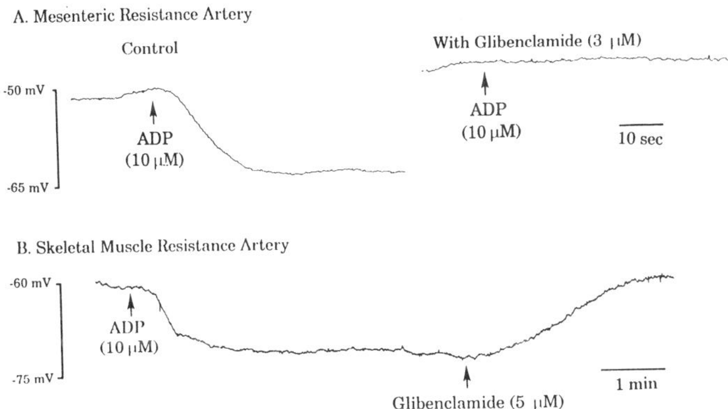

Figure 4. Effects of glibenclamide on adenosine diphosphate-induced membrane hyperpolarizations in isolated mesenteric and skeletal muscle resistance arteries. In (A), ADP was applied as a bolus directly to the myograph bath and glibenclamide was added via the superfusate five minutes before the second addition of ADP. In (B), ADP and glibenclamide were added via the superfusate.

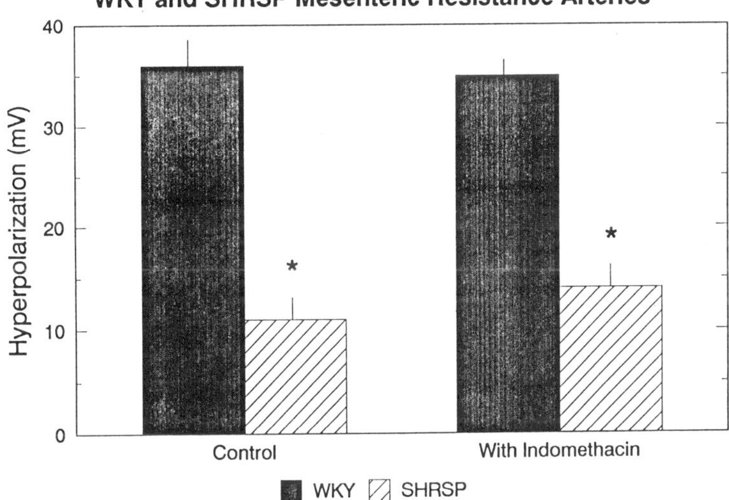

Figure 5. Reduced ACh-induced hyperpolarizations in arteries from Stroke-prone Spontaneously Hypertensive Rats vs. the WKY. Arteries were contracted and depolarized with norepinephrine ($1\,\mu$M) (V_m: -42 $\pm$ 1 mV). Effects of ACh ($0.3\,\mu$M) in the absence and presence of indomethacin ($10\,\mu$M) were then tested. Hyperpolarizations were significantly reduced in SHRSP arteries vs. control, with or without indomethacin ($p < 0.05$, n = 6).

Hyperpolarizations and relaxations were unaffected by indomethacin in arteries from any strain in these studies. Whether the smaller hyperpolarizations in arteries from hypertensive animals are due to reduced synthesis or release of EDHF, or to reduced effects of EDHF on smooth muscle from hypertensive animals has not yet been determined. However, it seems likely that depressed endothelium-dependent dilations in arteries from hypertensives may in part be explained by changes in the EDHF system.

Interestingly, although glibenclamide reduced the hyperpolarization to EDHF in mesenteric resistance arteries from SHRSP and WKY rats, the hyperpolarization was abolished only by exposure to a combination of glibenclamide, $BaCl_2$ and TEA (Figure 6). This observation suggests that in rat mesenteric resistance arteries, K_{ATP} channels may be only one of several potassium channels activated by EDHF.

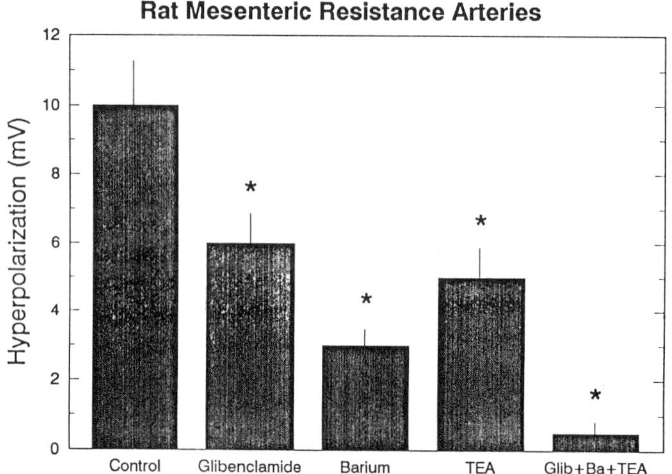

Figure 6. Effects of potassium channel blockers on ACh-induced hyperpolarizations of vascular smooth muscle in WKY mesenteric resistance arteries. Arteries were unstimulated and the average resting V_m was -61 ± 2 mV (n = 5). The magnitude of the hyperpolarization was significantly reduced in the presence of glibenclamide ($3\,\mu$M), barium chloride ($100\,\mu$M), TEA (1 mM), and in the combination of blockers (p < 0.05).

Adenosine. Cellular metabolites may also reduce arterial tone by activating K_{ATP} channels. Adenosine has long been thought to be an important vasodilator metabolite in coronary and skeletal muscle vascular beds (11). Adenosine hyperpolarizes bovine coronary vascular smooth muscle (71). This observation has been confirmed using rabbit coronary arteries where adenosine (20-100 nM) hyperpolarized vascular smooth muscle cells by up to 20 mV; this response was blocked by glibenclamide (J.E.B., unpublished). In addition, Daut et al. have reported that adenosine-induced vasodilator responses in isolated, perfused guinea-pig hearts are blocked by glibenclamide (26).

Hypoxic vasodilation. Short periods of hypoxia can induce a vasodilator response in intact vascular beds and in isolated arteries. Daut and colleagues (26) demonstrated that hypoxia-induced vasodilation in the intact guinea-pig heart is blocked by glibenclamide, suggesting a role for K_{ATP} channels in this response. Isolated cerebral arteries relax when the bath PO2 is reduced and this response is attenuated by glibenclamide (63). It is possible that hypoxic vasodilation is mediated by changes in intracellular concentrations of adenine nucleotides near K_{ATP} channels. This possibility is supported by recent observation by von Beckereth and colleagues (81). They found that blockade of glycolysis by substituting deoxyglucose for glucose in the perfusate caused vasodilation in isolated guinea-pig hearts and this vasodilator response was inhibited by glibenclamide. This observation suggests that disruption of pathways for formation of ATP may reduce the concentration of ATP or the ATP/ADP ratio (2) near the intracellular face of the membrane to the point where K_{ATP} channels are activated.

RELATIONSHIP BETWEEN MEMBRANE POTENTIAL AND FORCE

The preceding discussion argues strongly for a role of K_{ATP} channels in the vasodilator response to a variety of forms of stimulation. If the proposed mechanism is correct, i.e. that K_{ATP} channel-mediated membrane hyperpolarization results in closure of voltage-dependent calcium channels, then several predictions follow. First, steady state levels of force in the presence of various concentrations of a given vasodilator should be voltage-dependent. Second, the relationship between V_m and force should be independent of the particular vasodilator employed. Third, this relationship should be quantitatively similar to the relationship between V_m and open-state probability of voltage-dependent calcium channels from vascular smooth muscle. These predictions have been tested and have proven to be correct in recent studies using isolated rabbit cerebral arteries (19). Pre-contracted arteries were relaxed with varying concentrations of cromakalim and force and V_m were measured simultaneously. The relationship between V_m and force was exponential, with small hyperpolarizations associated with large changes in force. A similar relationship was observed for other vasodilators. These findings corresponded closely with data from single channel studies where the relationship between V_m and P_{open} of calcium channels in smooth muscle cells isolated from rabbit cerebral arteries had virtually the same steepness as the V_m/force relationship (83).

SUMMARY

Evidence presented in this review argues strongly that activation of K_{ATP} channels represents a powerful, general mechanism of vasodilation. Other potassium channels, although present in high density in many vascular smooth muscle cells, in most instances are not activated either by endogenous or by synthetic vasodilators. Over the physiological range of V_ms, vascular tone is strongly affected by small changes in V_m. This sensitivity is most likely mediated by the voltage sensitivity of calcium channels in the vascular membrane. Hyperpolarization in response to activation of K_{ATP} channels will close calcium channels, reducing calcium influx. The intracellular concentration of calcium will then arrive at a new, lower level as the various mechanisms of calcium handling exert their effects, and a lower level of tone will be attained.

REFERENCES

1.   Ashcroft, F.M., Ann. Rev. Neurosci. 11, 97-118 (1988).
2.   Ashcroft, S.J.H. and F.M. Ashcroft, Cellular Signaling 2, 197-214 (1990).
3.   Bean, B.P., Ann. Rev. Physiol. 57, 367-384 (1989).
4.   Bean, B.P., M. Sturek, A. Puga, and K. Hermsmeyer, Circ. Res. 59, 229-235 (1986).
5.   Beech, D.J. and T.B. Bolton, Br. J. Pharmacol. 98, 851-864 (1989).
6.   Benham, C. D. and T.B. Bolton, J. Physiol (Lond.) 381, 385-406 (1986.
7.   Benham, C.D. and R.W. Tsien, Nature 328, 275-278 (1987).
8.   Benham, C.D., P. Hess, and R.W. Tsien, Circ. Res. 61 (Suppl. 1), 110-116 (1987).
9.   Benham, C.D., T.B. Bolton, R.J. Lang, and T. Takewaki, J. Physiol. (Lond.) 387, 473-488 (1987).
10. Beny, J-L. and Brunet P.C., Blood Vessels 25, 308-311 (1988).
11. Berne, R.M., Circ. Res. 47, 807-813 (1980).
12. Blaustein, M.P., Am. J. Physiol. 232, C165-C173 (1977).
13. Bolton, T.B. and L.H. Clapp, Br. J. Pharmacol. 87, 713-723 (1986).
14. Bolton, T.B., Physiol. Rev. 59, 606-718 (1979).
15. Bolton, T.B., R.J. Lang, and T. Takewaki, J. Physiol. (Lond.) 351, 549-572 (1984).
16. Brayden, J.E. and G.C. Wellman, J. Cer. Blood Flow and Metab. 9, 256-263 (1989).
17. Brayden, J.E., Am. J. Physiol. 259, H668-H673 (1990).
18. Brayden, J.E., Circ. Res. (in press) (1991).
19. Brayden, J.E., Circulation 84 (Suppl. I), (in press) (1991).
20. Brayden, J.E., J. Mol. Cell. Cardiol. 23 (Suppl. III), S25 (1991).
21. Brayden, J.E., J.M. Quayle, N.B. Standen, and M.T. Nelson, Blood Vessels 28, 147-153 (1991).
22. Casteels, R., K. Kitamura, H. Kuriyama, H., and H. Suzuki, J. Physiol. (Lond.) 271, 41-61 (1977).
23. Cauvin, C., K. Saida, and C. VanBreeman, Blood Vessels 21, 23-31 (1984).
24. Chen, G.H., H.Suzuki, and A.H. Weston, Br. J. Pharmacol. 95, 1165-1174 (1988).
25. Cook, N.S., Trends Pharm. Sci. 9, 21-28 (1988).
26. Daut, J., W. Maier-Rudolph, N. von Beckerath, G. Mehrke, K. Gunther, and L. Goedel-Meinen, Science 247, 1341-1344 (1990).
27. Diederich, D., Z. Yang, F.R. Buhler, and T.F. Luscher, Am. J. Physiol. 27, H445-H451 (1990).

28.	Droogmans, G., L. Raeymaekers, and R. Casteels, J. General Physiology 70, 129-148 (1977).
29.	Edvinsson, L., B.B. Fredholm, E. Hamel, I, Jansen, and C. Verrecchia, Neurosci. Let. 58, 213-217 (1985).
30.	Edwards, F.R. and G.D.S. Hirst, J. Physiol. (Lond.) 404, 437-454 (1988).
31.	Edwards, F.R., G.D.S. Hirst, and G.D. Silverberg, J. Physiol.(Lond.) 404, 455-466 (1988).
32.	Feletou, M. and P.M. Vanhoutte, Br. J. Pharmacol. 93, 515-524 (1988).
33.	Fleckenstein-Grun, G. and A. Fleckenstein, Blood Vessels 27, 319-332 (1990).
34.	Fujii, K., K. Kobayashi, T. Koga, and M. Fujishima, Circulation 82 (Suppl. III), 344 (1990).
35.	Furchgott, R.F and P.M. Vanhoutte, FASEB J. 3, 2007-2018 (1989).
36.	Garland, C.J., J. Physiol. (Lond.) 392, 333-348 (1987).
37.	Gelband CH, Lodge NJ, van Breeman C., Eur. J. Pharmacol. 167, 201-210 (1989).
38.	Gibbins, I.L., J.E. Brayden, and J.A. Bevan, Neuroscience 13, 1327-1346 (1984).
39.	Haddy, F.J., Fed. Proc. 42, 239-245 (1983).
40.	Haeusler, G., Blood Vessels 15, 46-54 (1978).
41.	Hai, C.-M. and R.A. Murphy, Annu. Rev. Physiol. 51, 285-298 (1989).
42.	Hamilton, T.C., S.W. Weir, and A.H. Weston, Br. J. Pharm. 88, 103-111 (1986).
43.	Harder, D.R. and A. Waters, Eur. J. Pharmacology 93, 95-100 (1983).
44.	Harder, D.R., Circ. Res. 55, 197-202 (1984).
45.	Harder, D.R., R. Gilbert, and J.H. Lombard, Am. J. Physiol. 253, F778-F781 (1987).
46.	Hermsmeyer, K., Fed. Proc. 42, 246-252 (1983).
47.	Hirst, G.D.S. and D.F. Van Helden, J. Physiol. (Lond.) 333, 53-67 (1982
48.	Hirst, G.D.S. and F.R. Edwards, Physiol. Rev. 69, 546-604 (1989).
49.	Hirst, G.D.S., and F.R. Edwards, Physiol. Rev. 69, 546-604 (1989).
50.	Kauser, K., W. Stekiel, G. Rubanyi, and D.R. Harder, Circ. Res. 65, 199-204 (1989).
51.	Kuschinsky, W., M. Wahl, O. Bosse, and K. Thurau, Circ. Res. 31, 240-247 (1972).
52.	Langton, P.D., M.T. Nelson, Y. Huang, and N.F. Standen, Am. J. Physiol. 260, H927-H934 (1991).
53.	Lincoln, T.M. and T.L. Cornwell, Blood Vessels 28, 129-137 (1991).
54.	Linder, L., W. Kiowski, Buhler, R.R., and T.F. Luscher, Circulation 81, 1762-1767 (1990).
55.	Lockette, W., Y. Otsuka, and O. Carretero, Hypertension 8 (Suppl. II), II61-II66 (1986).
56.	McCarron, J.G., J.M. Quayle, W. Halpern, and M.T. Nelson, Am. J. Physiol. (in press) (1991).
57.	Mulvany, M.J., C. Aalkjaer, and T.T. Petersen, Circ. Res. 54, 740-749 (1984).
58.	Nelson M.T., J.B. Patlak, J.F. Worley, and N.B. Standen, Am. J. Physiol. 259, C3-C18 (1990).
59.	Nelson M.T., N.B. Standen, J.E. Brayden, and J.F. Worley, Nature 336, 383-385 (1988).
60.	Nelson, M.T., Y. Huang, J.E. Brayden, J. Hescheler, and N.B Standen, Nature 344, 770-773 (1990).
61.	Okabe, K., S. Kajioka, K. Nakao, K. Kitamura, H. Kuriyama, and A.H. Weston, J. Pharmacol. Exp. Ther. 252, 832-839 (1990).
62.	Osol, G., R. Osol, and W. Halpern in Essential Hypertension, K. Aoki, ed (Springer Verlag, Tokyo 1986) pp. 107-114.
63.	Pearce, W.J., J.E. Brayden, A.D. Hull, and D.M. Long, FASEB J. 5, 112 (1991).
64.	Post, J.M. and A.W. Jones, Am. J. Physiol. 260, H848-H854 (1991).
65.	Quast, U. and N.S. Cook, J. Pharmacol. Exp. Ther. 250, 261-271 (1989).
66.	Quast, U. and N.S. Cook, Trends Pharmacol. Sci. 10, 431-435 (1989).
67.	Quast, U., Br. J. Pharmacol. 93, 19P (1988).
68.	Quayle, J.M., J. McCarron, W. Halpern, and M.T. Nelson, Biophysical J. (Abstract) (In press) (1991).
69.	Rashatwar, S.S., T.L.Cornwell, and T.M. Lincoln, Proc. Natl. Acad. Sci. U.S.A. 84, 5686-5689 (1987).
70.	Rosenfeld, M., J. Mermod, S. Amara, L. Swanson, P.E. Sawchenko, J. Rivier, W.W. Vale, and R.M. Evans, Nature 304, 129-135 (1983).
71.	Sabouni, M.H., P.T. Hargittai, E.M. Lieberman, and S.J. Mustafa, Am. J. Physiol. 257, H1750-H1752 (1989).
72.	Smeda, J.S. and E.E. Daniel, Circ. Res. 62, 1104-1110 (1988).
73.	Standen, N.B., J.M. Quayle, N.W. Davies, J.E. Brayden, Y. Huang, and M.T. Nelson, Science 245, 177-180 (1989).
74.	Su, C., J.A. Bevan, R.C Ursillo, Circ. Res. 15, 20-27 (1964).
75.	Tare, M., H.C. Parkington, H.A. Coleman, T.O. Neild, and G.J. Dusting, Nature 346, 69-71 (1990).
76.	Tesfemarian, B. and W. Halpern, Hypertension 11, 440-444 (1988).
77.	Towart, R., Circ. Res. 48, 650-657 (1981).
78.	van Breemen, C. and K. Saida, Ann. Rev. Physiol. 51, 315-329 (1989).
79.	Van de Voorde, J., C. Cuvelier, and I. Leusen, Am. J. Physiol. 250, H711-H717 (1986).
80.	Vanhoutte, P.M. and J.C. Stoclet, Blood Vessels 27, 69-319 (1990).
81.	von Beckerath, N., S. Cyrys, A. Dischner, and J. Daut, J. Physiol. (Lond.) (in press) (1991).
82.	Winquist, R.J., L.S. Heaney, A.A. Wallace, E.P. Baskin, R.B. Stein, M.L. Garcia, and G.J.Kaczorowski, J.Pharmacol. Exp. Ther. 248, 149-156 (1989).
83.	Worley, J.F., J.M. Quayle, N.B. Standen, and M.T. Nelson, Am. J. Physiol. (in press) (1991).

ENDOTHELIUM-DERIVED HYPERPOLARIZING FACTOR (EDHF) -- A VASODILATING SUBSTANCE BY ACTIVATING K-CHANNELS

HIKARU SUZUKI and YOSHIMICHI YAMAMOTO

Department of Physiology, Nagoya City University Medical School, Mizuho-Ku, Nagoya 467, Japan

INTRODUCTION

Actions of acetylcholine (ACh) on vascular tissues were unique because of generation of muscle contractions with associated hyperpolarization of the membrane (34). This hyperpolarization was produced by activation of muscarinic receptors, since the response was blocked by atropine. Stimulation of these receptors by carbachol also hyperpolarized the smooth muscle membrane of the guinea-pig superior mesenteric artery, in an endothelium-dependent manner (5, 6). As ACh was shown to have potent actions to release the endothelium-derived relaxing factor (EDRF, 21) through activation of muscarinic receptors, this muscarinic receptor-mediated hyperpolarization was also considered the actions of EDRF (6). However, there were substances which release EDRF with no associated hyperpolarization of the membrane, and such involved substance P (5) or oxotremorine (32). The reasonable interpretation of these results was that the hyperpolarization was produced by substances other than EDRF. Contribution of different subtypes of muscarinic receptor for generation of the endothelium-dependent relaxation and hyperpolarization (i.e., M_1-receptor for hyperpolarization and M_2-receptor for relaxation, 31) also supported this concept. Such putative substance was termed an endothelium-derived hyperpolarizing factor (EDHF), although the hyperpolarization was not confirmed to be produced by a humoral substance (13, 55). Later, evidences in favor of an involvement of humoral substances in the endothelium-dependent hyperpolarization have been accumulated using the cascade experiments, nevertheless the chemical nature of the EDHF still remain to be unidentified. The release of EDRF from endothelial cells accompanies by a production of PGI_2 (43). The main physiological actions of PGI_2 on cardiovascular systems are considered to inhibit adhesion of blood cells to the vessel wall or platelet aggregation, and iloplost, a synthesized PGI_2, hyperpolarizes vascular smooth muscle membrane (10). Therefore, it is possible that PGI_2 is the mediator of the endothelium-dependent hyperpolarization. However, absence of any inhibitory action of indomethacin on the endothelium-dependent hyperpolarization suggests that metabolites of arachidonic acid may not be involved in the endothelium-dependent hyperpolarization. Although our knowledge about the endothelium-derived factors which hyperpolarize smooth muscle membrane is not enough, it would be worthwhile to overview here the physiological properties of the endothelium-dependent hyperpolarization updated.

PROPERTIES OF ENDOTHELIUM-DEPENDENT HYPERPOLARIZATION

Many types of EDRF-releaser such as ACh (9, 42, 51), bradykinin (4, 8, 42), or substance P (3, 8, 42) hyperpolarize endothelial cell membrane. There are gap junctions between endothelial and smooth muscle cells (48, 49), and such connections are probably working as electrical low resistance paths. The endothelial hyperpolarization is, therefore, possibly conducted to smooth muscles in an electrotonic manner (19). If this is the case, we do not need to assume contribution of humoral substance in the endothelium-dependent hyperpolarization of smooth muscle membrane.

Electrophysiological experiments demonstrated that in the rat aorta, both endothelial and smooth muscle cells are hyperpolarized by bradykinin with similar time courses, yet no electrical couplings are found to occur between these two cell layers (4). Injection of fluorescent dye which is permeable through gap junctions into an endothelial cell did not diffuse to smooth muscle cells (4). Thus, no functional connection between endothelial and smooth muscle cells was found to present in the rabbit aorta. Release of endothelial humoral substances which hyperpolarize smooth muscle membrane has been

Copyright 1991 by Elsevier Science Publishing Company, Inc.
Ion Channels of Vascular Smooth Muscle Cells and Endothelial Cells
Sperelakis and Kuriyama, Editors

demonstrated experimentally in the isolated blood vessels. Smooth muscle membrane of the endothelium-denuded canine coronary artery was hyperpolarized by ACh, if a segment of the endothelium-intact canine femoral artery was present in the same experimental chamber (20). In the rat femoral artery, smooth muscle cell membranes of the endothelium-denuded segment were hyperpolarized by ACh, when the endothelium-intact segment was connected at the upstream regions of the recording vessel (29). In the sandwich preparation made up from the endothelium-intact carotid artery and endothelium-denuded coronary artery of the guinea-pig (14), ACh hyperpolarized the membrane of the coronary smooth muscles (Figure 1). All these results strongly suggest that EDHF is indeed a humoral substance.

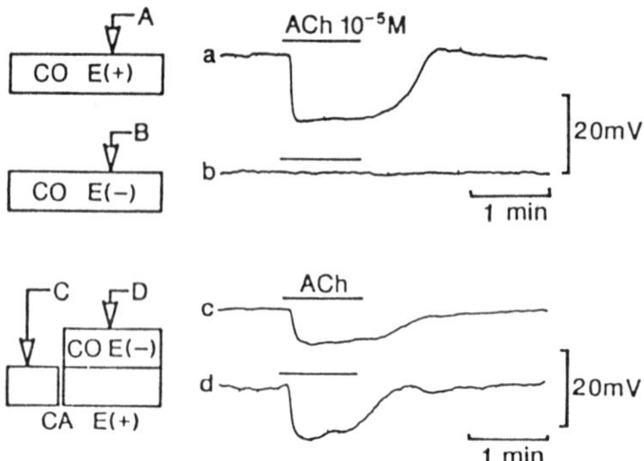

Figure 1. ACh-induced hyperpolarization of vascular smooth muscle membrane. In the guinea-pig coronary artery (CO), application of ACh (10^{-5} M) produces a membrane hyperpolarization in the endothelium-intact tissue (A,a) but not in the endothelium-denuded tissue (B,b). In the sandwich preparation made up with an endothelium-intact carotid artery (CA) and endothelium-denuded coronary artery (D), ACh hyperpolarizes the membrane in the endothelium-absent coronary artery (d). c, membrane hyperpolarizations produced by ACh in the intact coronary artery. From ref. 14.

Potency of endothelial cells to release EDHF differed between vascular beds, and also between agonists even in the same vascular bed. For example in the rat, ACh was a potent hyperpolarizer of the smooth muscle membrane in the main pulmonary artery but not in the intrapulmonary artery. Histamine produced a transient hyperpolarization in the aortic and intrapulmonary arteriolar muscles but no hyperpolarization in the main pulmonary artery muscles (Figure 2). Sensitivity of muscles to EDHF also differed between vascular beds, and EDHF released by ACh from endothelial cells of the guinea-pig carotid artery hyperpolarized the smooth muscle membrane by about 10 mV in the carotid artery and by about 20 mV in the coronary artery (Figure 3), i.e., the sensitivity to EDHF was higher in the coronary than in the carotid arteries (14).

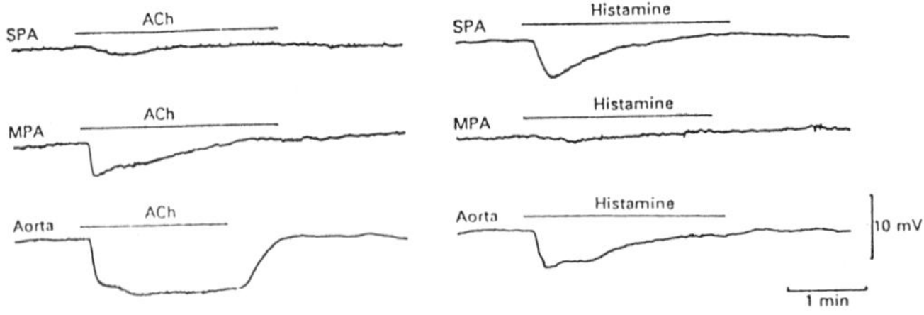

Figure 2. Hyperpolarizations produced by ACh (10^{-5} M) and histamine (10^{-5} M) recorded from smooth muscle cells of the rat intrapulmonary artery (SPA), main pulmonary artery (MPA) and aorta. From ref. 10.

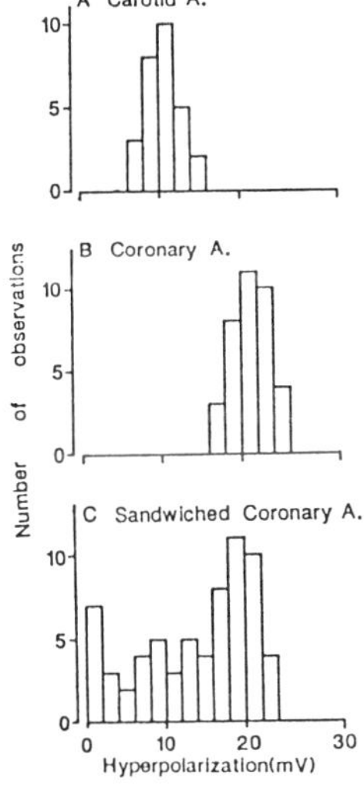

Figure 3. Histograms of the amplitude of ACh-induced hyperpolarization recorded from smooth muscle cells of the intact carotid artery (A), intact coronary artery (B) and sandwiched coronary artery (C) of the guinea-pig. The sandwich preparation was made up with an intact carotid artery and endothelium-denuded coronary artery. ACh: 10^{-5} M. The amplitude of the hyperpolarization was in the order of intact coronary artery > endothelium-absent coronary artery > intact carotid artery. From ref. 14.

IONIC MECHANISMS OF THE EDHF-INDUCED HYPERPOLARIZATION

In the rabbit saphenous artery, the membrane resistance estimated from the amplitude of electrotonic potentials produced by extra-cellularly applied current stimuli was decreased during the ACh-induced hyperpolarization (Figure 4), and such was interpreted by an increase in permeability of the membrane to K-ions (32). An endothelium-dependent increase in the efflux of the incorporated ^{86}Rb from the rat aorta also supported this concept (13). Amplitude of the endothelium-dependent hyperpolarization was related to the concentration of K ions in the external media ($[K^+]_o$), and it increased in low $[K^+]_o$ solution and decreased in high $[K^+]_o$ solution (Figure 5), and these phenomena were considered to be due to an increased K permeability of the smooth muscle membrane (10). However, the situation does not seem to be as they are, since change in $[K^+]_o$ also changes the endothelial membrane potential, and this causes the change in endothelial $[Ca^{2+}]_i$, as that the endothelial cell membrane has no voltage-dependent Ca-channels and in high $[K^+]_o$ solution amount of Ca ions being supplied from the external media for release of EDHF decreases (36). Thus, changes in $[K^+]_o$ indirectly modify the amount of releasable EDHF in endothelial cells and also the equilibrium potential for K ions in smooth muscles, i.e., an involvement of at least these two factors has to be considered in the change of the ACh-induced hyperpolarization in modified $[K^+]_o$ solutions.

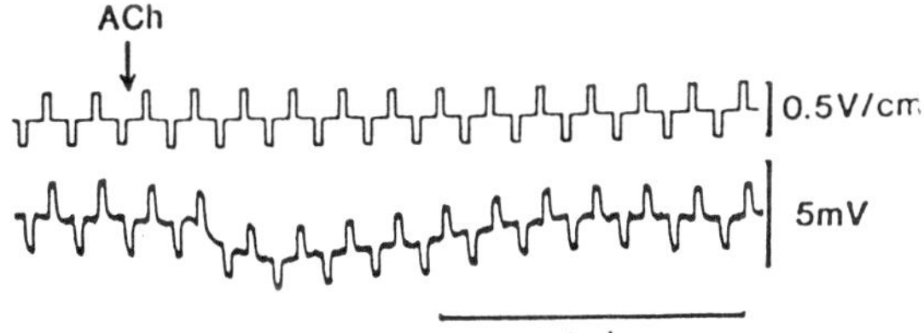

Figure 4. Effects of ACh on electrotonic potentials produced by alternate application of a constant intensity(0.3 V/cm) of inward and outward current pulses (1.5 s duration) at a constant rate (0.1 sec^{-1}) in the rabbit saphenous artery. Guanethidine (5×10^{-6} M) and tetrodotoxin (3×10^{-7} M) were present throughout. From ref. 32.

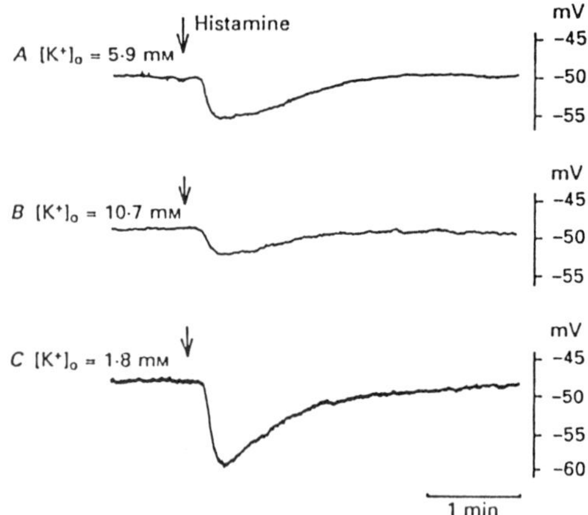

Figure 5. The histamine-induced hyperpolarizations recorded from the rat small pulmonary artery in different $[K^+]_O$ solutions. A, 5.9 mM; B, 10.7 mM; C, 1.8 mM. All responses from the same tissue. From ref. 10.

ACh-induced hyperpolarization of the canine coronary smooth muscle membrane was inhibited by ouabain, and this indicated an activation by EDHF of the electrogenic Na-K pump (20). However, no such evidence was obtained in the rabbit ear artery (56), dog coronary artery (12) and guinea-pig coronary artery (Figure 6). The membrane potential of endothelial cells is maintained by a high-permeability to K ions and also by an activity of electrogenic Na-K pump (18), and therefore inhibition by ouabain of the latter will depolarize the endothelial membrane. This may have causal relationship with the inhibition by ouabain of the endothelium-dependent hyperpolarization.

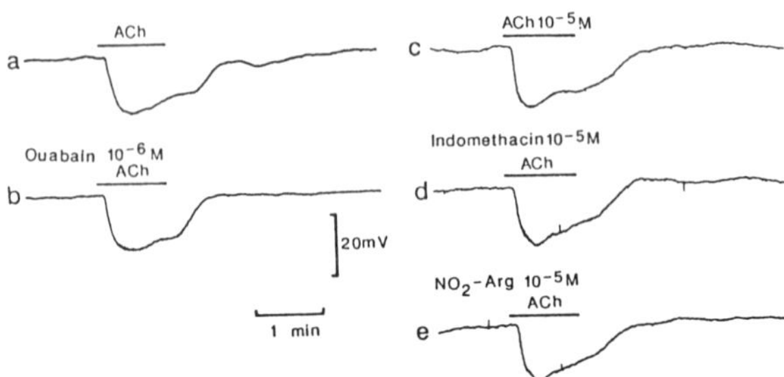

Figure 6. Effects of ouabain 10^{-6} M (b), indomethacin 10^{-5} M (d), and nitroarginine 10^{-5} M (e) on the ACh-induced hyperpolarization recorded from the endothelium-denuded coronary artery which was sandwiched with intact carotid artery. ACh, 10^{-5} M. From ref. 14.

Many types of K-channels are found to present in smooth muscle membrane (16). The type of K-channels involved in the EDHF-induced hyperpolarization was estimated from the effects of K-channel blockers on the hyperpolarization. In the rabbit basilar artery, ACh hyperpolarized the membrane in an endothelium-dependent manner, and the hyperpolarization was inhibited by glibenclamide (7, 54). This chemical has a property to block the ATP-sensitive K-channel, and possible contribution of this type of K-channel in the EDHF-induced hyperpolarization was considered (7, 54). The ATP-sensitive K-channel is also activated by a group of vasodilators which relax vascular smooth muscles by hyperpolarizing the membrane, i.e., the K-channel opener (23). Therefore, we may consider that EDHF has a property of an endogenous K-channel opener.

However in the sandwich preparation of the guinea-pig coronary artery prepared with carotid artery (14), the membrane hyperpolarizations produced by ACh were not blocked by glibenclamide, but was blocked by tetraethylammonium (TEA), while the hyperpolarizations produced by pinacidil, a K-channel opener, was blocked by glibenclamide but not by TEA (Figure 7). TEA is a non-specific K-channel inhibitor, but relatively high selectivity to the Ca-sensitive K-channel (16). However, the antagonism on muscarinic receptors, as was seen in the rat central nervous system (1), has also to be accounted for the interpretation of the actions of TEA. Thus, it is clear that EDHF hyperpolarizes the membrane by activating K-channels, but the type of K-channels involved remain to be determined.

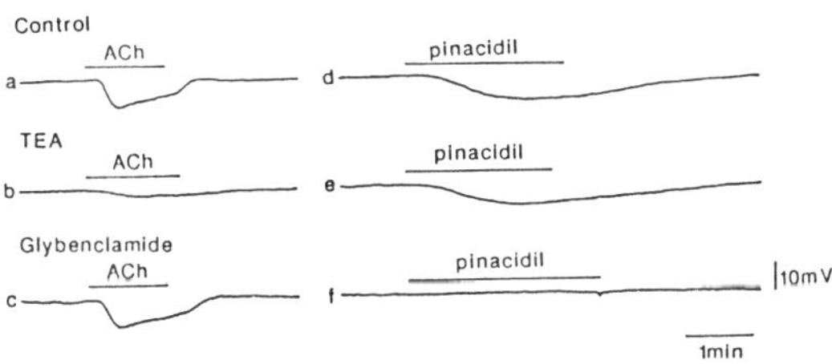

Figure 7. Effects of tetraethylammonium (TEA, 1 mM) and glybenclamide(10^{-5} M) on hyperpolarizations produced by ACh (a-c) and pinacidil (d-f) in smooth muscle cells of the intact guinea-pig coronary artery. From ref. 18. Note that the ACh-induced hyperpolarization is inhibited by TEA but not by glybenclamide, while the pinacidil-induced hyperpolarization is inhibited by glybenclamide but not by TEA.

DIFFERENCES BETWEEN EDRF AND EDHF

Possible actions by EDRF of the hyperpolarization of smooth muscle membrane are suggested in several vascular tissues (6, 29, 58). However, 1) many types of EDRF-inhibitors do not block the generation of the endothelium-dependent hyperpolarization (10, 13, 23), 2) Exogenously applied nitric oxide does not hyperpolarize the membrane (3, 33), 3) Drugs which relax vascular smooth muscles in an endothelium-independent manner, by producing nitric oxide in the cell, such as nitroglycerine or nitroprusside, do not hyperpolarize the membrane (35). These results suggest that the hyperpolarization is not the actions of EDRF.

Comparison of the endothelium-dependent relaxation with the hyperpolarization reveals differences between these responses in many points. The time course of the membrane hyperpolarization differs from that of the relaxation; in many vessels the former is a transient phenomenon and the potential disappears within 5-10 min or decays to a very small amplitude (Figure 8), while the latter can be maintained for up to 30 min with similar amplitude (Figure 14).

The transient nature of the endothelium-dependent hyperpolarization is not due to desensitization of receptors responsible to EDHF or due to depletion of releasable EDHF in the endothelial cells (Figure 9). The comparison between histamine- and ACh-induced hyperpolarization in the rat aortic smooth muscle membrane revealed that the rate of decay of the hyperpolarization was inversely related to the time required for the recovery of receptors on the endothelial cell membrane from the desensitization (10), i.e., transient generation of the hyperpolarization was partly due to the desensitization of receptors on the endothelial cell membrane. However, if this is the case, sustained generation of the relaxation by these agonists cannot be reasonably explained.

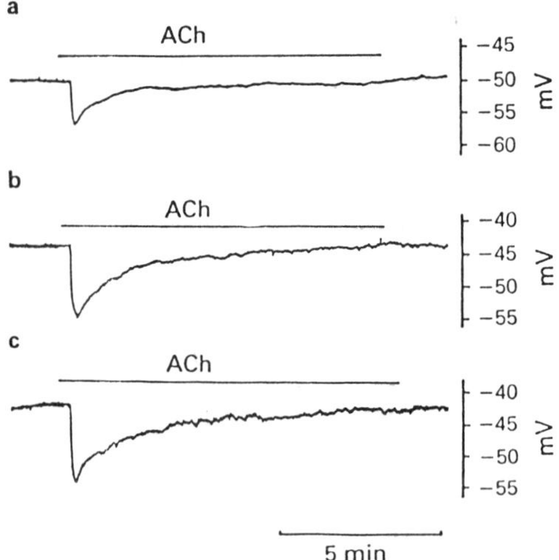

Figure 8. Effects of ACh (10^{-5} M) on membrane potential of smooth muscle cells in the rat main pulmonary artery. (a) Control, (b) in the presence of 10^{-7} M norepinephrine, (c) in the presence of 10^{-7} M norepinephrine plus 3×10^{-6} M methylene blue. From ref. 13.

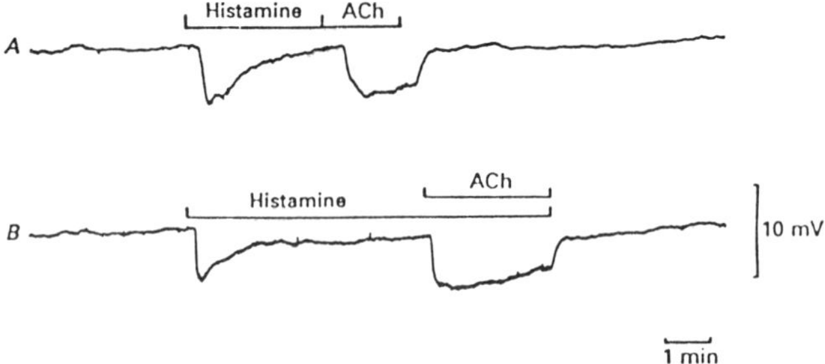

Figure 9. Hyperpolarization by ACh of the rat aortic smooth muscle cells generated after the histamine-induced hyperpolarization had ceased. ACh (10^{-6} M) was applied just after histamine (5×10^{-5} M) (A) or together with histamine (B). From ref. 10.

The time required for the recovery of the EDHF- and EDRF-actions was estimated from the amplitude of the ACh-induced hyperpolarization and that of relaxation of muscles pre-contracted with high-$[K^+]_o$ solution, respectively, in the canine coronary artery (12). In this artery, the recovery of the ACh-induced relaxation required about 15 min, whereas that of the hyperpolarizing response required over 30 min (Figure 10). When a fixed concentration of ACh was applied repetitively at a constant interval of 15 min, the decrease in amplitude of the hyperpolarizing response was faster than that of the relaxing response, and the 5th application of ACh produced hyperpolarization which was about 70% of that produced by the first application, whereas in the case of the relaxation, no significant decrease in the amplitude was observed (Figure 11). Thus, these results indicate that the recovery of the actions of EDRF is faster than those of EDHF.

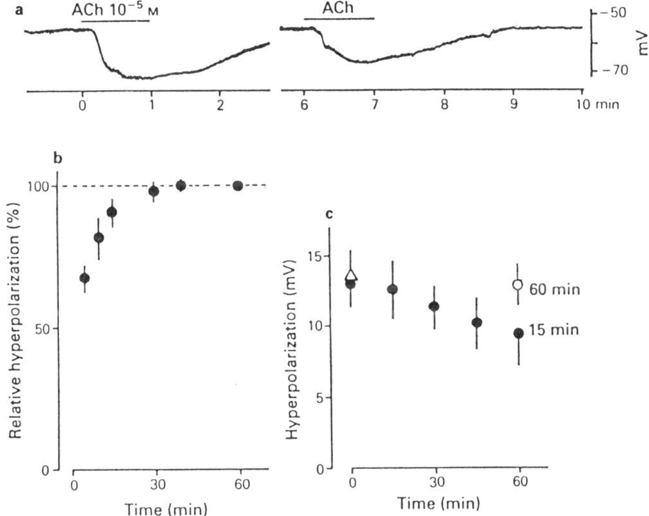

Figure 10. Tachyphylaxis of the ACh-induced hyperpolarization in smooth muscle cells of the canine coronary artery. **(a)** Hyperpolarizations produced by application of ACh (10^{-5} M) on two occasions separated by a 5 min interval (recordings were interrupted for about 3 min between two ACh applications). **(b)** Recovery of the ACh-response from desensitization. Amplitude of the hyperpolarization produced by the second ACh was measured as a percentage of the first, and plotted against the time between the two applications of ACh. About 40 min were required for the complete recovery of the ACh-induced hyperpolarization. **(c)** Tachyphylaxis of the ACh-induced hyperpolarization. Amplitude of the hyperpolarization produced by application of ACh (10^{-5} M) on 5 successive occasions each separated by a 15 min interval (filled circles), or on 2 occasions separated by a 60 min interval (open triangle and circle). Each ACh challenge was applied for 1 min, and the peak amplitude of the hyperpolarization was measured. From ref. 12.

Pharmacological properties of EDRF also differed from those of EDHF. The muscle relaxations produced by EDRF can be inhibited by methylene blue and by oxyhemoglobin (41). The former inhibition is due to reduced production of cyclic GMP as a result of inhibition of guanylate cyclase, while the latter inhibition is due to binding to and inactivate EDRF. Both of these EDRF-inhibitors did not block the generation of the ACh-induced hyperpolarization (Figure 8), indicating that the hyperpolarization may not be produced directly by EDRF or indirectly via increased production of cyclic GMP (10, 13, 23).

EDRF is probably nitric oxide(NO) produced from L-arginine in the endothelial cells (44, 47), and nitroarginine is a potent inhibitor of the production of EDRF, as estimated from the ACh-induced relaxation of the rabbit aorta (30). In the guinea-pig coronary and carotid arteries, the ACh-induced hyperpolarization was insensitive to nitroarginine (Figure 6), i.e., the production of EDRF was probably not directly related to the membrane hyperpolarization.

The release of EDRF requires a presence of Ca^{2+} in the external solution (37), and this suggests that the source of Ca^{2+} used for the release of EDRF is in the extracellular media. The experiments using Ca-sensitive fluorescent dye showed that many types of chemical stimulants which release EDRF elevate $[Ca^{2+}]_i$ in the endothelial cells (9, 15, 17, 27, 28, 38-40, 45). Electrophysiological experiments suggest that the release of EDHF also requires an increase in endothelial $[Ca^{2+}]_i$ (11). In the rabbit carotid artery, the ACh-induced hyperpolarization consisted of $[Ca^{2+}]_o$-insensitive and $[Ca^{2+}]_o$-sensitive components; the former appeared at the initial part of the hyperpolarization, and the latter formed the sustained part of the hyperpolarization. In

180

Ca-free solution, ACh produced only a transient hyperpolarization with the duration of up to 30s (Figure 12). The first component was decreased markedly by the pretreatment with A-23187, and was inhibited all but completely by procaine (11). A-23187 has a property of Ca-ionophore and, in vascular smooth muscles, acts at first to the membrane of the sarcoplasmic reticulum(SR) to deplete Ca, and then to the plasma membrane, to facilitate permeation of Ca^{2+} (26). Procaine inhibits the release of intracellularly stored Ca^{2+} (probably in SR) in vascular smooth muscles (25). Although no data are available about the actions of A-23187 on endothelial membrane, this agent hyperpolarizes the rabbit carotid arterial smooth muscle membrane in an endothelium-dependent manner, and once the tissues are exposed to the A-23187 containing solution, amplitude of the ACh-induced hyperpolarization decreases dramatically in an almost irreversible manner (Figure 13). Such allowed to estimate that A-23187 acts at first to the membrane of the endothelial endoplasmic reticulum to release stored Ca^{2+}. In the rabbit aorta, A-23187 relaxes the muscle by releasing EDRF, in an irreversible manner (22), and this strongly suggests that this agent has some irreversible actions on the endoplasmic reticulum of endothelial cells. Thus, these results strongly suggest that the increase in endothelial $[Ca^{2+}]_i$ is required for the release of EDHF. Two Ca sources are considered, one from the intracellular store sites and the release of them produces the initial transient component of the hyperpolarization, and another from the extracellular media to produce the sustained phase of the hyperpolarization.

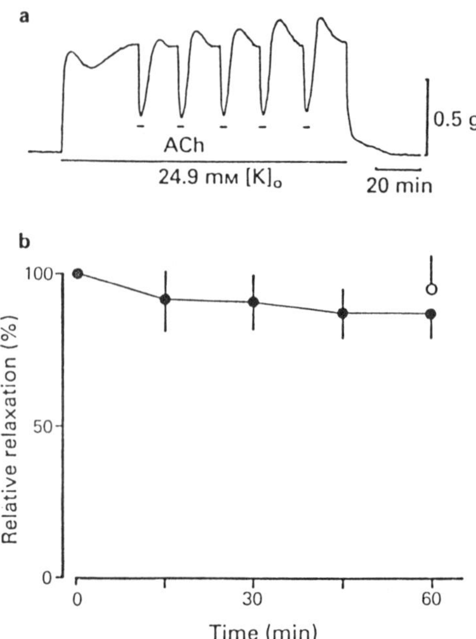

Figure 11. ACh-induced relaxation in the canine coronary artery. (a) A ring segment of the artery was contracted by 24.9 mM $[K^+]_o$ solution, and exposed to ACh (10^{-5} M) for 3 min on 5 occasions at 15 min intervals. (b) The relaxations produced by application of ACh as described in (a) (●) or on 2 occasions separated by a 60 min interval (o) were observed in tissues contracted with 24.9 mM $[K^+]_o$ solution, and the amplitude of the maximum relaxation relative to the first was measured. From ref. 12.

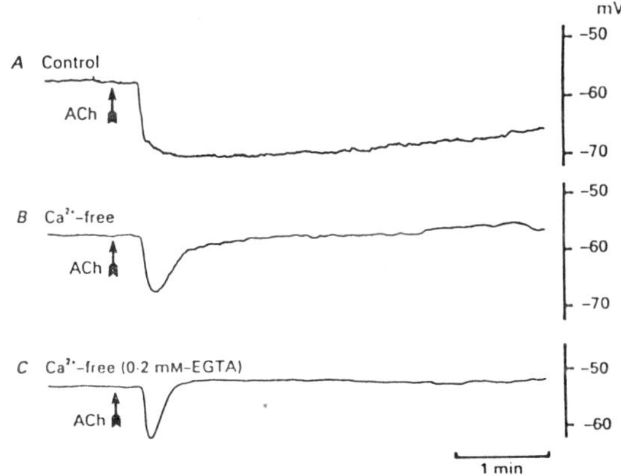

Figure 12. ACh-induced hyperpolarization of smooth muscle membrane of the rabbit carotid artery recorded in different $[Ca^{2+}]_o$ solutions. $[Ca^{2+}]_o$ = 2.5 mM (A), 0 mM (B) and 0 mM with 0.2 mM EGTA (C). ACh (10^{-5} M) was applied at the arrow in each trace. The sustained hyperpolarization was changed to a transient form in the absence of $[Ca^{2+}]_o$. From ref. 11.

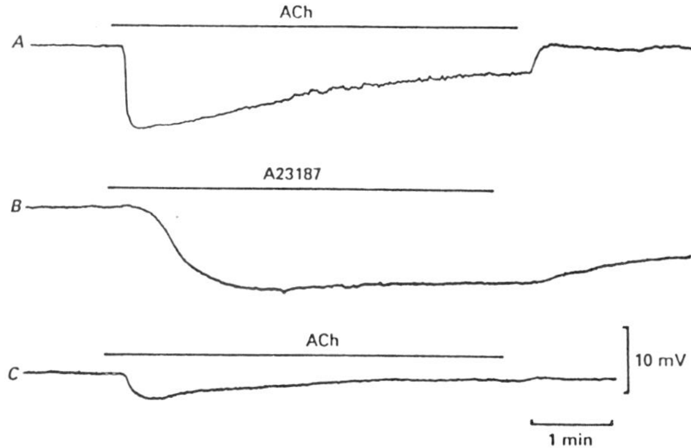

Figure 13. Endothelium-dependent hyperpolarizations produced by ACh (10^{-5} M) and A23187 (10^{-7} M) in the rabbit carotid artery. ACh was applied before (A) and after (B) application of A23187 (B). The latter response was diminished by A23187 in an irreversible manner. From ref. 11.

The cellular mechanisms of the EDRF-induced relaxation are mainly due to decrease in $[Ca^{2+}]_i$ by an increased production of cyclic GMP through activation of guanylate cyclase. Inhibition by cyclic GMP of phospholyration of contractile proteins may also contribute to the muscle relaxation (24). In the rat blood vessels, inhibition by methylene blue of the production of cyclic GMP did not alter the endothelium-dependent hyperpolarization and the efflux of the incorporated ^{86}Rb (13). Therefore, cyclic GMP may not be involved in the actions of EDHF.

HYPERPOLARIZATION AND MUSCLE RELAXATION

In the presence of EDRF-inhibitors, ACh or histamine has weak relaxing actions on rat aortic smooth muscles. The relaxations are mainly in a transient form, as in the case of the change in membrane hyperpolarization (Figure 14). Similar phenomena were seen in the guinea-pig carotid artery which was inhibited of the production of EDRF by nitroarginine (57). Therefore, the membrane hyperpolarization by EDHF is one of the contributor of the endothelium-dependent relaxation. It is demonstrated that in vascular smooth muscles, only a small amplitude of membrane hyperpolarization can decrease the open probability of Ca-channel to decrease the influx of Ca^{2+}, thus induces muscle relaxation (46). EDHF can hyperpolarize muscle membrane over 10 mV, and such a change in membrane potential would be large enough to relax muscles.

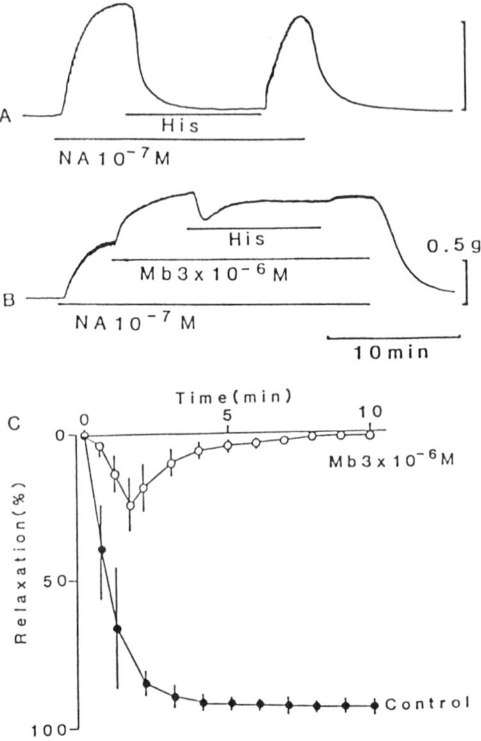

Figure 14. Sustained relaxation by histamine of smooth muscles precontracted with norepinephrine (NA, A), and the effects of methylene blue (Mb, 3 x 10⁻⁶ M) on the histamine-induced relaxation in the rat aorta. B, Mb enhanced the norepinephrine-induced contraction, and the histamine-induced relaxation took a transient form. C, time courses of the histamine-induced relaxation in the absence (filled circle) and presence of Mb (open circle).

The ratio of the EDHF-induced component within the endothelium-dependent relaxation can be estimated from the experiments using inhibitors of the actions of EDRF. For example in the rat aorta, comparison of the endothelium-dependent relaxations in the absence and presence of methylene blue or oxyhemoglobin indicated that the EDHF-induced relaxation amounted to 20-30% of the endothelium-dependent relaxation (10). In the guinea-pig carotid artery, inhibition by nitroarginine of the production of EDRF resulted a decrease of about 50% in the ACh-induced relaxation (57). However, the membrane hyperpolarizations are mainly of transient phenomena, and so are the nitroarginine-resistant relaxation. Therefore, EDHF contributes to the endothelium-dependent relaxation only at the initial phase. In certain conditions, ACh relaxes the rabbit aortic smooth muscles with two phases, an initial transient and following sustained relaxations, and they are produced by activation of muscarinic M_1- and M_2-receptors, respectively (50). Muscarinic M_1-receptor is involved in the ACh-induced hyperpolarization (31), and these results suggest that the initial phase of the relaxation is produced by EDHF.

SUMMARY AND CONCLUSION

Many types of physiologically active substances hyperpolarize the membrane of vascular smooth muscles, in an endothelium-dependent manner. This hyperpolarization is produced by a humoral substance derived from the endothelial cells, however the duration of the activity, pharmacological properties, and the cellular mechanisms seem to differ from those of EDRF or prostacyclin, and therefore identified as endothelium-derived hyperpolarizing factor (EDHF). The amount of released EDHF differ between agonists, and the releasing potency of endothelial cells and also the sensitivity of smooth muscle membrane to EDHF differ between vascular tissues. EDHF has properties of an endogenous K-channel opener, and hyperpolarizes the smooth muscle membrane by activating K-channel, although the type of the channel remains to be determined.

REFERENCES

1. Balduni, W., L.G. Costa, and S.D. Murphy, Neurochem. Res. 15, 33-39 (1990).
2. Beny, J.-L. and P.C. Brunet, Blood Vessels 25, 308-311 (1989).
3. Beny, J.-L. and P.C. Brunet, J. Physiol. (Lond.). 398, 277-289 (1988).
4. Beny, J.-L., Am. J. Physiol. 258, H836-H841 (1990).
5. Bolton, T.B. and L.H. Clapp, Br. J. Pharmacol. 87, 713-723 (1986).
6. Bolton, T.B., R.J. Lang, and T. Takewaki, J. Physiol. (Lond.) 351, 549-572 (1984).
7. Brayden, J.E., Am. J. Physiol. 259, H668-H673 (1990).
8. Brunet, P.C. and J.-L. Beny, Blood Vessels 26, 228-234 (1989).
9. Busse, R., H. Fichtner, A. Luckhoff, and M. Kohdhardt, Am. J. Physiol. 255, H965-H969 (1988).
10. Chen, G. and H. Suzuki, J. Physiol. (Lond.) 410, 91-106 (1989).
11. Chen, G. and H. Suzuki, J. Physiol. (Lond.) 421, 521-534 (1990).
12. Chen, G., H. Hashitani, and H. Suzuki, Br. J. Pharmacol. 98, 950-956 (1989).
13. Chen, G., H. Suzuki, and A.H. Weston, Br. J. Pharmacol. 95, 1165-1174 (1988).
14. Chen, G., Y. Yamamoto, K. Miwa, and H. Suzuki, Am. J. Physiol. 260, H1888-H1892 (1991).
15. Colden-Stanfield, M., W.P. Schilling, A.K. Ritchie, S.G. Eskin, L.T. Navarro, and D.L. Kunze, Circ. Res. 61, 632-640 (1987).
16. Cook, N.S., TIPS 9, 21-28 (1988).
17. Dabthuluri, N.R., M.I. Cybulski, and T.A. Brock, Am. J. Physiol. 255, H1549-H1553 (1988).
18. Daut, J., G. Mehrke, S. Nees, and W.H. Newman, J. Physiol. (Lond.) 402, 237-254 (1988).
19. Davies, P.P., S.-P. Olesen, D.E. Clapham, E.M. Morrel, and F.J. Schoen, Hypertension 11, 563-572 (1988).
20. Feletou, M. and P.M. Vanhoutte, Br. J. Pharmacol. 93, 515-524 (1988).
21. Furchgott, R.F. and J.V. Zawadzki, Nature 288, 373-376 (1980).
22. Furchgott, R.F., Circ. Res. 53, 557-573 (1983).
23. Huang, A.H., R. Busse, and E. Bassenge, Naunyn-Schmiedeberg's Arch. Pharmacol. 338, 438-442 (1988).
24. Ignarro, L.J., FASEB J. 3, 31-?? (1989).
25. Itoh, T., H. Kuriyama, and H. Suzuki, J. Physiol. (Lond.) 321, 513-535 (1981).
26. Itoh, T., Y. Kanmura, and H. Kuriyama, J. Physiol. (Lond.). 359, 467-484 (1985).
27. Jacob, R., J. Physiol. (Lond.) 421, 55-77 (1990).
28. Johns, A., T.W. Lategan, N.J. Lodge, U.S. Ryan, C. van Breemen, and D.J. Amams, Tiss. Cell. 19, 733-745 (1987).
29. Kauser, K., W.J. Stekiel, G. Rubanyi, and D.R. Harder, Circ. Res. 65, 199-204 (1989).
30. Kobayashi, Y. and K. Hattori, Jap. J. Pharmacol. 52, 167-169 (1990).

31. Komori, K. and H. Suzuki, Br. J. Pharmacol. 92, 657-664 (1987).

32. Komori, K. and H.Suzuki, Circ. Res. 61, 586-593 (1987).

33. Komori, K., R.R. Lorenz, and P.M. Vanhoutte, Am. J. Physiol. 255, H207-H212 (1988).

34. Kuriyama, H. and H. Suzuki, Br. J. Pharmacol. 64, 493-501 (1978).

35. Kuriyama, H., Y. Ito, H. Suzuki, K. Kitamura, and T. Itoh, Am. J. Physiol. 243, H641-H662 (1982).

36. Laskey, R.E., D.J. Adams, A. Johns, G.M. Rubanyi, and C. vanBreemen in Endothelium-Derived Relaxing Factors, G.M. Rubanyi and P.M. Vanhoutte, eds. (Karger, Basel 1990) pp. 128-135.

37. Long, C.J. and T.W. Stone, Blood Vessels 22, 205-208 (1985).

38. Luckhoff, A. and R. Busse, J. Cell. Physiol. 126, 414-420 (1986).

39. Luckhoff, A. and R. Busse, Pflugers Arch. 416, 305-311 (1990).

40. Luckhoff, A., U. Pohl, A. Mulsch, and R. Busse, Br. J. Pharmacol. 95, 189-196 (1988).

41. Martin, W., G.M. Villani, D. Jothianandan, and R.F. Furchgott, J. Pharmacol. Exp. Ther. 232, 708-716 (1985).

42. Mehrke, G. and J. Daut, J. Physiol. (Lond.) 430, 251-272 (1990).

43. Moncada, S., Br. J. Pharmacol. 76, 3-31 (1982).

44. Moncada, S., M.W. Radomski, and R.M.J. Palmer, Biochem. Pharmacol. 37, 2495-2501 (1988).

45. Morgan-Boyd, R., J.M. Stewart, R.J. Vavrek, and A. Hassid, Am. J. Physiol. 253, C588-C598 (1987).

46. Nelson, M.T., J.B. Patlak, J.F. Worley, and N.B. Standen, Am. J. Physiol 259, C3-C18 (1990).

47. Palmer, R.M.J., D.S. Ashton, and S. Moncada, Nature 333, 664-666 (1988).

48. Rhodin, J.A.G., in Handbook of Physiology, vol.II, Sec.2, The Cardiovascular System, D.F.Bohr, A.P.Somlyo, H.V.Sparks, and S.R.Geiger, eds. (Am. Physiol. Soc., Bethesda 1980) pp. 1-31.

49. Rhodin, J.A.G., J. Ultrast. Res. 18, 181-223 (1967).

50. Rubanyi, G.M., M. Mckinney, and P.M. Vanhoutte, J. Pharmacol. Exp. Ther. 240, 802-808 (1987).

51. Sakai, T., Jap. J. Pharmacol. 53, 235-246 (1990).

52. Schilling, W.P., A.K. Ritchie, L.T. Navarro, and S.G. Eskin, Am. J. Physiol. 255, H219-H227 (1988).

53. Siegel, G., J. Mironneau, F. Schnalke, G. Loirand, and G. Stock in Endothelium-Derived Relaxing Factors, G.M. Rubanyi and P.M. Vanhoutte, eds. (Karger, Basel 1990) pp. 265-280.

54. Standen, N.B., J.M. Quayle, N.W. Davies, J.E. Brayden, Y. Huang, and M.T. Nelson, Science 245, 177-180 (1989).

55. Suzuki, H. and G. Chen, NIPS 5, 212-215 (1990).

56. Suzuki, H., Eur. J. Pharmacol. 156, 295-297 (1988).

57. Suzuki, H., G. Chen, K. Miwa and Y. Yamamoto, Folia Pharmacol. Jap. 96, 44p (1990).

58. Tare, M., H.C. Parkington, H.A. Coleman, T.O. Neild, and G.J. Dusting, Nature 246, 69-71 (1990).

RECEPTORS AND SECOND MESSENGER MODULATION OF Ca^{2+} AND K^+ CHANNELS ACTIVITY IN VASCULAR SMOOTH MUSCLE CELLS

GHASSAN BKAILY

Department of Physiology and Biophysics, Faculty of Medicine, University of Sherbrooke, Sherbrooke, Quebec, Canada J1H 5N4.

INTRODUCTION

The vascular smooth muscle (VSM) consists of an assembly of small, narrow cells: these contiguous cells often come into close contact, forming various types of specialized junctions. Smooth muscle cells of different areas of the vascular tree are inherently different. Alternatively, it can be assumed that VSM cells are intrinsically homogeneous at some physiological level, but that fundamental similarities are masked by differences between vascular regions in their local environmental or architectural make-up (39). Because of the heterogenous responses of smooth muscle from different vessels to a wide variety of stimuli, it is difficult to make generalizations about the membrane properties of VSM as a group, as we do for heart muscle and skeletal muscle. Therefore, in this chapter we will focus on channels, of aortic VSM of the rabbit. The different types Ca^{2+} and K^+ currents in this muscle will presented as well as the regulation of these channels by pharmacological agents and their regulation by second messengers.

PRESENCE OF FAST Na^+ CURRENT IN AORTIC VSM CELLS

The presence of a TTX-sensitive Na^+ current was first been reported by Streck and Hermsmeyer (41) in neonatal rat portal vein and no other report confirmed these findings. Using the whole cell voltage clamp technique (25), we rarely detected the presence of a TTX-sensitive inward Na^+ current in aortic cells. In very few cases, we succeeded to record an inward current in normal Tyrode solution containing K^+ channels blockers and Figure 1 shows an example. As can be seen in this Figure, addition of 10^{-5} M, tetrodotoxin (TTX, a specific fast Na^+ channel blocker) rapidly decreased the peak inward current amplitude by 60% within 20 sec. A complete blockade of the inward current by TTX occurred at 100 sec in presence of the toxin (Fig. 1).

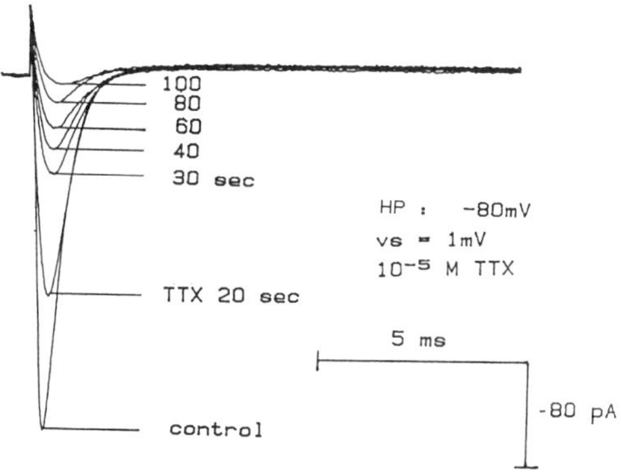

Figure 1. The inward current was recorded in Tyrode solution at 22^0 C. All records are from the same single aortic cell. HP is holding potential and VS is voltage step. (Bkaily and Peyrow, unpublished results).

These results suggests that a fast Na^+ current does also exist in aortic VSM cells. However, it is very difficult to ascertain the physiological role of this channel in aortic smooth muscle (SM) cells. The high difficulty to obtain this inward Na^+ current in aortic single cells makes further studies of this channel in VSM cells a very hard task.

Published 1991 by Elsevier Science Publishing Company, Inc.
Ion Channels of Vascular Smooth Muscle Cells and Endothelial Cells
Sperelakis and Kuriyama, Editors

PRESENCE OF THREE TYPES OF K$^+$ CHANNELS IN AORTIC VSM CELLS

At least four types of K$^+$ current (I_K) could be easily detected in aortic VSM cells of the rabbit. Two of these currents are normally expressed in normal Tyrode solution during whole-cell voltage clamp recording, an early outward K$^+$ current and a delayed outward rectifier. Some single cells show one type of K$^+$ current, however, some other group of single cells showed overlapping of both current types. Figure 2A shows current recordings taken from a single aortic VSM cell that show a predominant early outward K$^+$ current at different voltage steps, as well as the relationship between current and voltage (I/V curve, solid circles). This early I_K is outwardly rectifying, increase upon depolarization steps, then decreases slowly with a maintained depolarization, and reverses at about -80 mV. Fig. 2B, illustrates the I/V curve and current recordings (at different voltage steps) of a single cell that shows a predominant delayed rectifier I_K that increases slowly upon depolarization until it reaches a steady- state level. The reversal potential of this current measured from the tail current (holding potential, HP of -70 mV) was found to be between -65 and -55 mV. Both types of K$^+$ currents are greatly decreased by the K$^+$ channel blocker tetraethylammonium (12, 13). The fact that the delayed rectifier time-dependent I_K of aortic cells has a relatively low reversal potential compared with the expected reversal potential of a pure K$^+$ current, suggests that this current is not necessarily a pure I_K. The early outward current in rabbit aortic cells showed some similarities with the current reported in other tissues, such as heart muscle (22), drosophila flight muscle (37), neurons (1) and smooth muscle (23, 47), this current was called I_A. The difference between the current in a rabbit aortic VSM single cells and the current reported in neurons (1) lies in the fact that, in aortic single cells, both I_A and I_K (and only I_A) are decreased by TEA. Also, the I_A-like current in aortic cells seems to be different from those reported in smooth muscle of the stomach (47) and the urinary bladder because the K$^+$ currents in aortic cells were insensitive to Ca^{2+} blockers in normal Tyrode's solution (34).

RABBIT SINGLE AORTIC VSM CELLS CULTURE

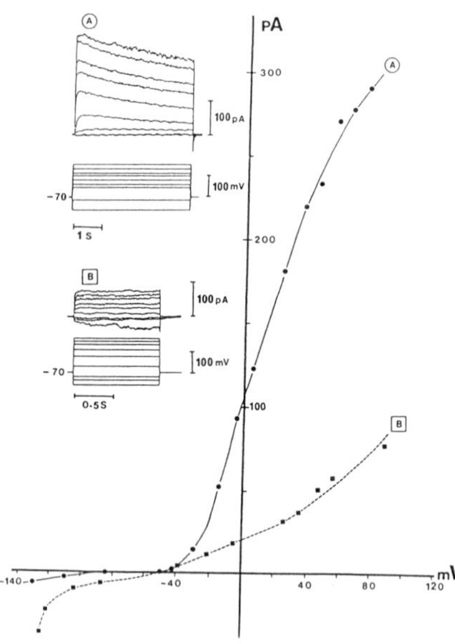

Figure 2. The relationship between current and voltage showed two different types of K$^+$ currents in rabbit single aortic voltage smooth muscle cell culture. Type one I_K (curve A, circles) showed outward rectification and decreased slowly with time (inset A). The second type of K$^+$ current (curve B, square) showed outward and inward rectifications and increased with time until it reached a steady-state level (inset B). The records illustrated in insets A and B were taken from two different cells. Holding potential was -70 mV. In inset A, voltage steps were to the following -130, -85, -30, -15, -3, +24, +37, +52, and +78 mV; and in inset B to the following -122, -105, -88, -38, -5, +27, +47, +56, and +87 mV. Currents were measured at peak amplitude (taken from ref. 12).

A delayed rectified type two K$^+$ current was recently reported in several types of smooth muscle cells including rat mesenteric arteries (14), rat vas deferens (30), rat tail venous and arterial cells (43), rabbit intestine smooth muscle cells (9), urinary bladder of guinea-pig (43), and stomach (47). These delayed rectifier K$^+$ currents seem to be different from the one found in the aorta of rabbit because (i) the outward-going rectification and K^+ current initiate differently; (ii) the delayed outward current in the aorta is not sensitive to calcium blockers in normal solution; and(iii) the type two I_K in rabbit aorta shows inward and outward rectification (34). The presence of two types of K^+ currents in the same single cell was reported by Salkoff and Wyman (37) and Bader et al. (1). Both I_K types were completely blocked when the pipette solution contained Cs^+, and when TEA was added to the extracellular medium.

Also a Ca^{2+}-activated K^+ current does exist in aortic VSM cells. However, this current does not seems to be active in relaxed aortic VSM superfused with normal Tyrode solution. This type of current, could be activated when the extracellular calcium concentration is increased from 2 to 10 mM or by addition of the L-type Ca^{2+} channel agonist, Bay-K-8644 (34). In single aortic cells that showed only a delayed outward I_K in normal Tyrode solution (containing TTX) addition of 10^{-8} M Bay-K-8644 increased the amplitude of this current, and addition of the Ca^{2+} blocker Mn^{2+} reversed the effect of Bay-K-8644 (34). The Bay-K-8644 activated I_K showed no rectification (34). The activation of $I_{K(Ca)}$ in aortic cells by Bay-K-8644 would be due to the increase of Ca^{2+} influx via the slow Ca^{2+} channels.

A fourth type of K^+ channel was also detected in aortic VSM cells (12). This type of channel had a single channel conductance between 80 to 100 pS (symetrical 150 mM K^+). The amplitude and the open channel probability are mostly seen at membrane potential (E_m) of $+75$ mV (KCl pipette solution = 150 mM). The channel behavior at E_m of $+75$ mV over a period of 3 min is illustrated in Figure 3Aa. This record shows three different types of fluctuation behavior exhibited by the K^+ channel. In type-I fluctuation pattern, the channel was mostly closed with brief openings; in type-II, the channel fluctuated rapidly in a burst-like pattern; in type-III, the channel remained opened with brief closures. The current traces Ab and Ac presented in Figure 3A, illustrate on faster time scales the fast transitions occurring while the channel was mostly closed (type-I). As can be observed in Figure 3Ab' and 3Ac', fast closures also took place during brief openings (type-II fluctuation). Figure 3B shows current amplitude histogram of single-channel recordings from Figure 3A. The mean single-channel current amplitude was 14 pA.

Figure 4A shows example of the effect of resting membrane potentials on single K^+ channel amplitude and open probability (P_o) using pipette solution containing 150 mM K^+. Figure 4A (a,b,c) shows single channel traces recording at three different E_M. As can be seen, the amplitude of the single channel was voltage dependent. The amplitude of the channel at E_M of $+75$ mV, was 14.9 pA (Fig. 4Aa) and decreased to 4.55 at E_m of $+19$ mV (Fig. 4Ab). At very low E_m of -160 mV, the channel amplitude was -6.93 pA (Fig. 4Ac).

Figure 4B shows the single-channel current/voltage relationship (I/V curve). The I/V curve was nearly linear over the entire voltage range studied. With patch pipettes containing 75 mM KCl (and 75 mM NaCl) instead of the standard 150 mM KCl, the I/V curve was shifted in the negative direction. This suggested that the observed channel was mainly permeable to cations, and the direction of the shift was not compatible with a Cl^--selective channel. Contribution of the Na^+ inside the cell is unlikely, since a Na^+ current down its electrochemical gradient, especially with patch electrodes containing solely KCl, would have resulted in current jumps with a polarity opposite to that observed experimentally. The I/V curves obtained with patch pipettes containing 75 mM KCl and 75 mM NaCl suggested that K^+ is the main charge carrier.

The effect of membrane potential on the channel open probability (P_o) is presented in Figure 4C. The P_o increases, as a function of E_m, especially within the depolarizing voltage range. This figure also shows that the probability that the channel is open is low at large hyperpolarizing potentials, and increasing the pipette potential to more depolarizing voltages increased the probability of the channel to open.

This channel seems not to be functional at the resting membrane potential of -70 to -80 mV. This channel may not contribute to generation of the normal resting potential, however it may play an important role, in depolarized cells and contribute to generation of the falling phase of the action potential. The open-channel probability (P_o) was greater at depolarizing potentials. At E_m level of $+75$ mV, which is not a physiological condition, fluctuation patterns of channel opening and closing were found, suggesting a complex kinetic behavior of the channel. This type of K^+ channel does not seem to be active in normal conditions. Activation of this particular K^+ channel would highly hyperpolarize the membrane and thus contribute to relaxation of vascular smooth muscle (9, 12). This possibility will be discussed later in this chapter.

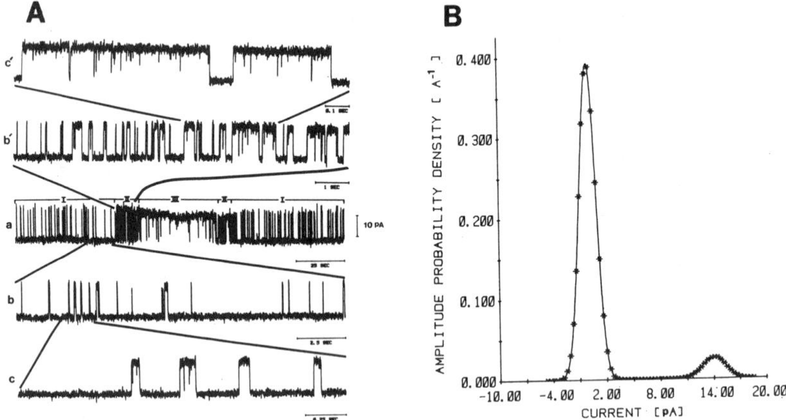

Figure 3. Single-channel records from cell-attached patch on an aortic vascular smooth muscle (VSM) cell in culture. Patch pipette contained 150 mM KCl and 5 mM EGTA. The resting membrane potential was held at +75 mV. **Aa:** the channel behavior over a period of 3 min is shown. Three types of fluctuation patterns, resulting from complex kinetic behavior can be seen: in type-I, the channel spends most of the time in the closed state, punctuated by brief openings; in type-II, the channel fluctuated rapidly in a bursting pattern; in type-III, the channel spends most of the time in the open state punctuated by brief closures. **Ab:** current traces on a much faster time scale to illustrate the type-I pattern. The channel spent most closed with brief openings. **Ac:** current traces on a much faster time scale of panel Ab, show that fast closures can also take place during the frief opening. **Ab':** current traces on a fast time scale to ullustrate the type-II pattern. **Ac':** current traces on a faster time scale than in panel Ab' to show that fast closures can also take place during the brief openings. **B:** current amplitude histogram of a very active single channel recording (from Fig. 3Aa). The amplitude of the current is 14 pA. Horizontal bar in Ac', b', a, b and c respectively: 0.1, 1.25, 2.5 and 0.25 sec. (Bkaily et al., unpublished results).

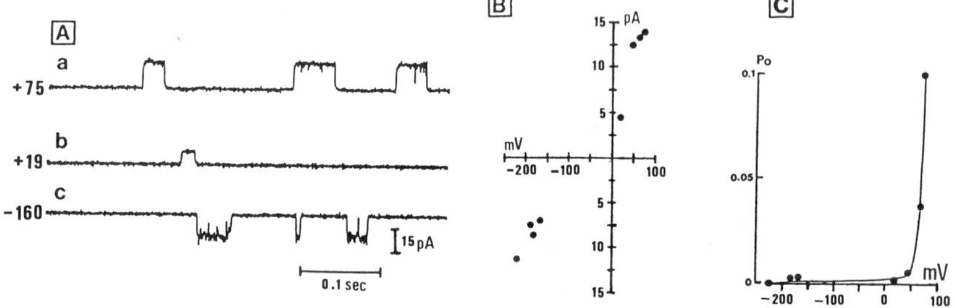

Figure 4. Effect of resting membrane potentials (Em) on single-channel current amplitude and opening probability (P_O) of K^+ channels in aortic vascular smooth muscle (VSM) cell in culture. Cell-attached patch configuration. The patch pipette contained 150 mM KCl. **A:** examples of the single-channel current observed at three different voltage levels. With depolarizing potentials, the current was outward; with hyperpolarizing potentials, the current became inward; the reversal potential was near E_K. **B:** single-channel current/voltage relationship (I/V). The I/V curve was essentially linear over the entire applied voltage range. The single-channel conductance was 100 pS. **C:** the channel opening probabilities were very low with hyperpolarization, and with depolarization, the probability of the channel being in the open state increased (taken from ref. 13).

PRESENCE OF THREE TYPES OF Ca^{2+} CHANNELS IN AORTIC VSM CELLS

Early in the 1980's, it became apparent that Ca^{2+} currents of different types coexist in several kinds of excitable cells (2-5, 9, 11-13, 16, 19, 21, 24, 26, 28, 31, 32, 35, 36, 40, 44). In vascular smooth muscle, at least two types of Ca^{2+} currents are present (5, 13, 17, 24, 40, 41, 48); L-type (for long lasting), and T-type (for transient) (3, 16, 44). Compared to L-type Ca^{2+} current, the T-type current activates and inactivates relatively rapidly at more negative test potential. The L-type Ca^{2+} channel inactivation is both voltage and Ca^{2+} dependent (i.e. biexponential), however, the T-type Ca^{2+} current inactivation is mono-exponential and slowly voltage dependent. The L-type Ca^{2+}-channels are about three times more permeable to Ba^{2+} than to Ca^{2+}, while the T-type Ca^{2+} channels are equally permeable to the two ions (2, 16, 26, 44). The L-type Ca^{2+} current is sensitive to classical Ca^{2+} channels blockers and the T-type I_{Ca} is sensitive to Ni^{2+}. The N-type Ca^{2+} current which has kinetic properties intermediated to the L-and T-type currents has been described only in nerve cells and this type of channel seems not to be functional in VSM cells in general (for review see 2, 3, 5, 11, 19, 28, 35).

As other types of vascular smooth muscle, aortic cells of the rabbit shows also L- and T-type Ca^{2+} currents (Figure 5). As can be seen in this figure, replacing Ca^{2+} by high concentration of Ba^{2+} (in order to better see and measure the slow inward current), two types of I_{Ba} are overlapping when the membrane potential of the single aortic cell is held at -70 and applying different depolarizing voltage steps induced an I_{Ba} that showed two peaks in the I/V curve. The first peak is near a potential of -40 mV and corresponds to the T-type I_{Ca}. The second peak of the I/V curve is at 0 mV and corresponds to the maximal inward L-type I_{Ca}.

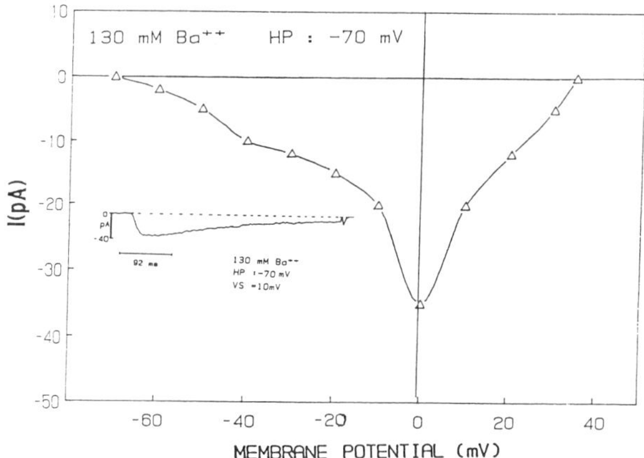

Figure 5. Current voltage relationship of the inward current recorded in extracellular solution containing 130 mM barium. The intracellular pipette solution contained (mM): NaCl 20, $MgCl_2$ 2, CsCl 120, glucose 5, EGTA 5 and HEPES 5. The holding potential is -70 mV. The inset current trace was induced from HP of -70 mV to VS to + 10 mV (Bkaily and Peyrow, unpublished results).

The T-type current is blocked by low concentration of Ni^{2+}, however, the L-type I_{Ca} is blocked by nifedipine as well as by the new Ca^{2+} channel blocker azelastine (29). Superfusion with low concentration of azelastine (10^{-10} M) decreased the peak amplitude of the L-type I_{Ba} by 20%. Increasing the concdentration of azelastine to 10^{-9} M further decreased this type of current to 60% of the control level. A concentration of 10^{-8} M of azelastine completely blocked the L-type I_{Ba}.

Very recently, a new type of Ca^{2+} channel called R-type (for resting, 10) was also found in aortic VSM of the rabbit as well as in human renal artery single cells (10, 11).

190

This type of channel is responsible for sustained increase of [Ca]$_i$ during tonic contraction of VSM (10).

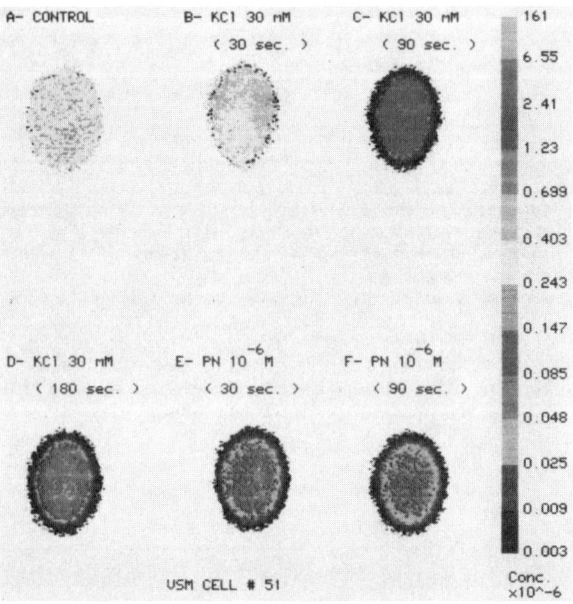

Figure 6. PN 200-110 blockade of the sustained component of [Ca^{2+}]$_i$ induced by 30 mM [K$^+$]$_o$ in single aortic vascular smooth muscle of rabbit. **A:** control (5 mM [K$^+$]$_o$) digital color image maps of free [Ca^{2+}]$_i$. **B and C:** record taken at 30 sec and 1.5 min after superfusion with 30 mM [K$^+$]$_o$ solution. **D:** sustained increase of [Ca^{2+}]$_i$ after 3 min in the presence of high [K$^+$]$_o$. **E and F:** superfusion with high K$^+$ solution containing 10^{-6} M PN 200-110 rapidly (within 30 sec) and progressively (up to 1.5 min) decreased the level of [Ca^{2+}]$_i$ distribution. The vertical colored bar shows calibration scales for [Ca] from 0.003 to 161 μM. (Taken from ref. 10).

This channel is activated during depolarization of the cell membrane of VSM cells with high extracellular concentration of potassium ([K]$_o$). Using microfluorometry and video imaging fura-2 Ca^{2+} measurement technique (18, 45), depolarization of the membrane of aortic VSM cell and raising the [K]$_o$ from 5 to 30 mM produce a rapid increase of [Ca]$_i$ from 109 $\pm$ 3.0 nM (n = 20) up to 346.3 $\pm$ 9.1 (n = 20) that decay slowly toward a sustained level of [Ca]$_i$ of 191.8 $\pm$ 5.0 nM (n = 20). The early phase of high [K]$_o$-induced increase of [Ca]$_i$ is blocked by L-type Ca^{2+} channel blockers nifedipine and PN 200-110 (10, 11). However, the sustained increase of [Ca]$_i$ is insensitive to nifedipine as well as to metabolic inhibitors (cyanide), to ouabain at increasing intracellular [cAMP]$_i$ (10, 11). Also the sustained increase of [Ca]$_i$ during sustained depolarization of the cell membrane is not sensitive to caffeine. Thus, the channel responsible of the tonic contraction of aortic VSM cells seems not to be due to sarcoplasmic reticulum -dependent (SR) Ca^{2+} release. However the R-type Ca^{2+} channel is blocked by PN 200-110 and with the Ca^{2+} chelator EGTA and Figure 6 shows an example using Ca^{2+} imaging technique. As can be seen in this figure, superfusion with 30 mM K$^+$ solution rapidly (at 30 sec, Fig. 6B) increased the level of [Ca]$_i$ near the cell membrane and reduced the size of aortic VSM cell. At 90 sec in the presence of high [K]$_o$, the [Ca]$_i$ near the cell membrane apparently become very low, however [Ca]$_i$ was greatly increased in the center of the single cell (Fig. 6C). As steady-state increase of [Ca]$_i$ took place at 3 min in the presence of high [K]$_o$ (Fig. 6D), superfusion with 10^{-6} M of PN 200-110 immediately decreased the level of [Ca]$_i$ and induced changes of [Ca]$_i$ distribution (Fig. 6E). No further decrease of [Ca]$_i$ and changes in distribution of [Ca]$_i$ were seen after 1.5 min in the presence of this drug (Fig. 6F).

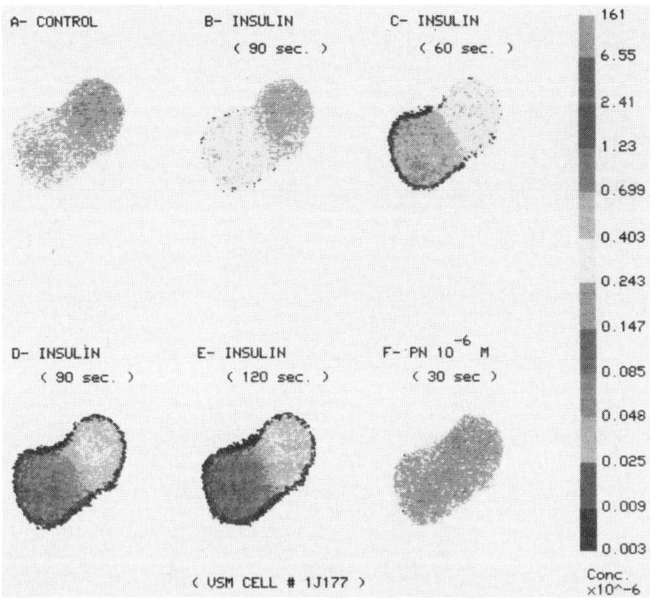

Figure 7. Time course effect of 80 μ U/ml on the level and distribution of [Ca]$_i$ in rabbit aortic single cells. **A:** steady- state basal resting intracellular free Ca^{2+} level and distribution in two attached single rabbit aortic VSM cells. **B-E:** changes of [Ca]$_i$ level and distribution at 0.5 and 1 min in presence of 80 μ U/ml of insulin. **D-E:** steady-state effect of 80 μ U/ml of insulin occurred at 1.5 min in the presence of the hormone. **F:** in the presence of insulin, PN 200- 110 (10^{-6} M) reversed the effect of insulin and [Ca]$_i$ distribution become homogenous in both single cells (at 30 sec). The vertical colored bar shows calibration scales for [Ca] from 0.003 to 161 μM. (Taken from ref. 10).

The R-type Ca^{2+} channel in aortic VSM was found to be activated by several hormones such as insulin. This hormone at a concentration of 20 to 40 μU/ml had no effect on the basal resting level of [Ca]$_i$ in aortic VSM cells. However, increasing the concentration of insulin to 60 and 80 μU/ml, respectively, caused an 18% (n = 7) and 73% (n = 11) increase over the basal resting [Ca]$_i$ level (10). No further increase could be seen at a concentration of insulin greater than 80 μU/ml. Figure 7 shows an example of insulin induced sustained increase of [Ca]$_i$ in VSM cell of rabbit aorta using the [Ca]$_i$ imaging technique. In this figure we could see the distribution of [Ca]$_i$ of two attached single VSM cells in control Tyrode solution (Fig. 7A). As can be seen in this figure, [Ca]$_i$ distribution and concentration are not the same in both single VSM cells. The lower single cell had a level of [Ca]$_i$ higher than the upper single cell. Superfusion with Tyrode solution containing 80 μU/ml of insulin progressively increased [Ca]$_i$ in both VSM cells (Fig. 7B-D) and no further increase of [Ca]$_i$ was observed after 90 sec in the presence of insulin (Fig. 7D and E). As in the control situation (Fig. 7A), the distribution and the level of the [Ca]$_i$ in the presence of insulin was not the same in both attached single cells (Fig. 7B-E). Addition of the Ca^{2+} blocker PN 200-110 (10^{-6} M) reversed the effect of insulin in both cells (Fig. 7F). The intracellular Ca^{2+} free distribution and concentration were similar in both VSM cells (Fig. 7F). The insulin-induced sustained increase of [Ca]$_i$ is not affected by the classical L-type Ca^{2+} channel blocker nifedipine (10).

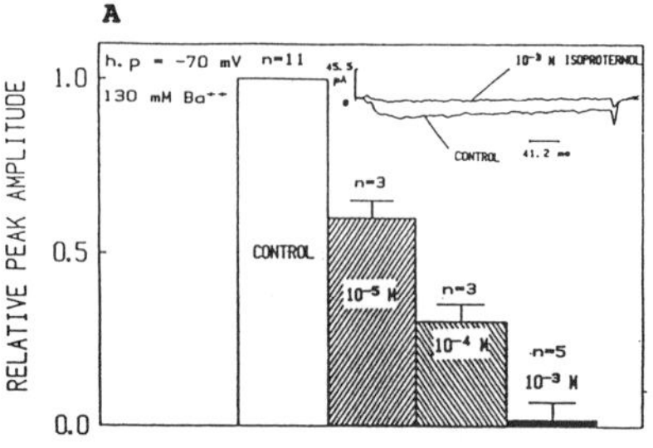
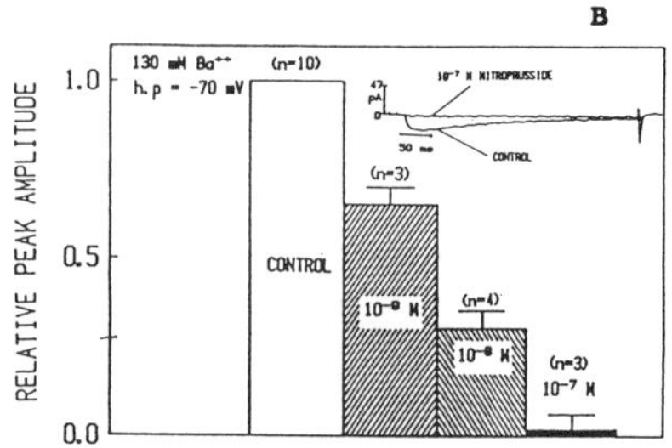

Figure 8. **A:** dose dependence of isoproterenol blockade of the slow inward current in VSM cells. Isoproterenol blocks the slow inward Ba^{++} current in a dose dependent manner the I_{si} was completely blocked by 10^{-3} M isoproterenol. **B:** dose dependence of nitroprusside blockade of the slow inward current in VSM cells. Nitroprusside blocks the slow inward Ba^{+} current in a dose dependent manner the I_{si} was completely blocked by 10^{-7} M nitroprusside. The holding potential was -70 mV and the current was induced by voltage step to 0 mV. (Bkaily and Peyrow, unpublished results).

In presence of insulin, depolarizing the cell membrane of aortic cells induced a transient increase of [Ca]i that decay back to the level of the insulin sustained component (10). This observation suggests that the R-type Ca^{2+} channels are fully activated by insulin and depolarization of the cell membrane would in this case induce activation of the L-type Ca^{2+} channel (10).

Using the whole-cell voltage clamp technique, insulin was found to be inactive on the L-type I_{Ca} and the delayed outward K^{+} current, thus supporting the hypothesis that the sustained increase of [Ca]i induced by insulin is not due to depolarization of the membrane and activation of the L-type Ca^{2+} current (10).

REGULATION OF THE L-TYPE Ca^{2+} CURRENT IN AORTIC CELLS BY CYCLIC NUCLEOTIDES

Calcium channels in heart and vascular smooth muscle were reported to be modulated by increasing [cAMP]i and [cGMP]i (for review see refs 7, 13, 38).

In aortic vascular smooth muscle cells, stimulation of β-receptors by isoproterenol induce activation of adenylate cyclase and increase [cAMP]i (for review see ref. 6). In heart cells, increasing [cAMP]i was reported to stimulate the L-type Ca^{2+} current (2, 7, 36, 38). However, in aortic VSM cells, superfusion with 10^{-5} M and 10^{-4} M of isoproterenol (ISO) decreased the peak of the L-type current respectively by 46% and 71%. A complete blockade of I_{si} takes place at 10^{-3} M of ISO (Fig. 8). This may suggests that the L-type Ca^{2+} current in aortic VSM cells is regulated by protein Kinase A and that phosphorylation of this channel by the Kinase may blocks it. Also, the vasorelaxation of aortic muscle by ISO could be due in part to blockade of the L-type Ca^{2+} channel.

As for ISO, nitroprusside (NP) is also known to relax VSM contracted by a number of agents (for review, see ref. 6). The relaxation of VSM by nitrogen oxide-containing compounds, such as NP, are thought to cause vasodilatation by formation of a reactive nitric-oxide free radical that activates guanylate cyclase and elevates [cGMP]i in VSM cells (6).

The vasorelaxation of VSM by NP seems to be due in part to the blockade of the L-type I_{Ca} via increasing [cGMP]i and stimulation of protein Kinase G' which seems as in heart muscle (7), to block the L-type Ca^{2+} current in aortic VSM cells (Fig. 8). As can be seen in this figure, superfusion with different concentrations of NP produced a dose-dependent blockade of the peak L-type Ca^{2+} current in aortic cells which an ID_{50} near 5 x 10^{-9} M (Fig. 8).

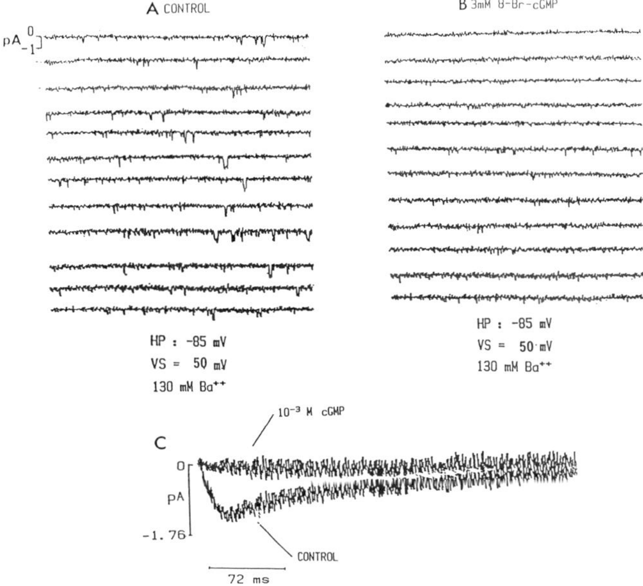

Figure 9. Regulation of single Ca^{2+} channel by $[cGMP]_i$. **A:** in cell attached configuration, depolarization of the cell membrane from -85 to -50 mV opens a 24 pS single channel. **B:** addition of 3 mM of 8-Br-cGMP decreased the channel activity. **C:** mean Ba^{2+} single channel current in the absence and presence of 8-Br-cGMP. All records are taken from the same patch. (Bkaily and Peyrow unpublished results).

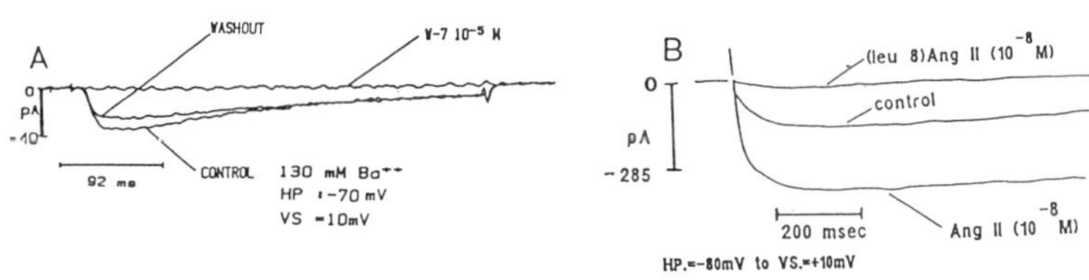

Figure 10. Blockade of the slow inward current by calmodulin inhibitor, W-7 and activation of I_{si} by Ang II in aortic single cells. **A:** superfusion with 10^{-5} M W-7 completely blocked the inward current. Washout with W-7 free solution restored the inward Ba^{2+} current. **B:** activation of slow inward current by Ang II. Superfusion with 10^{-8} Ang II rapidly increased the amplitude of I_{si}. Addition of Ang II antagonist completely blocked I_{si}. A and B are two different single cells. (Bkaily et al., unpublished results and ref. 13).

This effect of NP is mediated via increasing intracellular cGMP (6) and single Ca^{2+} channel studies using the cell attached configuration (Fig. 9) strongly support this idea. As can be seen in Figure 9, in patch pipette containing 130 mM Ba^{2+} and in extracellular solution containing 140 mM KCl, depolarization of the patch membrane from -85 mV to -30 mV induced opening of single Ba^{2+} channel (single channel conductance of 22 pS). Addition of 3×10^{-3} M 8-Br-cGMP largely decreased the channel openings (Fig. 9A and B) and the average of unitary L-type current (Fig. 9C).

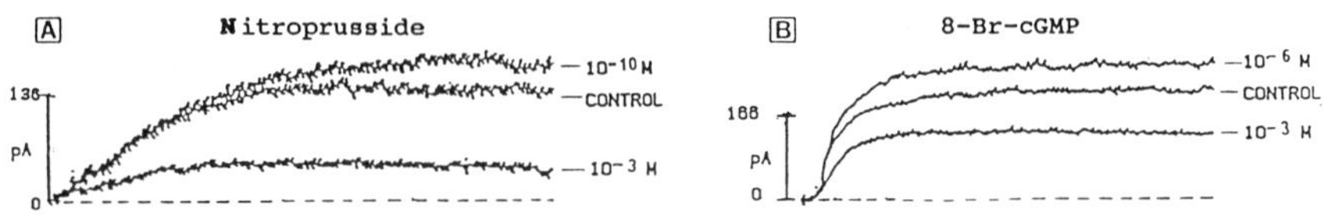

Figure 11. Effect of different concentrations of nitroprusside (A) and 8-Br-cGMP (B) on the delayed outward K^+ current in single aortic single cells. The K^+ currents were induced from holding potential of -70 to voltage step to +40 mV. **A:** addition of 10^{-10} M nitroprusside increased the peak amplitude of I_K. High concentration of nitroprusside largely decreased I_K. **B:** superfusion with 10^{-6} M 8-Br-cGMP increased I_K peak amplitude and high concentration (10^{-3} M) of 8-Br-cGMP largely decreased I_K. (Bkaily and Peyrow, unpublished results).

REGULATION OF THE L-TYPE I_{Ca} IN AORTIC CELLS BY CALMODULIN AND ANG II

Calmodulin was reported to potentiate the slow action potential in heart cells (8), however little is know about calmodulin -induced regulation of Ca^{2+} channel in aortic VSM cells. Using a well know calmodulin inhibitor, W-7 (20), this compound was found to block in a reversible manner the T- and L-types Ba^{2+} currents in aortic VSM cells and Figure 10A shows an example. As can be seen in this figure, superfusion with 10^{-5} M W-7 rapidly and completely blocked the peak and the late inward Ba^{2+} current. Washout of W-7 completely recovered the late inward current and partially recovered the peak inward current. These results suggests that as in heart cells (8), calmodulin may activates the L-type Ca^{2+} current in aortic VSM. It is possible that activation of the L-type Ca^{2+} by depolarization of the membrane by an hormone, would activate the L-type Ca^{2+} channel and increase $[Ca]_i$ which in turn activate calmodulin.

The Ca^{2+} -calmodulin complex than activates myosin light chain Kinase (MLCK), which phosphorilates the light chain of myosin to form myosin light- chain phosphate and smooth-muscle contraction (6). At the same time Ca^{2+} - calmodulin may further activates Ca^{2+} channels and allow further increase of $[Ca]_i$ which will contribute to further act in myosin interaction involving the thin filaments (27, 42) and tonic contraction of VSM.

Also Angiotensin II (Ang II) was found to increase the L-type Ca^{2+} current in aortic vascular smooth muscle (13). As can be seen in Figure 10B, the L-type inward current increased substantially within 3 min after the addition of 10^{-8} M Ang II. The addition of antagonist of Ang II, ([Leu[8]] Ang II, 10^{-8} M) rapidly reversed the Ang II effect on I_{Ba}, and also completely blocked the inward Ba^{2+} current (Fig. 10B). The addition of the [Leu[8]] Ang II alone (without Ang II) slightly decreased I_{si} amplitude but never blocked it. The increase of the L- type inward current by Ang II could be due: 1) to IP_3 formation and/or C-Kinase activation, or 2) to direct effect of Ang II on the slow channel, or 3) to activation of a G-protein that may stimulates directly the slow channel (15, 32) . However, the situation may be more complex, and more work must be done in order to determine the complex pathway by which Ang II affects I_{si} in aortic cells (13).

REGULATION OF K^+ CHANNELS IN AORTIC CELLS BY CYCLIC NUCLEOTIDES

As discussed in the previous section, Ca^{2+} channels are partly regulated by cyclic nucleotides. Phosphorylation of the aortic VSM cells Ca^{2+} channel by cyclic nucleotides seems to have an inhibitory action on this channel which may explain in part the vasorelaxation of VSM by compounds that stimulate these nucleotides. It is also possible that cyclic nucleotides may modulate the functioning of K^+ channels, that in turn modulate the resting membrane potential in VSM. Activation of some types of K^+ may hyperpolarize the cell membrane below the threshold of Ca^{2+} channel and thus relax VSM.

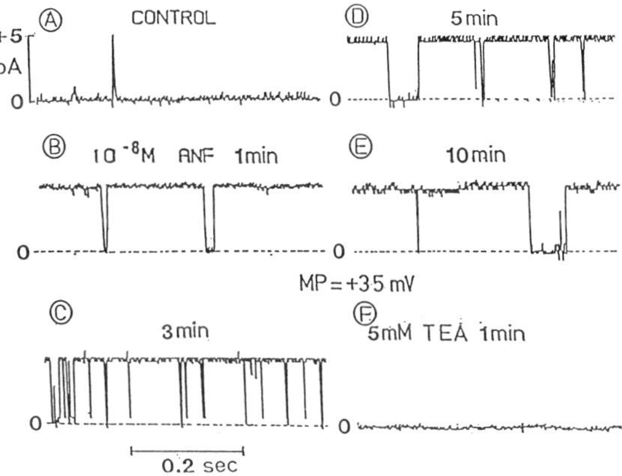

Figure 12. ANF increased the opening time of a potassium channel in aortic single cells **A:** cell-attached patch-clamp recording at a membrane potential of +35 mV. **B:** at 1 min after starting superfusion with 10^{-8} M ANF, and C, D, and E are the 3, 5 and 10 min respectively. **F:** superfusion with 5 mM TEA blocked the ANF effect. (Bkaily et al., unpublished results).

As Figure 11 shows, superfusion with a concentration of NP that does little affect on the L-type Ca^{2+} current in aortic cell (10^{-10} M) increased the macroscopic delayed outward K^+ current (Fig. 11A). A higher concentration than 10^{-9} M (which blocks I_{si}) largely decreased this K^+ current (Fig. 11A). As for nitroprusside, superfusion with 10^{-8} M to 10^{-6} M of 8-Br-cGMP largely increased I_K amplitude and higher concentration of cyclic-GMP ($> 10^{-6}$ M) blocked this type current (Fig. 11B).

Using single K^+ channel recording technique demonstrates that compound such atrial nutritic factor (ANF), which is known to increase intracellular cGMP, was found to increase the probability of opening of the 100 pS single K^+ channel conductance without affecting the I/V curve of this channel. Figure 12 illustrates an example of the effect of 10^{-8} M of ANF on the 100 pS single K^+ channel of aortic single cells. As can be seen in this figure, channel opening occurs at a very low holding potential of +30 mV (Fig. 12A). One min after the addition of ANF, (Fig. 12B) and throughout the incubation with ANF (Fig. 14C- E), the channel stays most of the time at the open state with few closing periods. The effect of ANF is completely blocked by the K^+ blocker TEA (Fig. 12F).

Similarly to ANF, 8-Br-cGMP also increased the probability of the 100 pS single K^+ channel to be in the open state. The cGMP sensitive K^+ channel was insensitive to the Ca^{2+} chelator EGTA, to the Ca^{2+}- activated K^+ channel blocker CTX (10^{-7} M) and to the ATP-sensitive K^+ channel blocker glybenclamide (10^{-6} M). These results suggest that the cGMP-sensitive K^+ channel in aortic vascular smooth muscle of the rabbit is not modulated by Ca^{2+} or ATP. Direct application of cGMP on the inner side of the membrane (using inside-out patch configuration) increased the probability of opening of the 100 pS single K^+ channel. This suggests again that ANF effect on this type of K^+ channel is mediated via increasing $[cGMP]_i$. The effect of cGMP on these channels were antagonised by increasing intracellular cAMP.

Thus, cyclic nucleotides seem to regulate the functioning of one type of K^+ channels in aortic VSM cells of the rabbit. It is postulated that phosphorylation of both L-type Ca^{2+} channel and the 100 pS K^+ channel by protein Kinase G would increase the probability of opening of a K^+ channel that hyperpolarize the membrane (and inactivate Ca^{2+} channels) and also at the same time would phosphorylate the L-type Ca^{2+} channels and thus decrease their activities. These dual effects of increasing cGMP may explain the long lasting vasorelaxation induced by the nitro compound. It is worth mentioning that increasing $[cGMP]_i$ would also affect Ca^{2+} release IP_3 production and contractile proteins (for review, see ref. 6).

Importantly, the vasorelaxation induced by b-receptor stimulation and increasing [cAMP]$_i$ could be mainly due to blockade of the L-type Ca^{2+} channel. Even if the K$^+$ channel is blocked by ISO, depolarization of the membrane cannot activate the L-type Ca^{2+} current which has already been blocked by the same second messenger.

The cGMP-sensitive K$^+$ channel seems to be also activated by bethanidine sulfate. This drug was reported to block some types of K$^+$ channels in heart and vascular smooth muscle (42) and to activate a K$^+$ channel that showed no rectification in aortic VSM of the rabbit (12). The addition of bethanidine (10^{-4} M) increased the probability of opening of the 80 to 100 pS single K$^+$ channel (9). In presence of bethanidine, ANF did not further increased the K$^+$ current in aortic single cell (9). This may suggests that bethanidine, hyperpolarize the membrane by activating the cGMP-sensitive K$^+$ channel in aortic cells.

CONCLUSION

Aortic vascular smooth cells of the rabbit as other vascular smooth muscle cells has at least three types of Ca^{2+} currents, T-, L- and R-types. The density of T-type Ca^{2+} current is lower than that at the L-type. The T-type Ca^{2+} current was reported to be responsible for pacemaker activity in heart muscle (2, 3). This current seems not to play a major role in normally quiescent VSM cells. However, it is possible that some hormones that may activates such a current are expected to induce spontaneous repetitive contraction. The pharmacology of this current is not yet clear and only Ni^{2+} at low concentration blocks specifically this type of current. Several organic Ca^{2+} blockers such as nifedipine, PN 200-110 and azelastine at a concentration that block completely the L-type I$_{Ca}$ decrease also the T-type I$_{Ca}$.

The T-type Ca^{2+} channel in heart muscle is not regulated by cyclic nucleotides (2), however, in VSM this channel, similarly to the L-type I$_{Ca}$, seems to be blocked by increasing [cAMP]$_i$ and [cGMP]$_i$. Also as the L-type I$_{Ca}$, the T-type current seems to be activated by calmodulin and Angiotensin II. Thus, both T- and L-type I$_{Ca}$ in aortic VSM cells of the rabbit appear to be regulated by the same second messengers.

Regulation of the L-type current in aortic cells by isoproterenol and [cAMP]$_i$ suggests that this channels once phosphorylated probably by cAMP-protein kinase' would decrease the mean open time and/or decrease the availability of the slow channels. As for cAMP, increasing [cGMP]$_i$ largely decreases the probability of opening of the L-type I$_{Ca}$. The regulation of the L-type by a cAMP in aortic VSM would be different from that of heart muscle. However, regulation of this type of channel by cGMP seems to be the same in both heart and aortic VSM cells.

The new R-type Ca^{2+} channel as been found to be highly active at normal resting potential of aortic cells (- 65 mV). However, this channel become highly active upon sustained depolarization of the cell membrane. This channel seems to not be modulated by second messengers and could be activated directly (or indirectly via a G-protein) at normal resting membrane by activation of a specific types of receptors such as insulin and endothelin receptors. Activation of this type of channel would induce raises of a sustained increase of [Ca]$_i$ which could be greatly responsible of the tonic contraction of VSM induced by some hormones. Coupling of a specific receptor to the R-type channel protein may fully open the channel. This, may explain why depolarization of the cell membrane in presence of an agonist did not further increased Ca^{2+} influx via the R-type channel. It is also possible that activation of a receptor by an hormone may directly (or indirectly via a G-protein) open the R-type channel by decreasing the sensitivity of the voltage sensor of the channel protein and make this channel to look like a receptor operated Ca^{2+} channel.

Whatever, the R-type Ca^{2+} channel may play a role on the tonic resting tension and could be implicated in the sustained increase of basal [Ca]$_i$ in VSM cells of hypertensive humans and animals, thus creating an opportunity for enhanced response to vasoactive compounds. Calcium accumulation due to R-type Ca^{2+} channels activation may also underlie the long-term ethiology of vascular deseases such as hypertrophy and proliferation of VSM cells seen in artheriosclerosis.

As for Ca^{2+} channels, at least three types of K$^+$ channels were detected in aortic VSM cells of the rabbit: 1) early outward K$^+$ current, 2) delayed outward K$^+$ current 3) Ca^{2+}-activated K$^+$ current and 4) cGMP-sensitive K$^+$ channel. The delayed outward K$^+$ current was found to largely contribute in maintening the resting potential of aortic cells. This current seems to be modulated by second messengers and Ang. II. The opening probability of cGMP-sensitive 100 pS K^+ channel is very low in normal physiological conditions. Nitro compound seems to act primarily via increasing cGMP which increase the probability of opening of the channel at the physiological range of membrane potential and hyperpolarize the membrane. This hyperpolarization of the membrane will not allow the L-type Ca^{2+} channel to be activated. This type of channel was not found in heart cells and seems to be unique for aortic VSM cells or maybe other types of VSM cells. The vasorelaxation of VSM by low concentrations of nitroprusside or ANF could be mainly due to the activation of the cGMP-sensitive K^+ channel. However, at high concentrations of these two compounds, the main affect could be mediated via cGMP induced blockade of L-type Ca^{2+} channels. The vasorelaxation of VSM by isoproterenol is mainly due to blockade of the L-type Ca^{2+} channel by cAMP.

The Ang II activated L-type Ca^{2+} channel could be mediated via protein Kinase C (46), however a possible stimulation of the L-type I_{Ca} by IP3 seems to be unlikely (33).

Finally, the controversy concerning drugs action and implication of receptors and second messengers on the regulation of different types of VSM is clearer than before. However, thin new knowledge still does not contribute to make a generalisation concerning VSM in general or classification of VSM types. Several questions must be answered in the future such as, how inositol phosphate, different G-proteins, different protein Kinase C, calmodulin, protein Kinase-A and G' regulate Ca^{2+} signaling and VSM tone.

Also more work must be done in order to understand the pathophysiology of different types of vascular smooth muscle and the role of endothelin-VSM cells coupling.

ACKNOWLEDGEMENTS

Supported by MRC MA 8920 to Dr Bkaily. Dr Bkaily is a Merck-Frosst-FRSQ Professor. The authors thank Ms Christiane Ducharme for her secretarial assistance.

REFERENCES

1. Bader, C.R., D. Bertrand, and E. Dupin, J. Physiol. (London). 366, 129-151 (1985).
2. Bean, B.P., Annu. Rev. Physiol. 51, 367-387 (1989).
3. Bean, B.P., J. Physiol. (London), 86, 1-30 (1985).
4. Bean, B.P., M. Sturek, A. Duga, and K. Hermsmeyer, Cir. Res. 59, 229-235.
5. Benham, C.D., P. Hess, and R.W. Tsien, Cir. Res. 61, I11-I16 (1987).
6. Bennett, B.M., C.R. Molina, S.A. Waldman, and F. Murad in: Physiology and Pathophysiology of the Heart, N. Sperelakis, ed. (Kluwer Acad, Press, Boston 1988) pp. 824-825.
7. Bkaily, G. and N. Sperelakis, Am. J. Physiol. 248, H745-H749 (1985).
8. Bkaily, G. and N. Sperelakis. J. Cyclic Nucleotide Protein Phosphorylation Res. 11, 25-34 (1986).
9. Bkaily, G. in: Frontier in Smooth Muscle Research, N. Sperelakis and J.D. Woods, eds (Allan R. Liss Inc. New York 1990) pp. 507-515.
10. Bkaily, G., D. Economos, L. Potvin, J.-L. Ardilouze, C. Marriott, J. Corcos, D. Bonneau, and C.N. Fong ,Mol. Cell. Biochem. In press.
11. Bkaily, G., in: In Vitro Methods in Toxicology, G. Jolles and A. Cordier eds. (Academic Press, London 1991). In press.
12. Bkaily, G., J.P. Caille, M.D. Payet, M. Peyrow, R. Sauve, J.-F. Renaud, and N. Sperelakis, Can. J. Physiol. Pharmacol. 66, 731-736 (1988).
13. Bkaily, G., M. Peyrow, A. Sculptoreanu, D. Jacques, D. Regoli, and N. Sperelakis, Pflugers Arch. 412, 448-450 (1988).
14. Bkaily, G., M. Peyrow, T. Yamamoto, A. Sculptoreanu, D. Jacques, and N. Sperelakis, Mol. Cell. Biochem. 80, 59-72 (1988).
15. Brown, A.M. and L. Birnbaumer, Am. J. Physiol. 254, H401-H410 (1988).
16. Fox, A.P., M.C. Nowycky, and R.W. Tsien, J. Physiol. (London), 394, 149- 172 (1987).
17. Friedman, M.E., G. Suarez-Kuratz, G.J. Kackorowski, G.M. Katz, and J.P. Reuben, Am. J. Physiol. 205, H699-H703 (1986).
18. Grynkiewzc, G., M. Poenie, and T.Y. Tsien, J. Biol Chem. 260, 3440-3450 (1985).
19. Hess, P., Can. J. Physiol. Pharmacol. 66, 1218-1223 (1988).
20. Hidaka, H., M. Asano, S. Iwadare, I. Matsumoto, T. Totsuka, and N. Aoki, J. Pharmacol. Ex. Ther. 207, 8-15 (1978).
21. Ice, K.S., M. Takahashi, B.M. Curtis, and W.A. Catterall, Cir. Res. 61, I24- I29.
22. Josephson, I., J. Sanchez-Chapula, and A.M. Brown, Cir. Res. 54, 157-162 (1985).
23. Klockner, U. and G. Isenberg, Pflugers Arch. 405, 340-348 (1985).
24. Loirand, G., P. Pacoud, C. Mironneau, and J. Moronneau, Pflugers Arch. 407, 566-568 (1986).

25. Marty, A. and E. Neher in: Single Channel Recording, B. Sackmann and E. Neher, eds (Plenum Press, New York 1983) pp. 107-172.
26. McCleskey, E.W., A.P. Fox, D. Feldman, and R.W. Tsien. J. Exp. Biol. 124, 177-190 (1986).
27. McGuffe, L.J., L. Hurwitz, S.A. Little, and B.E. Skipper, J. Cell Biol. 90, 201-210 (1981).
28. Miller, R.J., Science, 235, 46-52 (1987).
29. Molyvdas, P.A., F.W. James, and N. Sperelakis, Eur. J. Pharmacol. 164, 547- 553 (1989).
30. Nakazawa, K., N. Matsuka, K. Shigenobu, and Y. Kasuya, Pflugers Arch. 408, 112-119 (1987).
31. Nelson, M.T., J.B. Patlak, J.F. Worley, and N.B. Standen. Am. J. Physiol. 259, C3-C18 (1990).
32. Ohya, Y. and N. Sperelakis, Cir. Res. 68, 763-771 (1991).
33. Putney, J.W., H. Takemura, A.R. Hughes, D.A. Horstman, and O. Thastrup,. FASEB J. 3, 1899-1905 (1989).
34. Renaud, J.-F., G. Bkaily, M. Benabderrazik, D. Jacques, and N. Sperelakis, Mol. Cell. Biochem. 80, 73-78 (1988).
35. Reuter, H. and H. Porzig, Nature, 336, 113-114 (1988).
36. Reuter, H., Experientia, 43, 1173-1175 (1987).
37. Salkoff, L. and R. Wyman, Science, 213, 461-463 (1981).
38. Sperelakis, N. and G. Bkaily in Pathophysiology of Cardiovascular Injury, H.L. Stone and W.B. Weglicki, eds. (Martinus Nijhoff Publishing, Boston 1985) pp 109-144.
39. Sperelakis, N. in The Coronary Artery, S. Kalsner ed. (Croom Helm Ltd., London 1982) pp. 118-167.
40. Sperelakis, N.S. and Y. Ohya in: Physiology and Pathophysiology of the Heart, 2nd edition, N. Sperelakis, ed. (Kluwer Academic Press, Boston 1988) pp. 773-811.
41. Sturek, M. and K. Hermsmeyer, Science, 233, 475-478 (1986).
42. Sugi, H. and T. Daimon, Nature, 269, 436-438 (1977).
43. Toro, L., A. Gonzales-Kobles, and E. Stefani, Am. J. Physiol. 251, C763- C773 (1986).
44. Tsien, R.W., P. Hess, E.W. McCleskey, and R.L. Rosenberg, Annu. Rev. Biophys. Chem. 16, 265-290 (1987).
45. Tsien, R.Y. and M. Poenie, Trends Biochem. Sci. 11, 450-455 (1986).
46. Vivaudou, M.B., L.H. Clapp, J.V. Walshaud, J.J. Singer, Faseb, J. 2, 2497- 2504 (1988).
47. Walsh, J.V. and J.J. Singer, Pflugers Arch. 408, 83-97 (1987).
48. Wang, R., E. Karpinski, and P.K.T. Pang, Am. J. Physiol. 256: H361-H368 (1989).

ROLE OF Ca^{2+} CHANNELS ON THE RESPONSE OF VASCULAR SMOOTH MUSCLE TO ENDOTHELIN-1

GHASSAN BKAILY[1], SHIMIN WANG[1], DEMETRI ECONOMOS[1] and PEDRO D'ORLEANS-JUSTE[2]

Department of Physiology and Biophysics[1], Department of Pharmacology[2], Faculty of Medicine, University of Sherbrooke, Sherbrooke, Quebec, Canada, J1H 5N4.

INTRODUCTION

Endothelial cells in general synthetize and release substances that regulate contraction of vascular and non vascular muscles types. Among these substances, are the vasorelaxant agents such as endothelium-derived relaxing factor (EDRF) and prostacyclin as well as a vasoconstrictor peptide such as endothelin-1 (ET- 1) which has been recently been reported by Yanagisawa et al., (44-46).

Endothelin-1 (ET-1) was reported to act on both vascular smooth muscle (VSM) cells and endothelial cells. Interaction of ET-1 with its receptors in VSM cells produced increase of the muscle tone, however ET interaction with endothelial cells would release EDRF which attenuates the contractile response to ET itself (9, 23, 32, 39).

The responses of several types of muscles to ET is not homogenous and could be classified into three types (13): 1) group showing only monophasic sustained contractions and increases of intracellular free Ca^{2+} ([Ca]$_i$); 2) a group in which both rhythmic and slowly developing contractions occur and 3) a group that exhibit rapid rhythmic contractions alone devoid of monophasic developing contraction. The heterogenous responsiveness of vascular and non vascular cells to ET make it difficult to make a generalisation concerning the mode of action of this peptide in VSM cells.

The purpose of this paper is to study the effect of ET-1 on the intracellular level and distribution of [Ca]$_i$ using the Ca^{2+}- sensitive fluorescent dye, Fura-2 microfluorometry and digital imaging technique. Using the whole-cell voltage clamp technique, the effect of ET on K^+ current and the L and type calcium currents (I_{Ca}) was analysed in order to understand the effect of this peptide on these particular ionic channels.

DIFFERENCES IN SENSITIVITY OF VSM TO ET-1

Endothelin (ET) has been shown to produce prolonged contraction of smooth muscle in general from many mammalian species (6-8, 10, 12, 32, 33, 44, 46) including human (23). Also ET-1 was reported to induce a positive inotropic and chronotropic activity in isolated animal heart, atrial preparations (45, 46) as well as in single heart cells of human (26).

However, contrasting results of ET-1 effects on different types of vascular smooth muscles of different animal species was reported. Among the various arterial and venous preparations tested, rabbit jugular (RbJV) and mesenteric (RbMeV) veins displayed the greatest sensitivity to ET-1 (pD$_2$ of ET-1 on RbJV: 9.31; Fig. 1). This differential sensitivity was also observed using rat vena cava and aorta (Fig. 1). ET-1 was found to be much more potent that ET-3 as a vasoconstrictor in rat aorta. ET-3 was also much less potent in contracting the various venous vessels.

The contrasting results of the sensitivity of veins and arteries to ET cannot be entirely related to species differences but could be mainly due to differences with respect to the type of VSM used (23).

DIFFERENT PHARMACOLOGICAL PROFILES OF BAY-K-8644 AND ET-1

Early study of ET-induced contraction of VSM was due to opening of a Ca^{2+} channels that blocked by the L-type Ca^{2+} channel blockers nifedipine and nicardipine (14, 23, 25, 32, 36, 38, 44-46). In contrast, in many other reports, the ET-1 induced contraction was reported to be insensitive to nifedipine and nicardipine (5, 6, 11, 12, 29, 40, 41).

Published 1991 by Elsevier Science Publishing Company, Inc.
Ion Channels of Vascular Smooth Muscle Cells and Endothelial Cells
Sperelakis and Kuriyama, Editors

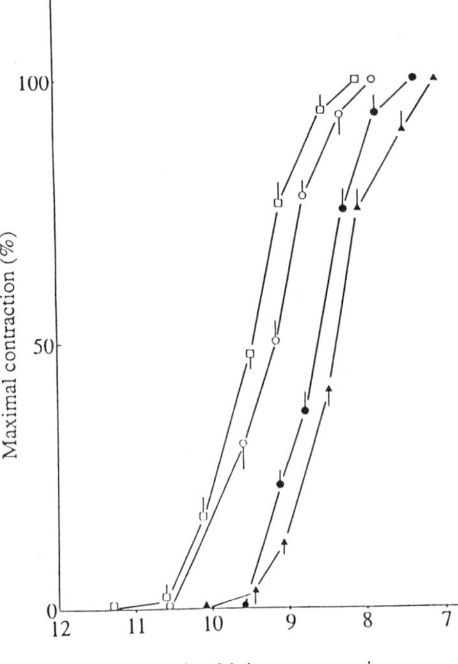

Figure 1. Concentration-response curves of porcine endothelin-1 (ET-1) on the rabbit jugular vein (RbJ) (□ - □) rabbit aorta (RbA) (▲ - ▲), rat vena cava (RVC) (O - O) and rat aorta (RA) (● - ●). Ordinate: contraction expressed as percentage of the maximal response of the agonist. Abcissa: -log of the molar concentration. Each point represents the mean ± SEM of at least four experiments. (Modified from D'Orleans-Juste et al. [10]).

In absence of endothelin, ET-1 (0.1-10 nM) and the L-type Ca^{2+} channel activator, Bay-K-8644 (0.025 - 1 μM) enhanced the tone of rat aorta. However, the contraction induced by Bay-K-8644 was different from that of ET-induced contraction (Fig. 2). ET-1 (1 nM) promoted a sustained contraction (Fig. 2) whereas Bay-K-8644 (100 nM) produced a tonic contraction with superimposed bursts of rhythmic contractions. Higher concentrations of Bay-K-8644 (1 μ M) increased the frequency of the phasic contractions (Fig. 2); this phenomena was not observed with ET-1 in rat aorta (10, 12) and various types of smooth muscle. However, ET-1 induced sustained contraction in VSM are not consistant, and there are reports in rat aortic cell line A7r5 (36), and porcine isolated aorta and ureter (11) where ET-1 as found to induce repetitive spiking in normally quiescent preparations which similarly to as Bay-K-8644-induced rhythmic contraction were nicardipine sensitive (11). However, the sustained increase of contraction induced by ET-1 in VSM seem mostly insensitive to the L-type Ca^{2+} channel blockers, nifedipine (6) and nicardipine (Fig. 2).

The inability of nifedipine to completely inhibit the ET-1 induced sustained contraction in VSM may suggests that the peptide may activate a non- inactivating Ca^{2+} channels (4, 17, 20, 24, 27, 36, 40) most probably the newly insulin-sensitive R-type Ca^{2+} channel (3).

ET-1 INDUCED INCREASES OF INTRACELLULAR DISTRIBUTION AND LEVEL IN VASCULAR SMOOTH MUSCLE CELLS

Recently, the development of Ca^{2+} sensitive dyes mainly fura-2 (15) made feasible to study the effects of substances on intracellular free $[Ca]_i$ as well as the its distribution in all types of single cells.

Several reports using Ca^{2+}-sensitive dyes demonstrate in various muscles and non muscles preparations that ET-1 induces different patterns of increase of $[Ca]_i$ such as: 1) transient increases of [Ca] ; 2) transient increase of [Ca] followed by a sustained component; 3) sustained increase of [Ca]$_i$ and 4) spike like increase of [Ca]$_i$ and wave like [Ca]$_i$ movement (32, 38, 39, 42).

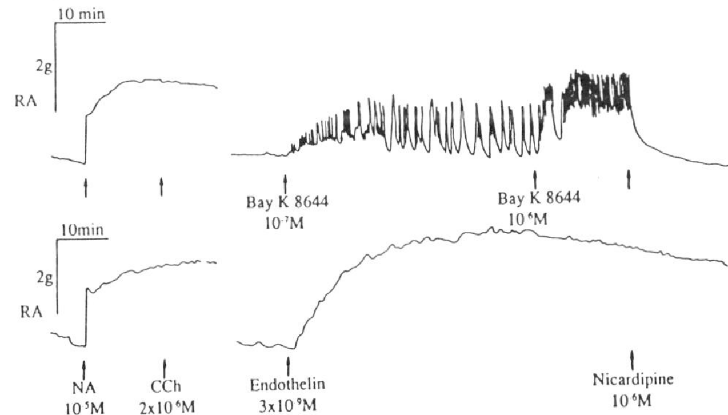

Figure 2. Porcine endothelin-1 (ET-1) and Bay-K-8644 contract the rat aorta via different channels or receptors; effect of nicardipine. ET-1 (3 nM) promotes a sustained contraction of the rat aorta whereas Bay-K-8644 (0.1 μM) induces a tonic and a phasic response on the same tissue. The frequency of the phasic response increases as higher concentration of Bay-K-8644 (1 μ M) is added. Nicardipine (1 μ M) completely inhibited the contractile response of Bay-K- (1.1 μ M) whereas the contractile response induced by ET-1 was not affected. (From D'Orleans-Juste et al. [10]).

As for isolated vascular smooth muscle (Fig. 1), single VSM cell respond to ET-1 in a dose dependent manner (Fig. 3). Superfusion of rabbit aortic single cell with 10^{-11} M of ET-1 induced a slowly transient increase of Ca^{2+} which the reached peak level within 100 sec and was followed by a sustained elevation. Increasing the concentration of ET-1 to 10^{-10} M induces only further increase of the [Ca]$_i$ sustained component. A higher concentration of ET-1 10^{-9} M induced further increase of both the transient and sustained component of [Ca]$_i$ whereas 10^{-8} M did not further increased these two components. A high concentration of ET-1 induced a large fast transient increase of [Ca]$_i$ followed by small further increase of the sustained component. The EC$_{50}$ of ET-1 induced sustained increase of [Ca]$_i$ in rabbit aorta single cells ($\sim$ 0.1 nM) is lower than the EC$_{50}$ reported for ET-1 induced sustained contraction in isolated rabbit aortic muscle (Fig. 1).

The higher sensitivity of single aortic cell to ET-1 when compared to isolated aortic muscle could be due to the capacity of ET-1 to reach its receptors site. The amplitude of ET-1 induced elevation of both components of [Ca]$_i$ is not the same in all the single cells tested. Some group of single cells showed the same pattern and amplitude of both ET-1 induced elevation of [Ca]$_i$, however, some other groups showed only a sustained increase of [Ca]$_i$. The heterogenity of elevation of [Ca]$_i$ induced by ET-1 could be due to the differences of the density of the ET-1 sensitive channel and/or to differences in ET-1 receptor density. The slow increases of [Ca]$_i$ by two concentrations of ET-1 (10^{-11} to 10^{-9} M is similar to that seen in ET-1 induced contraction (Fig. 2). Both the early and sustained increase of [Ca]$_i$ induced by endothelin are sensitive to the Ca^{2+} channels blocker PN 200-110 (10^{-6} M). As aortic VSM cells, 10^{-8} M of ET-1 slowly activated a sustained increase of [Ca]$_i$ in human renal artery single cells (Figure 4). As can be seen in this figure, addition of high concentration of ET-1 (10^{-8} M) slowly increased [Ca]$_i$ that reached a steady state level at 180 sec in presence of the peptide (Fig. 4C). There was no further increase of [Ca]$_i$ later (Fig. 4D). There was no development of pacemaker activity or Ca^{2+} spike in the human renal artery single cells. However, in several single aortic VSM cells of the rabbit, a high concentration of ET- 1 (10^{-8} M) induced within 20 sec a repetitive wave like [Ca^{2+}] movement in conjunction with a wave like contraction originated from a specific region adjacent to Ca^{2+} wave (Fig. 5). During the repetitive wave like the [Ca]$_i$ level increased by 40 to 60 nM (Fig. 5 D-E) above the resting control level (Fig. 5A). ET-1 seems not to only increase [Ca]$_i$ during the repetitive like wave of contraction but also changes the distribution of [Ca]ᵢ in aortic VSM cells of the rabbit (Fig. 5 A-F). The calcium channel blocker PN 200-110 blocks the Cawave as well as the repetitive movement of the cell. In several aortic single cell, [Ca]ᵢ and contraction waves were not accompanied by increase of [Ca]ᵢ. ET-1 in these single cells seems mainly to induce only a changes in [Ca]ᵢ distribution and PN 200-110 (10^{-6} M) again blocks both the [Ca]ᵢ and contraction waves without affecting the level of [Ca]ᵢ.

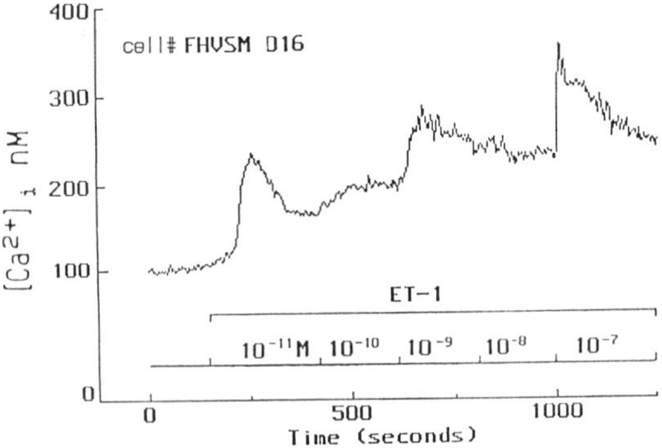

Figure 3. Time course of intracellular free Ca^{2+} changes in response to different concentrations of endothelin-1 (ET1) in rabbit single aortic cell determined with fura-2 (Bkaily et al., unpublished results).

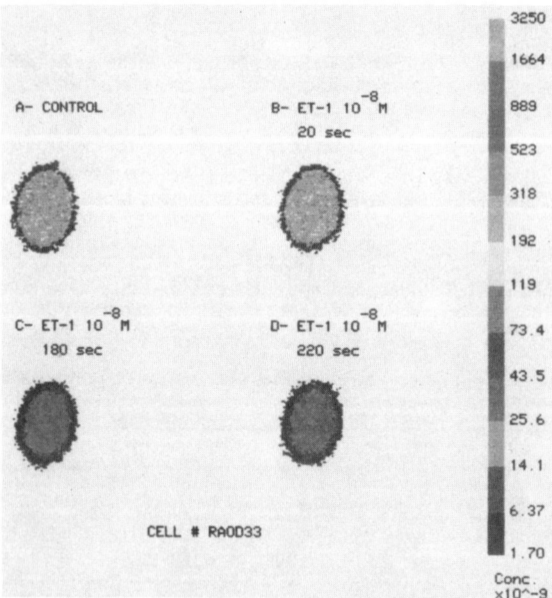

Figure 4. Effect of 10^{-8} M endothelin-1 (ET-1) on the distribution of intracellular [Ca^{2+}] in human renal artery single cell. **A:** intracellular Ca^{2+} distribution in normal Tyrode solution. **B-D:** intracellular distribution of [Ca^{2+}] following 20 (B), 180 (C) and 220 (D) sec in a solution containing 10^{-8} M endothelin-1 (ET-1). The vertical colored bar shows calibration scales for [Ca^{2+}] from 0.003 to 161 μM. (Bkaily et al., unpublished results).

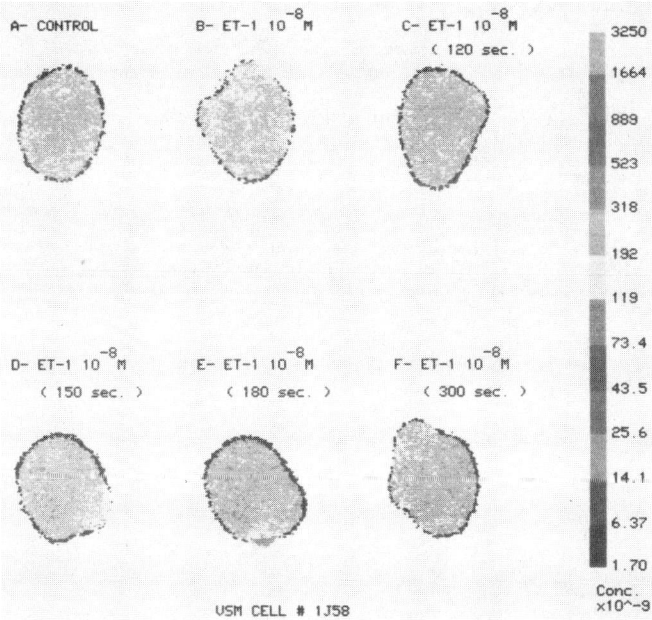

Figure 5. Endothelin-1 (ET) induced waves like repetitive contraction and changes in intracellular $[Ca^{2+}]$ of single aortic vascular smooth muscle cells. **A:** intracellular $[Ca^{2+}]$ distribution in normally quiescent single cell of rabbit aorta. **B-F:** changes of $[Ca^{2+}]_i$ distribution and induction of spontaneous contraction and Ca^{2+} waves 20 (B), 180 (C) and 220 (D) sec in a solution containing 10^{-8} M endothelin (ET-1). The vertical colored bar shows calibration scales for $[Ca^{2+}]$ from 0.003 to 161 μM. (Bkaily et al. unpublished results).

An ET-1 induced repetitive spiking alone devoid of monophasic developing contraction was reported in both rat aorta, aortic cell line A7r5 (36) and porcine isolated ureter (21). These ET-1 induced repetitive spiking seem to be blocked by nitrendipine and EGTA and due to activation of DHP sensitive Ca^{2+} channel (11).

The rate of progress and amplitude of $[Ca]_i$ waves seem to be independent of the concentration of the agonist concentration.

The peak transient increase of $[Ca]_i$ induced by ET-1 could be due to both Ca^{2+} influx following in the potential window current (13) an/or to release of Ca^{2+} from intracellular stores (24, 36, 40). The decline of the initial peak of the increase of $[Ca]_i$ could be due to inactivation of the VOCC and activation of Ca^{2+} -ATPase of the sarcolemma and the SR.

The sustained increase of $[Ca]_i$ as well as the sustained contraction induced by ET-1 are insensitive to nifedipine, nicardipine (10, 33), verapamil and diltiazem (33). Similar sustained increase of $[Ca]_i$ was reported to be induced by histamine, bradykinin (28), insulin and sustained depolarization of the membrane with high $[K]_0$ (3). The sustained increase of $[Ca]_i$ was reported to be insensitive to nifedipine (3, 28), nisoldipine, diltiazem, Bay-K-8644 (28), ouabain (3, 28), caffeine, cyanide, 8-Br-cAMP, 8-Br-cGMP, nitroprusside and ANF, but sensitive to PN 200-110, EGTA (3), Ni^{2+} and La^{3+} (28). This calcium channel is not permeable to Mn^{2+}, Co^{2+} and Ba^{2+} (28). All these observations suggests that agonist exposure (or sustained depolarization of the membrane) may activate Ca^{2+} influx through a pathway that has pharmacological and biophysical characteristics that are distinct from the classical L, T and N Ca^{2+} channels. The agonist (or sustained depolarization) activation of non activating Ca^{2+} channel ressemble curiously the R-type insulin sensitive Ca^{2+} channel in heart and VSM cells (3).

ET-1 ACTIVATES T-TYPE I_{Ca} WITHOUT AFFECTING K^+ CURRENT

The transient and sustained increase of $[Ca]_i$ was reported to be induced by depolarization of the cell membrane and, thereby, activating the VOCC (16, 36). It is possible that agonist may increase $[Ca]_i$ by blocking a K^+ current that is responsible of determining the resting membrane potential and activate the Ca^{2+} influx. Using the whole-cell voltage clamp technique, the effect of different concentrations of ET-1 was tested on the delayed outward K^+ current that mainly determined the resting membrane potential in aortic VSM of the rabbit. As can be seen in figure 6 superfusion with different concentrations of ET-1 (10^{-10} to 10^{-7} M) did not affect the shape and the amplitude of I_K. However, the K^+ channels blocker, barium (10 mM) completely blocks this outward current. These results may suggests that, ET-1 induce contraction and icnrease of $[Ca]_i$ is not due to depolarization of the cell membrane in rabbit aortic single cells. ET- 1 did not affected the I/V curve of the delayed outward K^+ current.

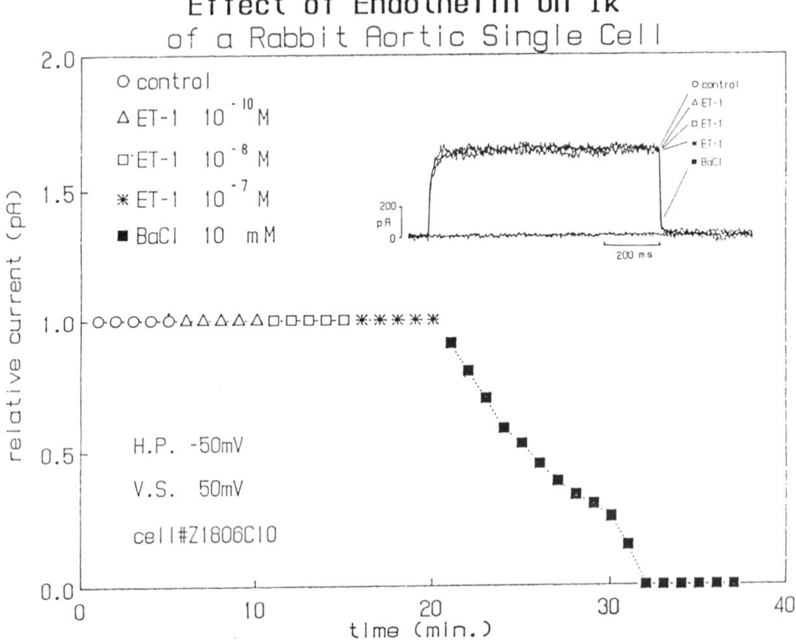

Figure 6. Absence of effect on endothelin-1 (ET-1) on the delayed outward K^+ current (I_K) in single aortic cells of the rabbit. **A:** time course effect of different concentrations of ET-1 on the peak amplitude of I_K measured at the end of a depolarization pulses (VS) to +50 mV from holding potential (HP) of -50 mV. The K^+ channels blocker, barium completely blocked the endothelin-insensitive I_K. **B:** current traces in absence and presence of different concentrations of ET-1. All records were taken from the same single cell. Temperature was 22° C. The experiments were carry out using a Tyrode solution containing tetrodotoxin (a blocker of the fast Na^+ channels) and Mn^{2+} (a blocker of the Ca^{2+} channels). (Bkaily et al., unpublished results).

The fact that once the nonactivating Ca^{2+} channel (or R-type) is stimulated by a hormone, and that K^+ depolarization did not further increase the sustained elevation of $[Ca]_i$ (3, 28) may suggest that the R-type Ca^{2+} channels are fully activated by the agonist opener.

A possible implication of the pacemaker Ca^{2+} current, T-type (2) was suggested to be involved in the action of ET, in the inhibition of ET-induced contraction by nickel, which inhibit T-type I_{Ca} (6). Using whole-cell voltage clamp technique, the effect of different concentrations of ET-1 was tested in both T- and L-type Ca^{2+} currents in chick embryonic heart single cells. Both types of Ca^{2+} currents were found to be stimulated by ET-1 in a dose-dependent manner and Figure 7 shows an example of the effect of this peptide on the T-type I_{Ca}.

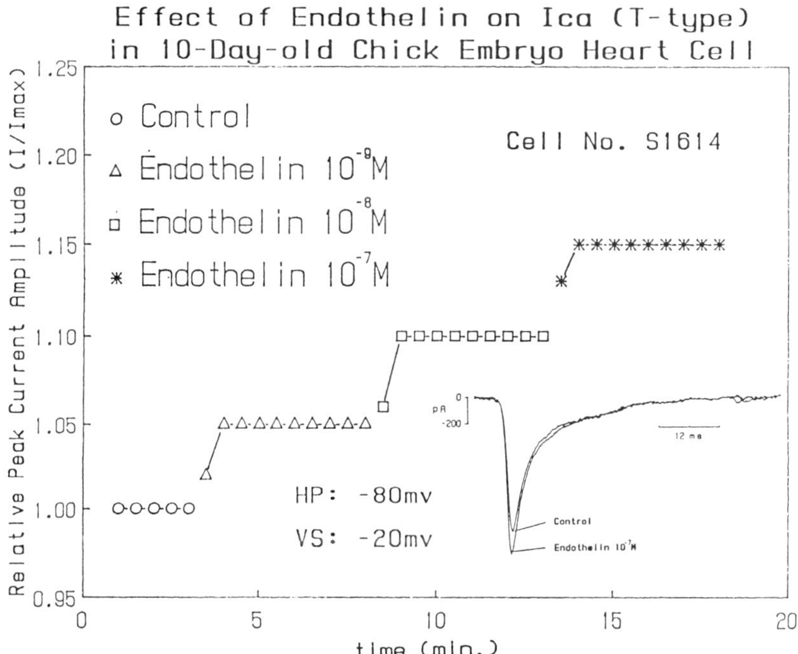

Figure 7. Increase of the T-type Ca^{2+} current by endothelin-1 in 10-day-old embryonic chick heart cells. **A:** time course effect of different concentrations of ET-1. **B:** current traces in absence and presence of 10^{-7} M ET-1. The T-type I_{Ca} was recorded in solution containing 2 mM Ca^{2+} and Na^+ and K^+ ions were replaced by the K^+ blockers, TEA and Cs^+. The experiment was carry out at 22° C and all records were taken from the same single cell. The capacitive current was eliminated using digital subtraction. The currents were measured at the peak. HP = holding potential and VS = voltage step. (Bkaily et al., unpublished results).

As can be seen in this figure, ET-1 increased the T-type in a dose- dependent manner (10^{-9} - 10^{-7} M). The increase of the T-type I_{Ca} as well the L- type I_{Ca} by ET-1 was highly significant.

These results may suggests that the transient increase of $[Ca]_i$ could be due in part to activation of the L-type I_{Ca}. However, the $[Ca]_i$ and the repetitive spikes induced by ET-1 could be due in part to activation of the pacemaker Ca^{2+} current, T-type.

SUMMARY AND CONCLUSIONS

All endothelial cells in addition to the synthesis and release of vasorelaxant agents EDRF and prostacyclin, also release a vasoconstrictor peptide: this peptide induces sustained increase of contraction of all VSM types of different animal species. Also ET-1 seems to induce repetitive spikes and $[Ca]_i$ waves. In general, ET-1 induced transient as well sustained increases of $[Ca]_i$. This rise of $[Ca]_i$ is an integral component of any agonist-induced smooth muscle contraction (19, 21, 34, 43). The transient increase induced by ET-1 seems to be due to activation of an L-type I_{Ca} as well as to Ca^{2+} release from the SR.

However, the ET-induced repetitive spikes and Ca^{2+} waves seem to be due in part to activation of a T-type Ca^{2+} channel and/or a release of Ca^{2+} from intracellular stores.

The activation of L-type and T-type by ET-1 could be to ET-1 stimulation of PLC induced activation of protein kinase C.

ET-1 seems to activate a nifedipine insensitive non inactivating Ca^{2+} channel that pharmacologicaly and biophysicaly seem to be similar to the R-type insulin sensitive Ca^{2+} channel. The activation of the sarcolemma Ca^{2+} channels (nifedipine sensitive and insensitive and Ni^{2+} sensitive) could be due to stimulation of different types of protein G that could be coupled directly to one type of Ca^{2+} channel or/and coupled to PLC that stimulates DG which in turn stimulates different types of protein kinase C (6, 8, 47). The transient increase of $[Ca]_i$ and the intracellular oscillation of $[Ca]_i$ could be due in part to ET-1 induced increased of IP_3 (22, 30, 36, 40).

In vivo or in isolated vascular tissues with intact endothelium, ET-1 effects could be attenuated by Et-1 induced release of endothelial vasorelaxant substances (23).

The physiological role of ET-1 is not yet very well defined and plasma levels in humans are low (1, 37, 40). However, in SHR and patients with essential hypertension were found to have increased plasma levels (31) and to increase sensitivity to ET-1 (35). Also, the presence of specific binding sites of $[^{25}I]$ endothelin on coronary tissues and the increased binding in atheromatous tissue suggest that ET-1 is a peptide which play a role in the maintenance of vascular tone and/or the pathogenesis of ischemic heart disease (7, 8).

ACKNOWLEDGEMENTS

Supported by MRC grant No MA 8920 to Dr Bkaily. Dr Bkaily is a Merck Frosst- FRSQ Professor. The authors thank Mrs Christiane Ducharme for her secretarial assistance.

REFERENCES

1. Ando, K., Y. Hirata, M. Shichiri, T.E. Mori and F. Marumo, Febs Lett. 245, 164-166 (1989).
2. Bean, B.P., Ann. Rev. Physiol. 51, 367-384 (1989).
3. Bkaily, G., D. Economos, L. Potvin, J.-L. Ardilouze, C. Mariott, J. Corcos, D. Bonneau, and C.N. Fong, Mol. Cell. Biochem. in press.
4. Bolton, T.B., Physiological Reviews 59, 606-718 (1979).
5. Charbier, P.E., M. Auguet, P. Roubert, M.O. Onchampt, V. Grillard, J.-M. Guillon, S. Delaflotte, and P. Braquet, J. Cardiovasc. Pharmacol. 13, S218-S219.
6. Cheng, H.C. and R.C. Dage, Neurochem. Int. 18, 497-501 (1991).
7. Dashwood, M.R., R.M. Sykes, M.J. Collins, S. Prehar, S. Theodoropoulos, and M.H. Yacoub, Neurochem. Int. 18, 439-444 (1991).
8. Dashwood, M., M. Turner, and M. Jacobs, J. Cardiovasc. Pharmacol. 13, S183- S185 (1989).
9. DeNucci, G., R. Thomas, P. D'Orleans-Juste, C. Walder, E. Antunes, D. Warner, and J.R. Vane, Proc. Acad. Sci. USA. 85, 9797-9800 (1988).
10. D'orleans-Juste, P., M. Finet, G. deNucci, and J.R. Vane, J. Cardiovasc. Pharmacol. 13, S19-S22 (1989).
11. Eguchi, S., M. Kozuka, S. Hirose, T. Ito, and H. Hagiwara, Biomedical. Res. 12, 35-39 (1991).
12. Eta, E. and C. Triggle, Neurochem. Int. 18, 559-564 (1991).
13. Ganitkevich, V.Y. and G. Isenberg, J. Physiol. (Lond). 435, 187-205 (1991).
14. Goto, K., Y. Kasuya, N. Matsuki, Y. Takuwa, H. Kurihara, T. Ishikawa, S. Kimura, M. Yanagisawa, and T. Masaki, Proc. Natl. Acad. Sci. USA 86, 3915-3918 (1989).
15. Grynkiewicz, G., M. Phoenie, and R.Y. Tsien, J. Biol. Chem. 260, 3440- 3450 (1985).
16. Hiley, C.R., S.A. Douglas, and M.D. Randall, J. Cardiovasc. Pharmacol. 13, S197-S199 (1989).
17. Himpens, B. and A.P. Semylo, J. Physiol. (Lond) 395, 507-530 (1988).
18. Huang, X-H., I. Takanayagi, and T. Hisayama, Gen. Pharmac. 21, 893-898 (1990).
19. Itoh, T., M. Ikebe, G.J. Kargacin, D.J. Hartshorne, B.E. Kemp, and F.S. Fay, Nature 338, 164-167 (1989).
20. Kai, H., H. Kanaide, and M. Nakamura, Biochem. Biophys. Res. Commun. 158, 235-243 (1989).
21. Kamm, K.E. and J.T. Stull, Annual Rev. Pharmacol. Toxicol. 25, 593-620 (1985).
22. Kasuya, Y., Y. Takuwa, M. Yanagisawa, S. Kimura, K. Goto, and T. Masaki, 161, 1049-1055 (1989).
23. Kiowski, W., T.F. Luscher, L. Linder, and F.R. Buhler, Circulation 83, 469-475 (1991).
24. Marsden, P.A., N.R. Danthuluri, B.M. Brenner, B.J. Ballermann, and T.A. Brock, Biochem. Biophys. Res. Commun. 158, 86-93 (1989).
25. Masaki, T., J. Cardiovasc. Pharmacol. 13, S1-S4 (1989).
26. Moody, C.J., M.R. Dashwood, R.M. Sykes, M. Chester, S.M. Jones, M.H. Yacoub, and S.E. Harding, Cir. Res. 67, 764-769 (1990).
27. Morgan, J.P. and K.G. Morgan, Pflugers Arch. 395, 75-77 (1982).

28. Murray, R.K. and M.I. Kotlikoff, J. Physiol. (Lond) 435, 123-144 (1991).
29. Pernow, J., A. Hemsen, A. Hallen, and J.M. Lundberg, Neurochem. Int. 18, 515-519 (1991).
30. Resink, T.J., T. Scott-Burden, and F.R. Buhler, Biochem. Biophys. Res. Commun. 157, 1360-1368 (1988).
31. Saito, Y., K. Nakao, M. Mukoyama, and H. Imura, New Engl. J. Med. 322, 205 (1990).
32. Sakata., K., H. Ozaki, S.-C. Kwon, and H. Karaki, Br. J. Pharmacol. 98, 483-492 (1989).
33. Sarria, B., E. Naline, E. Morcillo, J. Cortijo, J. Esplugues, and C. Advenier, Europ. J. Pharmacol. 187, 445-453 (1990).
34. Somlyo, A.P., Cir. Res. 57, 497-507 (1985).
35. Tomobe, Y., T. Miyauchi, A. Saito, M. Yanagisawa, S. Kimura, K. Goto, and T. Masaki, Eur. J. Pharmacol. 152, 373-374 (1988).
36. VanRenterghem, C., P. Vigne, J. Barhanin, A. Schmid-Alliana, C. Frelin, and M. Ladzunski, Biochem. Biophys. Res. Commun. 157, 977-985 (1988).
37. Vierhapper, H., O. Wagner, P. Nowothny, and W. Waldhausl, Circulation 81, 1415-1418 (1990).
38. Wagner-Mann, C., L. Bowman, and M. Sturek, Am. J. Physiol. 260, C763-C770 (1991).
39. Wagner-Mann, C. and M. Sturek, Am. J. Physiol. 260, C771-C777 (1991).
40. Wallnofer, A., S. Weir, U. Ruegg, and C. Cauvin, J. Cardiovasc. Pharmacol. 13, S23-S31 (1989).
41. Watanabe, T., K. Kusmoto, T. Kitayoshi, and N. Shimamoto, J. Cardiovasc. Pharmacol. 13, S108-S111 (1989).
42. Xuan, Y.-T., W.D. Watkins, and A.R. Whorton, Am. J. Physiol. 260, C492- C502 (1991).
43. Yagi, S., P.L. Becker, and F.S. Fay, Proc. Natl. Acad. Sci. USA. 85, 4109-4113 (1988).
44. Yanagisawa, M., H. Kurihara, S. Kimura, Y. Tombe, M. Kobayashi, Y. Mitsui, Y. Yazaki, G. Goto, and A. Masakai, Nature 332, 411-415 (1988).
45. Yanagisawa, M. and T. Masaki, Trend Pharmac. Sci. 10, 374-378 (1989).
46. Yanagisawa, M. and T. Masaki, Biochem. Pharmac. 38, 1877-1883 (1989).
47. Yang, Z., E. Bauer, L. VonSegesser, P. Stulz, M. Turina, and T.F. Luscher, J. Cardiovasc. Pharmacol. 16, 654-660 (1990).

A HYPOTHESIS FOR MECHANISM OF ENDOTHELIN-INDUCED VASOCONSTRICTION

TOMOH MASAKI[1,2], YOSHITOSHI KASUYA[2], MASASHI YANAGISAWA, and KATSUTOSHI GOTO[2]

Department of Pharmacology, Faculty of Medicine, Kyoto University, Kyoto 606, Japan[1] and Institute of Basic Medical Sciences, University of Tsukuba, Tsukuba, Ibaraki 305, Japan[2]

INTRODUCTION

Over the past decade, there has been an increasing body of evidence that endothelial cells are important in the regulation of vascular functions. Endothelial cells produce several vasoactive substances, such as prostacyclin, tissue specific plasminogen activator, thrombomodulin, etc.

The first report on the regulatory function of the endothelium on vascular tone was the discovery of the production of prostacylin by the vascular wall (40). This was followed by the discovery of endothelium-derived relaxing factor (EDRF) by Furchgott and Zawadki (10). The latter finding triggered a tremendous amount of research activity in this field. Subsequently, it was found that endothelium also produced vasoconstrictive factors. Since the endotheliumdependent constriction is partially blocked by inhibitors of cyclooxygenase, it has been thought that prostanoids are involved in this system (8). In addition to the prostanoids, peptidic vasoconstrictor was also thought to be one of the candidates (13, 45). However, the peptidic vasoconstrictor had remained unknown until the recent discovery of endothelin (64).

Endothelin (ET) is a potent vasoconstrictive peptide produced by endothelial cells (64). ET is composed of 21 amino acid residues with free amino- and carboxyl-termini, including four cystine residues at positions 1, 3, 11 and 15. The native form of ET has two disulfide bonds: 1-15 and 3-11. Theoretical isotypes having different disulfide bonds other than the native form show far lower levels of activity than the native peptide (27). They also consist of a hydrophilic amino terminal half and a strongly hydrophobic C-terminal half. The C-terminal hydrophobic amino acid cluster is very important in exerting vasoconstriction (21). The molecular weight of ET is estimated to be 2492 from its amino acid sequence.

Three sequences of structurally and pharmacologically distinct ET isotypes, designated ET-1, ET-2 and ET-3 are encoded in separate human genes (15) (Fig. 1). In addition to those ETs, sequences highly homologous to ET types have been found in sarafotoxins, which are the active components of the venom of the burrowing asp Atractapis engaddensis (56), suggesting the existence of ET-like sequences in the genes of various species.

ET exists ubiquitously in various tissues including endothelium, neuron, mesangium cell, liver cell, etc. (22). Expression of the three distinct ET isotypes has been observed in various tissues in different proportions (M. Yanagisawa, unpublished results). Vascular endothelial cells produce ET-1 exclusively. It was also demonstrated that ETbinding sites were distributed in various tissues (25), being particularly abundant in central nervous tissue. These results strongly suggest that ET acts differently on various tissues according to the ET subtype or the tissue type. Indeed, so far much evidence has accumulated pointing to diverse physiological functions of ET, not only in cardiovascular tissue, but also in non-cardiovascular tissues (35). The fact that the genes for the three distinct isoforms are located in different chromosomes (3), supports the conclusion that the ETs diverged evolutionally a long time ago, and that consequently each isotype has different functions.

Numerous papers on cellular mechanism of the production and release of ET in endothelial cells and on the cellular mechanism of ET-induced vasoconstriction have been published. However, they are still controversial. In this article, recent works on the cellular mechanism will be reviewed briefly.

Published 1991 by Elsevier Science Publishing Company, Inc.
Ion Channels of Vascular Smooth Muscle Cells and Endothelial Cells
Sperelakis and Kuriyama, Editors

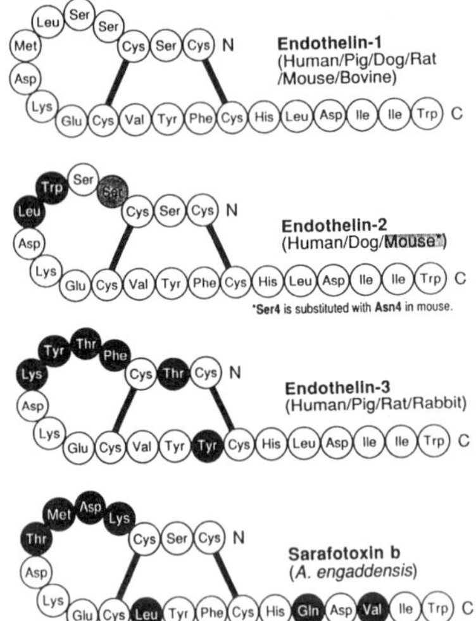

Figure 1. Structure of endothelins and sarafotoxins S6b

PRODUCTION AND RELEASE OF ET FROM ENDOTHELIAL CELL

ET-1 is synthesized and secreted by endothelial cells. In culture, endothelial cells release large amounts of ET. However, the basal release was not detectable when the cultured cells were incubated in phosphate buffered saline without calf serum. ET-1 was secreted in the presence of 10% calf serum, suggesting requirement of an essential factor in the calf serum (12). It was also demonstrated that many factors including thrombin, calcium ionophore, TGF-β, TNF-α, IL-1, adrenalin, histamine, angiotensin II and arginine vasopressin stimulate the production and release of ET from cultured endothelial cells (36). The mechanisms of basal production and stimulated production of ET may be different. Stimulation of production and release of ET by those agonists may be mediated by an increase in intracellular free calcium ion. On the other hand, production of ET upon stimulation with thrombin was potentiated by L-N^G-monomethyl arginine and methylene blue, and reduced by superoxide dismutase and 8-bromo cyclic guanosine 51-monophosphate (4). The inhibitory effects of nitric oxide on the production of ET are probably by mediated by a cGMP-dependent pathway. Oxyhaemoglobin also caused a significant increase in production of ET-1 (5). However, this effect may be ascribed to direct effect on endothelial cells, but is not mediated by abolishment of nitric oxide.

Sequence analysis of 203-amino acid porcine preproform ET-1 reveals that in cytoplasm, synthesized preproform of ET-1 penetrates the innermembrane as does preproform of insuline. The proform of ET-1 may then be cleaved by an endopeptidase specific for the paired dibasic amino acid residues to form an intermediate 39-amino acid peptide, designated big ET-1, which is in turn cleaved into mature ET-1 by a putative endothelin converting enzyme (64).

So far there has been no report on the isolation of the endothelin converting enzyme. However, the activity of conversion from big ET-1 to ET-1 is inhibited by EDTA, 0- phenanthroline and phosphoramidon, suggesting that the enzyme is a metalloproteinase (49). This activity is located in the membrane-bound fraction of endothelial cells.

MECHANISM OF ET-INDUCED VASOCONSTRICTION

In perfused blood vessels, ET elicits slow developing and long-lasting constriction. ET also elicits vasodilation as well as vasoconstriction. ET-3 is relatively potent in vasodilative action when compared with ET-1 (15). When a bolus low dose of ET-1 was injected into artery, vasodilation rather than vasoconstriction was prominent at the initial step (29). However, the artery ultimately showed constriction at the later stage. The vasoconstriction was sustained for a long time. Curiously, ET barely elicits relaxation in strips of isolated blood vessel. ET-induced vasodilation and vasoconstriction are dependent on the vessel employed. Vasodilation is prominent in hindquarter artery, particularly when the blood pressure of the animal was sustained at a low level under the anaesthetized and pithed state (29). In contrast, in mesenteric artery ET induces predominantly constriction. The vasoconstriction and vasodilation induced by ET may be mediated by different mechanisms. ET-induced vasodilation has been ascribed to EDRF released from endothelium by ET itself. Indeed, it was demonstrated that in perfused mesenteric artery, ET elicited EDRF release in perfusate (7). However, recent experiments demonstrated that in hindquarter skeletal muscle artery where ET predominantly induces vasodilation, ET-induced vasodilation was not abolished by pretreatment with methylene blue, suggesting the involvement of a different mechanism in this vasodilation.

To elucidate further the mechanisms of vasoconstriction and vasodilation, it is important to elucidate the receptor mechanism and intracellular signal transduction systems activated by ET. Biochemical analyses of the binding experiments of the fractionated membrane from various types of cells revealed that ET has multiple binding sites on the membranes of various kinds of cells including chicken cardiac muscle, rat lung, human placenta, rat mesangium cells etc. (37, 43, 54, 55, 63). In chicken heart muscle three binding proteins were detected in sodium dodecyl sulfate polyacryl amide gel electrophoresis (SDS PAGE)-autoradiographic experiments whose molecular masses were 53 kDa, 46 kDa and 34 kDa respectively (63), while in rat lung two binding sites were detectable with molecular masses of 44 kDa and 32 kDa (37). Smooth muscle has only one binding site for ET, with molecular mass of about 73 kDa (34).

According to experiments replacing membrane-bound iodine prelabeled ETs with the three distinct isotypes, there may be three different subtypes of ET receptors. The first type has a high affinity to ET-1 and ET-2, but a low affinity to ET-3. The second type has an equal affinity to all of the three different ETs. The third type has a high affinity to ET-3, but a low affinity to ET-1 or ET-2 (37). Indeed, two distinct CDNA clones for ET-receptors were isolated from bovine lung (2), rat lung (52), rat smooth muscle A10 cells (30), human liver (51) and human jejunum. Amino acid sequences of the first type ET-receptor from bovine lung and from rat A10 cell are 417 and 442 amino acid residues long respectively. The second type of ET-receptor, which was found in rat lung and human liver, consist of 426 and 441 amino acids respectively. The difference between the molecular sizes deduced from amino acid sequences and those deduced from the biochemical analyses may be ascribed to the glycosylation of the peptides. Two glycosylation sites were found in each ET-receptor sequence. Both of these two types of ET-receptor contain seven stretches of 20-27 hydrophobic amino acid residues, revealing a seven turned G-protein binding receptor which belongs to the rhodopsin-type receptor superfamily.

COS-7, COP-5 and CHO cells were transfected by constructs containing those CDNA clones. The expressed receptor by the cDNA for the first type receptor has a high affinity to ET-1 and ET-2 but a low affinity to ET-3. In contrast, the receptor expressed by the second type CDNA has an equal affinity to all of the three isotypes of ETs. Therefore, we name the first type receptor ETA, and the second type receptor ETB (52). Although the existence of an ET-3-specific third type was demonstrated in a binding experiment (34, 53), a cDNA clone for the third type receptor has not been cloned yet. The third type, if any, is quite different from the other two, as by molecular physiological methods, we were not able to detect the third type using a CDNA probe for ETB. This third type of ET-receptor was designated ETc. Smooth muscle cells probably have ETA receptor.

ET induces an increase in intracellular free calcium ions not only in smooth muscle cell but also in endothelial cell, firoblast cell, mesangial cell, cerebellar granule cell, etc. (14, 36, 57, 58). In A10 cell, ET-1 elicits a rapid and transient increase in cytosolic free calcium level followed by a prompt decrease to a lower level, which is significantly higher than the original basal level. This level is sustained for a long time.

On the other hand, ET clearly activates phospholipase C to produce phosphoinositol trisphosphate and diacylglycerol (20, 33, 42, 50, 58). This process was shown to be mediated by Gprotein, which is insensitive to pertussis toxin (20, 57, 58). However, this problem is still controversial. Several investigators demonstrated that ET-stimulated phosphoinositide hydrolysis was partially inhibited by pertussis toxin in vascular smooth muscle, mesangial cells and C-6 glioma cells, where pertussis toxin induced ADP ribosylation of about 40 kDa protein (31, 42, 58, 59). The initial transient increase in cytosolic free calcium ion induced by ET is ascribed to calcium ions from the intracellular caffeine-sensitive calcium store, which is stimulated by inositol 1,4,5-trisphosphate (20). Removal of exterior calcium ion did not affect the initial transient increase in cytosolic calcium ion (57). Repeated pretreatment of isolated porcine coronary artery in calcium free solution with caffeine and histamine completely depleted the intracellular calcium store (18), leading to a disappearance of the increase in intracellular free calcium ion induced by ET.

The latter sustained increase in concentration of the intracellular free calcium ions induced by ET may be ascribed to exterior calcium ions (57). Nickel ion, a calcium channel blocker, abolished the sustained phase of the increase in cytosolic calcium channel in fibroblast smooth muscle cells (58).

In porcine coronary artery smooth muscle, ET-1 clearly stimulates voltage-dependent, dihydropyridine-sensitive calcium channels (11). ET induces influx of exterior calcium ion without depolarization of the membrane. The activation of dihydropyridine-sensitive, voltage-dependent calcium channels by ET-1 in porcine coronary artery smooth muscle is mediated via a pertussis toxin-sensitive G-protein in a manner apparently independent of the activation of protein kinase C (Y. Kasuya unpublished data). It was also demonstrated that in guinea pig portal vein, ET-1 stimulates at least two types of calcium channels, which have different sensitivities to dihydropyridine calcium antagonist (16).

However, several reports have not been able to demonstrate ET-induced activation of calcium ions influx (38, 46). Moreover, in rat uterus and human bronchus, the contractile response occurs in two steps (1, 26). In the first step, constriction is induced by a lower concentration of ET, which requires calcium influx and is inhibited by dihydropyridinecalcium antagonist. The second step is characterized by a higher efficiency but a lower potency than those observed in the first step. The second step does not involve the dihydrophyridine-sensitive calcium channels. Very recently, Muldoon et al. demonstrated that in Rat-1 fibroblast cells, low concentration of ET-1 produced a significant increase in calcium influx without inducing comparable change in intracellular calcium mobilization or production of diacylglycerol level (41). In contrast, higher concentrations of ET-1 that were effective at stimulating phosphoinositide turnover were inhibitory for calcium influx. Muldoon suggested that in this case, following the mobilization of intracellular calcium pools by ET, the increase in concentration of intracellular free calcium ion inhibits calcium influx, as experimentally elevating intracellular calcium levels with the tumor promoter thapsigargin abolished ET-1 stimulated calcium influx. Thapsigargin inhibits calcium-activated ATPase, thereby releasing calcium ion from the calcium pools. The hypothesis is further strengthened by the fact that calcium influx was not inhibited by high concentrations of ET-1 but rather was stimulated in the presence of intracellular calcium chelator (1,2-bis(2-aminophenoxy) ethane-N,N,N',N'-tetraacetic acid) and its derivatives.

Since the plasma concentration of ET-1 is very low, subcellular phenomenon induced by low dose of ET-1 may be physiologically important. Therefore, the slow increase in cytosolic free calcium ions caused by low dose ET-induced calcium influx may be important in ET-induced vasoconstriction. The latter sustained increase in cytosolic free calcium level elicited by ET-1 may be ascribed to this calcium influx.

The two contrasting responses induced by low and high doses of ET-1 might not be mediated by two distinct receptors, as there seems to be no physiological difference between responses induced by both expressed receptors ETA and ETB (2, 30, 52). Probably they are activated by a single receptor acting through different G-proteins, as smooth muscle cell has only one type of ET-receptor.

The mechanism for opening the calcium channel is unclear. Cyclic AMP, cyclic GMP and inositol phosphate may not be candidate for factors affecting the action of ET on the calcium channel, as these factors did not activate the voltage-dependent calcium channel (23, 47, 48). One possibility is that inositol 1,3,4,5-tetrakisphosphate is involved in opening the calcium channel, as a local generation of this substance was reported to open plasma membrane calcium channels (28, 41) (Fig. 2).

Since the calcium channel is activated through the stimulation of protein kinase C (9), this pathway may also be activated (6) through an increase in concentration of diacylglycerol. ET-induced increase in diacylglycerol is biphasic, consisting of an early phase peak and a late sustained phase (39). The initial phase is ascribed to the hydrolysis of polyphosphoinositide, while the later phase is still unclear. Possibly the later phase is a peak of diacylglycerol produced by phospholipase D.

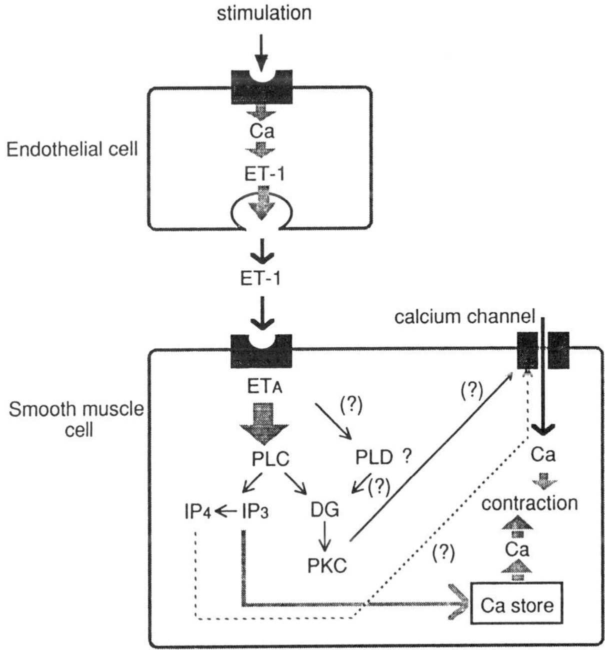

Figure 2. Hypothetical signal transduction systems of smooth muscle.

Van Renterghem et al. (60) demonstrated that, in cultured rat aortic smooth muscle cells, ET did not activate L-type calcium channels, but elicited transient hyperpolarization, followed by depolarization. They ascribed the long-lasting depolarization, at the later stage, to the ET-induced opening of the nonselective cation channel permeable to calcium ions. However, since the depolarization induced by maximum concentration of ET in porcine coronary artery is 8.4 mV at most (11), the depolarization might not activate the voltagedependent calcium channel in this case. ET also activated Na^+/H^+ exchanger, causing cytosolic alkalinization. The resulting intracellular alkalinization may enhance the sustained generation of diacylglycerol and may lead to a slowly developing contraction under certain conditions through activation of protein kinase C. However, Vigne et al. (61) demonstrated recently that the action of ET on Na^+/H^+ antiporter did not involve protein kinase C.

ET was shown to elicit vasoconstriction to some extent even in a calcium-free medium (24). Even after the complete depletion of calcium ion from caffeine-sensitive and histamine-sensitive intracellular calcium stores in calciumfree solution, a slow-rising tonic contraction with no increase in intracellular free calcium ion was elicited by ET. This constriction in calcium free solution is abolished by the protein kinase C inhibitor.

Since contractile action of ET-1 is very similar in many respects to that of phorbol esters, and ET-1 induces a sustained production of diacylglycerol, in smooth muscle, it could be that protein kinase C is involved in this mechanism (32). Protein kinase C possibly sensitizes contractile apparatus to intracellular free calcium ion as well as opening

214

calcium channels, leading to the slow developing and sustained contraction, suggesting the importance of the activation of proteins kinase C by ET in ET-induced vasoconstriction.

CONCLUSION

Vasoactive factors including angiotensin-II or arginine vasopressin may stimulate the production and release of ET-1 in vascular endothelial cells. Increase in the concentration of cytosolic free calcium ion may mediate those responses. ET-receptor is a seven-turned G-protein binding receptor. There are at least two subtypes of ET-receptor, designated ETA and ETB. Smooth muscle probably has ETA receptor. Low doses of ET released from endothelial cells activate ETA receptor on smooth muscle. Activated ET-receptor in turn induces phospholipase C activation to stimulate inositol phosphate turnover. The produced diacylglycerol stimulates protein kinase C. Activation of protein kinase C induces opening of calcium channels and in turn induces an influx of calcium ion from exterior medium. Pharmacologically, a high dose of ET induces production of inositol-trisphosphate. Produced inositol-trisphosphate stimulates the caffeinesensitive intracellular calcium store to elicit an increase in the concentration of intracellular free calcium ion, thus inducing muscle contraction.

REFERENCES

1. Advernier, C., B. Sarria, E. Naline, L. Puybasset, and L., Lagente, Br. J. Pharmacol. 100, 168-172 (1990).
2. Arai, H., S. Hori, I. Aramori, H. Ohkubo, and S., Nakanishi, Nature 348, 730-722 (1990).
3. Arinami, T., M. Ishikawa, A. Inoue, M. Yanagisawa, T. Masaki, M.C. Yoshida, and H. Hamaguchi, Am. J. Hum. Genet. 48, 990-996 (1991).
4. Boulanger, C. and T.F. Luscher, T. Clin. Invest. 85, 587- 590, (1990).
5. Cocks, T.M., E. Malta, S.J. King, R.L Woods, and J.A. Angus, Eur. J. Pharmacol. 196, 177-182. (1991).
6. Danthuluri, N.R. and T.A. Brock, J. Pharmacol. Exptl. Ther. 254, 393-399 (1990).
7. DeNucci, G., R. Thomas, P. D'Orleans-Juste, E. Autunes, C. Walder, T.D. Warner, and J.R. Vane, Proc. Natl. Acad. Sci. USA. 85, 9797-9800 (1988).
8. Eglen, R.M., A.D. Michel, N.A. Sharif, S.R. Swank, & R.L. Whiting, Br. J. Pharm. 97, 1297-1307 (1989).
9. Fish, D.R., G. Sperti, W.S. Colucci, and D.E. Clapham, Circ. Res. 62, 1049-1054 (1988).
10. Furchgott, R.F. and J.V. Zawadzki, Nature 288, 373-376 (1980).
11. Goto, K., Y. Kasuya, N. Matsuki, Y. Takuwa, H. Kurihara, T. Ishikawa, S. Kimura, M. Yanagisawa, and T. Masaki, Proc. Natl. Acad. Sci. USA 86, 3915-3918 (1989).
12. Hexum, T.D., C. Hoegar, J.E. Rivier, A. Baird, and M.R. Brown, Biochem. Biophys. Res. Commun. 167, 294-300 (1990).
13. Hickey, K.A., G. Rubanyi, R.F. Paul, and R.F. Highsmith, Am. J. Physiol. 248, C550-C556 (1985).
14. Hirata, Y., H. Yoshimi, S. Tanaka, T.X. Watanabe, S. Kumagai, K. Nakajima, and S. Sakakibara, Biochem. Biophys. Res. Commun. 154, 868-875 (1988).
15. Inoue, A., M. Yanagisawa, S. Kimura, Y. Kasuya, T. Miyauchi, K. Goto, and T. Masaki, Proc. Nat. Acad. Sci. USA 86, 2863-2867 (1989).
16. Inoue, Y., M. Oike, K. Nakao, K. Kitamura, and H. Kuriyama, J. Physiol. 423, 171-191 (1990).
17. Irvine, R.F. and R.M. Moor, Biochem. J. 240, 917-920 (1986).
18. Kai, H., H. Kanaide, and M. Nakamura, Biochem. Biophys Res. Commun. 158, 235-243 (1989).
19. Kasuya, Y., T. Ishikawa, M. Yanagisawa, S. Kimura, K. Goto, and T. Masaki, Am J. Physiol. 257, H1828-H1835 (1989).
20. Kasuya, Y., Y. Takuwa, M. Yanagisawa, S. Kimura, K. Goto, and T. Masaki, Biochem. Biophys. Res. Commun. 161, 1049- 1055 (1989).
21. Kimura, S., Y. Kasuya, T. Sawamura, 0. Shinmi, Y. Sugita, M. Yanagisawa, K. Goto, and T. Masaki, Biochem. Biophys. Res. Commun. 156, 1182-1186, (1988).
22. Kitamura, K., T. Yukawa, S. Morita, Y. Ichiki, T. Eto, and K. Tanaka, Biochem. Biophys. Res. Commun. 170, 497-503 (1990).
23. Klockner, V. and G. Isenberg, Pflugers Archiv. 405, 340- 348 (1985).
24. Kodama, M., H. Karaide, S. Abe, K. Hirano, H. Kai, and M. Nakamura, Biochem. Biophys. Res. Commun. 160, 1302-1308 (1989).
25. Koseki, M., Imai, Y. Hirata, M. Yanagisawa, and T. Masaki, Am J. Physiol. 256, R858-R866 (1989).

26. Kozuka, M., T. Ito, S. Hirose, K. Takahashi, and H. Hagiwara, Biochem. Biophys. Res. Commun. 159, 317-323 (1989).

27. Kumagaye, S., H. Kuroda, K. Nakajima, T.X. Watanabe, T. Kimura, T. Masaki, and S. Sakakibara, Int. J. Pept. Protein Res. 32, 519-526 (1988).

28. Kuns, M. and P. Gardner, Nature 326, 301-304 (1987).

29. LeMonnier de Gouville, A.C., S. Mondot, H. Lippton, A. Hyman, and I. Cavero, J. Pharmacol. Exp. Ther. 252, 300-311 (1990).

30. Lin, H.Y., E.H. Kaji, G.K. Winkel, H.E. Ives, and H.F. Lodish, Proc. Natl. Acad. 88, 3185-3189 (1991).

31. Lin, W.W., C.Y. Lee, and D.M. Chuang, Biochem. Biophys. Res. Commun. 168, 512-519 (1990).

32. Marsault, R., P. Vigne, and C. Frelin, Biochem. Biophys. Res. Commun. 171, 301-305 (1990).

33. Marsden, P.A., N.R. Danthuluri, B.M. Brenner, B.J. Ballermann, and T.A. Brock, Biochem. Biophys. Res. Commun. 158, 86-93 (1989).

34. Martin, E.R., B.M. Branner, and B.J. Ballermann, J. Biol. Chem. 265, 14044-14049 (1990).

35. Masaki, T., J. Cardiovascul. Pharmacol. 17 (suppl 7), Sl-S4 (1991).

36. Masaki, T., M. Yanagisawa, K. Goto, S. Kimura, and Y. Takuwa in: Cardiovascular Significance of Endothelium-Derived Vasoactive Factors, G.M. Rubanyi eds. (Mount Kisco, NY).

37. Masuda, Y., H. Miyazaki, M. Kondoh, H. Watanabe, M., Yanagisawa, T. Masaki, and K. Murakami, FEBS Lett. 257, 208-210 (1989).

38. Miasiro, N., H. Yamamoto, H. Kanaide, and M. Nakamura, Biochem. Biophys. Res. Commun. 156, 312-317 (1988).

39. Michitoshi, S., Y. Kawahara, K. Kariya, S. Araki, H. Fukuzaki, and Y. Takai, Biochem. Biophys. Res. Commun. 160, 744-750 (1989).

40. Moncada, S., R. Gryglewski, S. Bunting, and J.R. Vane, Nature 263, 663-665 (1976).

41. Muldon, L.L., H. Enslen, K.D. Rodland, and B.E. Magun, Am J. Physiol. 260, C1273-Cl281 (1991).

42. Muldoon, L.L., K.D. Rodland, M.L. Forsythe, and B.E. Magum, J. Biol. Chem. 264, 8629-8536 (1989).

43. Nakajo, S., M. Sugiura, R.M. Snajdar, F.H. Boehm, and T. Inagami, Biochem. Biophys. Res. Commun. 164, 205-211 (1989).

44. Nakamuta, M., R. Takayanagi, Y. Sakai, S. Sakamoto, H. Hagiwara, T. Mizuno, Y. Saito, S. Hirose, M. Yamamoto, and H. Nawata, Biochem. Biophys. Res. Commun. 177, 34-39 (1991).

45. O'Brien, R., R.J. Robbins, and I.F. McMurtry, J. Cell Physiol. 135, 263-270 (1987).

46. Ohlstein, E.H., S. Horohonich, and D.W.P. Hay, J., Pharmacol. Exp. Ther. 250, 548-555 (1989).

47. Ohya, Y., K. Kitamura, and H. Kuriyama, Pfluger Archiv. 408, 465-473 (1987).

48. Ohya, Y., K. Terada, K. Yamaguchi, R. Inoue, K. Okabe, K. Kitamura, M. Hirata, and H. Kuriyama, Pfluger Archiv. 412, 382-289 (1988).

49. Okada, K., Y. Miyazaki, J. Takeda, K. Matsuyama, T. Yamaki, and M. Yano, Biochem. Biophys. Res. Commun. 171, 1192- 1198 (1990).

50. Resink, T.J., T. Scott-Burden, and F.R. Buhler, Biochem. Biophys. Res. Commun. 157, 1360-1368 (1988).

51. Sakamoto, A., M. Yanagisawa, T. Sakurai, Y. Takuwa, H. Yanagisawa, and T. Masaki, Biochem. Biophys. Res. Commun., in press (1991).

52. Sakurai, T., M. Yanagisawa, Y. Takuwa, H. Miyazaki, S. Kimura, K. Goto, and T. Masaki, Nature, 348, 732-735 (1990).

53. Samson, W.K., K.D. Skala, B.D. Alexander, and F.S. Huang, Biochem. Biophys. Res. Commun. 169, 737-843 (1990).

54. Schwartz, I., 0. Ittoop, and E. Hazum, Endocrinology 126, 3218-3222 (1990).

55. Sugiura, M., R.M. Snajdar, M. Schmartzburg, K.F. Badr, and T. Inagami, Biochem. Biophys. Res Commun. 162, 1396-1401 (1989).

56. Takasaki, C., N. Tamiya, A. Bdolah, Z. Wallebing, and E. Kochva, Toxicon 26, 543-548 (1988).

57. Takuwa, N., Y. Takuwa, M. Yanagisawa, K. Yamashita, and T. Masaki, J. Biol. Chem. 264, 7856-7861 (1989).

58. Takuwa, Y., Y. Kasuya, N. Takuwa, M. Kudo, M. Yanagisawa, K. Goto, T. Masaki, and K. Yamashita, J. Clin. Invest. 85, 653- 658 (1990).

59. Thomas, C.P., M. Kester, and M.J. Dunn, Am. J. Physiol. 260, F347-F352 (1991).

60. Van Renterghem, C., P. Vigne, J. Barhanin, A. Shcmid-Alliana, C. Frelin, and M. Lazadunski, Biochem. Biophys. Res. Commun. 157, 977-985 (1988).

61. Vigne, P., A. Ladoux, and C. Frelin, J. Biol. Chem. 266, 5925-2928 (1991).

62. Vigne, P., C. Frelin, E.J. Cragoe, & M. Lazdunski, Biochem. Biophys. Res. Comm. 116, 86-90 (1983).

63. Watanabe, H., H. Miyazaki, M. Kondoh, Y. Masud, S. Kimura, M. Yanagisawa, T. Masaki, and K. Murakami, Biochem. Biophys. Res. Commun. 161, 1252-1259 (1989).

64. Yanagisawa, M., H. Kurihara, S. Kimura, Y. Tomobe, M. Kobayashi, Y. Mitsui, Y. Yazaki, K. Goto, and T. Masaki, Nature 332, 411-415 (1988).

EFFECTS OF AMMONIUM ON VASCULAR SMOOTH MUSCLE

TADAO TOMITA, HIROYUKI TOKUNO, MITSUAKI ITO, MOHSIN MD. SYED, and TOSHIHIRO MATSUMOTO

Department of Physiology, School of Medicine, Nagoya University, 466, Japan

INTRODUCTION

Intracellular pH (pH_i) is generally kept lower than extra- cellular pH by several regulatory processes. Whenever pH_i changes in muscle, nearly every step leading to contraction could be influenced, but there are still some uncertainties about its effect on these steps and the mechanism underlying the effects. For example, in the smooth muscle of rat caudal artery, skinned with Triton X-100, acidosis has been reported to potentiate contraction by shifting the Ca^{2+} concentration-tension curve to the left and slowing relaxation (13), whereas in glycerinated pig carotid artery, the opposite result has been obtained, *i.e*, acidification produces a rightward shift of the curve (23). In this article, effects of ammonium chloride (NH_4Cl) will be mainly described refering to the results on mechanical activities of intact vascular muscles and on electrical properties of single cells disparsed enzymatically. NH_4Cl has been used to alter pH_i since its application produces intracellular alkalinization and its removal acidification (*cf.* 27). Although it is known that Cl^- and HCO_3^- transports are involved in pH_i regulation in a complex fashion (2, 22, 27), this problem was not included in this article. In most of the experiments refered, HCO_3^- was omitted from the bathing solution, and the external pH was buffered either with trishydroxymethylaminomethane (Tris) or N- 2-hydroxyethylpiperazine-N'-2-ethanesulphonic acid (Hepes).

MECHANICAL ACTIVITY

Mechanical responses to NH_4Cl in isolated vascular muscles are complicated. In rat aorta, NH_4Cl (10-30 mM) produced slow contraction concentration-dependently and on wash-out there was a transient increase in contraction before slow recovery (9). NH_4Cl-induced contraction was dependent on the presence of external Ca^{2+}, but it was not affected by nifedipine (0.1 μM). When muscle tone was elevated by phenylephrine (0.1 μM) or KCl (20 mM), NH_4Cl induced relaxation for 10-15 min and then this gave way slowly to contraction.

In rabbit aorta, NH_4Cl (10 mM) applied for 10 min did not produce clear mechanical response, but when precontracted by noradrenaline NH_4Cl decreased the tone during application and transiently increased it after wash-out (12). Similarly, in canine coronary, femoral, and mesenteric arteries, precontracted with prostaglandin $F_{2\alpha}$, 4 mM NH_4^+ relaxed the muscle with slow recovery and upon removal a further transient contraction occurred (11). These responses were not affected by removal of endothelium and by chemical denervation by 6-hydroxydopamine. It has also been shown in the canine mesenteric artery that NH_4^+ (0.1-3 mM) did not significantly affect the resting membrane potential in the presence and in the absence of prostaglandin F_2 (11). In perfused rabbit ear vascular bed, vascular tone decreased during intracellular alkalinization and increased during acidification caused by NH_4Cl application and removal, respectively (28).

According to our own experimens on guinea-pig thoracic aorta, endothelium removed, NH_4Cl (10-20 mM) produced contraction reaching a peak in 5-10 min (Fig. 1A) and in some preparations a transient further increase in contraction on wash-out of NH_4Cl. Trimethylamine (($CH_3)_3N$) produced a similar response. Weak acids (butyrate or acetate) at a concentration of 10-20 mM also produced contraction, usually slightly smaller than that induced by NH_4Cl. When measured pH_i of muscle strips of the aorta using changes in absorption spectra of a pH indicator, dimethylcycloxy- fluorescine (Me_2CF), pH_i was increased with NH_4Cl and this was followed by slow recovery. On wash-out of NH_4Cl, pH_i decreased below the control value transiently, as shown in Fig. 1B. These results are similar to those found in other vascular muscles with a fluorescent dye, 2',7'-bis-(2-

218

carboxyethyl)-5 (and -6)- carboxyfluorescein (BCECF) (1, 9, 11, 16), or with nuclear magnetic resonance study (28). The weak acids produced intra- cellular acidification.

The NH$_4$Cl-induced contraction was strongly inhibited by phenoxybenzamine (PBZ, 1 μM, but after treatment with PBZ a transient contraction appeared on wash-out of NH$_4$Cl in most preparations (Fig. 1A). On average, the maximum contraction during NH$_4$Cl application was reduced to about 20%. Contractions induced by butyrate or acetate were not much affected by PBZ. The pH$_i$ changes caused by NH$_4$Cl as well as butyrate were not clearly affected by PBZ (Fig. 1B). Therefore, the NH$_4$Cl-induced contraction seems to be largely mediated by noradrenaline released from nerve fibres. Release of ^{3}H-noradrenaline by NH$_4^+$ has been demonstrated in the rat vas deferens and brain (31). When muscle tone was raised by noradrenaline or 30 mM K$^+$ after PBZ treatment, NH$_4$Cl induced relaxation during application and transient contraction on wash-out, as observed in canine arteries (11). It is likely that intracellular alkalinization leads to inhibition of contraction and acidification to contraction. Both NH$_4$Cl- and butyrate-induced contractions were resistant to verapamil, although butyrate-induced contraction was relatively more sensitive to verapamil.

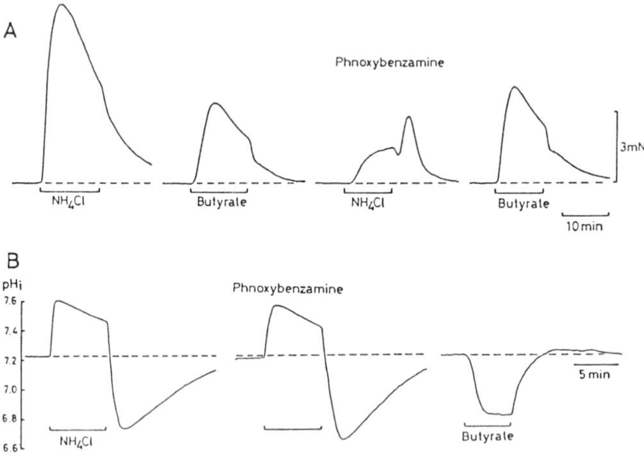

Figure 1. Effects of NH$_4$Cl (20 mM) and butyrate (20 mM) on tension development (A) and pH$_i$ (B) of muscle strip of guinea- pig aorta, in the absecne and presence of phenoxybenzamine (1 μM). pH$_i$ was estimated by the ratio of absorption of Me$_2$CF at 513 and 475 nm. pH$_i$ measurement was carried out in the presence of 0.1 mM Ca^{2+} to reduce mechanical artifact. A and B are obtained from different preparations.

When the external Na$^+$ was removed, a slow contraction developed, as in many other vascular muscles, and it is generally believed that Na$^+$-Ca^{2+} exchange is involved in this contraction (3, 21, 25, 26, 29). This was found to be accompanied with a decrease in pH$_i$ when measured with Me$_2$CF in the guinea-pig aorta (Fig. 2), as found in rat mesenteric arery (1). In rabbit aorta, the contraction by Na$^+$ removal is considered to be due to catecholamines released from nerve terminals, because the contraction is strongly inhibited by phentolamine (18), whereas in guinea-pig aorta, the contribution of this factor is known to be minor (25). This was confirmed in the present experiment.

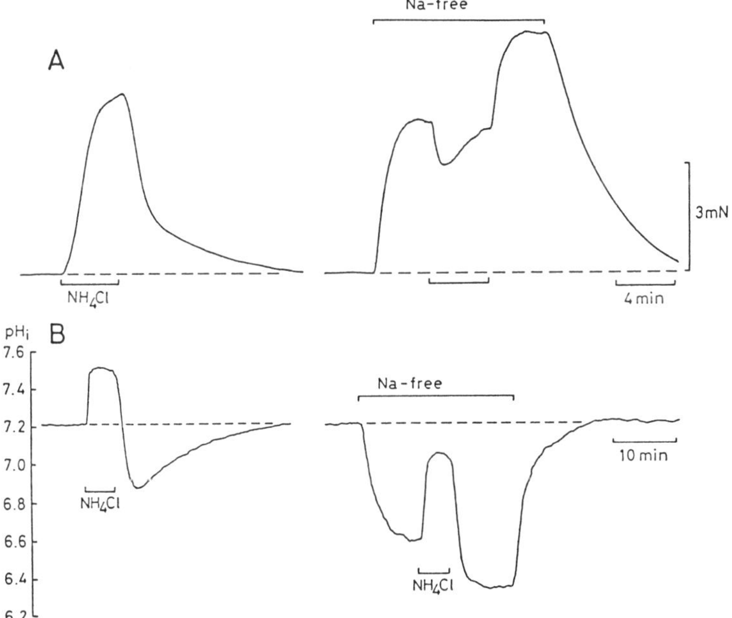

Figure 2. Mechanical response (A) and pH_i change (B) caused by NH_4Cl (20 mM) in guinea-pig aorta. Similar experiments to Fig. 1. After obtaining control response, NH_4Cl was applied in the absence of the external Na^+. Na^+ was replaced with N-methyl-D- glucamine. A and B are from different preparations.

During the contraction induced by Na^+ removal, NH_4Cl produced relaxation, and on wash-out of NH_4Cl followed a large sustained contraction until Na^+ readmission. Butyrate-induced contraction was potentiated in the absence of the external Na^+. It is possible that intracellular acidification caused by Na^+ removal is due to inhibition of Na^+-H^+ exchange (*cf.* 20) and that acidification is partly responsible for the contraction in the absence of the external Na^+, in addition to a contribution of the Na^+-Ca^{2+} exchange process. NH_4Cl-induced alkalinization counteracts acidification due to Na^+ removal, and this seems the reason for partial relaxation. Potentiation of contraction on NH_4^+ wash-out and also of butyrate-induced contraction in the absence of Na^+ can be explained by stronger intracellular acidification. In the crayfish neurone intracellular Na^+ concentration ($[Na^+]_i$) was found to increase markedly with more or less similar time course to acidification caused by NH_4Cl removal (22). If this is also true in smooth muscle, an increase in $[Na^+]_i$ during acidification may lead to activation of Na^+- Ca^{2+} exchange and to contraction.

MEMBRANE K^+ CURRENT

It has been found in generall that K^+ current is inhibited by intracellular acidification and increased by alkalinization. In human lymphocytes the voltage-dependent K^+ conductance is enhanced by alkalinization, without changing the threshold potential for activation and voltage dependency of steady state inactivation (10).

Ca^{2+}-activated (and ATP-dependent) K^+ channels in rat pancreatic B-cells are also inhibited by cytoplasmic acidifi- cation (7). In the epithelium of choroid plexus of amphibia a change in pH_i from 7.4 to 6.4 reduced the channel open probability of Ca^{2+}-activated K^+ channels mainly by increasing the channel closed time (6). Similar results have been obtained in cultured renal medullary cells (8). In membrane patches of smooth muscle cells dispersed from rabbit tracheal muscle, the main effect of decreasing pH_i is to decrease the open probability of the channel by reducing the sensitivity to Ca^{2+} and also shortening the open state (19).

In rabbit portal vein, we have also obtained the evidence with the whole-cell clamp experiment that intracellular acidification reduced and alkalinization increased steady state outward K^+ current, when pH_i was directly changed by perfusing patch pipettes with solution having different pH (Fig. 3). However, when intracellular alkalinization was achieved by external application of NH_4Cl, this current was found to be reduced. This apparent contradiction could be explained by the finding with the patch clamp experiment on single channels using inside-out configuration that NH_4Cl (20 mM) has a weak K^+ channel (probably Ca^{2+}-activated) blocking action in this cell. Trimethylamine which has a similar effect on pH_i to NH_4Cl also inhibited the K^+ channel.

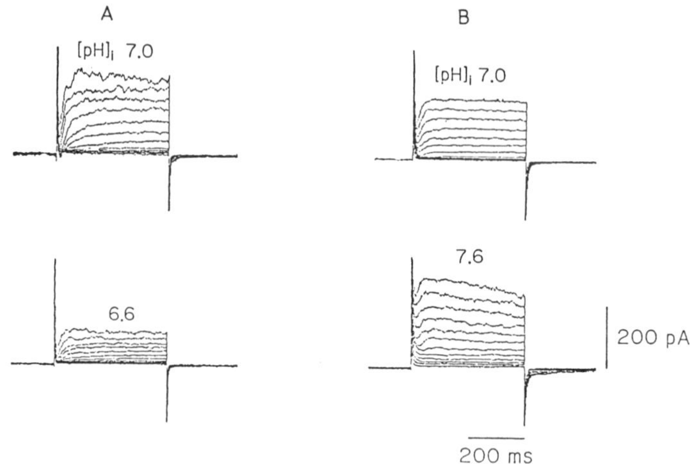

Figure 3. Effects of intracellular acidification (A) and alkalinization (B) on outward membrane currents obtained from a single cell of rabbit portal vein with the whole-cell voltage clamp. Electrode was filled with 130 mM KCl. Depolarizing command pulses (400 ms) were applied every 20 s from a holding potential of -80 mV. pH_i was changed with intrapipette perfusion. A and B are different cells.

When the membrane was voltage-clamped, spontaneous transient outward currents (STOCs) could be recorded in some sooth muscle cells (rabbit portal vein: 4; rabbit small intestine: 5, 24; rabbit ear artery: 5; guinea-pig ureter: 14). These currents are considered to reflect K^+ channel activation by Ca^{2+} released from intracellular stores. This type of currents could also be recorded from cells dispersed from guinea-pig portal vein (Fig. 4). The STOCs were inhibited by NH_4Cl and potentiated by butyrate. Our interpretation is that intracellular alkali- nization reduces intracellular Ca^{2+} and this in turn reduces the STOCs and that a direct potentiating effect of alkalinization on the K+ channel is masked by a decrease in Ca^{2+}. Intracelular acidification produced by butyrate exerts an opposite effect. These changes in intracellular Ca^{2+} are in accord with the relaxation produced by NH_4Cl and contraction by butyrate.

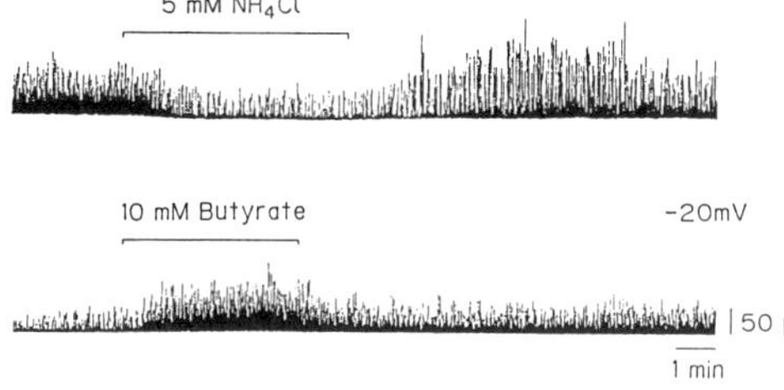

Figure 4. Effects of NH_4Cl and butyrate on spontaneous transient outward currents (STOCs) at a holding potential of -20 mV obtained with whole-cell clamp from guinea-pig portal vein (different cells). STOCs were inhibited by external application of NH_4Cl (5 mM) and potentiated by butyrate (10 mM).

INWARD Ca^{2+} CURRENT

In single cells dispersed from rabbit portal vein, effects of NH4Cl and butyrate on inward currents were studied in solution containing 2.4 mM Ca^{2+} under the condition in which outward currents were inhibited by 30 mM TEA in the external medium and 130 mM Cs$^+$ in pipettes. It was found that NH4Cl potentiated (Fig. 5) and butyrate reduced inward currents significantly (Fig. 6). In the presence of these agents, however, there was no significant changes in time course of the current and voltage- current relationship. It is very likely that these effects are mediated by intracellular alkalinization and acidification, respectively, because similar effects could be demonstarted by changing pH in intrapipette perfusion medium. When butyrate was washed out, inward currents were transiently increased before recovery to the control level. On the other hand, wash-out of NH4Cl produced no clear undershoot, although a decrease in inward currents would be expected, since pH$_i$ is decreased below the control value on removal of NH4Cl. It may be that pH$_i$ is well buffered against acid loads and that acidification on NH4Cl wash- out is weaker than that in the presence of butyrate under our experimental condition for the whole-cell clamp. Further experiments are necessary to clarify this point with simultaneous measurements of pH$_i$.

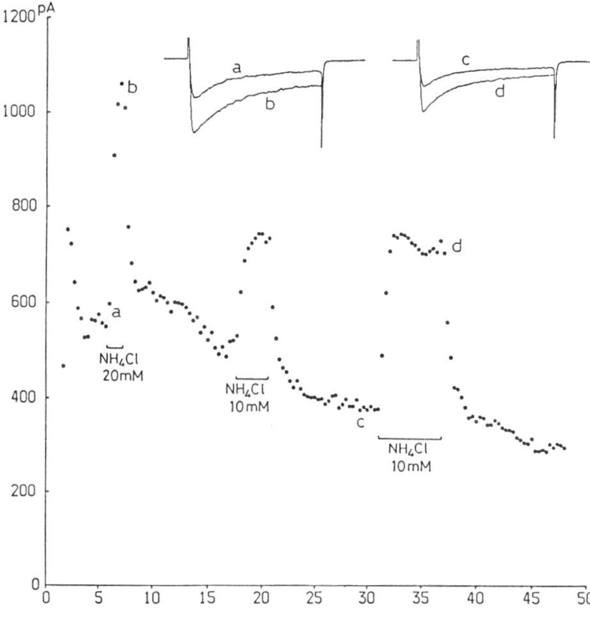

Figure 5. Potentiation of inward currents by NH4Cl in a single cells from rabbit portal vein, recorded with the whole cell clamp method in normal solution containing 2.4 mM Ca^{2+}. Peak currents are plotted which were obtained by command pulses applied from a holding potential of -80 mV to 0 mV. Current traces, a-d, shown above correspond to the currents having the same symbol.

222

The inhibition of Ca^{2+} current by lowering pH_i agrees with the result obtained in cardiac muscle (15, 17, 30). Effects of pH_i on Ca^{2+} channels are probably the same in many different types of cell, but it is not clear whether sensitivity to pH_i differs between "L-type" and "T-type" Ca^{2+} channels.

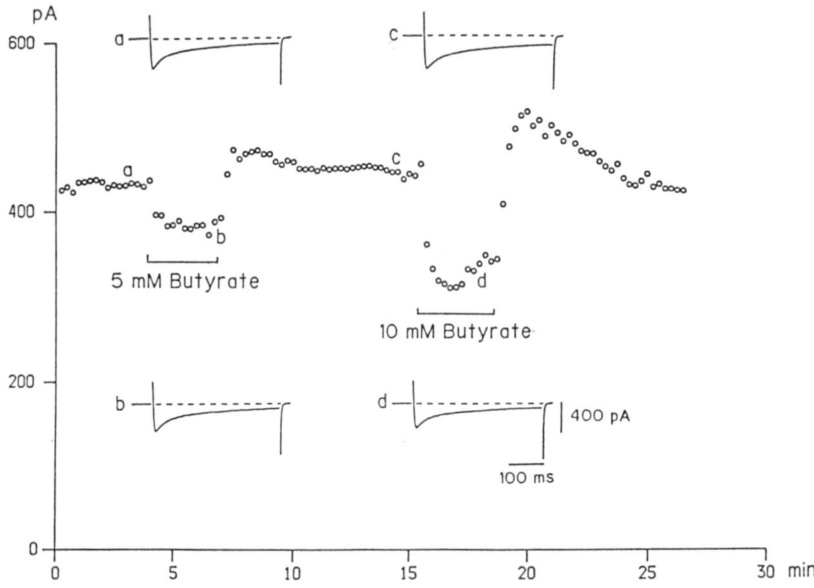

Figure 6. Effects of butyrate on inward currents. A similar experiment to that shown in Fig. 5.

CONCLUSION

Effects of NH_4Cl which produces changes in pH_i on vascular smooth muscle seem complicated. NH_4Cl may cause contraction or relaxation in different smooth muscles and depending on experimental conditions. One factor producing contraction in the presence of NH_4Cl is probably due to release of transmitters (and possibly also some autacoids), but there seems another factor producing contraction after treatment with phenoxybenzamine when muscle tone is low. Depolarization by NH_4Cl resulting from its blocking action of K^+ channel and potentiation of inward Ca^{2+} current may be responsible for the contraction in the presence of NH_4Cl. When muscle tone is kept high, however, NH_4Cl generally produces relaxation, but the mechanism underlying this process is not clear.

Withdrawal of NH_4Cl or application of butyrate are usually accompanied by contraction and these are potentiated in the absence of the external Na^+. A decrease in pH_i inhibits both outward and inward membrane currents, but the inhibition of outward current could be a major factor causing depolarization and in turn leading to contraction. However, since the contraction resulting from intracellular acidification is only partially reduced by verapamil, Ca^{2+} channels controlled by membrane potential may not be totally responsible. It could be that activation of Na^+-H^+ exchange increases $[Na^+]_i$ and that this secondarily increases intracellular Ca^{2+} through Na^+-Ca^{2+} exchange. This explains verapamil-resistant contraction. The complexity of the NH_4Cl-induced response may be due to involvement of these membrane transport systems, in addition to modification of both K^+ and Ca^{2+} channel activities.

REFERENCES

1. Aalkjaer, C. and E.J. Cragoe, J. Physiol. $\underline{402}$, 391-410 (1988).
2. Aickin, C.C. in: Proton Passage Across Cell Membranes, Ciba Foundation Symposium (John Wiley & Sons, Chichester 1988) $\underline{139}$, 3-22.
3. Ashida, T. and M.P. Blaustein, J. Physiol. $\underline{392}$, 617-635 (1987).
4. Beech, D.J. and T.B. Bolton, Biomed. Biochim. Acta $\underline{46}$, S673-676 (1987).
5. Benham, C.D. and T.B. Bolton, J. Physiol. $\underline{381}$, 385-406 (1986).
6. Christensen, O. and T. Zeuthen, Pflugers Arch. $\underline{408}$, 249-259 (1987).
7. Cook, D., M. Ikeuchi, and W.Y. Fujimoto, Nature $\underline{311}$, 269-273 (1984).
8. Cornejo, M., S.E. Guggino, and W.B. Guggino, J. Memb. Biol. $\underline{110}$, 49-55 (1989).
9. Danthuluri, N.R. and R.C. Deth, Am. J. Physiol. $\underline{256}$, H867-875 (1989).
10. Deutsch, C. and S.C. Lee, S.C. J. Physiol. $\underline{413}$, 399-413 (1989).
11. Feletou, M., C.T. Harker, K. Komori, J.T. Shepherd, and P.M. Vanhoutte, J. Pharmacol. Exp. Ther. $\underline{251}$, 82-89 (1989).
12. Furtado, M.R., Life Sci. $\underline{41}$, 95-102 (1987).
13. Gardner, J.P. and F.P.J. Diecke, Pflugers Arch. $\underline{412}$, 231-239 (1988).
14. Imaizumi, Y., K. Muraki, and M. Watanabe, J. Physiol. $\underline{411}$, 131-159 (1988).
15. Irisawa, H. and R. Sato, Circul. Res. $\underline{59}$, 348-355 (1989).
16. Kahn, A.M., E.J. Cragoe, J.C. Allen, R.D. Halligan, and H. Shelat, Am. J. Physiol. $\underline{259}$, C134-143 (1990).
17. Kaibara, M. and M. Kameyama, J. Physiol. $\underline{403}$, 621-640 (1988).
18. Karaki, H. and N. Urakawa, Europ. J. Pharmacol. $\underline{43}$, 65-72 (1977).
19. Kume, H., K. Takagi, T. Satake, H. Tokuno, and T. Tomita, J. Physiol. $\underline{424}$, 445-457 (1990).
20. Mahnensmith, R.L. and P.S. Aronson, Circul. Res. $\underline{56}$, 773-788 (1985).
21. Maseki, T., T. Abe, and T. Tomita, Europ. J. Pharmacol. $\underline{190}$, 355-363, (1990).
22. Moody, W., Ann. Rev. Neurosci. $\underline{7}$, 257-278 (1984).
23. Mrwa, U., I. Achtig, and J.C. Ruegg, Blood Vessels $\underline{11}$, 277-286 (1974).
24. Ohya, Y., K. Kitamura, and H. Kuriyama, Am. J. Physiol. $\underline{252}$, C401-410 (1987).
25. Ozaki, H. and N. Urakawa, Pflugers Arch. $\underline{390}$, 107-112 (1981).
26. Ozaki, H., H. Karaki, and N. Urakawa, Naunyn-Schmied. Arch. Pharmacol. $\underline{304}$, 203-209 (1978).
27. Roos, A. and W.F. Boron, Physiol. Rev. $\underline{61}$, 296-434 (1981).
28. Spurway, N.C. and S. Wray, J. Physiol. $\underline{393}$, 57-71 (1987).
29. Toda, N., Am. J. Physiol. $\underline{234}$, H404-411 (1978).
30. Vogel, S. and N. Sperelakis, Am. J. Physiol. $\underline{233}$, C99-103 (1977).
31. Wakade, A.R. and T.D. Wakade, Life Sci. $\underline{29}$, 2381-2389 (1981).

SMOOTH MUSCLE TONE OF AORTA OF SPONTANEOUSLY HYPERTENSIVE RATS DUE TO AN INCREASE IN THE VOLTAGE-DEPENDENT CA CHANNEL AND ITS MODIFICATION BY ENDOTHELIUM

SATORU SUNANO and KEIICHI SHIMAMURA

Research Institute of Hypertension, Kinki University, Osaka-Sayama, 589 Japan

INTRODUCTION

The contraction of smooth muscles has been known to be initiated through Ca-dependent and -independent mechanisms. Physiologically, the Ca-dependent contraction, especially the contraction via excitation-contraction coupling, plays a major role. This contraction is initiated by a Ca influx through voltage-dependent or receptor operated Ca channel, and/or by the Ca released from intracellular store sites such as sarcoplasmic reticulum (3, 61, 63). Among these mechanisms, the voltage-dependent Ca channels are opened by the depolarization of membrane, namely the membrane excitation.

In vascular smooth muscles of spontaneously hypertensive rats, changes in the membrane excitation have been reported. For example, we have observed using the smooth muscle of portal vein of stroke-prone SHR (SHRSP) that the manner of the spiking is altered when compared with that of control normotensive Wistar Kyoto rats (WKY) (47). Arterial smooth muscles of SHRSP often exhibit spiking of action potentials spontaneously or in response to the stimulation with certain agonists (14, 23, 24). These changes in the membrane electrical activities may lead to the change in the contractile activities including the change in the tone of vascular smooth muscles. In addition, Hermsmeyer et al. (43, communication at the 3rd International Symposium on Calcium and gene hypertension, Portland, 1991) have observed that the voltagedependent inward Ca currents are altered in the vascular smooth muscle cells of spontaneously hypertensive rats (SHR).

The vascular smooth muscles of normotensive rats, especially the smooth muscles of large arteries, are mostly quiescent and rarely show spontaneous electrical and mechanical activities. In addition, they exhibit no spontaneous tension development (active tone) in the absence of stimulation. Vascular smooth muscles of spontaneously hypertensive rats on the other hand, often exhibit spontaneously active tone. Even in smooth muscles of aorta, spontaneously developed active tone has been reported (11, 25, 37, 44, 45, 53, 55-57, 67). The active tone is sensitive to extracellular Ca and disappears by the removal of Ca or by the application of Ca-antagonists. Similar results have been obtained in the cerebral artery (67). The spontaneous tone of the smooth muscle from cerebral artery is accompanied with tension oscillations, indicating the causal presence of spontaneous electrical activities. Thus, it is suggested that the Ca permeability through voltage-dependent Ca channels is increased in vascular smooth muscle of spontaneously hypertensive rats.

In the present studies, relations between blood pressure of the rats and extracellular Ca-dependent active tone, influences of endothelium and effects of antihypertensive treatment of rats were examined using various strains of spontaneously hypertensive rats.

THE SPONTANEOUSLY HYPERTENSIVE RATS

Spontaneously hypertensive rats created by Okamoto and Aoki (39), are now reaching their 100th generation. Their blood pressure starts to elevate from a young age, reaching a peak level which is significantly higher than that of control normotensive Wistar Kyoto rats (WKY) at about 20 weeks of age. From these rats, Okamoto et al. (41) established a new strain of rats which exhibit a high rate of stroke incidence. These rats shows much higher blood pressure and were named stroke-prone spontaneously hypertensive rats (SHRSP). In our institute, lesions of stroke are found in more than 98% of SHRSP which die of diseases. Among the SHRSP, Okamoto et al. found some rats which showed much steeper elevation of blood pressure, died of stroke incidence within 13 to 14 weeks of age. After successive breeding, they finally (40) created a strain which was named malignant stroke-prone spontaneously hypertensive rats (MSHRSP).

Published 1991 by Elsevier Science Publishing Company, Inc.
Ion Channels of Vascular Smooth Muscle Cells and Endothelial Cells
Sperelakis and Kuriyama, Editors

226

We are now maintaining these strains of rats and control normotensive Wistar Kyoto rats (WKY) by successive breeding of animals. Great care in selecting males and females is required to maintain the respective blood pressures desired. The animals are fed with Japanese chow and tap water under circumstances of constant temperature (25° C), humidity (60%) and a light and dark cycle of 12 hours. At the time of mating, in M-SHRSP, antihypertensives are administrated to males and females. In females, the treatment is required during the pregnancy and the lactation.

Age-dependent changes in blood pressure of these rats are shown in Fig. 1. The blood pressure of WKY showed a tendency to elevate with aging and reached its peak at 20 weeks of age. The elevation however, was not prominent and the blood pressure at the age of 20 weeks was 135 ± 1.3 mm Hg (mean $\pm$ SE, n = 8). The blood pressure of SHR was significantly higher than that of WKY already at age of 5 to 6 weeks and raised steeply to attains its peak at 20 weeks of age. The blood pressure at the age of 20 weeks was 200 ± 1.8 mm Hg (n = 14). The elevation of blood pressure of SHRSP was steeper than that of SHR and the blood pressure at the age when reaching its peak (20 weeks) was 260.3 ± 1.3 mm Hg; the value being significantly higher than that of SHR. The steepest elevation of blood pressure was observed in M-SHRSP and attaining it peak at the age of 12 weeks. The blood pressure of ages after 12 weeks however, can not be measured because of the death of the rats or incidence of stroke which often lead to a decrease of blood pressure.

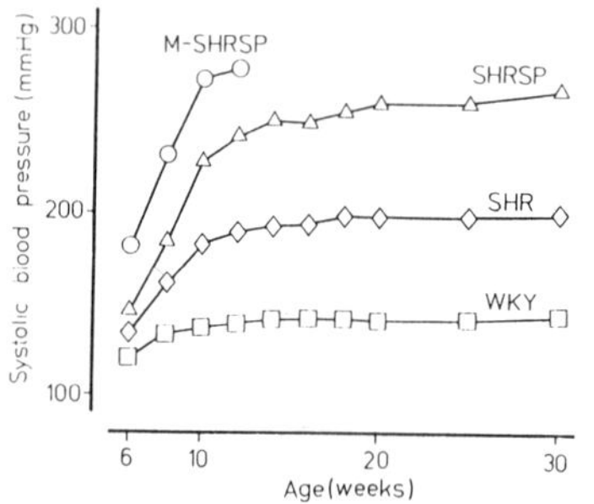

Figure 1. Age-dependent changes in blood pressure of various strains of spontaneously hypertensive rats. The blood pressure of M-SHRSP after 12 weeks could not be measured because of the stroke incidence.

The life span of SHRSP and M-SHRSP has been reported to be 212 ± 15 days and 91 ± 3 days, respectively (40). These values are markedly shorter than those of WKY and SHR (455 ± 20 days in WKY and 388 ± 21 days in SHR).

SMOOTH MUSCLE TONE OF AORTA
Extracellular Ca-Dependent Smooth Muscle Tone

The aortic smooth muscle of normotensive rats (WKY) does not exhibit any spontaneous active tone. On the other hand, the vascular smooth muscle of spontaneously hypertensive rats including all strains described above exhibit spontaneously developing active tone as has been reported previously (11, 25, 37, 44-46, 67). In the present experiment the relationship between the blood pressure of the animals and the spontaneous active tone of aortic smooth muscles were studied using thoracic aortae of the rats described above (WKY, SHR, SHRSP and M-SHRSP).

The experiments were performed at the age of 16 weeks except for M-SHRSP (12 weeks old). Endothelium-removed ring preparations of 1 mm width were made from thoracic aorta of the rats. From about half of the preparations, endothelium was removed by rubbing the inner surface of the lumen with a small instrument made of soft rubber. Preparations were incubated in a modified Tyrode's solution (48) and changes in tension were measured isometrically by force-displacement transducers (Shin-koh, Nagano, Japan). Prior to the measurement of the active tone, complete removal of endothelium was ascertained by applying 10^{-5} M acetylcholine to preparations precontracted in the presence of 5×10^{-7} M noradrenaline. After this observation, the preparations were soaked in the modified Tyrode's solution for 1 hour and then soaked in the solution from which all Ca was removed (Ca-free Tyrode's solution). The relaxation due to the removal of extracellular Ca was treated as an active tone of the preparation. We have measured the concentration of Ca remaining in the incubation medium after the incubation of preparations for 1 hour and found it to be around 10^{-5} M. However, EGTA was not added in the Ca-free Tyrode's solution in the present experiments, since it caused further increase in the active tone when Ca was reintroduced. The preparations soaked in Ca-free Tyrode's solution in the absence of EGTA were not relaxed further by the addition of papaverine $(10^{-4}$ M).

Fig. 2 shows the relaxing response to the removal of extracellular Ca in the preparations from 16 week old WKY and 12 week old M-SHRSP. Although most of the experiments were performed with preparations from rats with established hypertension (16 weeks old), M-SHRSP of this age could not be used because of stroke incidences as described above. However, hypertension in this strain of rats was established already at this age. The preparations from M-SHRSP of the age also exhibited a marked relaxing response to the removal of extracellular Ca.

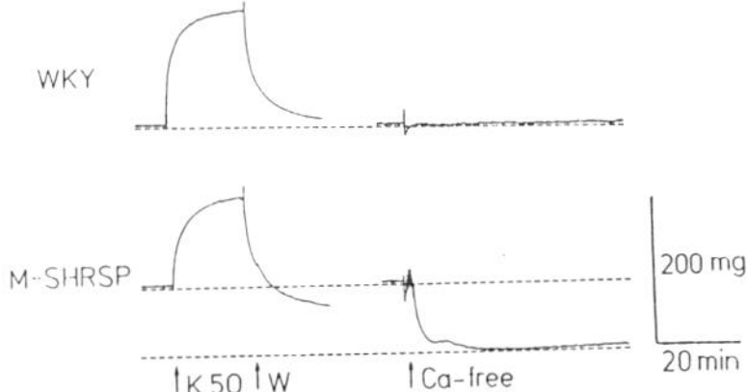

Figure 2. Effects of removal of Ca from a modified Tyrode's solution in the preparation from WKY and M-SHRSP aortae. The preparation from M-SHRSP showed a relaxing response to the removal of Ca, while no relaxation was observed in the preparation from WKY. The amplitude of the relaxation was treated as the active tone of the smooth muscle. (from 56 with permission).

A similar relaxation was observed when 10^{-6} M verapamil was applied in the presence of extracellular Ca. The relaxation by the removal of extracellular Ca or by the application of verapamil was similar in its amplitude to that of the contraction by 50 mM K^{+}. However, when the active tone presented before the application of high K^{+} was taken into account, the amplitude of the active tone was calculated as a half of the high-K^{+}-induced contraction. In the preparation from WKY on the other hand, no relaxing response to the removal of extracellular Ca or to the application of verapamil was observed.

The active tone which is blocked by the removal of extracellular Ca or by the application of Ca-antagonist was also observed in the preparations from SHR. The Ca-antagonist-sensitive active tone in SHR aorta has also been presented by Lindner and Heinle (25). However, the relaxation observed in the present experiment was not prominent although most of the preparations exhibited a relaxing response to Ca-removal or to Ca-antagonist. The preparations from SHRSP showed a marked relaxation in response to the Ca removal of extracellular Ca or to the application of verapamil, although the relaxation was significantly smaller than that observed in the preparation from M-SHRSP. The relationship between blood pressure of the rats and the amplitude of the active tone was illustrated in Fig. 3. As shown in this fig, the active tone increased as the blood pressure of the rats elevated, and good correlation between the blood pressure of the rats and the amplitude of the active tone was demonstrated.

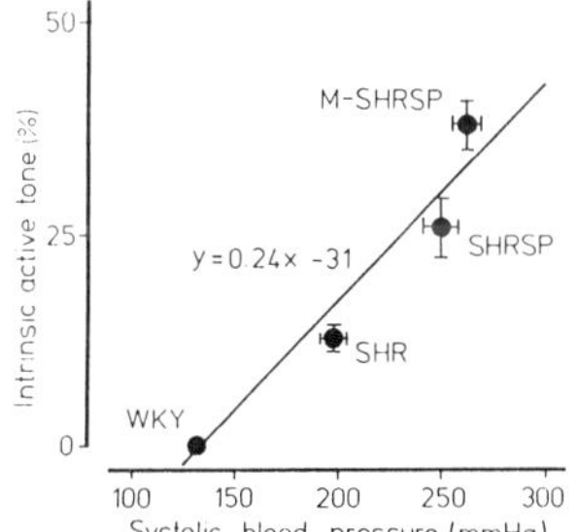

Figure 3. The relationship between the blood pressure of rats and the active tone of the smooth muscle. The amplitude of the active tone was expressed as percentages of the amplitude of high-K-induced contraction. The straight line in the figure was drawn by coefficient analysis using values obtained from all preparations.

228

Thus, it is suggested that Ca-influx through voltage-dependent channels under non-stimulated condition was greater as the blood pressure of the rats increased.

The relationship between the blood pressure and active tone can be studied by treating animals with antihypertensives (45). When animals are treated with drugs however, direct actions of drugs on smooth muscles should also be taken into consideration. Since rats which genetically showed different blood pressure were used in the present experiments, it can be thought that voltagedependent Ca channels of vascular smooth muscles is altered depending on the degree of hypertension.

It was also shown in the present experiment that the preparations relaxed in the absence of extracellular Ca contracted in response to reintroduction of Ca in a concentration-dependent manner (Fig. 4). The contraction was greater in the preparations from rats with higher blood pressure. Again, preparation from WKY did not respond to Ca at any concentration. The preparations from SHR showed a tendency to decrease in the developed tension at high concentration of Ca such as 5 mM or 10 mM.

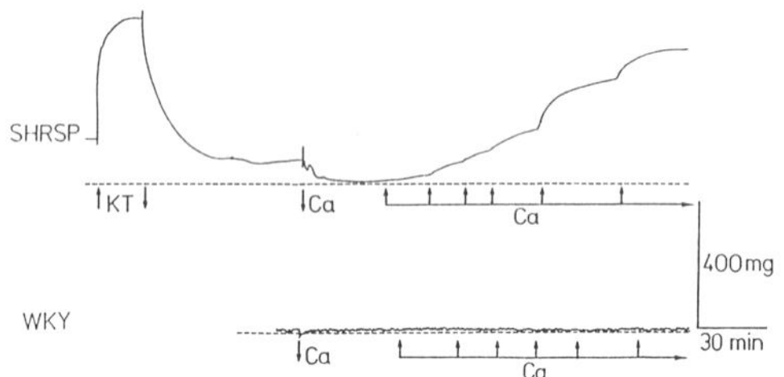

Figure 4. Contraction by readministration of Ca to Ca-removed preparation. Ca concentration was increased to 3×10^{-5} M to 10^{-2} M cumulatively. No contractile response was observed in the preparation from WKY at any concentration of added Ca.

The dependency of the active tone of vascular smooth muscle from spontaneously hypertensive rats on extracellular Ca has been observed in the aortic smooth muscle of SHR (11, 37). The present experiments were performed using rats which showed much higher blood pressure than conventional SHR. Unfortunately, the present experiment was performed with aortic smooth muscle and it can therefore not be concluded that the changes are directly contributed to the elevation of blood pressure. However, it can be expected that similar changes occurs in the smooth muscles of resistant arteries. To support this, an increase in the sensitivity to extracellular Ca (1, 32-34), abnormal sensitivity to Ca-antagonists (19, 33, 38) and increase in Ca-uptake (7, 62) has been presented in the smooth muscle of resistant mesenteric arteries. However, causal involvement of these changes in the initiation of hypertension is still less provable as will be described below.

Mechanisms of the Increase in Vascular Tone

In regard to the causes of the development of smooth muscle tone, following possibilities can be considered : 1) increase in the membrane excitability (depolarization and/or spike generation), 2) abnormality of Ca channels 3) abnormality of excitation-contraction coupling 4) decrease of relaxing activity 5) abnormality of regulatory factors and 6) abnormality of contractile proteins.

Among these possible mechanisms, 1) and 2) lead to an increase in Ca influx. In the experiment described above, it was suggested that Ca influx through voltage-dependent Ca channels is involved in the development of the active tone of the aortic smooth muscle of spontaneously hypertensive rats. However, it can not be simply concluded that the active tone of aortic smooth muscle of spontaneously hypertensive rats is brought about by possibility 1) and/or 2), since the tension development can be caused by possibility 3) to 6) even when Ca leak under nonstimulated condition is not different from that of WKY.

The extent of myosin phosphorylation in the relaxed and contracted state has been reported not differ between preparations from SHRSP and WKY (31). In addition, it has been reported that Ca sensitivity of skinned smooth muscle of tail arteries both from SHR and SHRSP were not also different from that of the preparations from WKY (31,36). A similar result has been presented by the experiment with the aortic smooth muscle of aldosterone hypertensive rats (28).Brayden et al., (4) reported that the actin and myosin contents of smooth muscles of SHR mesenteric and cerebral arteries were not significantly different between preparations from WKY. Sensitivity of contractile proteins to calmodulin has also been reported not to alter in vascular smooth muscles of SHR (2, 36), and of SHRSP (31). Thus, it is suggested that possibilities 5) and 6) are less probable. Although the contraction of smooth muscles through the activation of protein kinase C has been reported to be greater in the preparations from SHRSP (60), we could not observe the difference between hypertensive and control rats when extracellular Ca was also removed by the treatment with lanthanum. In any case, it can not be expected that the difference in this mechanism is involved in the initiation of active tone.

Results of experiments with Ca-indicators have revealed that intracellular Ca concentration is increased in the aortic smooth muscle of spontaneously hypertensive rats (46, 51, 52). The increase in intracellular Ca concentration can be brought about by possibilities 1) to 4). In regard to possibility 4), we have observed that both caffeine- and noradrenalineinduced Ca-releases from sarcoplasmic reticulum were greater in vascular smooth muscle from SHRSP (30, Moriyama et al, 1991 submitted; Sunano et al., communicated at FASEB meeting 1991). Since it has been revealed that caffeine induces the release of Ca from Ca-sensitive release sites of sarcoplasmic reticulum (18), the difference in the Ca-release may have been involved in the initiation of active tone under unaltered Ca-leak.

Even when Ca-influx or Ca-release is identical, increase in intracellular Ca concentration can be brought about when Ca-sequestration is decreased. in regard to this, we have observed that the rate of relaxation of the SHRSP mesenteric artery decreased (58). In addition, Ca-uptake by microsomal fraction of the SHR aortic smooth muscle (29, 35, 66) and by cell membrane fraction of the SHR mesenteric artery (20-22) have also been reported to be impaired. Thus, the decrease in Ca-sequestration can be involved in the elevation of basal tension (active tone) when leak influx of Ca is present.

The most probable possibility is however, the abnormality of voltagedependent Ca-channels, since the active tone of the aortic smooth muscle is highly dependent on extracellular Ca and blocked by Ca-antagonist as described above. This abnormality may be caused by or conversely lead to abnormalities of membrane excitation such as action potential generation (Ca-spike), as will be discussed below. The increase in the Ca influx through voltage dependent Ca-channels can be detected also by ^{45}Ca-uptake experiments. According to the observations by Cauvin et al. (7, 62), not only ^{45}Ca uptake stimulated by high K^+ or by noradrenaline but also the uptake under nonstimulated condition was increased in the mesenteric artery of SHR. However, it is still uncertain whether the Ca-influx under non-stimulated condition is brought about by influx through Ca antagonist-sensitive channels or through voltage-insensitive leak channels.

If membrane excitability is increased, namely when depolarization of membrane or spontaneous action potentials are present, the opening of the voltage-dependent Ca channels would be present. The possibility of membrane depolarization can however, be excluded, since most published papers agree that the resting membrane potentials of vascular smooth muscles in vitro are not significantly different between preparations from normotensive and hypertensive rats (10, 14-17, 23, 47, 49, 50). However, it has been known that the vascular smooth muscles of SHR easily generate action potentials, while these of WKY are quiescent under both stimulated and non-stimulated conditions (14, 23, 24). We have observed in the present experiment that the preparations from spontaneously hypertensive rats often showed spontaneous tension oscillations or twitch-like contractions. This result may indicate the occurrence of nonsynchronous spontaneous electrical activities in aortic smooth muscles from spontaneously hypertensive rats. Then, the decrease in the amplitude of the active tone observed at higher concentration of Ca can be explained by a stabilizing action of Ca (5).

Thus, it seems to be reasonable to assume that the opening of voltagedependent Ca channels are enhanced in the vascular smooth muscles of spontaneously hypertensive rats and the increase is greater as the blood pressure of rats increases. To support this, Rush and Hermsmeyer (43) and Hermsmeyer et al. (Communication at 3rd international symposium on calcium and gene hypertension, Portland, Oregon, 1991) have observed alterations of voltage dependent Ca currents in the smooth muscle of SHR, although the currents are not due to the spontaneous opening of the channels but initiated by the depolarization of membrane.

EFFECTS OF ANTIHYPERTENSIVE TREATMENTS ON THE SMOOTH MUSCLE TONE

As described above, the results of the present experiments demonstrated a close relationship between blood pressure of the rats and the amplitude of the active tone. This may indicate the close relationship between blood pressure and the opening of voltage-dependent Ca channels. However, it is uncertain whether the changes are genetically determined one or secondary changes to maintained hypertension. This problem can be solved by protecting the elevation of blood pressure of the rats (early treatment) and by maintaining the blood pressure to a level close to that of normotensive control rats (WKY).

In the present experiment, the blood pressure of SHRSP was kept at a similar level to that of WKY by treating the rats from the age of weaning with hydralazine or captopril. Figure 5 shows the effects of treatment of SHRSP with captopril. Although the blood pressure of SHRSP at the age of weaning (5 weeks) was already significantly higher than that of WKY, the difference was not marked when compared with that between these rats at an older age. The treatment was continued to the age when the hypertension was established (16 weeks), and the doses of the drug were controlled to maintain the blood pressure at a similar level to that of WKY.

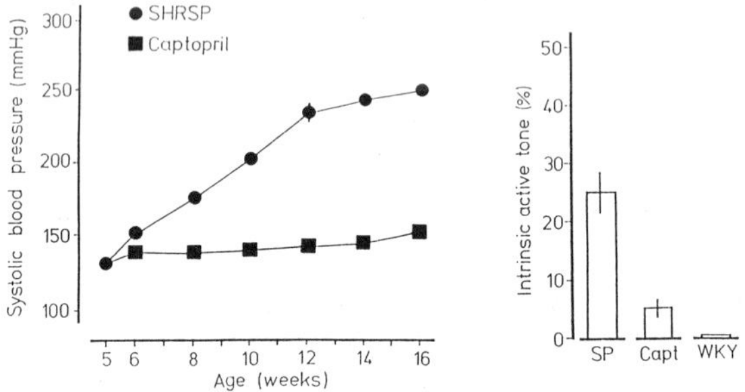

Figure 5. Prevention of the elevation of blood pressure by captopril in SHRSP and its effects on the active tone. **Left:** the treatment was started before the blood pressure of the rats was elevated (early treatment) and the dose of the drug was controlled to maintain the blood pressure at similar level to that of WKY. **Right:** The intrinsic active tone expressed as percentages of the contraction by high-K was markedly depressed in the preparation from captopril-treated rats.

The development of the active tone of the aortic smooth muscle of SHRSP was prevented by antihypertensive treatment and no relaxing response to the removal of extracellular Ca was observed in the preparation from treated rats (Fig. 5). Accordingly, no tension development was observed when Ca was reintroduced to Ca-removed incubation medium (Ca-free Tyrode's solution). The results indicate that the active tone observed in the aortic smooth muscle of SHRSP is a secondary change due to maintained hypertension, and can be prevented by the prevention of the development of hypertension. We have also observed that the active tone of the aortic smooth muscle was not observed in the preparation from young SHRSP This indicates that period exposed to hypertension is also a determinant factor of the development of the active tone in aortic smooth muscle of spontaneously hypertensive rats.

Can changes of vascular smooth muscles of hypertensive rats be restored by lowering blood pressure after the hypertension is established? This was examined by starting the treatment of SHRSP from the age (16 weeks) at which hypertension was established (late treatment). Similarly to the early treatment experiments, the dose of hydralazine was controlled to lower the blood pressure of SHRSP to the level close to that of WKY, although the SHRSP with established hypertension were relatively resistant to the treatment. The treatment was continued from 10 weeks as in the early treatment experiment.

The preparations from SHRSP which had received the late treatment also showed relaxing responses to the removal of extracellular Ca or to the application of Ca-antagonist indicating the existence of the active tone. The amplitude of the active tone however, was significantly lower than that observed in age matched WKY (26 weeks old). Thus, it was indicated that the changes in the aortic smooth muscle can be restored by lowering the blood pressure also after the hypertension is established. These results strongly suggest that the development of the active tone of aortic smooth muscle in spontaneously hypertensive rats is the secondary change due to maintained hypertension but is not a genetically determined one. However, details of the causal relationship between blood pressure and the increase in voltagedependent Ca-channel opening remains to be solved.

Similar results have been reported with SHR treated by angiotensinconverting enzyme inhibitors (44, 45). We have also observed that development of the active tone in aortic smooth muscle of SHRSP could be prevented by treating rats with captopril or SQ29852 (Sunano et al., to be published). In these experiments, it was demonstrated that the protecting action of the treatment with angiotensin-converting enzyme inhibitor was significantly more effective in preventing the development of the active tone than that with hydralazine. Sada et al., (44, 45) have observed similar results with conventional SHR. The results indicate the involvement of angiotensin in the alteration of Ca-channels in vascular smooth muscles of spontaneously hypertensive rats. At present however, molecular mechanism of the changes in smooth muscle membrane, hence molecular mechanisms of the effect of angiotensin, have not been studied.

However, it has been known that experiments described above were performed with endothelium-removed preparations. Endothelium controls vascular smooth muscle activity through the release of endothelium-derived relaxing factor (EDRF) and contracting factor (EDCF) (6, 12, 13, 64, 65). In addition, abnormalities of endothelium in blood vessels of spontaneously hypertensive rats have also been reported (26, 27, 42, 48, 54). Here, the influence of endothelium on the active tone of the aortic smooth muscle is described briefly.

The active tone of the aortic smooth muscle of all strains of spontaneously hypertensive rats was depressed in the presence of endothelium (Fig.6). As a result, no active tone was detected in the endothelium-intact preparation from SHR. Although the active tone of aortic smooth muscles of SHRSP and MSHRSP was also markedly depressed in the presence of endothelium, a small residual active tone could still be detected. Similar results were obtained in the development of the active tone induced by reintroduction of Ca to Caremoved solution. No tension development was observed in the endotheliumintact preparation from SHR at any concentration of added Ca. Again, only a small contraction was initiated in the preparation from SHRSP and M-SHRSP at a higher concentration of added Ca (1 to 10 mM). The inhibition by endothelium is thought to be due to a spontaneous release of EDRF.

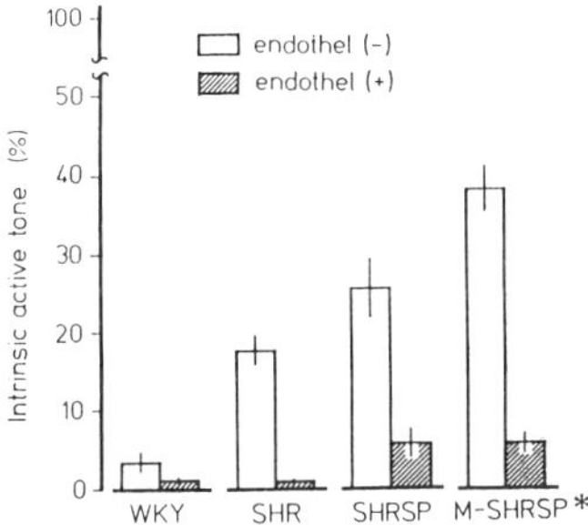

Figure 6. Depression of the active tone by endothelium. The active tone was expressed as percentages of the amplitude of high-K-induced contraction. endothel(-) and endothel (+) indicate the endothelium-removed and endothelium-intact preparations, respectively. It should be mentioned that the active tone in WKY aorta is due to the active tone of a few preparations which showed the active tone and majority number of the preparations did not show the active tone.

We have observed that the endothelium-dependent relaxations of aortae are impaired in the preparation from all strains of spontaneously hypertensive rats (54). The impairment of the relaxation was more prominent as the blood pressure of the rats increased. Therefore, it is possible to assume that the development of the active tone can not be abolished in the preparations from SHRSP and M-SHRSP because of the greater activity of smooth muscles and impaired activity of endothelium. It is uncertain how endothelium-derived factor depress the voltage-dependent Ca channels. Hyperpolarization of the smooth muscle membrane by EDRF (*NO*) (59) or another hyperpolarizing factor (EDHF) (8, 9) may be a cause of the depression.

CONCLUSION

Aortic smooth muscles from spontaneously hypertensive rats exhibit spontaneously developed active tone while those from control normotensive rats exhibit no such active tone. It is suggested in the present study that an increase in spontaneous opening of voltage-dependent Ca channels is the cause of the active tone. The active tone is greater in the preparation from rats with higher blood pressure, becoming greater with aging and is prevented by antihypertensive treatments. Thus it can be concluded that the changes in the Ca channels are secondary due to maintained hypertension and the degree of hypertension and the duration exposed to hypertension are the determinant factors of the degree of the active tone. Endothelium depresses the active tone releasing EDRF or EDHF; the activity of endothelium is impaired in the preparation from spontaneously hypertensive rats. Although these abnormalities are thought to be brought about by maintained hypertension, the changes would cause further elevation of blood pressure in a recurring manner. Therefore, early anti-hypertensive treatment is required to protect these changes in the vascular smooth muscle of the spontaneously hypertensive rats.

REFERENCES

1.   Aalkjaer, C. and M.J. Mulvany, in Calcium in Essential Hypertension, K. Aoki and E.D. Frohlich, eds., (Academic press, Tokyo 1989) pp.409-431.

2.   Asano, M., K. Masuzawa, M. Kojima, K. Aoki, and T. Matsuda, Jpn. J. Pharmacol. 48, 77-90 (1988).

3.   Bolton, T.B., Physiol. Rev. 59, 606-718 (1979).

4.   Brayden, J.E., W. Halpern, and L.R. Brann, Hypertension 5, 17-25 (1983).

5.   Bruner, C.A., R.C. Webb, and D.F. Bohr, in Calcium in Essential Hypertension, K. Aoki and E.D. Frohlich eds. (Academic press, Tokyo, 1989) pp.275-306.

6.   Busse, R., G. Trogisch, and E. Bassenge, Basic Res. Cardiol. 80, 475-490 (1985).

7.   Cauvin, C., in 6th International Adalat Symposium, P.R. Lichten ed. (Excepta Medica, Amsterdam, 1986) pp.225-231.

8.   Chen, G. and H. Suzuki, J. Physiology 410, 91-106 (1989).

9.   Chen, G., H. Suzuki, and A.H. Weston, Br. J. Pharmacol. 95, 1165-1174 (1988).

10.   Cheung, D.W., Can. J. Physiol. Pharmacol. 62, 957-960 (1984).

11.   Fitzpatrick, D.F. and A. Szentivanyi, Clin. Exp. Hypertension 2, 1023-1037 (1980).

12.   Furchgott, R.F. Ann. Rev. Pharmacol. Toxicol. 24, 175-197 (1984).

13.   Furchgott, R.F., Circulation Res., 53, 557-573 (1983).

14.   Harder, D.R., L. Brann, and W. Halpern, Blood Vessels 20, 154-160 (1983).

15.   Hermsmeyer, K. and D. Harder, Am. J. Physiol., 250, C1557-C562 (1986).

16.   Hermsmeyer, K. and S. Walton, Circ. Res. 40, Suppl.I, I-153-I-156 (1977).

17.   Hermsmeyer, K., Circ. Res., 38, 362-367 (1976).

18.   Iino, M., Biochem. Biophys. Res. Commun. 142, 47-52 (1987).

19.   Kannan, M.S., A.E. Seip, and D.J. Crankshaw, Can. J. Pharmacol. 64, 310-314 (1986).

20.   Kwan, C.Y., Can. J. Physiol. Pharmacol. 63, 366-374 (1985).

21.   Kwan, C.Y., L. Belbeck, and E.E. Daniel, Blood Vessels 16, 259-268 (1979).

22.   Kwan, C.Y., L. Belbeck, and E.E. Daniel, Mol. Pharmacol. 17, 137-140 (1980).

23.   Lamb, F.S. and R.C. Webb, Hypertension 13, 70-76 (1989).

24.   Lamb, F.S. and R.C. Webb, J.Hypertension 7, 457-463 (1989).

25.   Lindner, V. and H. Heinle, Europ. J. Pharmacol. 138, 147-149 (1987).

26.   Luscher, T.F. and P.M. Vanhoutte, in Relaxing and Contracting Factors, P.M. Vanhoutte, ed. (The Humana Press Inc., Clifton, 1988) pp.495-509.

27.   Martin, W., in Relaxing and Contracting Factors, P.M. Vanhoutte ed. (The Humana Press Inc., Clifton 1988) pp.159-177.

28.   McMahon, E.G. and R.J. Paul, Circ. Res. 56, 427-435, (1985).

29.   Moore, L., L. Hurwitz, G.R. Davenport, and E.J. Landon, Biochim. Biophys. Acta. 413, 432-443 (1975).

30.   Moriyama, K., S. Osugi, K. Shimamura, and S. Sunano, Blood Vessels 26, 280-289 (1989).

31.   Mrwa, U., K. Guth, C. Haist, M. Troschka, R. Herrmann, R. Wojciechowski, and M. Gagelmann, Life Sciences 38, 191-196 (1986).

32.   Mulvany, M.J. and N. Nyborg, Br. J. Pharmacol. 71, 585-596 (1980).

33.   Mulvany, M.J., in Essential Hypertension, Calcium Mechanism and Treatment (Springer Verlag, Tokyo 1986), pp.51-65.

34.   Mulvany, M.J., N. Korsgaard, H. Nilsson, and N. Nyborg, in Hypertension Mechanisms, W. Rascher, D. Clough, and D. Ganten eds. (Schattauer, Stuttgart 1982) pp.234-238.

35.   Mushlin, P.S., B.V.R. Sastry, R.C. Boerth, M.J. Surber, and E.J. Landon, J. Pharmacol. Exp. Therap. 207, 331-339 (1978).

36.   Nghiem, C.X. and J.P. Rapp, Clin. Exp. Hypertension A5, 849-856 (1983).

37.   Noon, J.P., P.J. Rice, and R.J. Baldessarini, Proc. Natl. Acad. Sci. 75, 1605-1607 (1978).

38.   Nyborg,N.C.B., J. Byg-Hansen, and M.J. Mulvany, J. Cardiovasc. Pharmacol. 7 Suppl 6, S43-S46 (1985).

39.   Okamoto, K. and K. Aoki, Jpn. Circ. J. 27, 282-293, 1963.

40.   Okamoto, K., K. Yamamoto, N. Morita, Y. Ohta, T. Chikugo, T. Higashizawa, and T. Suzuki, J. Hypertension 4 Suppl. 3, S21-S24 (1986).

41.   Okamoto, K., Y. Yamori, and A. Nagaoka, Circ. Res. 34, 35, suppl. 143-153 (1974).

42.   Osugi, S., K. Shimamura, and S. Sunano, Archives Int. Pharmacodyn. Therap. 305, 86-99 (1990).

43.   Rusch, N.J. and K. Hermsmeyer, Circ. Res. 63, 997-1002 (1988).

44.   Sada, T., H. Koike, and M. Miyamoto, Hypertension 14, 652-659 (1989).

45.   Sada, T., H. Koike, H. Nishino, and K. Oizumi, Hypertension 13, 582- 588 (1989).

46.   Sada, T., H. Koike, M. Ikeda, K. Sato, H. Ozaki, and H. Karaki, Hypertension 16, 245-251 (1990).

47.   Shimamura, K., N. Kurozumi, K. Yamamoto, and S. Sunano, Pfluger'S Arch. 414, 37-43, (1989).

48.   Shimamura, K., S. Osugi, K. Moriyama, and S. Sunano, J. Cardiovasc. Pharmacol. 17 suppl.3, S133-Sl36 (1991).

49.   Stekiel, W.J., in Blood Vessel Changes in Hypertension: Structure and Function, R.M.K.W. Lee ed. (CRC Press Inc., Boca Raton 1988) pp. 127-170.

50.   Stekiel, W.J., S.J. Contney, and J.H. Lombard, Am. J. Physiol- 250, C547 -C556 (1986).

51.   Sugiyama, T., M. Yoshizumi, F. Takaku, and Y. Yazaki, J. Hypertension 8, 369-375 (1990).

52.   Sugiyama,T., M. Yoshizumi, F. Takaku., H. Urabe, M. Tsukakoshi, T. Kasuya, and Y. Yazaki, Biochem. Biophys. Res. Commun. 141, 340-345 (1986).

53.   Sunano, S., S. Osugi, and K. Shimamura, Blood Vessels 27, 59-60 (1991).

54.   Sunano, S., S. Osugi, and K. Shimamura, Experientia, 45, 705-708 (1989)

55.   Sunano, S., S. Osugi, K. Yamamoto, and K. Shimamura, Arch. Int. Pharmacodyn. 305, 284 (1990).

56.   Sunano, S., S. Osugi, K. Yamamoto, and K. Shimamura, J. Cardiovasc. Pharmacol. 17 Suppl.3, S137-Sl4O (1991).

57.   Sunano, S., S. Osugi, K. Yamamoto, K. Moriyama, and K. Shimamura, Europ. J. Pharmacol. 183, 1802-1803 (1990).

58.   Sunano, S., T. Shimada, K. Moriyama, and K. Shimamura, Clin. Exp. Pharmacol. Physiol. 17, 413-425 (1990).

59.   Tare, M., H.C. Parkington, H.A. Coleman, T.O. Neild, and G.J. Dusting, J. Cardiovasc. Pharmacol. 17 Suppl.3, S108-Sll2 (1991).

60.   Turla, M. B. and R.C. Webb, Hypertension 9, Suppl III, III-150-III-154 (1987).

61.   Van Breemen, C. and K. Saida, Ann. Rev. Physiol. 51. 315-329 (1989).

62.   Van Breemen, C., C. Cauvin, A. Johns, P. Leijten, and H. Yamamoto, Fed. Proc. 45, 2746-2751 (1986).

63.   Van Breemen, C., P. Leijten, H. Yamamoto, P. Anderson, and C. Cauvin, Hypertension 8 suppl-II, II-89-II-95 (1986).

64.   Vanhoutte, P.M., G.M. Ruvanyi, V.M. Miller, and D.S. Houston, Ann. Rev. Physiol. 48, 307-320 (1986).

65.   Vanhoutte, P.M., J. Cardiovasc. Pharmacol., 12 Suppl.6, S21-S28 (1988).

66.   Webb, R.C. and R.C. Bhalla, J. Mol. Cell. Cardiol. 651-661 (1976).

67.   Winquist, R.J. and D.F. Bohr, Hypertension 5, 292-297 (1983).

SIGNAL TRANSDUCTION RESPONSES TO PRESSURE IN ARTERIAL MUSCLE: A POSSIBLE ROLE FOR ARACHIDONIC ACID METABOLITES.

DAVID R. HARDER, DEBEBE GEBREMEDHIN, YUNN-HWA MA, and RICHARD S. ROMAN

Medical College of Wisconsin, Department of Physiology, 8701 Watertown Plank Road, Milwaukee, WI 53226

Depending on the vascular bed being studied, arteries, either <u>in vivo</u> or <u>in vitro</u>, respond actively to elevation of transmural pressure (1, 5, 6, 11, 15). This myogenic response is thought to contribute to a basal level of vascular tone which is adjusted by metabolic and neurogenic influences. There has recently been a good deal of investigation aimed at understanding the mechanisms of pressure- induced tone. Some controversy has arisen regarding the role of the vascular endothelium in mediating the myogenic response to pressure, with earlier reports (7, 19) suggesting a requirement for an intact endothelium, and later reports demonstrating myogenic tone even in the absence of an endothelium (4, 14). The most plausible explanation for this discrepancy at present is that certain methods of endothelial removal, particularly those using enzymes, detergents and perhaps extended perfusion with air interfere with the normal mechanisms by which vascular muscle responds to a pressure load with an active response. Mild mechanical methods for removal of the endothelium do not eliminate pressure-induced myogenic tone in those vessels where harsher methods used to remove the endothelium do eliminate the active response to pressure (4, 8).

Elevation of transmural pressure in isolated arterial segments results in muscle cell depolarization that is often accompanied by Ca^{2+}-dependent action potentials (7, 20). While a pressure-induced elevation of active wall tension has been shown to be dependent on extracellular Ca^{2+}, we do not know if an increase in Ca^{2+} influx is a primary cause for generation of the myogenic response or if such influx is a result of the opening of potential-dependent Ca^{2+} channels resulting, in turn, from a depolarization mediated by other VSM membrane changes (e.g., such as a reduction in K^+ conductance). "Mechano-sensitive" ion channels have been described in a number of cell types where they usually have been recorded as single channel events in patch-clamp experiments (12, 17). However, the channels described thus far are non-selective, being broadly specific only to cations or anions (17). Mechanosensitive channels specific for Ca^{2+} have yet to be identified. Furthermore, a recent report by Morris and Horn (16) suggest that "single channel mechanosensitivity is an artifact of patch recording" in certain neuronal cells. Thus, while there is an electrophysiological mechanism involved in pressure-mediated myogenic activation of arterial muscle, only controversial evidence exists presently in support of the hypothesis for direct mechanical activation of ion channels with VSM membrane.

Several reports have demonstrated that inhibition of protein kinase C(PKC) can inhibit pressure-induced activation of arteries, and that phorbol esters augment myogenic tone (10, 13). PKC is activated by 1,2-diacylglycerol (1,2 DAG) formed via metabolism of phosphoinositide lipids (2). Activation of phospholipase C(PLC) through a membrane receptor G protein complex, or possibly mechanical deformation, cleaves phosphatidylinositol 4,5 biphosphate (Ptd-Ins 4,5 P_2) into the hydrophobic 1,2 DAG, and water-soluble inositol 1,4,5-triphosphate (IP_3). IP_3 elevates intracellular Ca^{2+} by its action on intracellular stores. 1,2 DAG serves as a second messenger to activate PKC, which has many cellular functions including phosphorylation of membrane ion channels (3, 9). At present, it is not known how pressure activates PKC; however, we have preliminary data demonstrating that IP_3 is elevated in dog renal arteries as transmural pressure is increased, suggesting that pressure stimulates PLC activity in the arterial wall (unpublished observation).

Recent studies from our laboratory have demonstrated that inhibition of cytochrome P-450 enzymes markedly attentuates pressure-induced myogenic tone, and that increasing substrate availability by supplying exogenous arachidonic acid, enhances the myogenic response. (Fig. 1).

Published 1991 by Elsevier Science Publishing Company, Inc.
Ion Channels of Vascular Smooth Muscle Cells and Endothelial Cells
Sperelakis and Kuriyama, Editors

236

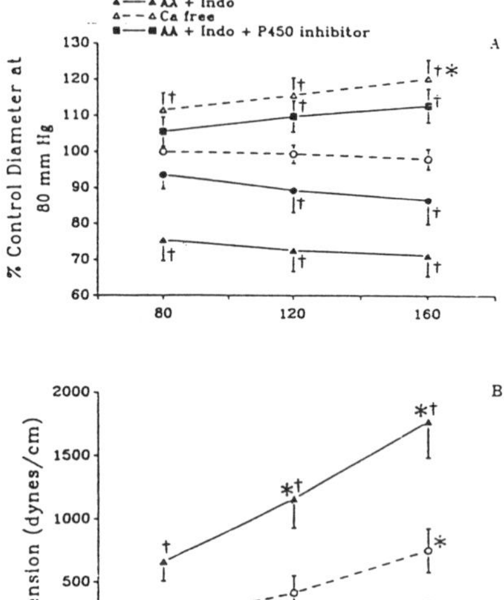

Figure 1. Effect of arachidonic acid (AA, 50 μM) in the presence and absence of indomethacin (10 μM), and indomethacin (10 μM) plus a P450 inhibitor ketoconazole (100 μM; n = 3) or SKF 525A (100 μM; n = 4) on the pressure-diameter (panel A) and pressure-active tension (panel B) relations of dog renal arcuate arteries at different levels of transmural pressure. Mean $\pm$ 1 SEM from seven vessels is presented. The response to both cytochrome P-450 inhibitors, ketoconazole and SKF 525A, was similar; therefore, the data from these two groups were combined and presented together. The control diameters of the vessels averaged 305 $\pm$ 25 μm at a pressure of 80 mm Hg. *Significant difference from the value measured at 80 mm Hg within a group. Significant difference from the control value measured at the same transmural pressure. Significant difference from the values obtained before administration of ketoconazole or SKF 525A in vessels pretreated with arachidonic acid and indomethacin.

We have also identified a cytochrome P-450 arachidonic acid (AA) metabolite from renal artery microsomal fractions which coelutes with 20-hydroxyeicosatetraenoic acid (20-HETE). (Fig. 2).

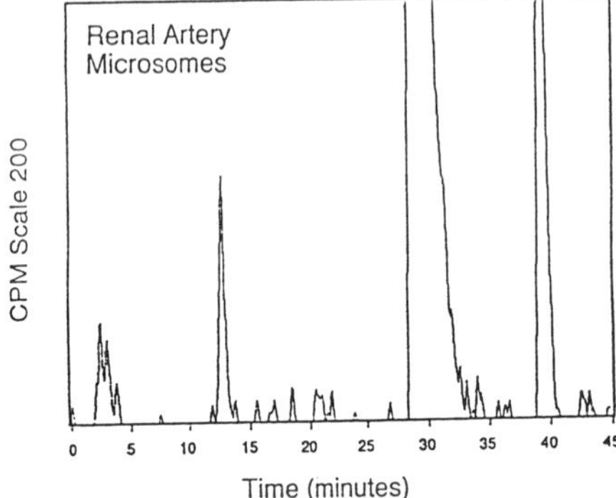

Figure 2. Metabolism of arachidonic acid by microsomes prepared from small renal arteries (200-450 μm). Microsomes (1 mg/ml) produced three products when incubated with [1-^{14}C]arachidonic acid (retention time, 30 minutes). The product with retention time of 12.5 minutes coelutes with 20-hydroxyeicosatetraenoic acid. Products were separated using a x 2.1mm x 25cm C18 reversed-phase high-performance liquid chromatography column and a linear elution gradient from acetonitrile:water:acetic acid (50/50/0.1) to acetonitrile:acetic acid (100/0.1) over 40 minutes.

There are two primary sources for the release of AA in arterial muscle, namely, from phosphatidylcholine by the action of phospholipase A_2 on phosphatidylcholine and from 1,2 DAG through the action of DAG lipase (18). Given our findings and those of others that PLC activity is enhanced upon pressure induced activation of arteries, it seems likely that the source of AA for production of 20-HETE is from 1,2 DAG.

When 20-HETE is applied to isolated dog renal arterial muscle cells loaded with the fluorescent Ca^{2+} probe FURA-2, we typically observe a large increase in signal indicating a large rise in $[Ca]_i$. (Fig. 3).

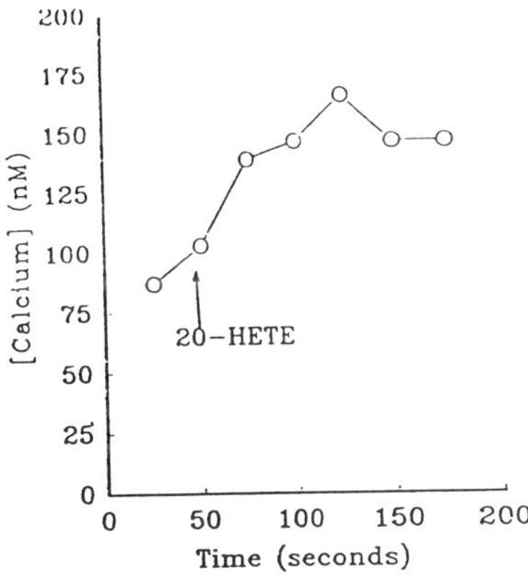

Figure 3. Effect of 20-HETE (1 μM) on intracellular calcium concentration in a suspension of vascular smooth muscle cells freshly isolated from renal arcuate arteries (200-300 μM) of a dog. Vascular smooth muscle cells were isolated by incubating several vessels for 30 minutes at 37° C in collagenase solution (0.3%) containing 1 mg/ml elastase and trypsin inhibitor. Cells were loaded by incubation with 2 μM Fura2-AM for 15 minutes and washed twice. Fluorescence intensifies at a wavelength 510 nm was measured in a 2 ml curette at 37° C using a Hitachi model F-2000 scanning fluorometer. Excitation was alternated between 340 and 380 nm at one second intervals.

In addition, we have a good deal of data demonstrating that isolated 20-HETE contracts and depolarizes renal arterial muscle by an action which is to inhibit K^+ channel activity. These effects of 20-HETE are most likely not due to non-specific actions of fatty acids in that the effect of AA on isolated renal arterial membrane patches is to markedly increase K^+ channel activity. (Fig. 4).

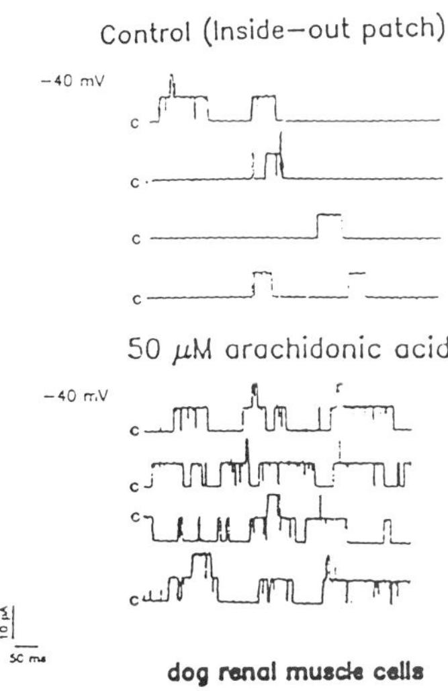

Figure 4. Arterial records of patch clamp recordings from excised, inside-out membrane patches of dog renal arterial muscle cells demonstrating the action of arachidonic acid is markedly increasing mean channel open time.

Thus, 20-HETE fulfills the requirements as a potential mediator of pressure-induced myogenic tone, i.e.: 1) inhibition of its formation blocks the myogenic response to pressure; 2) Increasing the concentration of the substrate (AA) for 20-HETE enhances the response to pressure, and 3) the product itself has an action on isolated vessels mimicking the response to pressure. An overall scenario encompassing many of the ideas described in this review is depicted in Fig. 5.

238

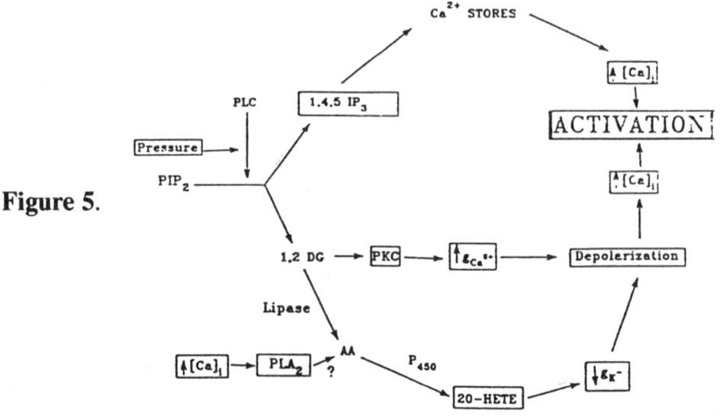

Figure 5.

REFERENCES

1. Bayliss, W.M., J. Physiol. (Lond.) $\underline{28}$, 220-231 (1902).
2. Berridge, M.J., Ann. Rev. Biochem. $\underline{56}$, 159-193 (1987).
3. DeRiemer, S.A., J.A. Strong, K.A. Alpert, D. Greengard, and L.K. Kaczmarek, Nature $\underline{313}$, 313-316 (1985).
4. Falcone, J.C., M.J. Davis, and G.A. Meininger, Am. J. Physiol. $\underline{260}$, H130-H135 (1991).
5. Folkow, B., Acta Physiol. Scand. $\underline{17}$, 289-310 (1949).
6. Harder, D.R., Circ. Res. $\underline{55}$, 197-202 (1984).
7. Harder, D.R., Circ. Res. $\underline{60}$, 102-107 (1987).
8. Harder, D.R., R.J. Roman, and J.H. Lombard in Proceedings of 3rd International Meeting on Resistance Arteries (Denmark May 1991) (In Press) .
9. Heagerty, A.M. and J.D. Ollerenshaw, J. Hypertension $\underline{5}$, 515-524 (1987).
10. Hill, M.A., J.C. Falcone, and G.A. Meininger, Am. J. Physiol. $\underline{259}$, H1586-H1594 (1990).
11. Johnson, P.C., J. Hypertension $\underline{7}$(Suppl.4), 533-540 (1989).
12. Kirber, M.T., J.V. Walsh, and J.J. Singer Pflugers Archiv. $\underline{412}$, 339-343 (1988).
13. Laher, I. and J.A. Bevan, Biochem. Biophys. Res. Commun. $\underline{158}$, 58-62 (1989).
14. McCarron, J.G., G. Osoi, and W. Halpern, Blood Vessels $\underline{26}$, 315-319 (1989).
15. Meininger, G.A., C.A. Mack, K.L. Fehr, and H.G. Bohlen, Circ. Res. $\underline{60}$, 861-870 (1987).
16. Morris, C.E. and R. Horn, Science $\underline{251}$, 1246-1249 (1991).
17. Morris, C.E., J. Membrane Biol. $\underline{113}$, 93-107 (1990).
18. Ohanian, J., J. Ollerenshaw, P. Collins, and A. Heagerty, J. Biol. Chem. $\underline{265}$, 8921-8928 (1990).
19. Osol, G., R. Osol, and W. Halpern in Resistance Arteries W. Halpern et al, eds., (Perinatology Press Ithaca, NY 1988) pp. 162-169.
20. Smeda, J.S. and E.E. Daniel, Circ. Res. $\underline{62}$, 1104-1110 (1988).

METHODS FOR SIMULTANEOUS VOLTAGE-CLAMP, MICROFLUOROMETRY, AND VIDEO OF CELLS. I. ELECTRONIC AND OPTICAL INSTRUMENTATION

M. STUREK, W.M. CALDWELL, D.A. HUMPHREY, and C. WAGNER-MANN

Department of Physiology, School of Medicine, and Dalton Research Center, University of Missouri, Columbia, Missouri 65211

Abstract

Instrumentation is described for measurement of ion concentrations with fluorescent indicators, which require rapidly alternating excitation with two wavelengths of light. Specific needs for measurement of intracellular free Ca (Ca$_i$) using fura-2 are considered primarily. Excitation light from a Xe arc lamp is passed to interference filters mounted in a light-tight enclosure and then to a microscope by liquid light guides secured in sleeves on the enclosure. The alignment of this optical configuration is simple, reliable, and permits isolation of light, electrical noise, and mechanical vibration from the microscope, thus permitting manipulation of the cell with microelectrodes for simultaneous electrophysiology. The use of transillumination with 600 nm light, 570 nm cut-on dichroic mirror, 510 nm wide-band bandpass filter, and video projection attachments on the fluorescence emission port of the microscope permit the simultaneous observation of the cell with no increase in background fluorescence. The timing signals necessary for synchronizing the measurement of fura-2 fluorescence emission at 510 nm with the appropriate excitation wavelengths originate from light reflecting and absorbing surfaces on a washer mounted concentrically on the interference filter wheel. Rotation periods of the filter wheel of 200, 100, 50, or 20 ms are precision controlled to within 1% of the selected speed. Separate signals resulting from 340 (or 360) and 380 nm excitation are derived from the composite fluorescence emission by averaging sample-and-hold circuits. The timing (position) signals sequentially switch sample-and-hold amplifiers such that the circuits sample composite photomultiplier output and then filter the separated signals by a variable time constant, depending on the rotation period of the interference filter wheel. Background fluorescence may be subtracted on-line from the separated fluorescence signals due to 340 (or 360) and 380 nm excitation and then analog ratio output provided. The signals can be fed into separate channels of any personal computer analog-to-digital converter to permit data acquisition and analysis with commonly used software; several examples are illustrated. The optical signal processing quantitation is valid, as shown by the linear relationship between output voltage and % optical transmission through neutral density filters. Measurement of fluorescence from microspheres indicates uniform sensitivity across the field of view in the video monitor. The uniformity should permit measurements from single cells or groups of ~6-10 cells to estimate population averaged Ca$_i$ responses. Fluorescence measurements are extremely reliable even when the alignment of the liquid light guides, etc. are manipulated in numerous ways. The errors introduced by incorrect background subtraction are determined. The innovations described enable determination of the temporal relationship between membrane electrical events, changes in Ca$_i$, and cell morphology (Refs. (41-43), this volume). The system is described in sufficient detail for investigators to replicate it and should be useful with fluorescent dyes for Ca, pH, Na, and Mg that require dual wavelength excitation.

INTRODUCTION AND BRIEF REVIEW OF THE LITERATURE

Numerous studies have shown that indicators of intracellular free ion concentrations employing ratio measurements of fluorescence have distinct advantages (e.g. 5, 11, 13, 14, 23-25, 53, 56) over fluorescent probes requiring only one excitation or emission wavelength (19). Dual wavelength measures largely cancel out variations in dye loading, optical path length (cell thickness), changes in absolute illumination intensity, changes in detector sensitivity, movement artifacts associated with light scattering, and bleaching of the dye (45,

Published 1991 by Elsevier Science Publishing Company, Inc.
Ion Channels of Vascular Smooth Muscle Cells and Endothelial Cells
Sperelakis and Kuriyama, Editors

47). The basics of instrumentation required for use of these dyes on single cells has been described in several reports (e.g. 14, 23, 47, 48) and an issue of *Cell Calcium* (February/March 1990).

Dual Excitation or Dual Emission Wavelength Microfluorometry?

Measurement of intracellular free Ca concentration (Ca_i) typifies the property of dual wavelength fluorescence measurements. The emission spectrum of indo-1 shifts and the excitation spectrum of fura-2 shifts upon binding of Ca (16). The main advantage of using indo- 1 is the relatively straightforward placement of a beam splitter/dichroic mirror and two photodetectors on the emission port of the microscope (28, 36, 44). The use of fura-2, in contrast, requires dual wavelength excitation and the resulting technical difficulty of synchronizing rapid (sometimes $\geq$ 50 Hz) switching of the excitation wavelength with the measurement of the appropriate fluorescence signal (5, 23, 59).

Although dual emission microfluorometry systems may be easier to construct, several technical reasons prompted us to design a dual wavelength excitation over dual wavelength emission system: 1) At the present time several fluorescent indicators of ion concentrations, including Ca, Na, K, H, and Mg, require and/or are well-suited to dual wavelength excitation (20, 24, 34). This property indicates a potentially more widespread need for instrumentation, thus making our system more versatile. 2) Because at the outset we were primarily interested in measurement of Ca_i, fura-2 was the preferred indicator. One potential disadvantage is that indo-1 bleaches more than fura-2; it is desirable to minimize bleaching because errors in estimation of Ca_i concentration may occur (4). Also, significant problems may exist in some preparations with autofluorescence of endogenous metabolic intermediates that overlaps the indo-1 emission spectra (33). Finally, most studies on Ca_i measurement in smooth muscle cells employ fura-2, rather than indo-1 (35). 3) Fura-2 allow an easier transition from the photometer recording described here to digital imaging, because fura-2 has no shift in the emission spectrum when Ca binds it. Fura-2, thus, requires only one camera and digital video acquisition system (23, 47, 48). Several reports indicate the usefulness of digital imaging of Ca_i with fura-2 (12, 14, 15, 22, 23, 54, 55, 57).

Overview of Other Instrumentation for Dual Wavelength Excitation Photometry

Chance et al. (10) introduced the application of time-sharing multiple wavelength photometry to the study of intracellular ion concentrations in the 1970's and various aspects of the method have been adapted by ourselves and other investigators (8, 26, 32, 45, 46, 49). Their basic technique utilizes optical monitoring of the position of interference filters located in a rotating wheel to synchronize with the optical signal emitted from the biological specimen (Fig. 1 of Refs. (10, 30)). The system has gained use in multicellular suspensions of cells (34), relatively large single skeletal muscle fibers (3) and multicellular tissue preparations of vascular smooth muscle, i.e. blood vessel segments (17). The sophistication of the sample-and-hold circuitry and mechanical configuration permits use of eight different wavelengths for exciting the indicator and biological specimen (30). In addition, fiber optic illumination make this system ideal for use with large skeletal muscle fibers (3). As a result, if absorbance dyes were suitable for use in small cells, this may be the superior system. However, fluorescent probes, with rare exception (2, 7), are required for small cells and use of the Chance et al. (10) instrumentation with smaller single cells has not been reported. Furthermore, only one fluorescent probe, rather than multiple probes, is likely to be used with more success in a single cell using epifluorescence microscopy because of more stringent requirements of dichroic mirrors, etc. This consideration makes 4 or 8 excitation wavelengths less important. The use of dual wavelength excitation is possible with the Chance et al. (10) system, but it is uncertain whether the vibrations resulting from the spinning air turbine wheel that alternates the excitation wavelengths (3, 10, 30) could be isolated from a microscope for voltage-clamp of a small cell with microelectrodes (5, 26, 32, 45). Simonson et al. (34) have used only single wavelength excitation with the Chance et al. (1) system and it involved the use of monolayer cultures of many small cells (> 10,000) on coverslips, rather than a single cell. Most applications of the system stem from the system made available through the Biomedical Instrumentation shop of the Johnson Research Foundation at the University of Pennsylvania group (3, 17, 30, 34). The U. of Pennsylvania will custom make this system for investigators outside of the University, but the system has not gained widespread use in many other laboratories since its inception in the 1970's.

Thomas (46) has provided a very detailed description of the electronic circuitry and some of the mechanical configuration was described for his time-sharing multiple wavelength photometry system. Technical factors arguing against the successful application of this system for simultaneous voltage-clamp and fluorescence measurements of small single cells are: 1) The optical configuration is not appropriate for use with small cells on a microscope (46). 2) The system is primarily used for light absorbance measurements, rather than fluorescence measurements (similar to the Chance et al. (10) system); therefore, the essential strengths of fluorescent dyes (16, 47) are not available. 3) Another consideration is that Thomas (46) uses a peak detector for obtaining the optical signal during illumination of the biological specimen at a desired wavelength. The peak detector only samples the optical signal for $10\,\mu$s prior to holding the signal. Although Thomas (46) finds that this method provides suitable signal-to-noise ratio for large cells (200 μm diameter), this is not likely to be the case with small cells having correspondingly smaller signals. Using an averaging time constant over most of the period of excitation would probably reduce noise more effectively and, thus, be more suitable for small cells.

Simultaneous fura-2 and voltage-clamp studies have been conducted on small single cells by several groups using the time-sharing mode of dual wavelength microfluorometry (21, 32). These investigators use modifications of commercially available systems. The convenient and flexible feature of the system is that monochromators are used for excitation, thus allowing the investigator to select between two ion-sensitive wavelengths (e.g. 340 and 380 nm for fura-2) or one ion-sensitive wavelength and the isosbestic wavelength of 360 nm for fura-2. In addition, entire excitation spectra can be conducted on a single cell. At the present time several disadvantages exist: 1) the isolation of cells from the vibrations of the excitation switching apparatus requires custom modification; 2) the computer interface is not especially compatible with electrophysiological software (e.g. pCLAMP or AxoBASIC software of Axon Instruments, Inc., Foster City, CA); 3) alignment of mirrors used in the diversion of light toward the microscope can be tedious and requires great precision, and; 4) the cost is almost prohibitive. Callewaert et al. (8) also used an oscillating mirror for switching excitation wavelengths, but do not indicate the degree of difficulty associated with the method. Their custom made electronic configuration demultiplexed the composite optical signal, but no circuit description of the method is available (8). Similarly, a microfluorometry system utilizing the monochromators of a dual beam spectrophotometer has been described (45). Neher (26) described the optical configuration for simultaneous voltage-clamp and fura-2 microfluorometry with reasonable detail, although it would be unlikely that the system could be replicated by a novice. It is especially important to note that this report stated "...a hard-wired instrument based on sample-and-hold circuits was used to track the fluorescence signals at the two excitation wavelengths." However, no details of the electronic configuration of the system were given.

Another main consideration in design of a microfluorometry system is the ability to simultaneously observe a cell during the experiment (49). One such system is that of Peeters et al. (28). The general configuration of optics for simultaneous measurement of Ca_i and motion in heart cells, but no interfacing with electrophysiology was described. A beam splitter permits passage of light to the video camera, while reflecting emitted fluorescence to a photomultiplier tube. A major difference is that their system utilizes the fluorescent indicator, indo-1, which involves dual wavelength emission, rather than dual wavelength excitation. An indo-1 is generally more convenient and simple to construct. The system developed at the University of Massachusetts (5, 23, 59) is closely related to the system that we have developed. Fura-2 has been used for measurement of Ca_i (5, 59) and SBFI for measurement of intracellular Na (24) in relatively small smooth muscle cells using time-sharing dual wavelength excitation. Furthermore, one of the excitation filters we use in our system is identical to theirs. The major difference is that Becker et al. (5) use an assembly language program for synchronization of emitted fluorescence with the appropriate excitation wavelength. Rather than the software approach of Becker et al. (5) our major emphasis has been the use of analog hardware. It should be noted, however, that the choice of computerized data acquisition and analysis system is totally the decision of the user. The analog signals from the optical processor we have designed can be fed into any commercially available analog-to-digital converter.

Purpose

The primary purpose of this chapter is to provide sufficient methodological detail for investigators to replicate this instrumentation. We have restricted this chapter primarily to technical design, theoretical considerations, construction, and optimization of dual wavelength microfluorometry, rather than use of the techniques on living cells. A separate chapter deals with the physiology and practical use of fura-2 primarily in smooth muscle and endothelial cells (43). The rationale for this detailed description is that all subsequent uses of fluorescent indicators in physiological experiments are absolutely dependent on the proven performance of this instrumentation. More specific purposes include:

1) Design accurate, valid, and reliable instrumentation that is also simple. Because scientists with widely different specialties and varying levels of expertise with optical and electronic instrumentation require such measurements for answering questions of physiological interest, this reinforces the need for a "turnkey" system. We have used this basic system over the past 3 1/2 years, have made revisions and are now using the fourth generation of the system. The technically expert and the novice have used the systems.

2) Enable simultaneous voltage-clamp, microfluorometry, and video of small cells. Endothelial cells are usually 5-15 μm in diameter (41) and vascular smooth muscle cells are less than 80 μm in length (43, 51). These dimensions are in contrast to, for example, large molluscan neurons, squid axons, or muscle cells of 200-1000 μm in diameter that can be easily manipulated (3, 30, 46). Microelectrodes that are used for voltage-clamp of small cells require isolation of mechanical vibrations from the cell that is observed through a microscope. Accordingly, the filter wheel turning mechanism must be different from the air turbine described previously (10). (In essence, we have chosen to "re-invent the wheel" of Chance et al. (10).)

3) Design analog circuitry for compatibility with existing data aquisition systems. Considering that many electrophysiologists already have an analog-to-digital converter, the analog signal output of our optical processor would be compatible with almost any data acquisition system currently in use.

To our knowledge this is the first full report describing the analog signal processing circuitry, including on-line ratio determination, and optical configuration for simultaneous fura-2 microfluorometry, voltage-clamp, and observation of single cells. Although the techniques described here were primarily developed for study of smooth muscle and endothelial cells, they are equally applicable to other relatively small cell types (31, 43).

METHODS

Figure 1 shows an overview of the instrumentation and experimental system. Several components, of course, are very similar to other systems previously described. For example, the photomultiplier tube (PMT), video camera and recorder, computer, epifluorescence microscope, and patch-clamp amplifier are considered almost "standard" equipment for some laboratories using electrophysiological and biophysical techniques. It is the optical signal processor, filter wheel enclosure, and projection system configuration described here that make this system a uniquely simple, versatile, and low cost method for simultaneous voltage-clamp, determination of Ca_i, and video monitoring in single cells. The emitted fluorescence of fura-2 is diverted from the video port of an inverted microscope to the PMT by means of a dichroic mirror. The mirror allows transillumination of the cells with 600 nm light and passage to a video camera for display on a video monitor. A wide band interference filter with center wavelength of 510 nm, selected for the peak of the fura-2 emission spectrum, minimizes background fluorescence during simultaneous video monitoring and microfluorometry. The optical signal processor synchronizes the high frequency switching of the 340 (or 360) and 380 nm excitation wavelengths with the measurement of the appropriate fluorescence signal and presents the separated signals resulting from 340 (or 360) and 380 nm excitation to the analog-to-digital converter of the computer. Each major component is described below in detail.

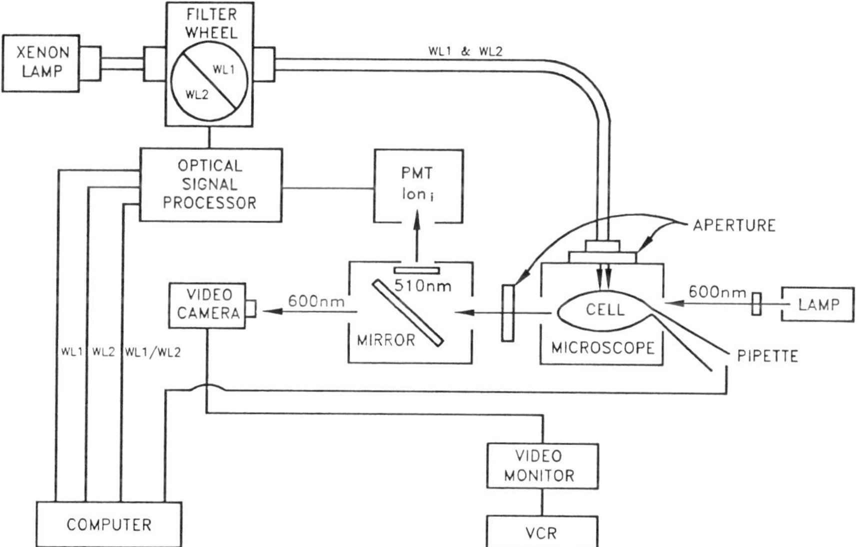

Figure 1. Schematic overview of instrumentation and experimental system. See text for details.

Instrumentation Components

Configuration for ultraviolet excitation. A 150 W Xe arc lamp and accessories (Oriel, Corp., Stratford, CT) are used for excitation of the fura-2 contained by the cell (Fig. 1), unless otherwise indicated. The Xe lamp is preferred over a Hg lamp because of the more uniform spectral irradiance across the relevant wavelengths of 300 to 400 nm, specifically, 340, 360, and 380 nm (Ref. (27), p. 42). The Hg lamp has greater than three-fold the spectral irradiance than does the Xe lamp at 340 and 380 nm, but the Hg lamp also has a peak irradiance at 365 nm that is almost 100-fold greater than the Xe lamp (Ref. (27), p. 42). We were concerned that the use of a rejection band filter might be necessary for blocking the 365 nm excitation (if 340 and 380 nm are used) and, therefore, selected the Xe lamp. Some of our studies have involved the use of the isosbestic wavelength of 360 nm (16) as a monitor of Mn influx (43) and intracellular fura-2 concentration. Future studies may also involve complete excitation spectra in a single cell using a monochromator (Oriel Corp.), thus making uniform spectral irradiance highly desirable. An ultraviolet grade condensing lens (f/1.0) composed of fused silica is included in the lamphouse. The procedure for proper adjustment of the condenser is provided by Oriel Corp. Welding goggles are highly recommended for this procedure because of the intense ultraviolet light. The main conveniences of this lamphouse arrangement are that it is self-standing and other optical components are easily coupled to it as needed. A self-standing lamphouse, separate from the microscope, is essential for vibration isolation (45). We have also briefly tested an Optiquip 150 W Xe arc lamp, lamphouse, and ignitor and found it to ignite reliably, but it creates excessive 60 Hz noise that may interfere with electrophysiological measurements. The lamphouse is not free-standing and does not have a fan for cooling of the Xe lamp. The lamp becomes excessively hot after several hours use. An Oriel fiber optic input assembly (#1, Fig. 2) is coupled to the condenser on the Oriel Xe lamphouse and focuses the light onto the end of a liquid light guide. A custom-machined fitting is necessary to attach the Oriel fiber optic input assembly to the Optiquip lamphouse. The alignment of the liquid light guide (#2, Fig. 2) in the input assembly is trivial. A manually controlled shutter on the input assembly permits excitation of the cell for specific time intervals to minimize photobleaching of fura-2, thereby reducing artifacts in measurement of Ca_i (4).

244

The shutter also reduces bleaching of interference filters due to high intensity illumination by the arc lamp. In practice the shutter is opened continuously during experiments, because of minimal bleaching with this optical configuration, as shown in the companion chapter (43). Microprocessor controlled shutters could be implemented if frequent shutter control is necessary.

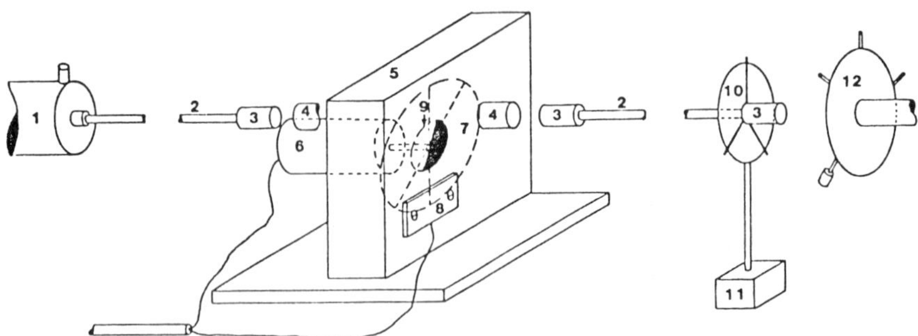

Figure 2. Configuration for ultraviolet excitation: interference filter wheel enclosure and liquid light guide alignment. Major components of the system are indicated by circled numbers as follows: 1) fiber optic input assembly; 2) liquid light guides; 3) collimating beam probes; 4) metal sleeves for alignment of collimating beam probes; 5) filter wheel enclosure; 6) rotary motor; 7) interference filter wheel (4-sector fura-2 wheel); 8) infrared emitting diode and phototransistor; 9) light reflecting and absorbing washer; 10) adjustable flange mount; 11) magnetic base; 12) Nikon epifluorescence illuminator. See text for additional details.

Liquid light guides (1 m long x 3 mm diameter; #2, Fig. 2) are used for transmission of light to the interference filter wheel and microscope instead of ultraviolet grade fused silica fiber optics. The light guides are shown in Fig. 2 as being broken for clarity in the limited space available for illustration. The main reasons are that the liquid light guides have equally good spectral transmission at the relevant wavelengths (Ref. (27), p.197) at less than half the cost. The transmission of the liquid light guide is 0.53 at 340 nm and 0.68 at 380 nm (Ref. (27), p.197); this relatively small difference in transmission does not present any quantitative problems because a calibration curve is generated on this optical system and shows an acceptable quantitative relationship between $[Ca^{2+}]$ in known Ca-EGTA buffers and the measured fluorescence ratio (F_{340} or F_{360} / F_{380}) (Fig. 18) (29,37,40,51). In order for the interference filters to transmit light at 340, 360 and 380 nm accurately, the incident light must be collimated. If the light strikes the filter at an angle other than perpendicular to the filter, the transmitted light will be of a different wavelength. The light exiting the liquid light guides has the path of a diverging cone, but it is easily collimated into parallel beams by attachment of a collimating beam probe (#3, Fig. 2).

The filter wheel enclosure, constructed of 4 mm thick aluminum plate, is basically a light-tight housing in which the filter wheel is mounted and collimated beam probes are aligned. In particular, the sleeves for beam probe mounting (#4, Fig. 2) are essential for easy alignment of the beam probes. The sleeves are pulled away from the surface of the enclosure in Fig. 2 for clarity of illustration. The beam probes are positioned in the sleeves by inserting them carefully until they contact the filter wheel and then backing up 0.5 to 1.0 cm. This places the beam probes sufficiently close for proper illumination and far enough to avoid contact with the filter wheel, in the event that it would not spin exactly at a 90° angle to the long axis of the beam probe. The positions of the beam probes in the sleeves are then marked with indelible pen or etching. The beam probes can then be removed, for example, to provide full spectrum irradiance to the microscope when an exciter filter is in place in the epifluorescence cube of the microscope, rather than using an interference filter wheel in the enclosure as the exciter filter. This is especially helpful for us because we have the need for fluorescence microscopy requiring the Nikon G-cube for rhodamine excitation to identify vascular endothelial cells (50). The collimating beam probes can then be replaced in the same alignment in the sleeves without affecting quantitation of fluorescence. The reliability test procedure is explained in *Procedures for Verifying Technical Acceptability of Operations.*

The light-tight property of the enclosure is perhaps conducive to proper transmission of light through the interference filter, although we have not systematically tested this notion. The enclosure mainly permits proper operation of the photosensitive transistor (#8, Fig. 2) which detects the position of the filter wheel (#7, Fig. 2). Also, the enclosure shields the user from high intensity ultraviolet light that could damage skin and eyes. The 4-sector fura-2 wheel (#7, Fig. 2) is mounted on an adjustable speed servo motor (#6, Fig. 2; Electro-Craft Corp., Hopkins, MN) that turns the wheel, thereby allowing the changing of the illuminating wavelengths at the appropriate frequency (more characteristics of the servo motor will be considered with the optical processor).

Excitation in a previous report was provided by the diffraction grating of a dual beam spectrophotometer (45). The alignment was difficult and the availability of such instruments at a reasonable cost is unlikely. Instead, an interference wheel originally designed and used successfully by Fay and coworkers (4, 23, 59) and manufactured by Omega Optical, Inc. (Brattleboro, VT) is used here (#7, Fig. 2). The filter wheel is 9 cm diameter and comprised of four sectors of alternating composition. Two pie-shaped sections each comprise 144° of the circle and are the interference filter sections of interest. One of the filters is 340 nm (11 nm HB) and the other provides 380 nm (13 nm HB) illumination as shown in Fig. 5. Two opaque sections each comprising 36° of the circle separate the sections providing light transmission. Surprisingly, these sections do not have complete optical density; instead, they pass a substantial amount of light. This problem is alleviated by taping the opaque sections with two layers of black electrical tape, which reduce light transmission to zero at the amplification used in these studies. One primary reason for the use of the Omega Optical filter wheel is the success of the University of Massachusetts group (4, 23, 59). An additional reason is that the large sectors allow a longer time of illumination with the 340 (or 360) and 380 nm light, thus permitting analog averaging of the fluorescence signal over a longer time and consequent reduction in noise by averaging with a variable time constant (See Figs. 9, 10). We also use in 2 of our 4 systems, a 2-sector fura-2 wheel shown in Fig. 1. The 2-sector increases the interference filter wheel sampling time per rotation. The 4-sector wheel was originally used for digital instrumentation to provide a zero light level and enable the summing of photon counts during exactly 180° of the rotation, rather than delay because of the transition from one to another of the ultraviolet light-emitting sections (4, 23, 59). It is trivial for us to impose a delay with our timing circuitry, thus the 2-sector wheel is useful. The position of the filter wheel is monitored by photosensitive transistors placed on the filter wheel enclosure (#8, Fig. 2) that detect the light reflecting and absorbing surfaces of a washer (#9, Fig. 2) placed on the shaft of the servo motor that rotates the filter wheel. The phototransistors are indicated below the washer in Fig. 2 for clarity; they are actually at the same level as the washer. The filter wheel position detection circuitry will be described in more detail below.

Another collimating beam probe is placed immediately after the interference filter wheel to focus the light onto the end of another liquid light guide (#3 in Fig. 2), which directs light to the epifluorescence illuminator (#12, Fig. 2) of the Nikon Diaphot microscope (Fig. 1). A third collimating beam probe immediately in front of the epifluorescence illuminator is secured in place by an adjustable flange mount (#10, Fig. 2) suspended on a post attached to a magnetic base (#11, Fig. 2). The post is made longer than that provided by Oriel Corp. to reach the height of the epifluorescence illuminator in place on the microscope. The magnetic base permits placement and securing of the beam probe at the appropriate location as close to the microscope as possible, without touching the microscope. The flange mount is coarse adjusted to position the excitation light in the center of the epifluorescence illuminator and then fine adjustments of the alignment made as described in the *Procedures for Verifying Technical Acceptability of Operations* section below. The advantage of the magnetic base (#11, Fig. 2) for positioning the beam probe is that the beam probe never touches the microscope, so it is less likely to transmit vibrations from the rotating filter wheel. A more convenient method is to insert the probe into a custom-made sleeve (not shown in Fig. 2) attached to the epifluorescence illuminator (#12, Fig. 2). We have had no difficulties with vibrations interfering with microelectrode work while using this attachment.

The epifluorescence illuminator (#12, Fig. 2) contains an iris diaphram, which we find is essential for controlling the area of ultraviolet excitation of the cell. This is one of five methods for reduction of background fluorescence (See companion chapter, Ref. (43)). The area of excitation can be adjusted from almost the total observation area of approximately 425 μm diameter (using a 40x objective) to less than 10 μm diameter, only 0.06% of the total area, for excitation of portions of a cell. The standard area used in these studies on smooth muscle and endothelial cells is a 100 μm diameter circle and is shown in Fig. 3.

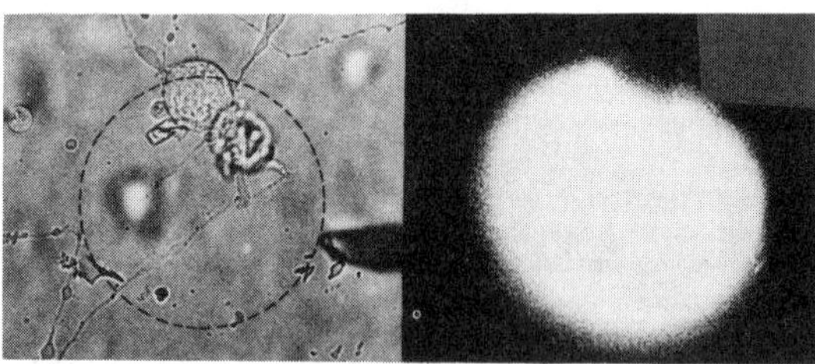

Figure 3. Area of ultraviolet excitation. Left panel: A wide bore patch pipette is located out of focus in lower right 1/3 of photo, touching dashed line circle. This circle approximates the area of ultraviolet illumination. Right panel: Ultraviolet excitation of fura-2 showing circular area of approximately 100 μm diameter.

A patch pipette containing 100 μM fura-2 pentapotassium salt is broken off and positive pressure of about 10 cm H_2O is applied to the pipette to cause continuous expulsion of fura-2 from the pipette in a large stream (Fig. 3, left). In the left panel of Fig. 3 the field is simultaneously illuminated by 600 nm light and ultraviolet excitation with 340 and 380 nm light originating from 4-sector fura-2 interference filter wheel. The resulting fluorescence from ultraviolet excitation alone (no 600 nm light) is shown in the right panel of Fig. 3. Note that the fluorescence from the patch pipette is negligible, while the 100 μm diameter area of ultraviolet excitation is obvious. In practice, we utilize the standard 100 μm diameter area for excitation, because this area will completely encompass a single cell when using the 40x objective and yields low background fluorescence. The area of excitation can be moved in the X-Y plane of the observation area by movement of the positioning controls for the iris diaphragm on the epifluorescence illuminator (#12, Fig. 2). The illuminator has a sliding shutter, but this is not used in our system. The Nikon lamphouse, power supply, etc. are also not used because the interference filter wheel must be isolated physically from microscope, since the vibrations occurring at wheel rotations of 50 Hz would prohibit simultaneous patch-clamp studies. The consideration with this type of optical arrangement is that a more substantial loss of excitation light intensity may occur than if the beam of light were collimated directly from the arc lamp condenser to the interference filter wheel and then to the epifluorescence illuminator (4, 23, 59). In contrast, in our system the approximate 50% transmittance of the liquid light guides is likely to have reduced the light intensity by 25% of other excitation configurations. The fluorescence emission from fura-2 loaded cells, however, is certainly large enough to resolve (See companion chapter, Ref. (43)) and the reduced excitation intensity may decrease the amount of photobleaching to insignificant levels (See companion chapter, Ref. (43)). From the light pipe of the epifluorescence illuminator the excitation light (340 or 360 and 380 nm) passes to the optics of the Nikon Diaphot inverted microscope.

Inverted microscope optics. The 340 or 360 and 380 nm excitation light first strikes a DM400 dichroic mirror (Nikon, Inc.), which reflects through the objective. The exact spectrum of the DM400 is not provided by Nikon, but our data suggest that the spectral transmission is adequate. The UVF 40x oil immersion objective (Nikon, Inc.) has a numerical aperture of 1.3, thus permitting high quality transmission of the light in the near ultraviolet range (340, 360 and 380 nm) and optimizing collection of fluorescence emitted from the cell (18). For future studies involving complete excitation spectra ranging from 320-400 nm, this high quality objective is essential. The adjustable aperture on the objective permits the reduction of light collected and is another method of reducing background fluorescence. The Diaphot has a prism configuration, which permits 100% of the light collected by the objective to be directed toward the eyepieces or 100% to be diverted to the video port and on to the PMT and the video system. In reality, of course, less than 100% of the fluorescence emitted from the cell is captured by the objective; some of the fluorescence is lost in directions away from the objective. Nonetheless, given that the optical configuration is similar during experiments, the amount of fluorescence lost should be a constant function of the total fluorescence emitted from the cell. This is a normal occurrence with epifluorescence microscopy (18) and only extreme measures would prevent such losses of light. Specifically, ellipsoid mirrors could be constructed above the microsuperfusion chamber, as done for studies with aequorin (6). This maneuver, however, would make simultaneous measurements of ionic currents using patch pipettes impossible, thus ruining a main purpose of this technique. The epifluorescence microscope is mounted on a vibration isolation table (Technical Manufacturing Corp., Peabody, MA) within a light-tight Faraday cage with the Xe light source isolated electrically, as well as mechanically, outside the cage. A 480 nm barrier filter (Nikon, Inc.) is used to eliminate fluorescence emission below that wavelength, thereby selecting primarily for the fura-2 fluorescence, instead of autofluorescence from the cell (33), background fluorescence, etc. The exact spectrum again is not provided by Nikon, but our indications are that the filter is acceptable when combined with the other optical components described here.

Projection system configuration for simultaneous microfluorometry and video monitoring. When the prism of the inverted microscope causes the light captured by the objective to be diverted to the video port depicted as item #1 in Fig. 4, the fluorescence of the cell in most microscopes is normally relayed directly to the PMT. In this system, however, light microscopic and fluorescence signals are projected to the video port in order to obtain a visual image of the manipulation of the patch pipette and contraction of the cell. The additional task of obtaining a visual image is made possible by also illuminating the cell with light at a center wavelength (CWL) of 600 nm. This is acheived by insertion of a narrow band (half-band width, HBW = 10 nm) interference filter (600DF10, Omega Optical, Inc.) in a filter holder slot of the Diaphot lamphouse. The transmission spectrum of the filter is shown in Fig. 5 (labelled 600). Note that the maximal transmission is 75%, which is sufficient to furnish the low light intensities detectable by the video camera. The optical density is approximately 6.0 at wavelengths from 200 to about 580 nm and from about 620 nm to 1200 nm. A 610 nm CWL narrow band filter is used in 2 of our 4 microfluorometry systems; essentially any bandpass filter having a CWL within the transmitting region of the dichroic mirror will be suitable. In the case of the 570 nm dichroic spectrum shown in Fig. 5, anything $\geq$ 600 nm is suitable. The filter is placed in the slot closest to the lamp to minimize stray white light (multichromatic) from the lamp that might contribute to background light. Also, aluminum foil is placed on the downward-facing side of the lamp if needed to further decrease stray light. The normal, 12 V 50 W tungsten-halogen illumination is of sufficient intensity and is regulated by a DC power supply, rather than the power supply in the Diaphot microscope.

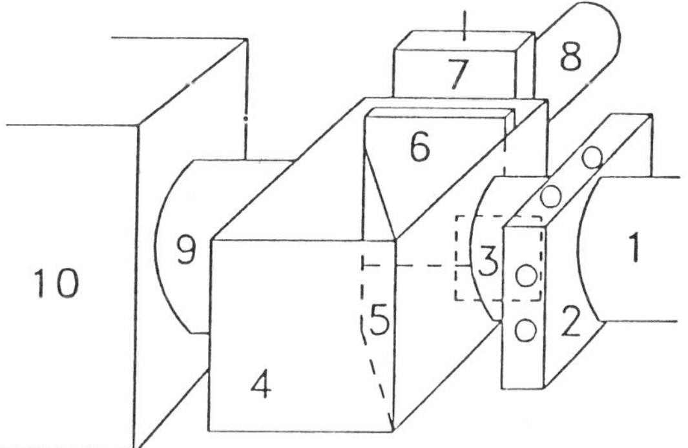

Figure 4. Projection system configuration. Major components of the system are indicated by circled numbers as follows: 1) video port of inverted microscope; 2) adjustable rectangular aperture; 3) 4x projection lens; 4) mirror enclosure; 5) 570 nm center cut-on wavelength dichroic mirror; 6) 510 nm center wavelength wideband bandpass filter [See Fig. 5 for transmission spectra of #5 and #6]; 7) PMT shutter; 8) PMT; 9) 8x zoom lens; 10) video camera. See text for additional details.

An adjustable rectangular aperture (Ealing Electro-Optics, Inc., South Natick, MA) is used to spatially define the cell for visual observation and recording of fura-2 fluorescence; hence, this aperture is also referred to as the measurement aperture. The distinct advantage of this aperture over the aperture available through Nikon is that each side moves independently. This degree of spatial resolution is necessary because it is desirable to acheive a measure of the averaged Ca_i in the whole cell. The variable size of the cell necessitates the rectangular aperture arrangement. The main purpose of these measures is not to spatially resolve Ca_i gradients optically in these cells; instead, the image

248

processing systems described by several groups is more appropriate for that purpose (14, 22, 23, 44, 47, 48, 55, 57). In this regard, recent evidence suggests that sampling of Ca_i from the middle portions of the cells may not represent the mean of the whole cell (15), although Cannell et al. (9) make this assumption for cardiac myocytes. Until a more definitive analysis exists we conservatively obtain measurements from at least greater than 75% of the cell area that could be visualized. The holder for the rectangular aperture (#2, Fig. 4) is fitted into the video port of the microscope and also is fitted to secure a 4x projection lens (#3, Fig. 4). The lens magnifies the images obtained from the microscope prior to passage to the video camera and PMT. At the mirror box (#4, Fig. 4) light from both the 600 nm interference filter and the fura-2 fluorescence are presented to a dichroic mirror (#5, Fig. 4; DC570LP, Omega Optical, Inc.) that primarily reflects fura-2 fluorescence to the PMT (#8, Fig. 4) and allows the 600 nm light to pass to the zoom lens (#9, Fig. 4) and the video camera (#10, Fig. 4). Alignment of the dichroic mirror at a 45° angle to the incident fluorescent light and 600 nm light is ensured by the slots machined in the mirror box and small set screws on each side and corner of the dichroic mirror for fine adjustment of the angle. This alignment requires careful attention and must be done for each microfluorometry system despite uniform machining of parts, etc. The alignment is described in more detail in the section *Procedures for Verifying Technical Acceptability of Operations*.

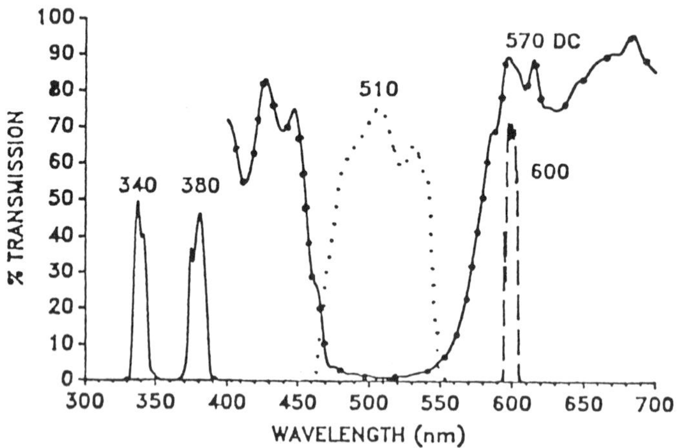

Figure 5. Transmission spectra of interference filters. Spectra are labelled according to the center wavelength. Solid lines = sections of the 4-sector fura-2 wheel providing narrow band transmission with center wavelengths of 340 and 380 nm light; dotted line = wide band filter with center wavelength of 510; dashed line = narrow band filter with 600 nm; dotted/solid line = dichroic mirror with cut-on wavelength of 570 nm at 50% transmission (labelled 570 DC). Spectra were kindly provided by Omega Optical, Inc.

Note several key features of the transmission spectrum of the dichroic mirror in Fig. 5 (labelled 570DC). First, the transmission of the dichroic is only 2-3% at wavelengths near 510 nm. This means that at least the remaining 97% of the light is reflected to the PMT and/or absorbed by the dichroic mirror. Second, the transmission at 600 nm is only about 85% and, therefore, the remaining 15% is also reflected and/or absorbed by the dichroic. Third, substantial transmittance of greater than 50% occurs at wavelengths less than 460 nm. In summary, the dichroic mirror (like any dichroic mirror) is not a perfect transmitting and reflecting filter and additional barrier filters, etc. must be used to acheive the final desired transmittance. Inclusion of the wide-band interference filter (#6, Fig. 4; Omega Optical, Inc.) with a center wavelength of 510 nm (the peak of the fura-2 emission spectrum) and a half-band of 60 nm reduces background optical noise to the same level noted in the absence of illumination of the cell with 600 nm light. Note the transmission spectrum of the 510 nm wide-band filter in Fig. 5. Regardless of the absence of exact estimations of the amount of fura-2 fluorescence lost due to the less than perfect optical properties of the dichroic mirror and wide band interference filter, the detection of large amplitude fura-2 fluorescence from cells and consistency of any fluorescence losses makes such losses manageable. A 585 nm cut-on wavelength dichroic mirror is used on 1 of the 4 microfluorometry systems in our laboratory. The utility of this dichroic is that with the higher cut-on wavelength more light between 550 and 570 nm is reflected to the PMT. Therefore, this dichroic could also be used for other indicators of free ion concentrations, such as BCECF that emits light with a center wavelength maximum of 540 nm for pH measurements (34). The shutter (#7, Fig. 4) allows changing of the dichroic mirror and bandpass filter in normal room light without over-exposing the PMT to high intensity light. Prior to attaching the shutter to the mirror enclosure and PMT, a 1-2 mm diameter white dot should be placed at exactly the center of the shutter on the side which receives light by reflection from the dichroic mirror. This marks the best approximation to the center point of the PMT photocathode.

The video detection system consists of a Panasonic WV-D5010 or WV-D5100 CCD camera (#10, Fig. 4) equipped with a Panasonic WV-LZ14/8AF 8x auto focus zoom lens (#9, Fig. 4). The images of the cell were observed on a video monitor (Panasonic CT-1301 or CT-1381Y) during fura-2 microfluorometry experiments. The entire experiment, including real time and other annotation provided by a character generator (WV-KB12A), can be recorded on VHS cassette tape with a video cassette recorder (AG-1230, Panasonic). Only the high wavelength, narrow bandwidth light (600 or 610 nm) was allowed to pass from the halogen lamp of the Diaphot through the dichroic mirror to the video camera. Another very important feature of the projection system configuration (Fig. 4) is that the distances between the dichroic mirror (#5, Fig. 4) and both the video camera (#10, Fig. 4) and the photocathode of the PMT (#8, Fig. 4) were similar in an attempt to equate the area of photodetection by the PMT equal to the area of observation in the video monitor.

Photodetection and optical signal conversion overview. The general principles of the optical signal detection and processing are shown schematically in Fig. 6. The electrical signals derived during 1/2 of a filter wheel rotation period are shown starting with the PMT output (top of Fig. 6) and ending with output to the analog-to-digital converter (A-D CONVERTER, bottom of Fig. 6). The PMT output is characterized by pulses corresponding to photons and a variable baseline noise as shown in the top of Fig. 6. The two different types of photodetectors used are similar to that described previously (45). The PMTs used for detection and amplification of photon emission have high sensitivity at the 510 mn peak of the fura-2 emission spectrum and also have a high frequency response for high gain photon counting. Low background counts are ensured because the photocathode diameter is electrostatically reduced to < 10 mm. One of the PMTs (#9893A/350, Thorn EMI Gencom, Inc., Plainview, NY) has been in operation for 3 1/2 years in this laboratory. The power supply (3000R, Thorn EMI) is adjusted to provide a voltage of -2500 V to the PMT. The emission signal from this PMT is amplified and threshold detected by a pulse discriminator amplifier (APED-II, Thorn EMI). The other PMT in use is a Hamamatsu R647-04 using an input voltage of -1000 V (Hamamatsu Corp., Bridgewater, NJ) in a radio frequency shielded housing (#PR1401RF005, Products for Research, Inc., Danvers, MA). The housing is constructed to yield a cathode position only 0.5 inches from the front of the housing (instead of the normal 1 3/16 inches) and fitted for a shutter (#PR305). The cathode position is very important to ensure proper focal distance from the dichroic mirror in Fig. 4. The fluorescence emission transmitted through the 4x projection lens (#3, Fig. 4) is diverging and, therefore, if the cathode distance from the dichroic mirror were greater, then larger portions of the image (cell) would strike the PMT peripheral to the area of the photocathode. The result would be an area of fluorescence emission measurement that might be substantially smaller than the image displayed on the video monitor. With larger smooth muscle cells (1,4) or groups of cells this could be a limitation. Procedures to circumvent this limitation involve alignment of the dichroic mirror to equate photosensitive regions of the PMT with images displayed on the video monitor. Details are given in another section (*Procedures for Verifying Technical Acceptability of Operations*). The Hamamatsu PMT is used in conjunction with a Hamamatsu C3866 photon counting unit and Bertan high voltage power supply (#230-03R, Bertan Associates, Inc. Hicksville, NY). We greatly prefer the Hamamatsu over the Thorn EMI systems because of the lower cost, greater availability, and comparable performance for our purposes.

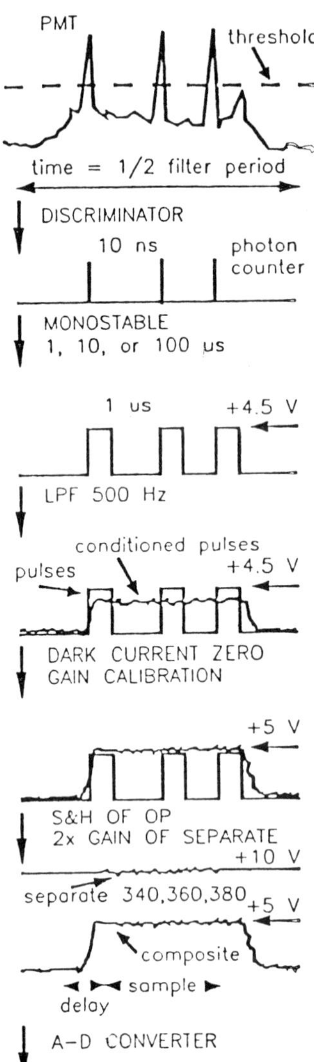

Figure 6. Overview of optical signal conversion

A threshold is set by the amplifier/discriminator of the photodetection systems (- - -, top of Fig. 6). Pulses of 10 ns duration are generated each time the annode current of the PMT exceeds the threshold. Note in Fig. 6 that low amplitude pulses within the noise of the PMT do not elicit 10 ns pulses from the discriminator. Quantification is possible with high speed counters, i.e. photon counter (Fig .6); however, we convert the pulses (counts) to the analog mode for signal processing. The pulses are applied to a monostable multivibrator (74LS123) circuit similar to that described previously (45), which generates user-selectable pulse widths of 1, 10, or 100 μs (MONOSTABLE, Fig. 6). The amplitude of these TTL pulses is typically +4.5 V (Fig. 6). The monostable is shown in more detail in Fig. 7 as the IC1 and IC2 sections of the present frequency-to-voltage converter circuit. An improvement of this circuit (45) is the use of trim potentiometers for precise adjustment of pulse widths. A voltage proportional to pulse frequency is generated from the constant-width, constant-amplitude monostable output pulses (LPF 500 Hz, Fig. 6) by the IC3-IC4 4-

pole, 500 Hz, active lowpass filter (Fig. 7). This filter substitutes for the previously applied commercial lowpass filter (45). Dark current output zero adjustment, amplification, and calibration are provided by amplifier IC5 (Fig. 7). The low-pass filter basically integrates the area under the pulses, thus yielding an analog signal (conditioned pulses, Fig. 6). The overall gain of the PMT instrument chain is proportional to the selected monostable pulse width. The 1, 10, and 100 ls pulse widths (PMT gain factors) are selectable with a front panel switch. Because the monostable pulses are +4.5 V, we provide a gain calibration to increase the analog conditioned pulse signal to +5 V (GAIN CALIBRATION, Fig. 6).

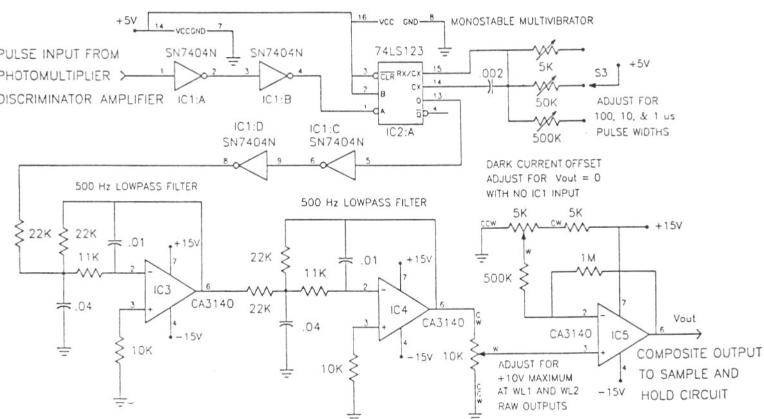

Figure 7. PMT signal frequency-to-voltage converter. Constant-amplitude pulses of constant switch-selected widths are generated at the PMT output frequency by a monostable multivibrator. The pulses are low-pass filtered to produce a composite DC output voltage proportional to fluorescence intensity. See text for details.

The analog voltage output of the above monostable and lowpass filter frequency-to-voltage converter general design was shown previously to be directly proportional to light intensity at the PMT by use of a series of neutral density filters (45) and is verified here in the RESULTS, Fig. 13. This configuration provides an analog frequency counting device that retains the increased signal-to-noise ratio associated with photon counting. The method also allows convenient presentation of an analog signal to the optical processor sample-and-hold circuitry (S&H OF OPTICAL PROCESSOR, Fig. 6 and see below). The sample-and-hold amplifiers separate the composite output into its components resulting from 340 or 360 and 380 nm excitation (separate 340, 360, or 380 nm; bottom of Fig. 6). Because the analog-to-digital converters that we use accept +10 V maximum input, we provide a 2-fold gain of the separated signals (2x GAIN OF SEPARATE, Fig. 6). This is shown in Fig. 6 by the 2-fold higher amplitude of the separated signal than the composite signal (composite, bottom of Fig. 6). For simplicity and to indicate the general use of this technique for any 2 wavelengths, we refer to 340 and 360 nm excitation as wavelength 1 (WL1) and 380 nm as wavelength 2 (WL2). The separated signals are then presented to the analog-to-digital converter of the computer (A-D CONVERTER, bottom of Fig. 6).

Optical filter wheel period and position measurement. The timing signal necessary for separation of the frequency-to-voltage converted PMT signal into discrete 340 and 380 signals is generated by detection of light reflecting and absorbing surfaces on a washer attached to the filter wheel drive motor. This washer, interference filter wheel, and rotary motor are indicated functionally in Fig. 8 (See also Figs. 1 and 2). The washer is illuminated by an infrared light-emitting diode (LED, Texas Instrument TIL-99). The reflected light from the polished half activates a phototransistor (Texas Instrument 906-1), which reduces the transistor collector voltage at test point TP1 to almost zero. The non-reflecting painted half places the transistor in the off state, producing 5 volts at TP1 (test point 1). The waveform "smoothing" due to the angular transition time from washer flat black to polished surfaces is removed by buffer IC1:C and inverter IC1:D. The washer is positioned so that the 340 nm filter section is active when the light reflecting half of the washer passes the infrared LED. This optical filter section is indicated by a filter position pulse low voltage state (L). Because the LED and phototransistor are located diametrically opposite to the incident ultraviolet light (Fig. 2), the 340 or 360 section of the interference filter wheel must be placed opposite of the polished half of the washer; specifically, at the position of the number 7 in Fig. 2. The resultant motor rotation square wave IC1:D output, hereafter designated as the "filter position pulse", is used for filter wheel position synchronization by the 340 and 380 signal separation sample-and-hold circuitry.

Fig. 8 also shows the filter wheel rotational period measurement circuit for monitoring motor speed control and accuracy. The counting time interval of an Intersil 7225 four-digit counter is controlled, via dual monostable multivibrator IC4 (4528B), by the filter position pulse waveform period. The counter input is a 10 Khz signal that is digitally derived from a 1 Mhz crystal by the IC1:A, IC1:B, IC2, and IC3 oscillator-divider chain. This counter, and associated LED numerical readout, continuously displays the filter rotational period in milliseconds on the instrument front panel. The complete wiring connections between the 7225 counter and the LED numerical displays are not detailed in Fig. 8; this information may be obtained from the 7225 manufacturer's specification sheet or from the authors. The four switch-selected motor speeds provide filter periods of 20, 50, 100 and 200 ms. Sufficient 1 ms monitoring accuracy is obtained by using only three of the four available 7225 counter output digits; the 0.1 ms digit may also be employed if more critical measurement is desired. A tachometer feedback servo controller described below maintains the selected motor speed within 1% accuracy.

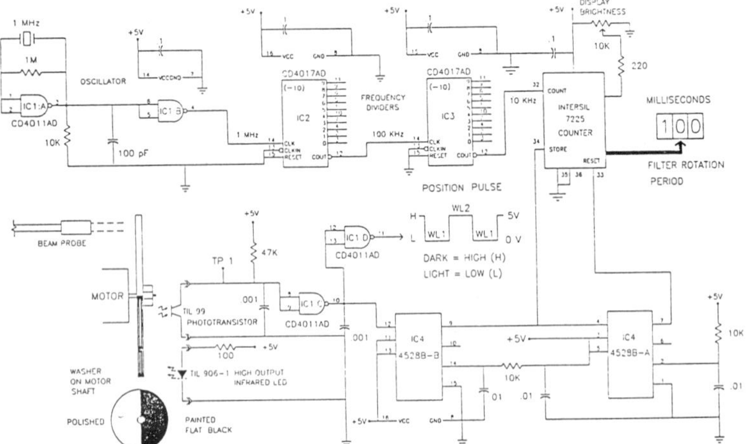

Figure 8. Filter position pulse generator and filter rotation period indicator. Infrared light reflected from the polished half of a washer on the filter wheel drive motor alternatively switches a phototransistor on and off to form the filter position pulse signal. Motor speed accuracy and stability are monitored by resetting a fixed-frequency 10 KHz counter with the position pulse. A digital readout displays the filter wheel rotation period in milliseconds. See text for details.

Functions of optical signal processor circuitry. Fig. 9 shows the optical wavelength signal separation and ratio circuit. The complete circuit for the wavelength 1 (WL1, either 340 or 360 nm) signal is presented; identical circuit sections for the wavelength 2 (WL2, 380 nm) signal are indicated by text blocks. Circuitry and components common to both wavelengths are fully detailed. An idealized frequency-to-voltage converted PMT signal resulting from 1 1/2 rotations of the 4-sector filter wheel is shown at the circuit input. The actual signal, as supplied by the Fig. 7 frequency-to-voltage converter output, is a continuous train of sequential WL1 and WL2 signal amplitudes resulting from constant filter wheel rotation. The WL1 and WL2 sections of the composite signal in Fig. 9 are separated by a brief zero voltage filter transition period, which is caused by passage of an opaque area positioned between the two filter wheel wavelength-selective (light transmitting) sections. The 0-5 volt optical filter position pulse supplied for circuit synchronization is buffered, amplified, and inverted by Q1-Q4. This position pulse then, in its inverted (Q2) and re-inverted (Q4) form, triggers separate WL1 (IC1A, IC1B) and WL2 monostable multivibrator chains to generate sample-and-hold circuit control pulses for the resultant separation of the PMT composite optical signal into WL1 and WL2 components.

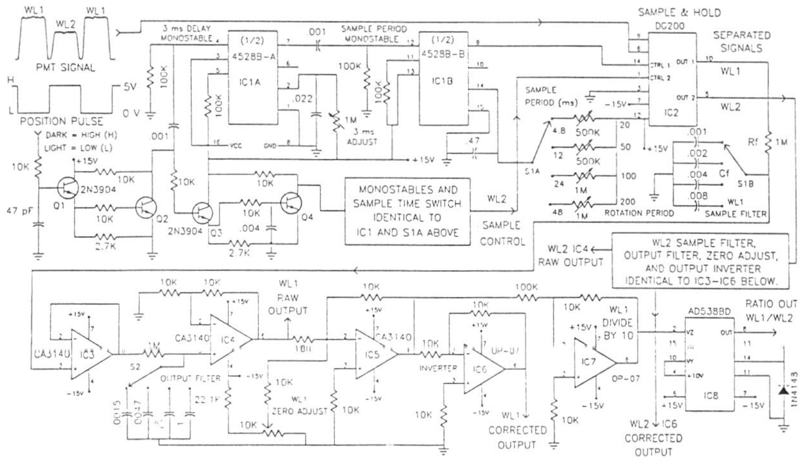

Figure 9. Optical wavelength signal separation and ratio circuit. Complete circuitry for wavelength 1 (WL1) is shown; text blocks indicate identical circuits for wavelength 2 (WL2). All resistors in the IC3-IC7 circuits are metal-film 1%. The composite PMT signal resulting from excitation by WL1 and WL2 is sampled by dual electronic switch IC2 at times controlled by the filter position pulse to generate separate output voltages for each wavelength. These signals are filtered and background fluorescence voltage subtracted before the analog divider calculates the WL1/WL2 ratio. See text for details.

Sampling and holding functions are performed by a DG200 electronic switch IC2 in Fig. 9, that has two individual switches that are turned on and off by voltage levels at the control inputs CTRL 1 and CTRL 2. A zero control voltage turns an individual switch on and a positive control voltage turns it off. The switch inputs, pins 6 and 9, are both connected to the PMT composite WL1 and WL2 input signal. Sequential WL1 and WL2 control multivibrator pulses are applied, to CTRL 1 and CTRL 2, respectively. These control pulses occur synchronously with the corresponding input signal wavelength time slot so that the WL1 and WL2 voltages are switched to the proper IC2 output. Therefore, the composite input signal alternately appears as separated, discontinuous WL1 and WL2 signals at the individual switch outputs OUT 1 and OUT 2.

Separation of the WL1 signal begins by monostable IC1A generating a 3 ms delay pulse at a filter position pulse voltage fall. A 15-turn trim potentiometer is used for precise calibration of the 3 ms delay. This action is triggered by an inverted position pulse voltage rise at the collector of Q2. The delay pulse postpones signal sampling until the PMT WL1 output signal plateau is obtained, i.e., until the light is shining only on the light-transmitting section of the interference filter and not on the transition between the opaque section in the case of a 4-sector filter wheel. The delay is shown schematically in Fig. 6 (near bottom). We found it simplest to fix the delay at 3 ms, rather than change the delay with each of the 4 different rotation periods. The delay of 3 ms is also used with the 2-sector filter wheel; we have not determined whether shortening the delay and increasing the sampling time would be beneficial.

After the 3 ms delay, monostable IC1B is triggered to generate a sample time length pulse that corresponds to the filter wheel rotation period currently in use. These 4.8, 12, 24, and 48 ms sampling times are selected by switch S1A for the motor and filter wheel rotation periods of 20, 50, 100, and 200 ms, respectively. To facilitate timing calibration, the sample period monostable circuit uses 15-turn trim potentiometers for the variable resistances shown at IC1B and S1A. The sample time length control pulse applied to IC2 CTRL 1 by IC1B is a normally high, inverted waveform that falls to zero volts during the sample period. This zero level turns on the CTRL 1 switch, connecting the composite signal WL1 level present at this time to OUT 1. This output signal is averaged during the sampling period by the passive RC low-pass filter (Rf-Cf). The 1, 2, 4, and 8 ms filter time constants are selected by S1B for the 4.8, 12, 24, and 48 ms sample times, respectively. At least 4 filter time constants lapse during the sampling period, thus providing a reasonable average of the signal without distorting the time resolution. At the end of the WL1 sampling period, the CTRL 1 switch opens and the average WL1 value at that time is held by capacitor Cf. It is held constant until the next WL1 sampling period starts, which begins 3 ms after the beginning of the next complete filter wheel cycle. Wavelength sampling time plus the initial 3 ms delay is not equal to the total time for a filter wheel half-rotation; a dead time is made available so that the sampling time ends before the next filter transition. For example, during the 20 ms rotation period (with half cycle time of 10 ms) the filter position pulse initiates the 3 ms delay before sampling the WL1 signal and then sampling occurs for 4.8 ms, thus totalling 7.8 ms; the remaining 2.2 ms is not sampled because this time is again a transitional period. This stopping of sampling well before the filter transition assures that the end sample value is completely within the interference filter active region. The sample period is shown schematically in Fig. 6 (near bottom).

A position pulse rise signals the WL2 second half of the optical filter period. The twice-inverted position pulse at the collector of Q4 triggers an identical pair of monostable multivibrators, indicated in the text block, for generation of WL2 delay and sampling time pulses. The WL2 sample pulse is applied to IC2 CTRL 2 for switching of the WL2 data to IC2 OUT 2. Therefore, the WL1 and WL2 information is sampled during the time that the proper section of the interference filter wheel is in place and held when the other wavelength of light is being transmitted. The sample time and sample filter switches, S1A and S1B, are 2 separate switch sections of a single 4-position rotary switch deck wafer. The identical WL2 channel sample time and filter switch, as well as the motor speed switch, are also included on this 3-deck, 4-position rotary switch. The multiple switch, mounted on the instrument front panel, allows all signal processor operational parameters to be changed simultaneously.

Discharge of the WL1 signal voltage held by sampling filter capacitor Cf between samples is negligible due to the extremely high input impedance of voltage follower IC3. This amplifier also provides low driving impedance for the multiple-frequency low-pass output filter at the non-inverting (+) input of amplifier IC4. This filter determines the overall system frequency response. Switch-selected filter cutoff frequencies are 100, 34, 16, and 0.16 Hz, corresponding to 1.5, 4.7, 500, and 1000 ms time constants, respectively. These values provide sampling visualization, switching transient suppression, signal smoothing, or heavily averaged output frequency responses, respectively.

Fig. 10A shows actual fluorescence signals recorded from a piece of fluorescent uranyl glass (18) with the pCLAMP data acquisition system (See below). Two complete cycles of filter rotation using the 20 ms period are shown by the position pulse. The conditioned PMT output that is a composite of fluorescence emission resulting from 340 (WL1) and 380 (WL2) nm excitation is shown. The fluorescence emitted as a result of 340 nm excitation being about two-fold greater than the fluorescence due to 380 nm excitation. In this case the filtering time constant was 1 ms, thus yielding an average over the 4.8 ms sampling period. The separated, discrete 340 and 380 nm signals from the optical processing circuitry are superimposed on the composite PMT output, thus displaying the effective sampling of the signal during the appropriate period and then holding of the voltage until the next cycle is initiated. A longer record of 1000 ms (50 filter wheel rotations) is shown in Fig. 10B.

Although low-pass filtered, the IC4 output is termed "raw" due to the presence of a non-zero background fluorescence signal resulting from non-specific fluorescence, etc. This background value is set to zero by adding an adjustable negative voltage to the WL1 signal with summing amplifier IC5. The setting of IC5 output to a zero voltage baseline is accomplished by using front-panel mounted 10-turn adjustment potentiometers and a zero detector circuit described below. The resulting zeroed signal, negative-going due to inversion by IC5, is again inverted by amplifier IC6 to produce a background subtracted output signal which is directly proportional to changes in fluorescence resulting from excitation by WL1 light. WL2 signal processing after the sample-and-hold separation is identical to the above WL1 signal filtering, amplification, background subtraction, and inversion. This circuitry duplication is outlined by the WL2 text block in the IC2 OUT 2 signal line.

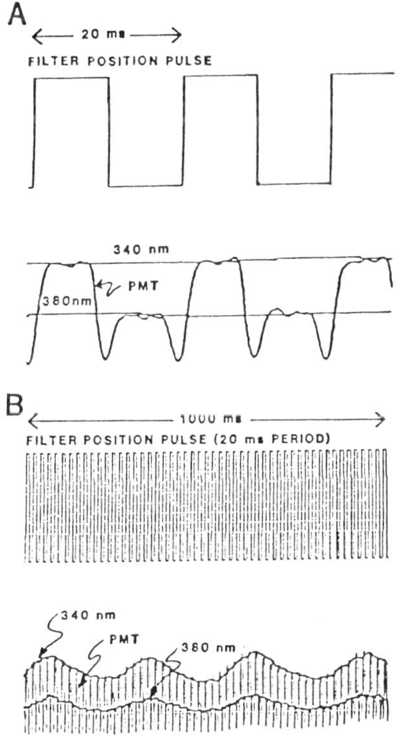

Figure 10. Composite, conditioned signal from PMT and associated filter wheel position pulse and sample-and-hold outputs of the optical processor. Fluorescence was recorded from a piece of uranyl glass using excitation from a 4-sector interference filter wheel. **(A)** The rotation period of 20 ms is shown by the 0-5 volt excursions of the position pulse. Two and one half rotation periods are shown. Immediately below is the conditioned PMT output (PMT) that was low-pass filtered at 500 Hz. When the position pulse is high (5 V) the fluorescence resulting from 340 nm excitation (WL1) is sampled and subsequently held after the position pulse goes low (0 V), as shown by the flat line (340). Similarly, the fluorescence resulting from 380 nm excitation (WL2) is sampled when the position pulse is low and subsequently held after the position pulse goes high, as shown by the flat line (380). Note also the transitional periods where the PMT output goes to zero. Data were digitized by the A-D converter at 96 μs intervals. **(B)** Tracking of the oscillating conditioned PMT output by the sample-and-hold circuits over a 1000 ms segment of data acquisition. The fluorescence intensity measured by the PMT was caused to change by manually turning the focus knob on the microscope, thus changing the focal plane. A 20 ms rotation period was used.

After background fluorescence subtraction (zero correction) by IC5, the WL1 and WL2 signals are applied to analog divider IC8 for calculation of the WL1/WL2 ratio. The ratio output signal, and the WL1 and WL2 output signals, are developed within a 1-10 volt range. Computation of the ratio of the WL1 output voltage divided by the WL2 output voltage is performed by an Analog Devices AD538BD computational module. This integrated circuit performs the operation VZ/VX; these variables are the WL1 and WL2 analog voltages supplied to the device input ports. The WL2 output signal is applied directly to the denominator input VX. Since neither of the numerator inputs can be zero, input VY is connected to the highly accurate +10 volt reference output provided by the module. This necessitates reducing the WL1 numerator signal by a factor of 10 so that the AD538BD 10 volt maximum output specification is not exceeded. IC7 is, therefore, employed to divide the IC5 WL1 output voltage by 10 before application to the VZ numerator input. The inverted IC5 output is used as the input to IC7 due to a second inversion by IC7. Therefore, numerator, denominator, and the ratio output are all positive.

Optical filter motor speed controller and accessory circuits. Fig. 11 illustrates the servo motor controller (#9086-0007, Electro-Craft Corp., Hopkins, MN). The controller is on a printed circuit board which, along with the included power transformer, conveniently installs in the system cabinet with the signal processing boards. The manufacturer provides a speed potentiometer adjustment on the board, which is capable of controlling motor rotational periods from 8 ms to 12 seconds. Since only switch selected fixed periods of 20, 50, 100, and 200 ms were desired, this potentiometer was bypassed and replaced by the four potentiometers as shown. These internal 15-turn trim potentiometers are adjusted only during initial system calibration. As stated above, the motor period selector (switch S1C) is ganged with the master sample period (switch S1A) and sample filter time constant (switch

256

S1B) rotary switch. An analog output proportional to motor speed is calibrated from the motor tachometer signal by amplifier IC3. This speed output signal, inversely proportional to sampling period, is recorded as an indicator of the period in use.

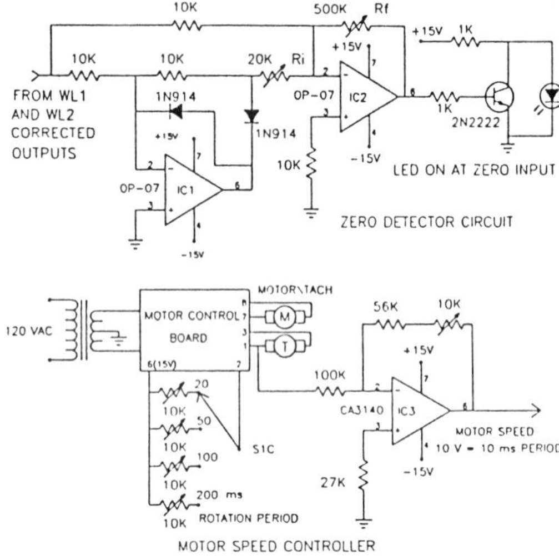

Figure 11. Wavelength signal zero detector and motor speed controller. The zero voltage indicator facilitates wavelength signal zero baseline adjustment. The filter wheel motor speed is maintained within 1% by a tachometer feedback controller. IC3 generates a proportional analog voltage for monitoring and recording the selected motor speed.

A zero voltage detector for precisely adjusting the WL1 and WL2 output signal baselines is also shown in Fig. 11. This circuit is included in the system due to the importance of zero voltage signal baseline (background fluorescence subtraction) at the analog divider inputs. A separate zero detector circuit is used for each wavelength output voltage. The circuit is an amplified precision rectifier that produces a positive DC output voltage for either positive or negative mV level inputs from the main WL1 and WL2 outputs. IC2 output turns the transistor on, causing the front-panel mounted LED to remain off. When a zero input is achieved, IC2 output also becomes zero and the LED is illuminated. Ri balances the circuit to obtain identical IC2 output for positive and negative inputs; gain adjustment Rf sets the LED indicator on-off threshold to a 10 mV input level.

The complete optical processor instrument is housed in a Bud HC-14103 13 x 17 x 5.25 inch benchtop or rack mountable cabinet. This ample cabinet size allows all circuit boards and other internal components to be mounted horizontally on the cabinet floor by 1" stand-off threaded posts. All circuitry except the filter period readout, motor controller, and zero detectors are on a single 11.5 X 23 cm single-sided printed circuit board. Power requirements, which are easily furnished by standard commercial modular power supplies, are 5 V at 300 mA and + 15 and -15 V at 150 mA.

Procedures for Verifying Technical Acceptability of Operations

The section on *Instrumentation Components* described the characteristics of the optical and electronic components and the basic configurations. Setting up the microfluorometry system according to that section should enable recordings, but the system would not be optimized. This section gives more of a step-by-step, ordered procedure for testing the microfluorometry system before its ultimate use on cells. These procedures should be performed on each system to verify its acceptable operation with fluorescent standards, etc.

Alignment. Three major types of alignment must be done.
1) Collimation of excitation light. True collimation of the ultraviolet light beam proximal to the Xe arc lamp transmitting light toward the filter wheel enclosure (Fig. 2) must be confirmed by shining the light on a flat surface about 15 cm from the beam probe. Use of welding goggles is again essential to avoid eye damage. A circular beam the size of

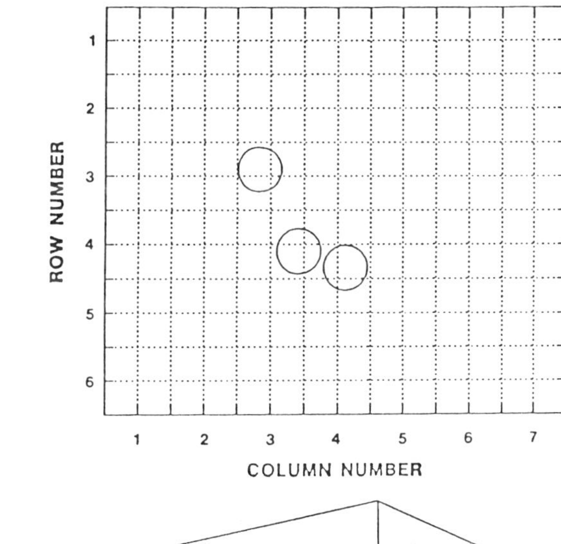

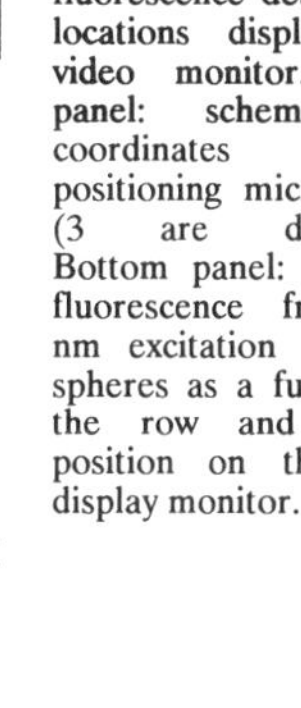

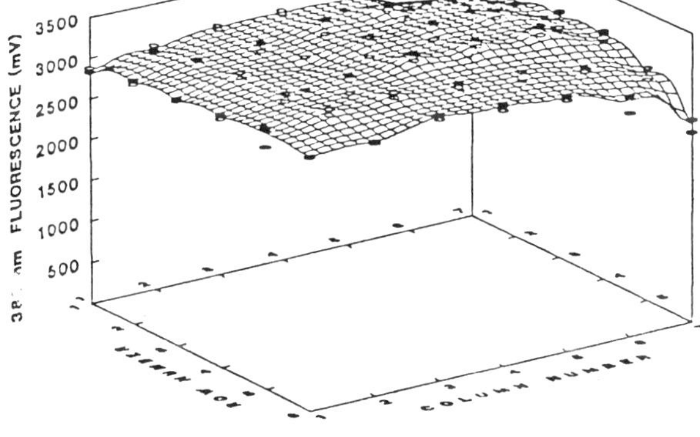

Figure 12. Uniformity of fluorescence detection at locations displayed on video monitor. Top panel: schematic of coordinates for positioning microspheres (3 are displayed). Bottom panel: Absolute fluorescence from 380 nm excitation of 6 lm spheres as a function of the row and column position on the video display monitor.

Optimizing fluorescence recording. In addition to the alignments described above the following considerations will optimize fluorescence recordings and minimize artifiacts.

1) Dark current zero. The zero adjust 10-turn potentiometers on the front panel are first turned to zero. The PMT shutter (#7, Fig. 4) is closed and a 10-turn potentiometer on the front panel associated with IC5 in Fig. 7 is then adjusted until the front panel LED of the zero detector circuit (Fig. 11) is turned on. This procedure nullifies intrinsic dark current (noise) of the PMT. PMT dark current may also need to be re-zeroed when changing optical processor gain (monostable pulse width).

2) Adjust gain for maximum signal amplitude. This procedure should be done only after the alignments discribed above and the dark current zero have been completed. The full scale of the analog-to-digital converter is used by adjusting the maximum amplitude of a saturating fluorescence signal. Any highly fluorescent material is satisfactory. The microspheres we have used (Duke Scientific) are 6, 10, 16, 22, and 48 μm diameter; saturating fluorescence can easily be achieved with diameters $> 10\,\mu$m. The adjustment is made by an internal trim potentiometer at the input of IC5 in the circuit shown in Fig. 7. As shown schematically in Fig. 6, the maximum amplitude of the composite output becomes $+5$ V; the maximum amplitude of the separated signals becomes $+10$ V because of a 2-fold gain in the sample-and-hold circuitry (contributed by IC4 in Fig. 9). Typical analog-to-digital converter input maximums are $+10$ or $+5$ V; the maximum output should be set close to these limits for best analog-to-digital resolution.

258

3) Choice of objectives. The Nikon UVF 40x ojective is recommended for single cells, rather than 100x, because the absolute fluorescence is about 40% greater with 40x and there is almost no difference in background fluorescence between the two objectives. In addition, immersion oil should always be used because it increases the absolute fluorescence by about 30-50%. The UVF 20x objective can also be used, but gives higher background fluorescence.

4) Determining depth of field. Microspheres of varying diameters are used to provide a fluorescent standard of varying thickness. The measurement aperture (#2, Fig. 4) is set at only 2 μm^2, which is only 1/3 the diameter of the smallest (6 μm) microsphere. An etch is made on the coverslip with a diamond pencil so that the microscope may be easily focused on the top of the coverslip, which is the bottom of the microsphere. This procedure is repeated on at least 5 microspheres each of 6, 10, 16, and 48 μm in diameter because the absolute fluorescence of all microspheres of a given diameter is not equal. Means of absolute fluorescence resulting from 380 nm excitation, rather than 340 or 360 nm excitation, are compared.

5) Determining X-Y (2 dimensional area) selectivity of measurement aperture. The effectiveness of the measurement aperture (#2, Fig. 4) in restricting the area of fluorescence measurement to only the cell(s) in view through the aperture and displayed on the video monitor is determined using microspheres. Microspheres are placed on a coverslip. Several microspheres are often found close together randomly, similar to those shown schematically in Fig. 12; also, microspheres can be positioned close to each other with a patch pipette. Pairs of microspheres are most extensively studied. One microsphere is defined spatially by placing all 4 sides of the measurement aperture at the largest diameter of the microsphere, while the view of the other microsphere is completely masked by the aperture sides, and the absolute fluorescence resulting from 380 nm excitation is then recorded. The process is repeated on the adjacent microsphere and then the aperture is placed around both microspheres to determine the additivity of their fluorescence. Multiple pairs of microspheres are tested.

6) Determining PMT averaging of fluorescence across photosensitive area. Similar to procedure 5 above, the fluorescence from individual microspheres in a field is determined. The main difference from procedure 5 is that $\geq$ 7 microspheres are measured in a single field such that the measurement aperture is almost entirely opened when all 7 microspheres are measured by the PMT simultaneously. The sum of the fluorescence from individual microspheres is compared to the combined fluorescence of all 7 determined simultaneously by the PMT to determine the additivity of measurement by the PMT.

7) Background fluorescence reduction. The effect of background fluorescence on the accuracy of analog ratio calculation is determined using standard solutions with calibrated free Ca concentrations of 100, 400, and 1000 nM that were made as described previously (37, 51) and in other chapters (41, 43). The 510 nm wide-band filter (#6, Fig. 4) was removed from the mirror enclosure (#4, Fig. 4) prior to conducting the experiments. Without the wide-band filter in place the 600 (or 610 nm) transillumination must be turned off, otherwise the background flourescence may typically be 30-50% of the total absolute fluorescence signal; with no transillumination the background is zero. We utilize this property to add varying amounts of background light to the fluorescence recordings. A drop of the Ca calibration solution is placed on a coverslip and the fluorescence ratio (F_{340} / F_{380}) determined. A stable ratio is monitored for about 1 min and then, the transillumination is turned on, adjusted to be either 10% or 30% of the absolute fluorescence recorded from 380 nm excitation, and the effect of background fluorescence on the ratio is then recorded for 1 min.

Reliability. The reliability (reproducibility) of fluorescence recordings from the microfluorometry system are established on an intra-day and inter-day basis.

1) Microsphere fluorescence stability. Because the fluorescent microspheres should show no decrease in fluorescence due to bleaching, unlike some reports on fura-2 (4) and indo-1 especially (1, 28, 36, 44), these are ideal for determining intra-day reliability. The tests on microspheres are quite appropriate because one is working with a microscopic object and the inherently more precise technique required, rather than work with a large piece of fluorescent uranyl glass or other fluorescence sources (18). The 6 μm microsphere is placed on a coverslip, all 4 sides of the measurement aperture (#2, Fig. 4) reduced to the maximum diameter of the sphere, and the fluorescence is then recorded continuously for greater than 90 min. The focal point (plane) on the microsphere is maintained constant

throughout the recordings. Similar recordings are done on microspheres on different days to determine inter-day variability.

2) Manipulation of liquid light guides, etc.. One major manipulation of the optical configuration involves changing excitation wavelength, which requires several steps. The liquid light guide (#2, Fig. 2) is removed from the fiber optic input assembly (#1, Fig. 2) and placed aside. The beam probe (#3, Fig. 2) is removed from the sleeve (#4, Fig. 2) exiting the filter wheel enclosure (#5, Fig. 2), and then the liquid light guide is removed from this beam probe. The liquid light guide is then placed in the fiber optic input assembly to deliver ultraviolet and visible excitation light directly to the epifluorescence illuminator (#12, Fig. 2) of the microscope. A different exciter filter, dichroic mirror, and barrier filter are then placed in the Nikon Diaphot microscope. We use rhodamine excitation filters extensively (41). All these components are then returned to their configuration for excitation by the interference filter wheel in the reverse order of that just described. Another very simple manipulation of the optics is removal and replacement of the 510 wide-band filter (#6, Fig. 4) in the mirror enclosure (#4, Fig. 4).

3) Inter-investigator variability on fura-2 Ca calibration standards. Reliability of the fluorescence ratio (F_{340} / F_{380} or F_{360} / F_{380}) of Ca calibration standards containing fura-2 is determined as for the microspheres. Variability between investigators is established by each of 7 laboratory personnel performing fura-2 Ca calibration tests on the same day and test standards.

Validity. These procedures verify that electrically processed signals are directly related to light emission and ion concentration.

1) Output voltage of circuitry proportional to percent optical transmission. The 600 (or 610) nm narrow-band interference filter is removed from the microscope lamphouse to provide multichromatic illumination. The neutral density filters are then placed individually on the stage of the microscope above the objective in the path of the multichromatic light. The 510 nm wide-band interference filter (#6, Fig. 4) is kept in place in the mirror enclosure (#4, Fig. 4) to ensure that green fluorescence is detected by the PMT (#8, Fig. 4), as would occur during usual microfluorometry experiments on cells. The fluorescence emission is set to nearly maximal at about 9.5 V on the analog-to-digital converter with no neutral density filter in place and, thus, is 100% transmission. The percent of transmission is then set to several values between 10% and 100% transmission with neutral density filters.

2) Accuracy of analog divider. Several different voltages from 0-10 V are applied to the VZ input of IC8 in Fig. 9 to represent the input from the corrected WL2 output of IC6 (Fig. 9). Several different voltages ranging from 0-1 V are applied to the VX input of IC8 in Fig. 9 to represent the corrected WL1 output divided by 10 from IC7 in Fig. 9. These inputs to the divider integrated circuit are from highly accurate DC power supplies, calibrated on a digital voltmeter, and the ratio output from the analog divider is recorded. The output from the analog divider is compared to the ratio calculated from the inputs.

3) Fura-2 fluorescence ratio vs Ca concentration calibration. The details of this procedure are in published manuscripts (37,51) and in other chapters (41, 43). The fluorescence ratio (F_{340} / F_{380} or F_{360} / F_{380}) is compared between 3 different microfluorometry systems in our laboratory constructed as described here, which have different objectives (20x vs 40x) or different excitation wavelengths.

Computerized Data Acquisition and Analysis

Data are presented to an analog-to-digital converter, displayed on the computer terminal, and saved to disk for off-line analysis. Two analog-to-digital converters have been used. The Labmaster analog-to-digital board (Scientific Solutions, Inc., Solon, OH) has been used with pCLAMP software (versions 4.1 through 5.5) and AxoBASIC 1.0 (Axon Instruments, Inc., Foster City, CA) and the ADAC 5500 board has been used with Labtech Acquire software (ADAC Corp., Woburn, MA). pCLAMP and Acquire were used in 80286 processor-based computers, but AxoBASIC requires 80386 computers. Select fluorescence values are often placed into an ASCII file and transferred into a spreadsheet program (Quattro, Borland, Inc.). The companion chapter provides a more thorough description of computerized data acquisition and analysis (43).

Statistical analysis. Systat version 5.01 is used for 3-dimensional graphics. Sigmaplot 3.1 is used for graphics and descriptive statistics (mean, standard deviation, standard error).

The coefficient of variation [(SD / X) x 100, stated as percent)] is used as an index of reliability.

RESULTS AND DISCUSSION

The qualitative assessment of equipment were described in the METHODS, while this section deals primarily with more quantitative evaluation and results of tests. These tests should be performed for any microfluorometry system to verify the quality of its operation.

Photosensitive Area of Video Monitor

Fig. 12 shows the uniformity of fluorescence detection across a majority of the area displayed on the video monitor. The 6 μm microsphere was used as an approximation to a point source of fluorescence. Note that the display maximum 380 nm fluorescence is only 3500 mV, which is only 35% of the full scale 10,000 mV fluorescence. The variability across all locations expressed as the coefficient of variation of the fluorescence is only 4%. Thus, whether one is studying large cells spanning the entire ~420 μm^2 area (60 μm vertical x 70 μm horizontal) of the video monitor or small cells comprising only 0.1 the area of the monitor, the measured fluorescence signal should represent an average of the fluorescence throughout the cell. This averaged signal is not grossly distorted by spatial inhomogeneities in fluorescence recording resulting from instrument characteristics. These results contrast with those obtained using a Nikon PFX projection system, which showed reductions in fluorescence to about < 10% of the maximal value when the point source of fluorescence is moved only about 20 μm from center (data not shown).

Areas of low fluorescence, however, do exist and must be documented for each microfluorometry system. Note that the fluorescence is least in the right bottom corner, designated by row coordinate 6 and column coordinate 7. Similarly, the right top corner shows low fluorescence. The fluorescence drops to essentially zero when the microsphere is placed to the right of column 7 outside of the video monitor display. The reason for this reduction is twofold: 1) the excitation area is restricted to only the 100 μm diameter area shown in Fig. 3 and; 2) any fluorescence from these peripheral areas are reflected from the 570 nm dichroic mirror (#5, Fig. 4) and likely to have diverged to outside the perimeter of the photocathode of the PMT (#8, Fig. 4). The great utility in this steep fluorescence decrease is in reduction of background fluorescence from the patch pipette during simultaneous voltage-clamp and microfluorometry experiments, as illustrated in the companion chapter (43).

These results suggest that multiple cells can be studied simultaneously to yield population averaged Ca$_i$ responses (43). A strong implication of these results is that microfluorometry systems employing indo-1 (1,28,36,44) need to closely document the uniformity of fluorescence emission because mirror/beam splitters are essential to quantification of the indo-1 ratio and 2 separate PMTs must be used.

Output Voltage of Circuitry Proportional to Percent Optical Transmission

Photon counting of the 10 ns pulses generated by the amplifier/discriminator shown schematically in Fig. 6 (photon counter, near top) would be certain to yield a linear relationship between counts per unit time and light intensity, because the pulses represent threshold discriminated photon amplification by the PMT (Fig. 6, top). However, we have further processed the 10 ns pulses through numerous circuits (Figs. 7, 9) that convert the digital signal to analog form and then the data were re-digitized by the analog-to-digital converter for presentation to the computer. The fidelity of this signal processing must be verified. Fig. 13 shows that the output voltage of optical processing circuitry is proportional to percent transmission of light. The analog voltage out of the frequency-to-voltage converter was shown previously to be directly proportional to light intensity at the PMT over a range of 0 to +5 V. This was verified by the use of a series of neutral density filters (Fig. 2 in Ref. (45)). We improved the linearity of the relationship to +10 V set by a trim potentiometer (associated with IC5, Fig. 7) and 2-fold gain provided by IC4 in Fig. 9. The 1 and 10 μs pulse widths most commonly used gave a linear relationship with a correlation coefficient (r) of 0.99. The data show a simple improvement over the previous report (45) and satisfy one of the specific criteria for the validity of the microfluorometry instrumentation.

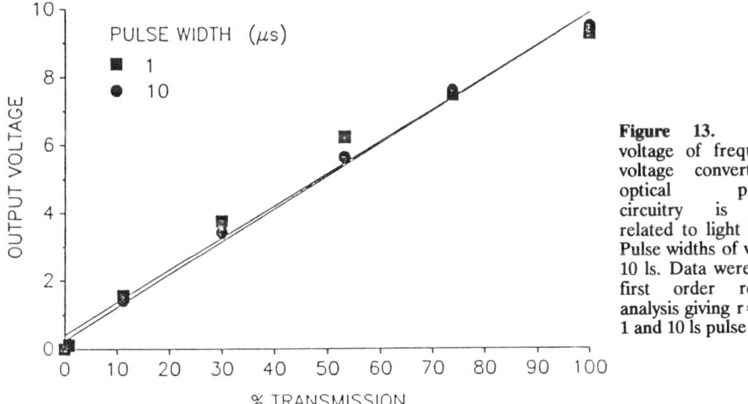

Figure 13. Output voltage of frequency-to-voltage converter and optical processing circuitry is directly related to light intensity. Pulse widths of were 1 or 10 ls. Data were fit by a first order regression analysis giving r=0.99 for 1 and 10 ls pulse widths.

The ratios out of the analog divider typically are within <1% of the calculated ratios and, therefore, satisfy another criterion for validity of the system. It should be noted, however, that the accuracy of the AD538BD divider output (IC8, Fig. 9) depends on the precision of the inputs to the divider. The AD538BD device accuracy is 0.5%; this degree of precision is maintained by using low input offset voltage operational amplifiers for the WL1 and WL2 outputs and the divider numerator input.

Single vs. Multiple Cells

Fig. 14 indicates the effectiveness of the measurement aperture (#2, Fig. 4) in masking the fluorescence emission of one object from the recording of fluorescence emission from an adjacent object, similar to the arrangement in Fig. 12 (top panel). The 6 μm diameter microspheres (#1-#4) showed similar fluorescences, as expected. It is important to note that the individual fluorescences of about 700 mV for these adjacent microspheres in Fig. 12 is similar to the fluorescence of individual 6 μm microspheres isolated entirely in the video monitor and measured during the same experimental session. The sum of these fluorescences from #1 and #2 or #3 and #4 (SUM) differed from that measured by the PMT by only <4.5%.

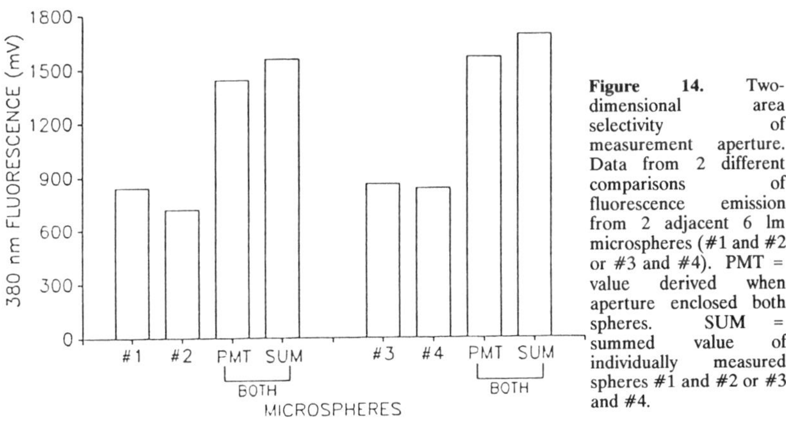

Figure 14. Two-dimensional area selectivity of measurement aperture. Data from 2 different comparisons of fluorescence emission from 2 adjacent 6 lm microspheres (#1 and #2 or #3 and #4). PMT = value derived when aperture enclosed both spheres. SUM = summed value of individually measured spheres #1 and #2 or #3 and #4.

262

The ability of the the microfluorometry system to sum accurately the fluorescence of several objects depends on the data in Fig. 11 showing uniformity of fluorescence across the video monitor and the data in Fig. 14 showing the ability to define spatially the area of measurement. Fig. 15 shows that the individual fluorescence values of 7 microspheres randomly dispersed across the video monitor (similar to Fig. 12, top panel) were similar. The summed fluorescence of all 7 (SUM) differed from that measured by the PMT by only <4.5%. Furthermore, differing ratios from different size spheres were also accurately averaged by PMT recordings to yield values comparable to the mathematical average. The implication of these findings is that averaged Ca_i responses might be derived from groups of 7 or more dispersed cells simultaneously. Comparison to single cell Ca_i dynamics might then be possible, similar to that done with digital imaging (22, 58).

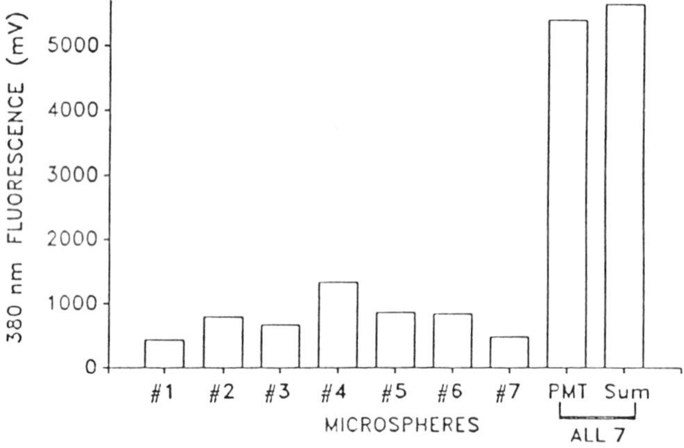

Figure 15. Additive effect of multiple microspheres in field on fluorescence. Values recorded individually from 7 microspheres. PMT = value derived when aperture enclosed ALL 7 spheres. SUM = summed value of individually measured spheres #1-#7.

The data in Figs. 12, 14, and 15 show the uniform photosensitivity across the video monitor and ability to accurately define the area of measurement in the horizontal plane within at least 6 μm diameter areas. We determine from the data in Fig. 16 that the limits of resolution in the vertical plane are somewhere between 16 and 48 μm. The absolute fluorescence at 380 nm increases in direct proportion for sphere diameters from 6 to 16 μm and then saturates beyond 16 μm. The results are not likely due to the saturation of the circuitry, because Fig. 13 indicates that the output voltage of the optical processor was linear up to +10 V. None of the 48 μm microspheres had fluorescences of 10 V. One implication of these results is that the limit of specimen (cell) thickness appears to be between 16 and 48 μm. In addition, background fluorescence from a pipette loaded with fluorescent indicator is likely to be minimal if the measurement aperture is limited to the area of the cell and if the focal plane is below the cell and pipette.

Frequency Response

We have crudely tested the frequency response of the system by monitoring fluorescence using the 20 ms rotational period as in Fig. 10A and have induced rapid changes in fluorescence intensity by turning the focus knob on the microscope. Note in Fig. 10B that the sample-and-hold amplifiers of the optical processing circuitry closely track the oscillations in PMT fluorescence intensity. In addition, we have monitored Ca_i in field stimulated cardiac myocytes and tracked peak changes in Ca_i occurring in less than 75 ms (31) as shown in the companion chapter (43). The frequency response is limited entirely by the rotation period, since the analog divider, electronic switch, etc. all have frequency responses in the 10 KHz range. If greater than 20 ms resolution were required, a maximum 20 ms rotational period and a 4-sector wheel with 2 sectors of each excitation wavelength, thus providing 10 ms ratios, is highly recommended. This would require modification of our circuitry, but would provide 10 ms ratios. In limited attempts to use a 10 ms rotational period, the vibrations were excessive and may create problems for voltage-clamp of single cells and stability of the wheel configuration. An alternate strategy for maximizing time resolution during voltage-clamp is to switch wavelengths only between identical voltage-clamp steps as done for cardiac myocytes (9). The only limitation would then be the speed of the dye in binding Ca. Indo-1 and other dual wavelength emission dyes certainly hold a decisive advantage over fura-2 and other dual wavelength excitation indicators for the above reasons (1, 36, 44).

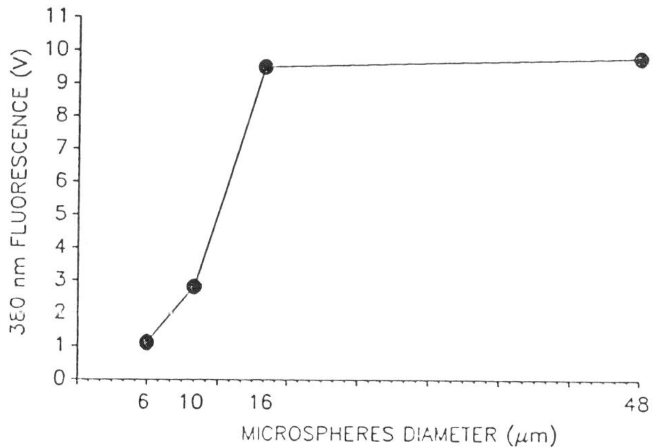

Figure 16. Effect of specimen thickness on fluorescence recording.

Validity: Fluorescence Ratio is Proportional to [Ca^{2+}]

The empirical determination of fluorescence ratio as a defined function of free ion concentration is a clear indicator of the validity of the microfluorometry instrumentation. Calibrations of fluorescence ratios take into account optical alignment, detector sensitivity, and electronic differences, etc. that may influence recordings of fluorescence independent of the physiological status of the cell(s) studied. Fig. 17 shows that despite some differences, the fluorescence ratios show a predictable relationship to the free Ca concentration in known standards set with Ca-EGTA buffers (29, 37, 41, 45, 51). Calibration using a 20x objective yields yet another predictable relationship, though not identical to those in Fig. 17 (data not shown). In addition to reasons of physiological variability, the relatively frequent calibration of the system is necessary to ensure accurate compensation for instrumentation drifts, etc. We noticed, for example, that over a period of 12 months the ratio for 100 nM Ca (using Ca-EGTA buffers) drifted from about 0.90 to only 0.70. Similar variations were reported by Neher (26).

Effect of background fluorescence. A potentially large source of background fluorescence arises from the transillumination with 600 (or 610) nm light during microfluorometry. Specifically, the 600 nm light causes a five to ten-fold elevation in background fluorescence when illuminating the cell during fura-2 fluorescence measures. Inclusion of the wide-band interference filter (#6, Fig. 4) with a center wavelength of 510 nm (the peak of the fura-2 emission spectrum) and a half-band of 60 nm reduces background optical noise to the same level noted in the absence of illumination of the cell with 600 nm light. High background fluorescence that is not compensated has been shown to influence the quantitation of fluorescence recordings (23). The data in Fig. 18 clearly indicate that when measuring the ratio from a 1000 nM Ca standard and the background fluorescence was increased by 30% the percent error in the ratio was >20%. Fig. 19 shows the adverse effect of uncompensated background fluorescence is greatest as the fluorescence ratio ([Ca^{2+}]) increases. These conditions would exist with cell activation and a rise in Ca$_i$. Clearly, these data indicate the necessity to minimize background fluorescence during microfluorometry on cells, as detailed in the companion chapter (43).

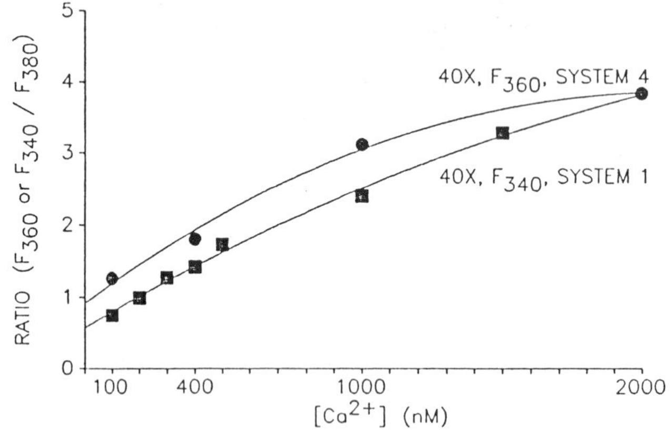

Figure 17. Validity of instrumentation: fluorescence ratio (F_{340} or F_{360}/F_{380}) vs. $[Ca^{2+}]$ calibration standards. N = 5 repeat measurements per point. Smooth line was least squares fit to the data points by a second order polynomial. Excitation on System #1 was provided by 340 and 380 nm light and on System #4 was provided by 360 and 380 nm light.

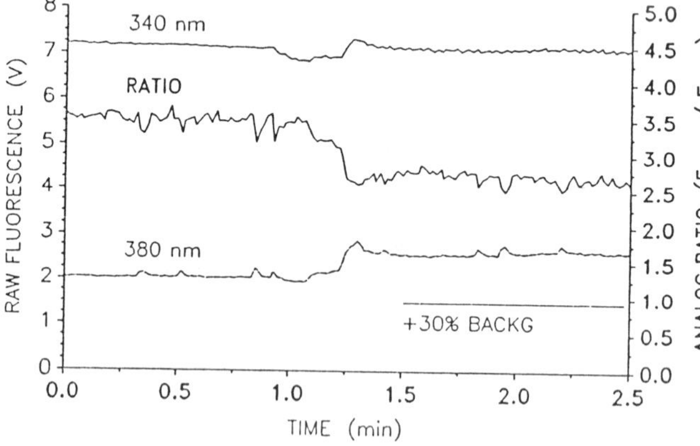

Figure 18. Effect of high background fluorescence on the accuracy of ratio ($[Ca^{2+}]$) determination.

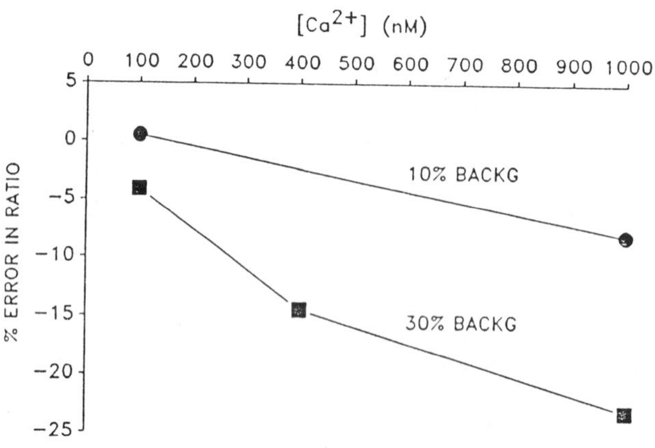

Figure 19. Percent error of the analog ratio circuit as a function of background light and $[Ca^{2+}]$.

Reliability

The reliability of calibration constants during the same experimental period of several hours (intra-assay variation) was very high. The coefficient of variation [(SD / X) x 100, stated as percent)] for 10 repeat measurements using a Ca-EGTA buffer of 100 nM was only 4.2%; inter-assay variation was measured by determination of the ratio of the 100 nM Ca-EGTA buffer over several days within a week yielded a coefficient of variation of <5%. Also, the average coefficient achieved for 7 different users on 10 repeat measures of a Ca standard was only 3.7%. The excellent stability of fluorescent measures derived from a 6 μm microsphere is shown in Fig. 20; the coefficient of variation was only 0.9% during 90 min of measurement. Manipulation of the alignment of the beam probes at the filter wheel enclosure housing a 340 nm / 380 nm filter wheel and fiber optic input assembly described in the METHODS yielded variability in measurement of Ca standards similar to inter-assay variability of about 5%. No manipulation of the alignment of the beam probes at the filter wheel enclosure holding the 360 nm / 380 nm interference wheel was attempted because of the more precision alignment necessary. Indeed, it was necessary to align the beam probes to yield an <u>uncollimated</u> light beam incident on the filter wheel, thereby changing slightly the wavelength transmitted by the filter to achieve an isosbestic wavelength for fura-2 (16).

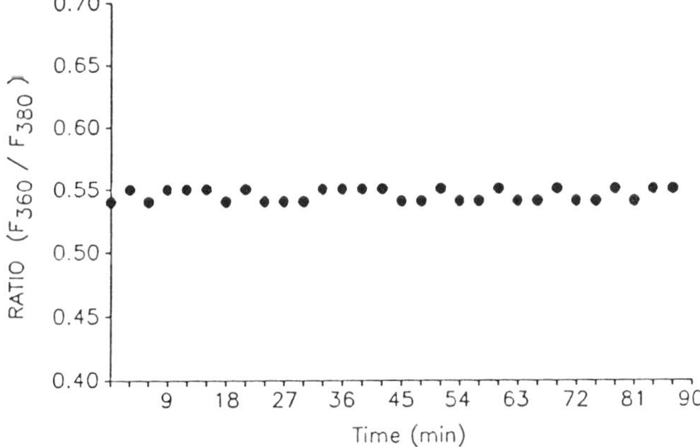

Figure 20. Reliability (stability) of 6 μm microsphere fluorescence ratio recording.

SUMMARY AND CONCLUSIONS

The previous methods for synchronization of the fluorescence emission with the appropriate excitation wavelength required computer programming, in addition to mechanical components and hardware for rotating the interference filters (45). Although computer programming was used successfully, there always remains an unavoidable need for hardware. Accordingly, our strategy was to use sample-and-hold circuitry synchronized with the rotation of the filter wheel motor to process PMT output, rather than computer programming. The analog optical processing hardware then provided separate analog signals corresponding to 340 or 360 and 380 nm excitation to the analog-to-digital converter of the computer. The main advantages of the improvements are ease of interface to any typical A-D converter, essentially no computer programming, optical arrangement simplicity and, in contractile cells such as smooth muscle cells, the ability to monitor contraction provides another measurement parameter. On-line background subtraction and ratio calculation have greatly increased the convenience. We addressed the major points regarding the optical configurations and analog processing of fluorescence signals from dual wavelength excitation systems. Although the instrumentation was originally designed for fura-2 measurements of Ca_i, the techniques are immediately applicable to measurement of K, Na, and Mg.

The proven validity, reliability, relative simplicity and convenience, low cost, and flexibility of this microfluorometry system make it extremely useful for the study of intracellular ion concentrations and electrophysiology in vascular smooth muscle and endothelial cells (38-43, 51, 52, 52a).

ACKNOWLEDGEMENTS

Supported by a University of Missouri Alumni Fund for Faculty Development, Institutional Biomedical Research Support Grant RR 07053, the American Heart Association-Missouri Affiliate, Marion Laboratories, Marion Merrell Dow, Inc., NIH HL41033, and NIH Training Grant T-32 HL07094. DAH was a Dalton Research Center Cardiovascular Fellow. The authors thank Tom Erdal and Charles Jorgenson of the Dalton Research Center for help in construction of electronics and printed circuit board/artwork preparation, respectively. We greatly appreciate the loan of neutral density filters from Dr. Virginia Huxley. We also thank the Science Instrument Shop of the U. of Missouri for expert craftsmanship of the mechanical components and Omega Optical, Inc., Brattleboro, VT for permitting the publication of transmission spectra for interference filters. Additional information on this system is available from the authors.

REFERENCES

1. Aaronson, P.I. and C.D. Benham, *J. Physiol. (Lond)* 416, 1-18 (1989).
2. Adams, S.R., A.T. Harootunian, Y.J. Buechler, and S.S. Taylor, *Nature* 349, 694-697 (1991).
3. Baylor, S.M. and S. Hollingworth, *J. Physiol. (Lond)* 403, 151-192 (1988).
4. Becker, P.L. and F.S. Fay, *Am. J. Physiol.* 253, C613-C618 (1987).
5. Becker, P.L., J.J. Singer, J.V. Walsh Jr., and F.S. Fay, *Science* 244, 211-214 (1989).
6. Blinks, J.R., *Methods Enzymol.* 172, 164-202 (1989).
7. Bolsover, S.R., *J. Gen. Physiol.* 88, 149-165 (1986).
8. Callewaert, G., L. Cleeman, and M. Morad, *Proc. Natl. Acad. Sci. USA* 85, 2009-2013 (1988).
9. Cannell, M.B., J.R. Berlin, and W.J. Lederer *Science* 238, 1419-1423 (1987).
10. Chance, B., V. Legallais, J. Sorge, and N. Graham, *Anal. Biochem.* 660, 498-514 (1975).
11. Chandra, S., D. Gross, Y.-C. Ling, and G.H. Morrison, *Proc. Natl. Acad. Sci. USA* 86, 1870-1874 (1989).
12. Cheung, J.Y., D.L. Tillotson, R.V. Yelamarty, and R.C. Scaduto,Jr., *Am. J. Physiol.* 256, C1120-C1130 (1989).
13. Golchini, K., M. Macjivuc-Basic, S.A. Gharib, D. Masilamani, M.E. Lucas, and I. Kurtz, *Am. J. Physiol.* 258, F438-F443 (1990).
14. Goldman, W.F., S. Bova, and M.P. Blaustein, *Cell Calcium* 11, 221-231 (1990).
15. Goldman, W.F., W.G. Wier, and M.P. Blaustein, *Circ. Res.* 64, 1019-1029 (1989).
16. Grynkiewicz, G., M. Poenie, and R.Y. Tsien, *J. Biol. Chem.* 260, 3440-3450 (1985).
17. Himpens, B. and A.P. Somlyo, *J. Physiol. (Lond)* 395, 507-530 (1988).
18. Kawamura, A., Jr. and Y. Aoyama, in *Immunofluorescence in Medical Science.* (Springer-Verlag Berlin 1983) p. 75-98.
19. Minta, A., J.P.Y. Kao, and R.Y. Tsien, *J. Biol. Chem.* 264, 8171-8178 (1989).
20. MOLECULAR PROBES, *Bioprobes* 9, (1988).
21. Mollard, P., P. Vacher, B. Dufy, B.P. Winiger, and W. Schlegel, *J. Biol. Chem.* 263, 19570-19576 (1988).
22. Monck, J.R., E.E. Reynolds, A.P. Thomas, and J.R. Williamson, *J. Biol. Chem.* 263, 4569-4575 (1988).
23. Moore, E.D.W., P.L. Becker, K.E. Fogarty, D.A. Williams, and F.S. Fay, *Cell Calcium* 11, 157-179 (1990).
24. Moore, E.D.W. and F.S. Fay, *Physiologist* 32, 147 (Abstract) (1989).
25. Murphy, E., C.C. Freudenrich, L.A. Levy, R.E. London, and M. Lieberman, *Proc. Natl. Acad. Sci. USA* 86, 2981-2984 (1989).
26. Neher, E., *J. Physiol. (Lond)* 395, 193-214 (1988).
27. Oriel Corporation, *Vol. II. Light Sources, Monochromators, Detection Systems* (1985).
28. Peeters, G.A., V. Hlady, J.H.B. Bridge, and W.H. Barry, *Am. J. Physiol.* 253, H1400-H1408 (1987).
29. Poenie, M., J. Alderton, R. Steinhardt, and R.Y. Tsien, *Science* 233, 886-889 (1986).
30. Scarpa, A., in *Techniques in Cellular Physiology. Vol. 1, Part II, Techniques in Life Sciences,* P.F. Baker, ed. (Elsevier/North Holland, Amsterdam 1982) p. 1-40.
31. Schaefer, M.E., M.H. Laughlin, and M. Sturek, *Med. Sci. Sports Exercise* 21, S17 (Abstract) (1989).
32. Schlegel, W., B.P. Winiger, F. Wuarin, G.R. Zahnd, and C.B. Wollheim, *J. Receptor Res.* 8, 493-507 (1988).
33. Sick, T.J. and M. Rosenthal, *J. Neurosci. Methods* 28, 125-132 (1989).
34. Simonson, M.S., S. Wann, P. Mene, G.R. Dubyak, M. Kester, Y. Nakazato, J.R. Sedor, and M.J. Dunn, *J. Clin. Invest.* 83, 708-712 (1989).
35. Somlyo, A.P. and B. Himpens, *FASEB J.* 3, 2266-2276 (1989).

36. Spurgeon, H.A., M.D. Stern, G. Baartz, S. Raffaeli, R.G. Hansford, A. Talo, E.G. Lakatta, and M.C. Capogrossi, *Am. J. Physiol.* 258, H574-H586 (1990).
37. Stehno-Bittel, L., M.H. Laughlin, and M. Sturek, *J. Appl. Physiol.* (In press) (1991).
38. Stehno-Bittel, L., M.H. Laughlin, and M. Sturek, *Am. J. Physiol. Heart Circ. Physiol.* 259, H643-H647 (1990).
39. Stehno-Bittel, L. and M. Sturek, (Submitted) (1991).
39a. Stehno-Bittel, L. and M. Sturek, (Submitted) (1990).
40. Sturek, M., K. Kunda, and Q. Hu, (Submitted) (1991).
41. Sturek, M., P. Smith, and L. Stehno-Bittel, in *Electrophysiology and Ion Channels of Vascular Smooth Muscle Cells and Endothelial Cells*, by N. Sperelakis and H. Kuriyama, eds. (Elsevier New York, 1991) (in press).
42. Sturek, M., L. Stehno-Bittel, and P. Obye, in *Electrophysiology and Ion Channels of Vascular Smooth Muscle Cells and Endothelial Cells*, N. Sperelakis and H. Kuriyama, eds. (Elsevier New York, 1991) (in press).
43. Sturek, M., L. Stehno-Bittel, and P.K. Obye, in *Electrophysiology and Ion Channels of Vascular Smooth Muscle Cells and Endothelial Cells*, N. Sperelakis and H. Kuriyama, eds.(Elsevier New York 1991) (in press).
44. Takamatsu, T. and W.G. Wier, *Cell Calcium* 11, 111-120 (1990).
45. Thayer, S.A., M. Sturek, and R.J. Miller, *Pflugers Arch.* 412, 216-223 (1988).
46. Thomas, M.V., in *Techniques in Calcium Research* (Academic Press London 1982) p. 139-165.
47. Tsien, R.Y., *Trends Neurosci.* 11, 419-424 (1988).
48. Tsien, R.Y. and A.T. Harootunian, *Cell Calcium* 11, 93-109 (1990).
49. Tsien, R.Y., T.J. Rink, and M. Poenie, *Cell Calcium* 6, 145-157 (1985).
50. Voyta, J.C., D.P. Via, C.E. Butterfield, and B.R. Zetter, *J. Cell Biol.* 99, 2034-2040 (1984).
51. Wagner-Mann, C.C., L. Bowman, and M. Sturek, *Am. J. Physiol. Cell Physiol.* 260, C763-C770 (1991).
52. Wagner-Mann, C.C. and M. Sturek, *Am. J. Physiol. Cell Physiol.* 260, C771-C777 (1991).
52a. Wagner-Mann, C.C., Q. Hu, and M. Sturek, *Brit. J. Pharmacol.* (1991).
53. Weiner, I.D. and L.L. Hamm*Am. J. Physiol.* 256, F957-F964 (1989).
54. Wier, W.G. and L.A. Blatter, *Cell Calcium* 12, 241-254 (1991).
55. Williams, D.A., P.L. Becker, and F.S. Fay, *Science* 235, 1644-1648 (1987).
56. Williams, D.A. and F.S. Fay, *Cell Calcium* 11, 75-83 (1990).
57. Williams, D.A., K.E. Fogarty, R.Y. Tsien, and F.S. Fay, *Nature* 318, 558-561 (1985).
58. Winiger, P. and W. Schlegel, *Biochem. J.* 255, 161-167 (1988).
59. Ygi, S., P.L. Becker, and F.S. Fay, AY, *Proc. Natl. Acad. Sci. USA* 85, 4109-4113 (1988).

METHODS FOR SIMULTANEOUS VOLTAGE-CLAMP, MICROFLUOROMETRY, AND VIDEO OF CELLS. II. PHYSIOLOGY

M. STUREK, L. STEHNO-BITTEL, and P.K. OBYE

Department of Physiology, School of Medicine, and Dalton Research Center, University of Missouri, Columbia, Missouri 65211

Abstract

Physiological data are presented on voltage-clamp and fura-2 measurements of intracellular free Ca concentrations (Ca_i) in smooth muscle cells that were derived using the instrumentation described in another chapter (59). Contractile responses of freshly dispersed cells from the coronary artery to K depolarization and agonists were observed with a video monitor. Ratio measurements of Ca_i were valid and reliable, as shown by in vitro and myoplasmic calibrations. Myoplasmic fura-2 in fura-2/AM-loaded smooth muscle cells showed decreased Ca_i sensitivity; therefore, accurate calibration of fluorescence ratios with Ca_i should be conducted under approximately "in situ" conditions. There was no interference from intrinsic birefringence and background fluorescence was less than 5% of the total fluorescence signal of cells loaded with fura-2. In cells loaded with fura-2/AM up to 97-99% of fura-2 was free in the myoplasm. With brief (15 min) loading periods there was no adverse buffering of Ca_i or sequestration of fura-2 by organelles in smooth muscle cells. Some fura-2 was lost from cells during constant superfusion, but did not photobleach significantly. The Ca_i transients elicited from single smooth muscle cells had variable time courses and amplitudes, but the averaged responses closely resembled the Ca_i transient measured from groups of 5-10 cells. The rate of quenching of fura-2 fluorescence by Mn estimated unidirectional divalent cation influx. The excellent time resolution (20 ms/ratio) of the microfluorometry system was shown by measurement of Ca_i transients in cardiac myocytes; versatility was shown by measurement of intracellular Na with SBFI in smooth muscle. Fura-2 pentapotassium salt in Ca-EGTA buffers dialyzed into smooth muscle cells via a patch pipette within a maximum of 3 min, allowed clamping of Ca_i and yielded ratios similar to those found with in vitro (cuvette) calibration curves. These methods for simultaneous recordings in single smooth muscle cells made it possible to determine the precise temporal relationship between Ca_i transients caused by Ca release from the sarcoplasmic reticulum and the activation of ion channels during voltage-clamp. It is hoped that researchers recognize the limitations and the strengths of the methods described here for using simultaneous voltage-clamp with fura-2 and will utilize the full potential of the instrumentation with other ion-sensitive dyes.

INTRODUCTION

Fura-2 has gained widespread use as an indicator of intracellular free calcium concentration (Ca_i) in vascular smooth muscle, endothelial, and other cells types for several reasons (For reviews see Refs. 1, 53, 65). First, there is a shift in the excitation spectrum of fura-2 when fura-2 binds Ca, which allows one to calculate a ratio of the fluorescence intensities that is related to Ca_i in a manner unaffected by differences in dye concentration or optical path length (65). Second, fura-2 can be loaded into cells as the cell permeant ester form (55, 56, 60-62, 68, 69) or through the patch pipette as the membrane impermeant pentapotassium salt form (8, 15, 49, 56, 58, 62). This avoids the use of hyperpermeabilization of smooth muscle cells that is necessary for loading aequorin (13, 44), but necessitates knowing the extent of fura-2 de-esterification and cellular distribution for appropriate quantification of fura-2 fluorescence (2, 32, 41, 60, 61, 68, 71). Third, epifluorescence measurements offer a better signal-to-noise ratio than absorbance dyes because the light emitted is usually compared to a very low background. In contrast, a small absorbance difference between two relatively large signals must be measured with metallochromic indicators (48). Fourth, it is assumed that there are fewer artifacts due to light scattering in contractile cells when using fura-2 rather than, for example, absorbance dyes (3, 37, 48). In small cells fluorescent Ca_i indicators are used almost exclusively

Published 1991 by Elsevier Science Publishing Company, Inc.
Ion Channels of Vascular Smooth Muscle Cells and Endothelial Cells
Sperelakis and Kuriyama, Editors

because they are the most technically feasible method for measurement of Ca_i (65). Fifth, the combined use of voltage-clamp techniques and fura-2 microfluorometry yields data on regulation of Ca_i and membrane electrical properties that neither technique can provide alone (9, 15, 30, 56, 58, 62-64).

The companion chapter established the quality of optical and electronic operations of the dual wavelength excitation microfluorometry system in use in this laboratory by using fluorescent standards (59). The purpose of this chapter was to validate the use of the instrumentation for fura-2 microfluorometry on single smooth muscle cells by determining: 1) the behavior of intracellular fura-2, including myoplasmic distribution, calibration of fura-2 fluorescence ratios, Ca_i buffering, and photobleaching; 2) fluorescence properties of smooth muscle cells at rest and during contractions observed through a video monitor; 3) estimation of divalent cation influx with Mn; 4) single cell vs. population averaged Ca_i responses, and; 5) simultaneous whole-cell patch-clamp and microfluorometry for measurement of ionic currents and Ca_i.

The types of experiments described here validate the use of fura-2 as an indicator of Ca_i in the bulk myoplasm of coronary smooth muscle and are relevant to the study of other cell types. We have also used the microfluorometry system with endothelial cells (36, 57, 61) and cardiac myocytes (28). Furthermore, the feasibility of using other ion-sensitive fluorescent indicators requiring dual wavelength excitation was demonstrated (SBFI for measurement of Na).

METHODS
Electronic and Optical Instrumentation

The electronic and optical instrumentation is described in greater detail in the companion chapter (59). We add more detail on the instrumentation in this chapter as it pertains to physiological techniques and interpretation of data in later sections. We also refer frequently to various figures in the instrumentation chapter (59) for details.

Fresh Isolation of Cells From the Coronary Artery

Single cells from bovine or porcine left circumflex, left anterior descending, and right coronary arteries are enzymatically isolated (66, 68). Sterile procedures are used throughout the collection of the arteries and dispersion of the cells. At an abattoir, bovine, but not porcine arteries, are cut longitudinally to expose the lumen while still attached to the heart. The arteries are then grossly dissected from the heart by cutting into the myocardium and removing 1-5 mm of the adjacent myocardium beneath the arteries. The length of each main artery is typically about 10 cm for bovine arteries and less than 5 cm for porcine arteries. The arteries are rinsed with about 100 ml sterile physiological salt solution (PSS) composed of (in mM): 2 $CaCl_2$, 138 NaCl, 1 $MgCl_2$, 5 KCl, 10 Hepes, 10 glucose, and 0.02 (vol:vol) penicillin/streptomycin, pH 7.4 with NaOH. This procedure maintains sterility of the arteries, which are then placed in a container with ice-cold cell culture media containing (in mM): 2 $CaCl_2$, 135 NaCl, 1 $MgCl_2$, 5 KCl, 0.44 KH_2PO_4, 0.34 Na_2HPO_4, 2.6 $NaHCO_3$, 20 HEPES, 10 glucose, pH 7.4 with NaOH. Also, the following components are added in the dilutions (vol:vol) indicated: 0.02 amino acids, 0.01 vitamins, 0.002 phenol red, 0.01 penicillin/streptomycin (GIBCO, Grand Island, NY), and 0.02 horse serum (Fisher Biotech). This buffer is similar to Eagle's Minimal Essential Medium, except that 20 mM HEPES is included as an additional pH buffer; hence, it is referred to as EMEM/HEPES storage media. The bottle containing the arteries is transported on ice to the laboratory. In a laminar air flow cabinet (Forma Scientific, Marietta, OH) the arteries are rinsed again with PSS containing 2% penicillin/streptomycin. The myocardial tissue, connective tissue, and fat are dissected from the artery while in a solution similar to the above EMEM/HEPES, but containing only 0.5 mM Ca (Low Ca) and no horse serum. During dissection care is taken not to stretch the arteries excessively or damage the endothelium, which may result in the release of vasoconstrictor agents such as endothelin. The artery is cut into segments and then placed in a 30 ml bottle. Routinely, the left anterior descending artery is cut into its proximal, middle, and distal thirds because there is evidence for functional differences (50). Cells are dispersed from the arteries immediately or the arteries could be stored in EMEM/HEPES plus 2% horse serum at 4° C, typically for up to 7 days prior to dispersion.

For dispersion the arteries are rinsed in Low Ca solution and a smaller 1-2 cm long segment is cut for the dispersion. This amount of tissue usually gives enough cells for voltage-clamp and microfluorometry experiments on single cells. The 1-2 cm segment of artery is placed into a bottle containing a cured layer of Sylgard elastomer (Dow Corning, Inc., Midland, MI) and about 1-2 ml of Low Ca. The arterial segment is pinned onto the elastomer with the lumen facing up. The Low Ca solution is removed and 1.5 ml of enzyme solution is added, which contains the following components in Low Ca solution: 294 U/ml collagenase (CLS II, Worthington, Freehold, NJ), 2 mg/ml bovine serum albumin (Fraction V, Sigma Chemical Co., St. Louis, MO), 1 mg/ml soybean trypsin inhibitor (Worthington), 0.4 mg/ml DNAase I (Type IV, Sigma). The artery is then placed in a 37° C water bath and shaken in a reciprocating motion at 100 strokes/min for 60 min. The supernate fraction at this step contains mostly endothelial cells, which are easily identified morphologically as sheets of cells or "grape- like clusters", as shown in Fig. 1 of another chapter (61). Few smooth muscle cells like those in Fig. 1 are seen in this fraction. It is very important to remove a small droplet of the enzyme solution and observe the solution under the microscope (Nikon TMS, Nikon, Inc., Garden City, NY) to monitor the progress of the dispersion. The supernate is removed and 1.5 ml enzyme solution again added and the artery placed in the water bath for 60 min. Vigorous pipetting of the enzyme solution onto the lumen of the artery greatly facilitates dispersion of cells from the artery. The next fraction usually contains a substantial number ($\sim 10^6$) of relaxed smooth muscle cells. This dispersion procedure is repeated several times if more cells are needed; the fractions could be pooled or kept separate. Also, we have had good success with dispersing cells by incubating the artery at room temperature (typically $22\text{-}25^O$ C) in 3 ml enzyme solution for >8 h.

Loading of cells with fura-2 is done as described previously (55, 60, 68, 69). For experiments on cells loaded with the membrane permeant acetoxymethylester form of fura-2 (fura-2/AM), cells are incubated with 2.5 μM of fura-2/AM for only 15 min at 37° C. Typically, 0.5 ml of pre-warmed PSS that contains 0.2% bovine serum albumin (Fraction V, Sigma Chemical Co.) is added to the cells suspended in enzyme solution to bring the volume to 2 ml; 5 μl of a 1 mM stock fura-2/AM solution (made in 100% dimethylsulfoxide). This is the "normal" loading procedure. Following incubation, the cells are centrifuged at 1500 x g for 4 min to form a loose pellet of cells so that the fura-2 containing solution could be aspirated. Cells are rinsed once by suspending the cells into 2 ml of pre-warmed storage media at 37° C for 30 min. The cells are then centrifuged, the storage media aspirated, and the cells resuspended in 2 ml of PSS containing 2% bovine serum albumin with no added fura-2. The "excessive" fura-2/AM loading procedure is identical to that described above, except that the fura-2 concentration is 5 μM and the incubation is 60 min. For patch-clamp experiments gigaseal formation is often enhanced by dispersing the cells in serum-containing enzyme solution. These cells usually are not loaded with fura-2/AM, because fura-2 pentapotassium salt is loaded into the cells via the patch pipette.

Smooth muscle cells remain relatively relaxed for up to 6-8 h after the dispersion and cells contract when exposed to vasoconstrictors. Experiments described in this chapter and our previous work (38, 55, 60, 68, 69) are conducted on relatively relaxed, freshly isolated smooth muscle cells distinguished by contractile responses to depolarization (80 mM K) or caffeine (5 x 10^{-3} M) (Fig. 1). Contractile responses to endothelin (5 x 10^{-8} M, data not shown) are another criterion for identifying cells as smooth muscle. The video system described in the companion chapter (59) is used to verify contraction during each experiment.

Calibration of Fura-2 Fluorescence Ratios with Ca$_i$

A calibration curve is generated for fura-2 pentapotassium salt in the range of Ca concentrations likely to be encountered in experiments on these smooth muscle cells. Eleven different solutions of various Ca$_i$ concentrations are used with fura-2 pentapotassium salt in a mock intracellular solution containing (in mM): 126 KCl, 10 NaCl, 20 HEPES, 1 MgCl$_2$ 2H$_2$O, 0.1 fura-2 pentapotassium salt, pH 7.1 with KOH. Free Ca concentrations ranging from 0 to 1500 nM are achieved by using appropriate ratios of H$_2$K$_2$EGTA and CaK$_2$EGTA calculated using the apparent stability constant of 3.969 x 10^6 (temperature 22° C , pH 7.1) (14) as done for other calibrations (56, 60, 61, 64, 68). The final EGTA concentration is always 10 mM to ensure a high Ca buffering power.

In vitro calibration constants for fura-2 fluorescence signals are obtained using the mock intracellular solutions indicated above and this microfluorometry system, thus accounting for the unique optical arrangement of this system. The affect of the unique optical system (excitation wavelengths, spectral properties of the dichroic mirrors, etc.) on the exact ratio values derived was shown in Fig. 17 of the companion chapter (59). The same type of coverslip (25 mm diam., 0.17 mm thickness; Fisher Scientific) serving as the bottom of the superfusion chamber (See ref. 64 for an illustration) is used in these calibrations. Three to five 2 μl droplets of the final calibration solution are placed on the coverslip then the investigator focuses on the edge of a droplet for proper optical alignment. The top panel of Fig. 2 shows schematically how we compare fluorescence ratios recorded from each of the droplets. Fluorescence is recorded from the periphery of the droplet to achieve optical path lengths of < 10 μm. The droplet is then moved so that the measurement aperture is near the center of the drop and, therefore, represents > 300 μm depth (path length) of the $[Ca^{2+}]$ calibration solution.

"Myoplasmic" calibration of fura-2 fluorescence ratios is conducted to estimate whether fura-2 was properly hydrolyzed to the pentapotassium salt (47, 60, 67, 68); in other words, whether the myoplasmic calibration for the salt is similar to the *in vitro* measurements. Cells from bovine coronary arteries are dispersed as described above and then divided into two different fractions; one fraction is loaded normally (See above) with fura-2/AM and the other fraction is not loaded, but subjected to the same incubation and centrifugation steps. The two fractions are each divided: the fraction loaded with fura-2/AM divided into 4 fractions and the fraction not loaded is divided into three fractions to which fura-2 pentapotassium salt (fura-2 K_5) is added later. The 7 cell fractions are centrifuged at 1500 x g for 4 min, the supernate removed, and cell pellets (volume of approximately 50 μl) resuspended in 500 μl of final fura-2 calibration solution. The calibration solution contains either 100, 400, 1000, or 2000 nM free $[Ca^{2+}]$ for fractions loaded with fura-2/AM, or 100, 400, or 1000 nM for fractions not loaded with fura-2/AM; the fura-2 K_5 salt is excluded from the calibration solutions for all fractions at this stage. This procedure strongly buffers the external Ca to the desired free $[Ca^{2+}]$. The cells are centrifuged, the 500 μl of calibration solution removed, and 25 μl of the same calibration solutions containing 100 μg saponin/ml added to each cell pellet. For cells that had not been loaded with fura-2/AM, 40 μM of fura-2 pentapotassium salt is now included in the saponin solution. Saponin has been shown to effectively permeabilize the sarcolemma of cells without affecting the intracellular organelles (75). In the case of cells loaded with fura-2/AM the cytoplasmic fura-2 is released from the cells into the suspension in a manner very similar to that done in cuvette measurements in cell suspensions (18). Buffering the free Ca^{2+} concentration to known levels permits determination of the Ca sensitivity of the myoplasmic fura-2 that the cells process (presumably hydrolyzed) from the fura-2/AM that entered the cells during loading. Addition of fura-2 pentapotassium salt in different cell pellets was a control for instrument calibration, possible turbidity of the suspension because of membrane lysis, and interaction of fura-2 with myofilaments and other cell structures (4, 72). The data for all the calibration analyses are fit with a non-linear least squares method in Sigmaplot 3.1 graphics program (Jandel Scientific, Corte Madera, CA).

Computerized Data Acquisition and Analysis

Data are presented to an analog-to-digital (A-D) converter, displayed on the computer terminal, and saved to the fixed disk for off-line analysis. Two A-D converters have been used: the Labmaster (Scientific Solutions, Inc., Solon, OH) and the ADAC 5500 (ADAC Corp., Woburn, MA). Numerous data acquisition and analysis programs and hardware are now available and all would probably be satisfactory for the microfluorometry data. We have used 3 types of software.

Labtech Acquire software (ADAC Corp.) is used in 80286 or 80386 processor-based computers. The simplicity of this system makes it ideal for long-duration microfluorometry experiments with sampling rates of less than about 2 Hz. The A-D converter samples at a maximum rate of 50 Hz, but memory is soon exhausted. Obviously, this software is inadequate for voltage-clamp. The data are immediately placed in an ASCII file during the experiment. For analysis the ASCII file is imported by a spreadsheet program (Quattro, Borland, Inc.). The great flexibility in execution using macros makes the subsequent task of background fluorescence subtraction and division of hundreds of data points trivial. Graphics are of low resolution, but the flexibility in display, labelling, etc. is helpful.

One software package used for voltage-clamp is pCLAMP (versions 4.1 through 5.5) from Axon Instruments, Inc. (Foster City, CA), which is used with 80286 and 80386 processor-based computers. The pCLAMP versions 4.1 through 5.5 are very good for immediate data acquisition, as there is absolutely no programming necessary. The serious drawback, however, is the extreme difficulty in programming needed options. First, there is no option for division of one channel of data by another, thus making simple ratio calculations impossible. To overcome this problem data are output in ASCII format. The second problem is that earlier versions of pCLAMP did not demultiplex the 3 channels of data (ionic current, fluorescence from 340 (or 360) and 380 nm excitation) before sending the data to an ASCII file, thus making it necessary to create multiple files. The multiple files are extremely cumbersome.

The AxoBASIC 1.0 software (also from Axon Instruments, Inc.) requires an 80386 computer. This is overwhelmingly the preferred system because of the flexibility and relative ease of programming. AxoBASIC is currently the main system used in this laboratory, along with some use of Acquire software. The drawback is the investment of time in program development, but the convenience of data acquisition and analysis programs designed specifically for our needs has outweighed the time investment in program development. AxoBASIC is excellent for long-duration protocols involving only microfluorometry and ideal for simultaneous voltage-clamp and microfluorometry. For example, with a program written for acquisition of microfluorometry data (no voltage-clamp) 1024 data points of fluorescence at 2 wavelengths can be collected over a 200 ms period, background subtraction and ratio calculations performed, and data from the on-line calculations displayed every second (we have not attempted faster acquisition with this type of protocol). One of the main strengths of the software is the ability to read the keyboard while waiting execution, thus allowing single character keystrokes to change test potentials, stimulus interval, and gain. The resulting output is assembled in one file for straightforward analysis. Speed and convenience of analysis of all data by AxoBASIC is at least 2-fold greater than by pCLAMP or Acquire. Original records for publication-quality graphics are exported as an HPGL (Hewlett Packard Graphics Language) file and then imported by Corel Draw! 2.0.

Microfluorometry

This section describes a typical experiment in which microfluorometry alone (no voltage-clamp) is performed on smooth muscle cells. We have intended this section to be very "how-to", step-by-step instructions on doing an experiment. Please consult the companion chapter (59) throughout this section for details of data output, instrument components, etc. Given the slow Ca_i responses of most coronary vascular smooth muscle and other tonic smooth muscle (11, 53), microfluorometry experiments typically involve responses that may require minutes to develop fully. We use sample intervals of 48 ms and found that the most brief time-to-peak response of the coronary smooth muscle cells is 5-10 s in response to a maximal, 5×10^{-3} M concentration of caffeine. Accordingly, a sample interval of 1 s and low-pass filter cutoff of 0.16 (time constant = 1 s) are adequate. Note the difference in noise of the fluorescence signals with different low-pass filtering in Figs. 8 and 10. The reader should keep in mind that many of the procedures described in this section also apply to preparation for later procedures in the section *Simultaneous Whole-Cell Patch-Clamp and Microfluorometry* and will not be repeated in that section.

All equipment are turned on, except optical processor, which we have typically left off until excitation of cells with ultraviolet light is needed. We do this to avoid wear on the DC rotary motor. The Xe arc lamp is ignited first and allowed to warm about 15 min to achieve stability of output before use. A compelling reason for igniting the arc lamp first is that it requires voltage pulses of about 15,000 volts for ignition and, therefore, all integrated circuits should be off to avoid damage. It should be made certain that the shutter at the fiber optic input assembly (#1, Fig. 2, Ref. 59) is closed after the arc lamp is ignited and while the optical processor is turned off. When the optical processor is turned off the filter wheel (#7, Fig. 2, Ref. 59) is not turning and, therefore, intense ultraviolet light will be incident only on one spot of the interference filter and may cause bleaching of the interference filter wheel. The computer is checked for proper data acquisition parameters as described above.

The cell superfusion chamber is prepared and the flow of solution through the chamber checked during the 15 min required for instrument warm-up. The coverslip serving as the bottom of the superfusion chamber is changed daily. The solutions are typically room temperature (22-25° C), but we have a series of experiments that indicates warming an aluminum plate on the stage can warm the solution in the bath to 37° C. The solution must flow at a rate of 1-2 ml/min in order to exchange the bath solution in 15-30 s. Cells are placed into the superfusion chamber by turning off the valve system connecting the 4-7 fluid reservoirs with the superfusion chamber (Anspec, Ann Arbor, MI). The cells settle to the bottom and adhere to the coverslip within 5 min. Final checks are made on the instrumentation then the investigator is free to find a relaxed cell isolated from other cells. Once the cells settle, the superfusion is re-started; some cells float away, but typically an abundance of cells remain attached to the coverslip.

The optical processor is now turned on and the shutter on the fiber optic input assembly opened so that the cells containing fura-2 are excited with 340 (or 360) and 380 nm light. Observation of the cell through the binocular eyepieces of the microscope gives an indication of the degree of loading of the cells with fura-2 by the relative brightness of the green fluorescence. The cell is then moved by adjustment of the movable microscope stage to place it in the point in the field previously determined to be the midpoint of the uniform field of photosensitivity (section *Procedures for Verifying Technical Acceptability of Operations. Alignment.* in Ref. 59). If the cell is relaxed, then the fluorescence and 600 (or 610) nm light is diverted to the video port of the microscope (#1, Fig. 4, Ref. 59). The measurement aperture (#2, Fig. 4, Ref. 59) may initially be wide open to the full video monitor dimensions. The cell will appear out of focus initially in the video monitor because the video camera is not parfocal with the microscope. This is a trivial matter and requires only a minor adjustment of the focus knob on the microscope. The measurement aperture is closed around the cell and then the photomultiplier tube shutter (#7, Fig. 4, Ref. 59) is opened.

Composite photomultiplier output, representing the fluorescence emission from the photomultiplier before separation of 340 (or 360) and 380 nm signals, is always monitored on an oscilloscope during experiments. The position pulse from the optical processor is used to trigger the oscilloscope, thus one can observe composite output during one full rotation period continuously. The resulting signals are similar to those in the lower section of Fig. 10A of the companion chapter (59). Of course, the display of the separated signal has been decreased by half in Fig. 10A, thus making the separated 340 and 380 signals track the corresponding components of the composite output (compared with the signals near the bottom of Fig. 6, Ref. 59). Constant observation of the composite output provides assurance that the rotation period is sufficiently fast to monitor the Ca_i dynamics of the cell, because any transient change in fluorescence during the sampling period would be easily noticed on the composite output. Composite output should be of "sufficient" amplitude. Any voltages between 0 and $+10$ V are satisfactory because of the linearity of the amplification processes, etc. (Fig. 13 of Ref. 59). However, for display purposes an initial voltage of between $+5$ and $+10$ volts is desired. Several procedures allow adjustment of the fluorescence signal amplitude: 1) The aperture on the Nikon 40x UVF Fluor fluorescence objective may be adjusted. This usually does not cause drastic movement of the objective or superfusion chamber to cause the cell to move out of the field of view. 2) The pulse width of the optical processor may be changed between 1, 10, or 100 μs. With "normal" loading of fura-2/AM (described above) a pulse width of 1 is sufficient with the smooth muscle and endothelial cells we have studied (61, 62). 3) If a clear focus on the surface of the cell is not needed, then changing the focal plane to yield an out-of-focus image actually yields about a 10-20% increase in fluorescence over that measured with the cell in focus on the video monitor. It is extremely important to note, however, that the cell must be in focus for masking the cell from adjacent cells to reject the fluorescence from the adjacent cells as effectively as shown in Fig. 14 of the instrumentation chapter (59). A combination of any of these 3 procedures may be used to increase or decrease the fluorescence signal as needed.

After obtaining a suitable fluorescence signal from the cell, background fluorescence may be determined and subtracted from the absolute fluorescence of the cell. This is performed for every cell. The cell is moved out of the rectangular measurement aperture and fluorescence recorded or zeroed with the 10-turn potentiometers on the front panel; the associated LED is activated when zero reached. The analog ratio then becomes very erratic and eventually saturates (> 10 V) because of the very low denominator (380 nm signal) and, ultimately, division by zero. A front panel switch shorts the ratio output to zero in this case if needed to avoid confusion in the recorded data; the ratio output is then turned on again when the cell fluorescence is observed and the experimental protocol involving drug application, etc. begins after a stable baseline (resting) fluorescence is obtained. Our AxoBASIC data acquisition program prompts the user for background and the program proceeds to the experimental protocol after collection of these background fluorescence data. The analog recorded ratio then is a "corrected" ratio. Alternatively, background fluorescence may be collected by moving the cell out of the opening specified by the measurement aperture and proper subtraction of background fluorescence and corrected ratio are calculated on-line or during off-line analysis. All data are saved to the fixed disk of the computer. The background fluorescence was only $4.2 \pm 0.4\%$ ($X \pm SE$) of the absolute level of fluorescence in 20 representative cells. This corresponds to less than 315 mV, given an average absolute fluorescence of 7500 mV on the cell. Although background fluorescence is minimal and is compensated, the raw and/or background subtracted fluorescence resulting from 340 (or 360) and 380 nm excitation are also saved to disk for every experiment.

Our constant superfusion of solutions holds a decided advantage over a stagnant bath, because complete washout and exposure of the cell to several different drugs is possible. Also, the minimal amount of fura-2 leaking from the cells is swept away by the superfusion, rather than accumulating in the bath near the cell to artifactually increase the fura-2 fluorescence ratio to give the misleading view that there is a small elevation of Ca_i (43). After the experimental protocol is completed the cells are wiped off the coverslip bottom of the superfusion chamber and the chamber is rinsed with distilled water and/or PSS. The average time for cleaning the superfusion chamber and preparing new cells for an experiment is usually 5-10 min, unless extensive rinsing of the chamber is necessary to remove drugs that are irreversible and stick to the chamber. For example, thapsigargin appears to be irreversible and adheres to the superfusion chamber excessively; rinsing with ethanol, distilled water, and then PSS is necessary after each application of thapsigargin (R. Woodward, unpublished observation). Often 5-10 cells can be studied daily using individual cells for each experiment lasting 30-40 min each. Depolarization with 80 mM K in experiments described in this chapter was achieved by equimolar substitution of K for extracellular Na. Experiments were conducted at room temperature ($22\text{-}25^{\circ}$ C).

Simultaneous Whole-Cell Patch-Clamp and Microfluorometry

The whole-cell configuration of the patch-clamp technique (21) is used to record ionic currents from single smooth muscle cells. Fig. 11 shows the configuration, including dialysis of the cell with the contents of the pipette. This configuration permits control of the intracellular ionic environment, including dialysis of fura-2 pentapotassium salt into the cell. The external solution for recording of ionic currents in this report is PSS (with no penicillin/streptomycin). One internal (pipette) solution is termed KAsp,FURA and composed of (in mM): 135 K, 75 aspartate, 10 Na, 47 Cl, 1 Mg, 2 MgATP, 0.5 Tris GTP, 0.2 EGTA, 0.1 fura-2 pentapotassium salt, pH to 7.1 with KOH. Aspartate (Asp) is the main internal anion to allow normal smooth muscle cell Cl concentrations of about 47 mM (26). The calculated reversal potentials for the relevant ions are: E_K = -80 mV, E_{Cl} = -30 mV, E_{Na} = +67 mV, and E_{Ca} = > +120 mV. Another basic type of pipette solution is used to clamp Ca_i to desired nM concentrations (Ca-EGTA buffer solution). The solutions are termed KCl,FURA,x nM to indicate that these are high K and Cl concentration solutions containing "x" concentration of Ca_i; mainly 200 and 400 nM Ca_i in these studies. Ionic currents are monitored during 250-350 ms periods at intervals from every 1.5 s to minutes at appropriate holding and test potentials when the data are digitized at 600 μs intervals (1666 Hz). For longer duration, continuous recordings the digitizing (sample) interval is 60 ms and the low-pass filter cutoff frequency is 100 Hz. The currents are amplified by a List EPC-7 patch-clamp amplifier with a headstage having switchable feedback resistors of 0.5 and 50 GOhm. The currents are filtered by an 8-pole low-pass filter at a cutoff frequency of 400 Hz when digitized at 600 μs intervals. Because the smooth muscle Ca_i transients are so slow, we achieve sufficient time resolution by using a rotational period of 100 ms and output filter low-pass cutoff frequency of 100 Hz for the optical processor. Experiments on cardiac myocytes use a rotation period of 20 ms. All data are stored and analyzed on 80286 or 80386 processor-based personal computers with software describe in the section above (*Computerized Data Acquisition and Analysis*).

276

Gigaseal formation and reduction of background fluorescence. A droplet of cell suspension is placed in the superfusion chamber on the inverted microscope and rinsed thoroughly with PSS for several minutes, similar to that done for microfluorometry experiments when voltage-clamp is not done. The cells selected are typically 50-90 μm long. The cell is approached with a low resistance (typically 2-7 MOhm) fire-polished pipette containing KAsp,FURA or KCl,FURA,x nM internal solutions described above. Smaller electrodes (higher resistance) do not allow adequate perfusion of the cells with the pipette solution and larger pipettes (lower resistance) are not conducive to seal formation. Positive pressure of 4-6 cm H_2O is always applied prior to submerging the pipette in the cell superfusate until a high resistance seal starts to form. The positive pressure causes mild flow of the pipette solution (intracellular solution) out of the pipette. Indeed, the importance of positive pressure is reinforced as the visual observation (or observation of PMT output) indicated the complete decrease of fluorescence to zero with a latency of only less than one second if the positive pressure is removed. This indicates that extracellular buffer is aspirated into the pipette by capillary action in the absence of constant positive pressure. One difficulty arising from such capillary action is the possible elevation of the free Ca concentration in the pipette.

After a cell is chosen for study it is moved to the center of the photosensitive area of the video monitor, but 5-10 μm of the cell is left outside of the photosensitive area, if possible. In the case of the photosensitive area in Fig. 12 of the companion chapter (59), the area displayed on the video monitor is almost uniformly photosensitive. The areas not photosensitive are outside the field of the video monitor and, therefore, the pipette is aligned on the cell near column #7. The pipette is then manipulated over the cell, but placed out of focus. The light is diverted to the video port of the microscope for viewing of the cell and pipette in the video monitor. Positioning the pipette out of focus above the cell ($>10\,\mu$m above the cell) is very important because there is substantial vibration of the microscope when the light is diverted to the video port and the pipette would break if it were closer to the cell. Once the image of the cell is in the video monitor, manipulation of the pipette is similar to the typical seal formation procedure for patch-clamp when not also using microfluorometry. Monitoring cell contraction is a great advantage of the video projection system, but equally advantageous is the ability to observe the cell during manipulation of the pipette. In previous studies after the light was diverted to the video port and photomultiplier the investigator "blindly" manipulated the pipette to the cell (64). After obtaining a gigaseal the edge of the measurement aperture (#2, Fig. 4, Ref. 59) closest to the pipette tip is moved to mask the pipette. This is the greatest advantage of the Ealing Optical aperture over the Nikon aperture, which has a minimum of two sides moving on each adjustment. In addition, the Ealing aperture is less expensive. Movement of the edge of the aperture is easily accomplished by a knurled finger screw and does not mechanically disturb the seal. This ability to define spatially the measurement area with such flexibility by independent movement of all 4 sides is necessary because it is desirable to achieve a measure of the averaged Ca_i in the whole cell. The somewhat variable size of the cells necessitates the rectangular aperture arrangement. Pinhole apertures available commercially would be inadequate for our purposes.

Background fluorescence is minimized by several procedures, which are attributed primarily to close attention to the following details of the gigaseal formation procedure. First, a small (5-10 μm) area of the cell with the pipette attached is outside the photosensitive area, if possible. Second, the measurement aperture is closed 5 μm beyond the pipette tip. Third, the 2-7 MOhm pipettes results in only a little "spray" of fura-2 out of the pipette. In addition, the constant flow of the extracellular superfusate solution is removed from the cell avoiding "sticking" to the outside of the cell or loading the cell. Fourth, the focal plane is adjusted to beneath the cell and the pipette is at a 60^0 angle to the coverslip of the cell superfusion chamber, thus largely out of the focal plane. Fifth, the illumination aperture on the epifluorescence attachment is adjusted to be only about 100 μm diameter of the entire microscope visual field.

After a gigaseal is formed in the PSS external solution, the cell membrane at the pipette tip is ruptured by suction to allow whole-cell voltage-clamp. The extent of dialysis of fura-2 in to the cell is closely monitored (See below).

RESULTS AND DISCUSSION

Functional State of Cell Preparations Determined by Video Monitoring

The top panel of Fig. 1 shows a relaxed smooth muscle cell that was freshly isolated from bovine coronary artery. This is typical of the smooth muscle cells that were used for these studies. The yield of relaxed cells such as this was usually about 30-50%. The presence of functional voltage-gated Ca channels is shown by the contraction to high K (80 mM K) saline solution (Fig. 1, middle panel) and the sarcoplasmic reticulum is able to release Ca as shown by further contraction to 5×10^{-3} M caffeine (Fig. 1, bottom panel). The relaxed length of 39 cells in which this was quantified was $75 \pm 2.5 \mu$m ($X \pm$ SE); 80K and 5×10^{-3} M caffeine contracted the cells to $83.8 \pm 2.3\%$ and $90 \pm 2.2\%$ of resting length, respectively. The contraction to 80K was half the magnitude reported by Van Dijk and Laird (66) for cells from the coronary artery that were dispersed by similar methods and is probably due to the longer initial length of $92 \pm 16.3 \mu$m of their cells. Similar to earlier reports (66), relaxation of the cells is slow and incomplete. Note also in Fig. 1 (bottom panel) that these cells also display "blebs" when contracted, similar to other reports of isolated vascular smooth muscle cells (12). There was no obvious, systematic difference in contraction (or Ca_i and ionic currents) in cells isolated from the artery the same day the tissue was harvested from the animal or up to 7 days of storage at 4° C. These criterion for selection of cells were considered very important in order to ensure that the cells studied were fully functional, since the differentiated state of the smooth muscle cells may modulate the expression of voltage-gated Ca channels, as suggested for cultured rat aorta (46) and clearly demonstrated for contractile proteins (38). Another important aspect of the contractions demonstrated here is that the cell does not contract so vigorously as to move out of the area of fluorescence measurement. We also observed under video analysis during simultaneous Ca_i measurements, that endothelin (5×10^{-8} M) caused contraction during the Ca_i transient in porcine cells. The Ca_i returned fully to resting levels within 2 min of peak, as we have previously documented (69). Most striking was the finding that in the continued presence of endothelin, the cell <u>actively shortened</u> for 3 more min <u>after</u> the return of Ca_i to the resting level. The contractions are largely dependent on extracellular Ca, thus the data suggest that Ca influx may match Ca efflux (42). The data in Fig. 8 indicate that Mn may be used to estimate unidirectional divalent cation flux, i.e. Ca influx, in these cells, similar to that done for tracheal smooth muscle (34). Another reason for the importance of video monitoring of contraction during an experiment is to document the identity of smooth muscle cells. This would be especially important for cultured cells, which can show de-differentiation and loss of smooth muscle specific actin and myosin when subcultured (38). It is possible that changes in Ca_i in subcultured cells might be associated with de-differentiation (46). Note how different the morphology of these cells is from endothelial cells in another chapter (61). In summary, some quantification of contraction is possible, but the greatest utility of observing contraction is for the identification of the dispersed cells as healthy smooth muscle cells.

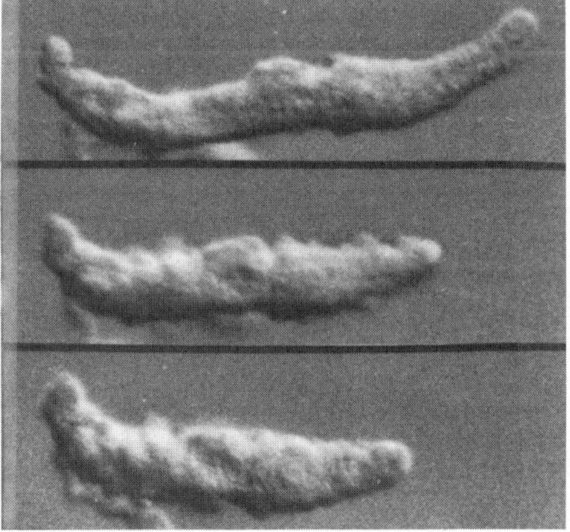

Figure 1. Freshly dispersed smooth muscle cells from bovine coronary artery. A relaxed cell (top panel) contracted when depolarized with 80 mM K (middle) further contracted by exposure to 5×10^{-3} M caffeine (bottom). Length of the cell in the top panel is 74μm.

Behavior of Intracellular Fura-2

Calibration of fluorescence ratios to Ca_i. The method of Grynkiewicz et al. (18) employs the equation $[Ca^{2+}]_i = K_D\,(max_{380}\,/\,min_{380})\,(R - R_{min})\,/\,(R_{max} - R)$, where K_D is the dissociation constant for fura-2, min_{380} and max_{380} are the minimum and maximum fluorescence at 380 nm, R is the measured ratio (the independent variable, 340/380 fluorescence intensity), and R_{max} and R_{min} are the maximum and minimum ratios of (340 fluorescence) / (380 fluorescence). In this widely used method the R_{max} and R_{min} are obtained at saturating amounts of Ca and at 0 Ca, respectively. This method may not be appropriate for some microfluorometry systems for several reasons: 1) relative background fluorescence increases greatly in mM Ca concentrations because the fluorescence resulting from 380 nm excitation becomes extremely low; the variability of these measures is unacceptable (See RESULTS); 2) it is usually not possible to attain mM Ca concentrations in a living cell, unless the cell is lysed or permeabilized; problems, such as dye leakage, occur with single cells and are difficult to compensate for (Fig. 4) and; 3) on a theoretical basis a more appropriate method may be to use $[Ca^{2+}]$ calibration standards within the range that would be expected in the cell (40).

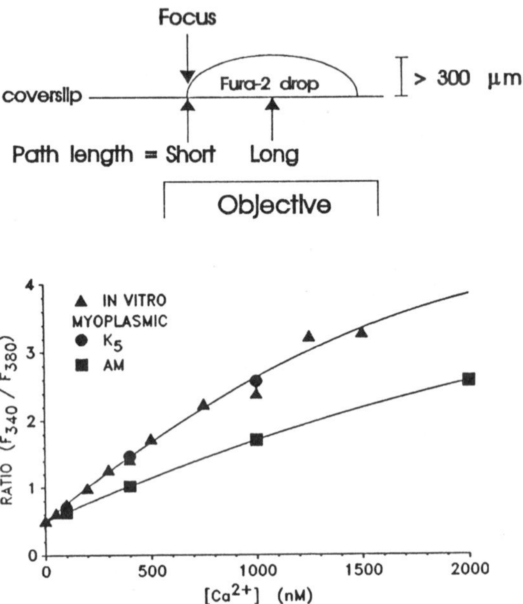

Figure 2. Calibration of fura-2 fluorescence ratios with $[Ca^{2+}]$. Ratios were obtained as described in METHODS. Triangles = *in vitro* calibration solutions (N = 2 per point). Circles = myoplasmic ratios from permeabilized cell suspensions with pentapotassium salt added exogenously (N = 5). Squares = myoplasmic ratios from cell suspensions loaded normally with fura-2/AM (N = 5).

The curve fit to the open circles in Fig. 2 is very similar to that shown previously (56, 60, 68) and is also in another report using very different optics (64). Furthermore, the ratio was unaffected by path length (Fig. 2, top panel). This relationship between ratio and Ca is quite appropriate to use for determination of maximum and minimum fluorescence for insertion into the equation of Grynkiewicz et al. (18) or for using simple regression prediction to derive Ca_i directly from the curve (64) if an essential prerequisite is met. The essential prerequisite is that the fura-2 behaves in the myoplasm as it does *in vitro* (in a cuvette), which is true for some cell types (64). The solid circles in Fig. 2 indicate that the pentapotassium salt of fura-2 does not bind any organelles, undergoes no quenching, etc. by myoplasmic contents in the skinned cells used. In contrast, when cells were loaded with fura-2/AM by the normal loading procedure and then allowed to cleave the molecule with endogenous esterases, the apparent Ca sensitivity of the fura-2 released by saponin skinning was reduced. Saponin alone had no fluorescent properties. At low Ca concentrations the differences were not easily detected, but the ratio corresponding to 1000 nM free Ca following lysate of fura-2/AM-loaded cells was only 61% of the value when fura-2 pentapotassium salt was added to saponin lysed smooth muscle cells. In addition to reasons of physiological variability, the relatively frequent calibration of the system is necessary to ensure accurate compensation for instrumentation drifts, etc. We noticed, for example, that over a period of 12 months in the *in vitro* calibrations the ratio for 100 nM Ca (using Ca-EGTA buffers) drifted a few tenths of a ratio value. Similar variations were reported by Neher (35). The excellent reliability of calibration constants was indicated in the companion chapter (59).

Fluorescence changes in cells. Fig. 3 shows typical changes in fluorescence emitted as result of excitation of smooth muscle cells that have been stimulated by 80 mM K depolarization and 5×10^{-3} M caffeine (CAF). The data shown by the open squares in Fig. 3B are the calculated Ca_i values using ratios shown in Fig. 3A and the calibration constants derived from the curve fitted to the triangles in Fig. 2. The main point is that Ca_i values are readily derived from the data in Fig. 3A, but Ca_i obtained using *in vitro* curves are less than those reported using other instrumentation systems, however when ratio values are converted to Ca_i using *in situ* calibration the Ca_i values are similar to other reports (45). Furthermore, by observation with the video system it was found that loaded smooth muscle cells contracted to $89 \pm 2\%$ and non-loaded cells contracted to $90 \pm 2\%$ of control as a result of 80K and caffeine exposure. Although buffering of Ca_i in smooth muscle cells is likely to occur to some extent as a result of introduction of fura-2, it is reasonable to assume that the buffering is not too extreme, since this most relevant physiological function of the cells seems normal. Normal function despite the introduction of the dye is an excellent indicator of minimal buffering effects (65). The lack of any fluorescence change in the cells not loaded with fura-2 in Fig. 3A conclusively shows that these cells have virtually no autofluorescence under resting conditions and no birefringence changes upon depolarization of stimulation with caffeine; therefore, these light signals are not a source of artifact in recordings on smooth muscle cells of the coronary artery. This is in contrast to some other cells where autofluorescence (51) and birefringence changes (10) are pronounced. Fig. 3B also shows Ca_i calculated using the raw data in Fig. 3A to derive ratios and the calibration constants derived from curve fitted to the squares in Fig. 2.

These data strongly indicate that in bovine smooth muscle cells fura-2/AM is not hydrolyzed completely. Incomplete cleavage of the ester would decrease values of measured Ca_i. Scanlon et al. (47) found that the excitation spectrum of fura-2 hydrolysis reaction intermediates peaks at 380 nm, therefore, 360 nm might be better wavelength to use. Our data on intracellular behavior of fura-2 in smooth muscle and endothelial cells is similar to that shown by Scanlon et al. (47), which indicates incomplete cleavage of fura-2/AM by endogenous esterases. These properties of the dye make calibration of fura-2 fluorescence ratios with absolute levels of Ca_i more difficult. We would suggest using calibration standards derived from permeabilized cells as in Fig. 2, which allow monitoring empirically the Ca sensitivity of the fura-2 that has been "processed" by the cells, i.e. de-esterified. In addition, the suggestion made by Scanlon et al. (47) that 360 nm be used to avoid the excitation peak of partially de-esterified and Ca-insensitive intermediates of fura-2 is prudent, also as a better monitor of fura-2 concentration (Figs. 3-5). Such effects have been shown to be less of a problem in, for example, neurons (64). Another important aspect of Fig. 2 to note is that the loading procedure was briefer and at a lesser concentration (2.5 μM for 15) than typically used for smooth muscle cells.

Accurate calibration of fura-2 fluorescence ratios in nM concentrations of Ca_i must involve calibration parameters closely mimicking intracellular conditions. We emphasize that species and vessel differences must be taken into consideration. Recent data of Wagner-Mann in our laboratory indicated that cells from porcine coronary arteries have similar myoplasmic fura-2 calibrations to bovine cells, therefore suggesting that esterases, ionic environment, etc. are similar (67). Nonetheless, differences in Ca_i regulation exist (67-69) and it would be reasonable to suspect that cells from other blood vessels might have different esterase activity. Our data in Fig. 3C using calibration constants derived under "myoplasmic" conditions of Fig. 2 may closely approximate, but are not absolute indicators of the myoplasmic milieu. The most accurate measurements are most likely *in situ* calibrations such as those reported by Peeters et al. (40). These measures involve clamping the Ca_i concentration in living cells to several known values (similar to Fig. 2). *In situ* calibrations are absolutely dependent on a consistent means for equilibrating intracellular and extracellular Ca in cells loaded via the ester. Ionophores such as ionomycin or bromo A23187 have been used successfully in some cell types (39, 40), but in cells from bovine coronary artery (data not shown) and porcine coronary artery (25) ionomycin does not cause equilibration. We tried to circumvent this problem by the use of saponin. The saponin technique has been well-documented to selectively solubilize the sarcolemma of smooth muscle cells, while leaving intracellular membranes, such as the sarcoplasmic reticulum and mitochondria, intact (75). These data are shown in Fig. 4.

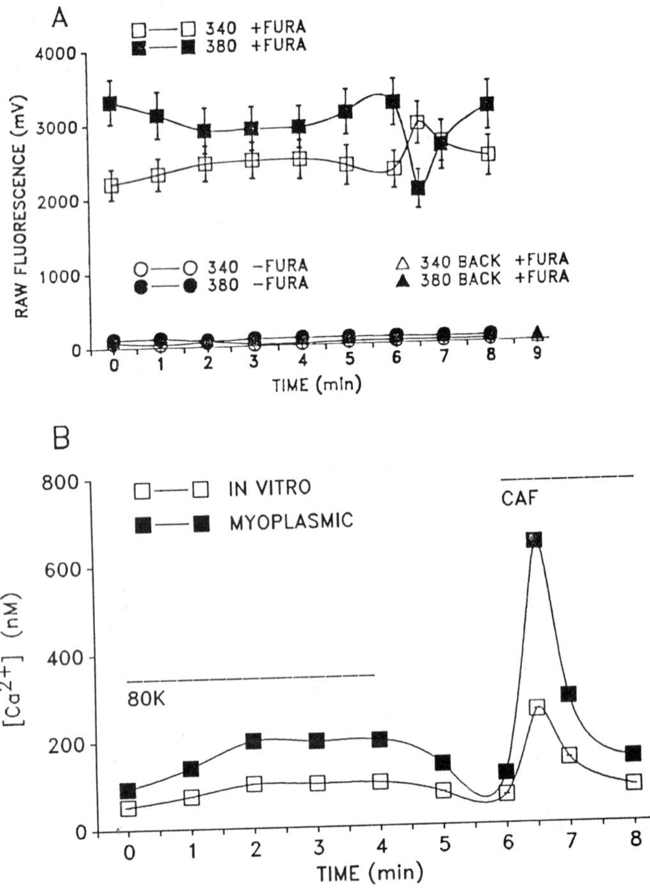

Figure 3. Fluorescence changes in smooth muscle cells. (A) Absolute (raw) fluorescence changes in cells loaded "normally" with fura-2/AM are shown in open (340 +FURA) and solid (380 +FURA) squares. Fluorescence changes in cells that were not loaded with fura-2/AM, but were exposed to the same washing and centrifugation steps (See METHODS) are shown in open (340 -FURA) and solid (380 -FURA) circles. Both groups were depolarized with 80 mM K for min 0-4, returned to PSS for min 4-6, then exposed to 5×10^{-3} M caffeine (CAF) in PSS for min 6-8 (durations indicated by horizontal bars, shown in panel B for clarity). One min after the protocol was finished (at min 9) a measure of background fluorescence was determined and shown in open (340 BACK +FURA) and solid (380 BACK +FURA) triangles. (B) Open squares = Ca$_i$ calculated from fluorescence values in panel A and "*in vitro*" calibration constants derived from Fig. 2 (triangles). Solid squares = Ca$_i$ calculated from the fluorescence values in panel and "myoplasmic" calibration constants derived from Fig. 2 (solid squares). All values are X $\pm$ SE, N=13 cells/group.

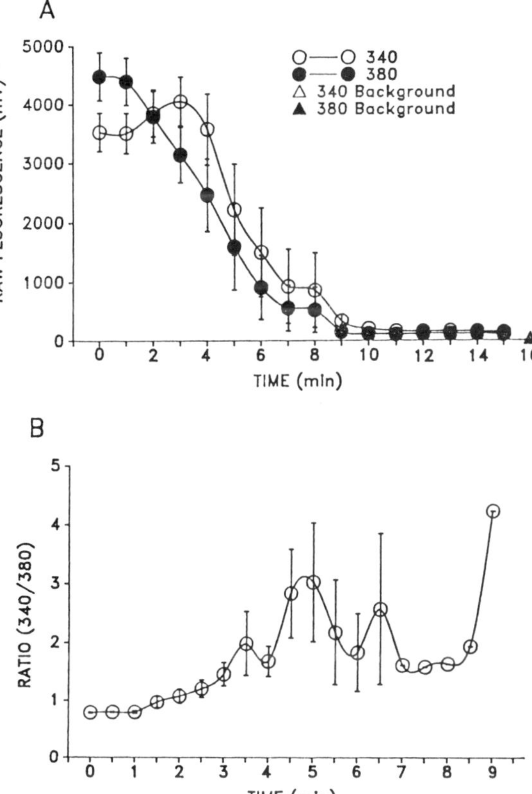

Figure 4. Changes in absolute fura-2 fluorescence during permeabilization of sarcolemma. Cells were loaded normally with fura-2/AM and then exposed to saponin (50 μg/ml) starting at min 1 and continuing through the entire protocol to remove the sarcolemma. (A) Absolute (raw) fluorescence changes in cells shown in open (340) and solid (380) circles. Background fluorescence is shown in open (340 Background) and solid (380 Background) triangles after the protocol was completed. All values are X $\pm$ SE, N = 8 cells/group. (B) Ratio of fluorescence obtained by 340 and 380 nm excitation of cells corresponding to data in panel A. The ratios were not calculated after 9 min into the protocol because the fluorescence was < 2-fold higher than background, resulting in great variability in ratio calculation. Values are X $\pm$ SE.

Myoplasmic distribution and Ca$_i$ buffering. The rationale for using saponin permeabilization was that the extracellular fluid certainly would be equilibrated with the myoplasm (75) because of the large pore size formed by saponin. The intracellular fura-2 would also leak from the cell, but there would be a short period when the maximum fluorescence ratio was achieved. Fig. 4A shows the change in absolute fluorescence recorded during 340 and 380 nm excitation, while the sarcolemma of smooth muscle cells is being selectively permeabilized with saponin (75). Saponin caused a progressive decrease in the absolute fluorescence measured in the entire cell, which after 9-10 minutes of constant exposure reached its lowest value. It was hoped that this method would permeabilize the sarcolemma rapidly enough to allow Ca to rush into the myoplasm from the extracellular space and saturate the fura-2 contained there to obtain a maximum value for the ratio. Unfortunately, the simultaneous loss of fura-2 during the permeabilization procedure made the ratios extremely variable as shown in Fig. 4B. After approximately 9 min of saponin exposure the ratios became erratic, ranging from <1 to >20. This is largely explained by the low absolute fluorescence relative to the background fluorescence. The ratio of about 4 obtained at 9 min, although near the maximum obtained from *in vitro* calibration standards in Fig. 2 (triangles), was not used as the maximum fluorescence ratio because it was deemed most desirable to keep background fluorescence less than 10% of total fluorescence. Background fluorescence at 340 and 380 nm was 17 ± 6 and 35 ± 6 mV, respectively; this corresponded to 11.6% and 33.3% of the minimal fluorescence values obtained during saponin exposure at 15 min. (Fig. 4A). These experiments yielded important information concerning the behavior of fura-2 in the myoplasm of cell. The minimal fluorescence values at 340 and 380 nm obtained after 14 min of saponin exposure were only 3.7% and 1.6%, respectively, of initial fluorescence, thus indicating that less than about 3-4% of the dye was sequestered in intracellular organelles. This is in favorable contrast to the cardiac muscle cells, which contain almost 50% of the total cellular dye content in the mitochondria or other organelles (28, 54).

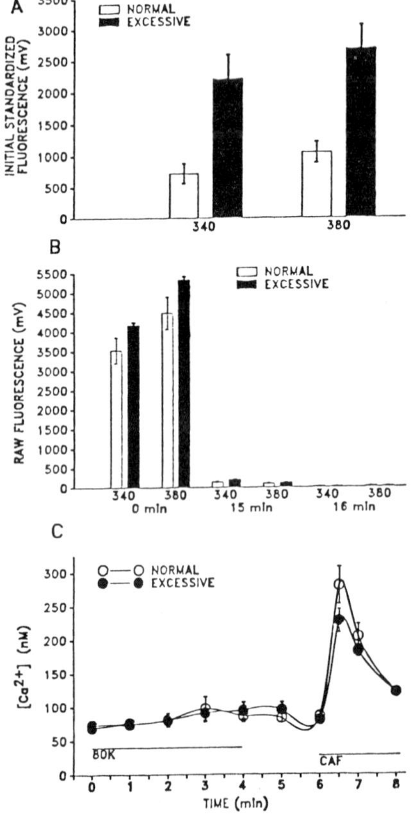

Figure 5. Effect of normal vs. excessive fura-2/AM loading of smooth muscle cells on intracellular fura-2 distribution and Ca$_i$ transients. (A) Cells were loaded with fura-2/AM "normally" (2.5 μM, 15 min, 37° C) or excessively (5 μM, 60 min, 37° C). Values are X $\pm$ SE, N=7 cells/group. (B) The sarcolemma of the cells was removed as described in Fig. 4. Raw fluorescence values for 340 and 380 nm excitation are shown only for the start (min 0) of the protocol, the end (min 15), and the background measurement (min 16). The measurement aperture was opened to encompass the entire cell for this protocol. (C) Ca$_i$ levels in cells as a result of depolarization with 80 mM K (80K) and caffeine exposure (CAF), same as the protocol in Fig. 3.

It has been suggested that long durations of loading with high concentrations of fura-2 may result in saturation of endogenous esterases and trapping of fura-2 in intracellular organelles (11, 72). This possibility was investigated in coronary artery smooth muscle cells. The normal loading procedure was compared to excessive loading of smooth muscle cells by use of 5 μM fura-2/AM for 1 hr. First, a measure of fluorescence was standardized within an area only 2 x 2 μm as delineated by the rectangular aperture. Variations in cell thickness were reduced by sampling an area adjacent to the nucleus of the cell (Figs. 1 and 11). This technique has been used quite successfully in studies on cells with quin-2 as the indicator (27), but requires measurements from a sufficiently large number of cells to reduce the variability. As shown in Fig. 5A the initial fluorescence was almost three-fold higher in the cells that were excessively loaded. However, we saw no significant elevation of the fluorescence remaining in the excessively loaded cells after myoplasmic fura-2 was released by saponin treatment (Fig. 5B), thus indicating no compartmentalization in these cells. The Ca_i values calculated from *in vitro* calibration constants were the same in the normal and excessively loaded cells during the slow rises in Ca_i normally accompanying stimulation with 80 mM K. Also, the more rapidly changing increase in Ca_i during caffeine exposure was mildly attenuated in the excessively loaded cells, as predicted (65).

Photobleaching. As Becker and Fay (5) have noted, the ratio method for calculation of Ca_i in cells does not entirely normalize for effects of photobleaching. In addition, since the interference filter wheel used in the present studies was essentially the same as that used by Fay and coworkers (5, 6, 32, 70, 74), it was necessary to determine whether photobleaching was a problem. Fig. 6 shows raw fluorescence obtained from cells that were either continuously exposed to ultraviolet illumination or were exposed for only 10 s every 2 min (intermittently). There was no difference in the rate of decay of the raw fluorescence signal at either 340 or 380 nm between the two groups. Lack of photobleaching in this system may be due to the liquid light guide arrangement, which would tend to decrease ultraviolet illumination intensity more than the system used by Fay and coworkers (5, 6, 32, 70, 74), which involves illumination of the filters by a collimated beam directly from the Xe arc lamp.

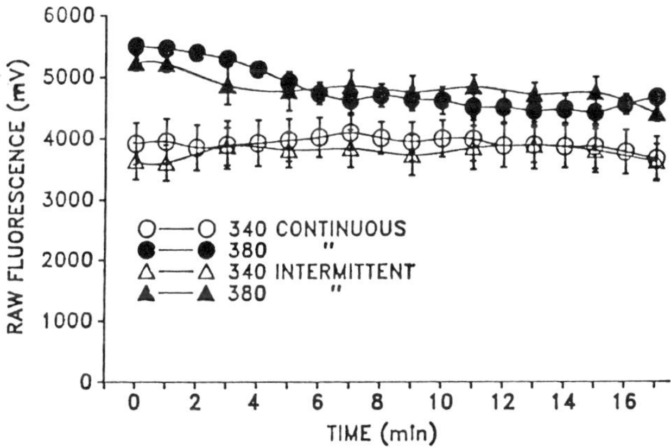

Figure 6. Changes in absolute fura-2 fluorescence in intact smooth muscle cells during continuous vs intermittent ultraviolet excitation. Cells were loaded normally with fura-2/AM. Absolute fluorescence with continuous excitation is shown in open (340 CONTINUOUS) and solid (380 ") circles; fluorescence changes with intermittent excitation of 10 s every 2 min is shown in open (340 INTERMITTENT) and solid (380") triangles. Values are X ± SE, N = 8

Single Cell vs. Population Averaged Ca_i Responses

Evidence from digital image analysis of Ca_i in pituitary cell lines (73) and the A10 smooth muscle cell line (31) has indicated the heterogeneity of single cell responses to various agonists. It was suggested that data derived from cuvette measurements of Ca_i responses from suspensions of cells represented a population average of the variable single cell Ca_i responses. This is especially striking considering that these are presumably homogeneous cell populations. The convincing aspect of these experiments was that several cells could be studied simultaneously and, thus, yield single cell and averaged Ca_i responses from the group in only one agonist exposure. Data from our laboratory also suggest that distinct Ca_i response patterns of smooth muscle cells of bovine coronary artery to endothelin (5×10^{-8} M) may be the result of distinct cell types in the artery (68).

In the companion chapter on instrumentation (59) we showed that the photosensitivity of the field of measurement in our microfluorometry system was uniform and average fluorescence data could be derived from multiple fluorescent microsphere standards. Thus, we proposed that smooth muscle cell population averaged Ca_i responses could be derived and compared to single cell Ca_i responses. We hypothesized that the caffeine-induced Ca_i responses from groups of cells would be an average of all the single cell responses within the group of cells. Fig. 7 illustrates the difference between Ca_i responses of single smooth muscle cells or a group of cells to caffeine (5×10^{-3}). Either single cells or groups of up to 10 cells were selected for microfluorometry of Ca_i responses to caffeine. The time of onset of Ca_i responses of single cells to caffeine was more variable than in groups of cells. Also, single cells showed a steeper increase in Ca_i to its peak value in only about 10 s and more rapid decrease of Ca_i to resting levels. Single cell ratios returned to resting values within 2 min of caffeine application. Groups of cells had a somewhat more gradual increase in Ca_i to the peak value in 15-25 s and decrease in Ca_i to baseline within 2 min. The response of the group of cells is similar to an average of several single cell responses. Therefore, when measuring groups of cells we are, in fact, measuring an average of the responses from the single cells within the group. These data indicated that single cell Ca_i responses can be compared to group average Ca_i responses using this microfluorometry instrumentation. The method is not as elegant, nor as definitive, as digital image analysis (31, 73), but provides a good estimate for comparison of single cell vs. group Ca_i responses if sufficient numbers of experiments are conducted (also our instrumentation is a fraction of the cost of digital image analysis). Estimates of whole artery Ca_i responses might ultimately be achieved using groups of cells from the artery. Preliminary experiments have been conducted on intact dog coronary arteries using our instrumentation (Magliola, L. and Jones, A. Unpublished observations).

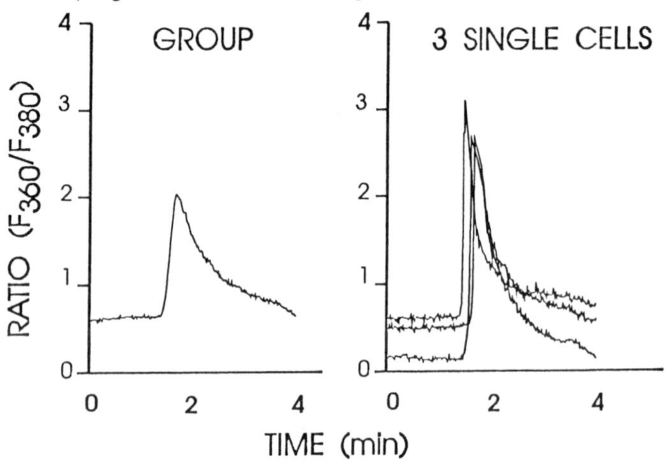

Figure 7. Smooth muscle cell responses to caffeine (5×10^{-3} M) exposure from min 1 to 3. Ca_i was measured as the ratio of 360 nm / 380 nm fluorescence (F_{360} / F_{380}). Right panel: 3 individual tracings from single cell responses. Left panel: response from a group of approximately 7 cells.

Estimation of Divalent Cation Influx with Mn

Previous data from this laboratory clearly indicated that the Ca_i response to endothelin was transient (69). However, contraction persisted even after Ca_i had returned to resting levels. We considered the possibility that Ca influx was matched by Ca efflux, thus giving rise to a sustained "cycling" of Ca across the sarcolemma that transduced a signal to the contractile proteins to sustain contraction (43). One limitation in fura-2 measurements of Ca_i is the inability to distinguish clearly the origin of the increase in Ca_i; specifically, whether Ca is entering the myoplasm from the sarcoplasmic reticulum or via influx. Furthermore, fura-2 only measures directly the net accumulation of Ca_i, not rates of influx, efflux, or release.

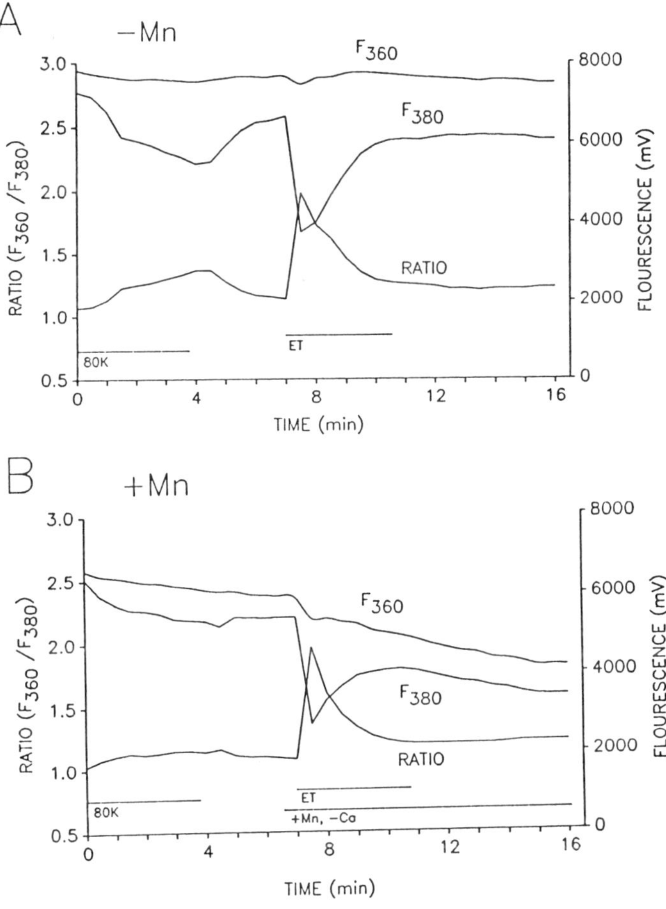

Figure 8. Single smooth muscle cell fluorescence responses to endothelin (5 x 10^{-8} M). Raw fluor- escence values from excitation at 360 nm and 380 nm are shown as well as the ratio of F_{360} / F_{380}. (A) Response to endothelin in the presence of extracellular Ca without Mn. (B) Response to endothelin in the presence of extracellular Mn (no Ca).

A method used by others to study ligand-gated divalent cation influx into endothelial cells is Mn quench of fura-2 fluorescence (20). Extracellular Ca was replaced with Mn, which has a 40 times greater affinity for fura-2 than Ca (18) and quenches the florescence of fura-2 when bound (19, 20). We compared the change in the 360 nm florescence in the presence of Ca and in the presence of Mn. The importance of using the 360 nm filter is that 360 nm excitation is very near the isosbestic wavelength for the fura-2:Ca emission spectrum. Therefore, as Ca concentration changes, ideally there should be no change in the raw florescence value due to 360 nm excitation. However, when Mn is present the 360 nm florescence will be quenched (as will the 380 nm), thus making it possible to differentiate the contributions made by each of the two ions in a whole-cell microfluorometry experiment. In Fig. 8 single smooth muscle cells were exposed either to endothelin in the presence of Ca, or endothelin in the presence of Mn. In Fig. 8A the smooth muscle cell was bathed with PSS (no Mn present) and there was very little change in the 360 nm florescence. In Fig. 8B in the presence of Mn (no Ca present), there is a sharp and sustained decrease when endothelin is added to the bath. This indicates that

286

there is influx of this Mn upon application of endothelin to smooth muscle cell. These data provide more evidence for divalent cation (Mn and probably Ca) influx and suggest that Mn (and Ca) influx is sustained after Ca_i has returned to resting levels. Murray and Kotlikoff have thoroughly characterized the use of Mn for the study of ligand-gated divalent cation influx in airway smooth muscle (34).

The data in Fig. 8 also illustrate several technical points with practical application. First, the data were collected using ideal data acquisition parameters for maximizing signal/noise ratio and, also, having appropriate time resolution. The sample interval of the A-D converter was 1 s (1 Hz) and the output filter low-pass cutoff frequency of the optical processor was 0.16 Hz (time constant of 1000 ms). Second, this is an example of the usefulness of adjusting the absolute fluorescence to a predetermined level. In this case the absolute fluorescence resulting from 360 nm excitation is needed, by itself, for quantitative purposes. Because the absolute fluorescence depends on dye loading in the cell, the size of the cell, etc. the absolute fluorescence may vary excessively. Adjustment of the fluorescence to 6000 mV, specifically, enables compilation of group data with less variability. Third, the data were collected using on-line background subtraction and ratio calculation, leaving export of the data for graphics almost the only analysis task.

Examples of Time Resolution and Versatility

We now digress briefly from our emphasis on microfluorometry of Ca_i in smooth muscle cells to illustrate 2 important aspects of our microfluorometry methods that were dealt with using fluorescent standards in the companion chapter (59). One aspect is the time resolution of the system. We first crudely tested the frequency response by monitoring fluorescence using the 20 ms rotational period by rapidly turning the focus knob on the microscope, as shown in Fig. 10B of the companion chapter (59). We sought to provide a physiological example here.

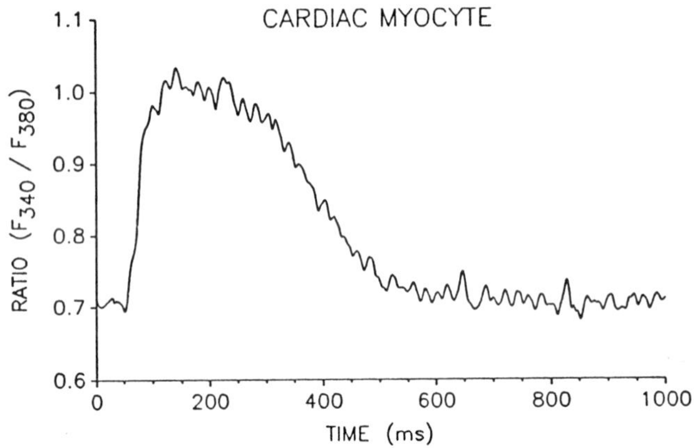

Figure 9. Ca_i response of cardiac ventricular myocyte to electrical field stimulation. Record is the average of 6 individual Ca_i transients synchronized by triggering from the stimulator. Electrical stimulation was at 0.2 Hz, 2 ms duration, 10% above the threshold voltage needed to cause contraction.

Accumulating data on depolarization-induced increases in Ca_i in smooth muscle cells indicates that the response is almost an order of magnitude slower (53, 55, 56, 58, 60, 62, 67-69) than e.g., neurons (64) or cardiac muscle cells (8, 28, 40, 54, 63). Little change is expected in Ca_i in tonically contracting cells during voltage-clamp depolarizations of 1-2 seconds. The exception are phasically contracting smooth muscle cells, such as the classical Bufo Marinus stomach muscle cell preparation (6,70). Caffeine causes the most rapid change in Ca_i in coronary smooth muscle cells, but the time to peak is at least 5-10 s (Figs. 14 and 15); therefore, Ca_i responses to caffeine will not provide a good test of the time resolution on a physiologically relevant preparation. We, therefore, monitored Ca in cardiac ventricular myocytes stimulated with platinum field electrodes. Cardiac myocytes were dispersed by methods described previously (16) and loaded with fura-2 by the "normal" loading procedure for fura-2/AM described above. As shown in Fig. 9 the peak changes in Ca_i occurred in less than 75 ms (28). Note also the relatively small change in the fluorescence ratio from about 0.7 to 1.0. This would imply either a decreased sensitivity of the myoplasmic fura-2 (as we have shown in Fig. 2 for coronary smooth muscle cells) or a sequestration of fura-2 in organelles (54). "Myoplasmic" calibrations performed on suspensions of cardiac myocytes indicated comparable sensitivity of the myoplasmic fura-2 to *in vitro* fura-2 pentapotassium salt (28), thus indicating that the problem is in sequestration of fura-2.

The versatility of the microfluorometry system is shown by the measurement of intracellular Na with SBFI in Fig. 10. The pore-forming antibiotic, nystatin (NYS, 60 μg/ml), makes the cell membrane permeant to all monovalent cations; therefore, Na ions flow down their electrochemical gradient to increase the intracellular Na concentration. We show this crude experiment simply to indicate that the instrumentation has the capability of measuring intracellular Na in near physiological conditions; other reports provide more detail in the use of SBFI (22, 24). The main point is that no changes in the instrumentation were needed for this recording. In addition, several fluorescent indicators of ion concentrations, including Ca, Na, K, and Mg, require and/or are well-suited to dual wavelength excitation (29, 33, 52). Our microfluorometry system can be used for all of these indicators with no alterations; this property of the indicators suggests a potentially more widespread need for instrumentation, thus making our system extremely versatile.

We also indicate here a technical point with practical application. The data in Fig. 10 were not collected using ideal data acquisition parameters for maximizing signal/noise ratio, in contrast to Fig. 8. The sample interval of the A-D converter was 5 s (0.2 Hz), which was certainly adequate for the slow response is Fig. 10. However, the output filter low-pass cutoff frequency of the optical processor was 100 Hz (time constant of 1.5 ms) and thus, essentially provided no filtering and yielded the noisy record in Fig. 10.

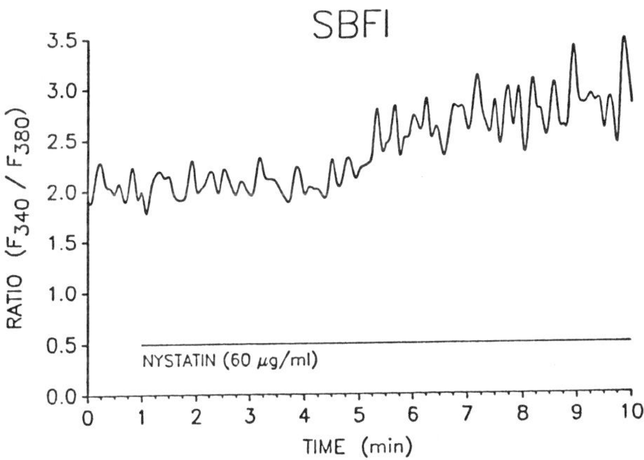

Figure 10. SBFI measurements of intracellular Na in coronary smooth muscle cell. Cell was loaded with 5μM SBFI/AM for 80 min at 37° C, then rinsed in storage media similar to fura-2/AM loading (See METHODS).

Simultaneous Whole-Cell Patch-Clamp and Fura-2 Microfluorometry

Dialysis of cell and behavior of intracellular fura-2 pentapotassium salt. The conventional whole-cell recording configuration of the patch-clamp technique was used (21). One of the notable features of this method is the dialysis of the cell interior with the contents of the pipette. We have capitalized on this property to manipulate the intracellular ionic milieu as described below. Fig. 11 shows a patch pipette and smooth muscle cell. The pipette contained KAsp,FURA intracellular solution and the external solution was PSS. In the left panel the cell was simultaneously illuminated by 600 nm light and ultraviolet excitation with 340 and 380 nm light from a 4-sector fura-2 interference filter wheel. The rectangle indicates the area of fluorescence measurement specified by the adjustable rectangular aperture (#2, Fig. 4 of Ref. 59) and the dashed circle is the area of ultraviolet excitation and photosensitive area of the photomultiplier. The patch pipette is oriented horizontally in the lower half of the photograph and almost totally excluded from the ultraviolet illumination area, fluorescence measurement area, and photosensitive area of the photomultiplier. This spatial configuration of pipette and cell reduced background fluorescence appropriately, as explained in greater detail in the section *Gigaseal formation and reduction of background fluorescence*. In the right panel the ultraviolet excitation causes bright fura-2 fluorescence of the cell, indicating adequate dialysis of the cell with KAsp,FURA pipette solution. Note that fluorescence from the patch pipette is negligible. The actual background fluorescence of the pipette measured by the optical processor, while in the cell-attached configuration (21) prior to achieving a whole-cell recording, is still less than 10% of the total fluorescence signal when quantitative measures of Ca_i are desired. In addition, when the focus was adjusted up the pipette as much as 20 μm the fluorescence was decreased to almost undetectable levels, consistent with results from fluorescent microsphere standards that determined the depth of field to be about 20 μm (Fig. 16 of Ref. 59).

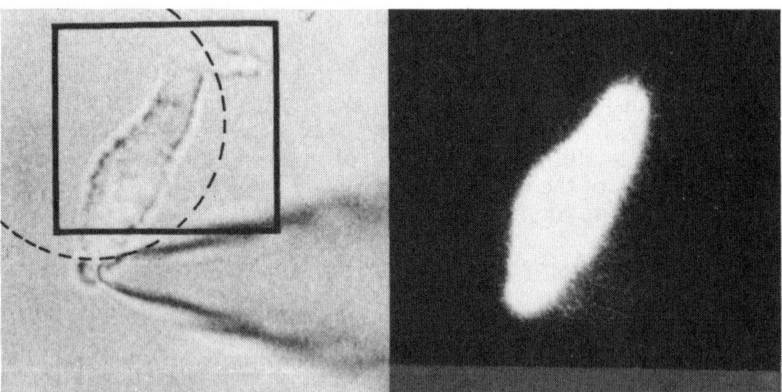

Figure 11. Configuration of pipette for simultaneous whole-cell patch-clamp and fura-2 microfluorometry. Left panel: light micrograph, dashed circle indicates area of ultraviolet excitation, interior of square indicates area of measurement aperture opening. Right panel: fura-2 fluorescence.

Manipulating the pipette solution allows for changing of the intracellular ionic equilibrium concentrations, clamping of Ca at a given concentration, and introduction of membrane impermeant factors into the cell. Again, this is a strength of the conventional whole-cell recording configuration. Therefore, the time required for the pipette solution to dialyze maximally into the cell is an important variable when completing experiments. We dialyzed cells with pipette solutions containing fura-2 to measure the time course of increasing fluorescence as the dye entered the cells. The cells had not been previously loaded with fura-2. Since we were interested in increases in raw fluorescence values rather than changes in the ratio, we included 10 mM EGTA to clamp intracellular Ca at constant concentrations (as described in METHODS). Fig. 12 illustrates the increase in the 340 (solid line) and 380 (dashed line) as the fura-2 entered a typical porcine smooth muscle cell. At min 0 access to the internal milieu of the cell was gained by applying suction to a gigaseal and rupturing the patch of membrane directly under the pipette, thus allowing the pipette solution to enter the cell. The micropipette tip resistance prior to attachment on the cells was 4.8 MOhms and the pipette solution was clamped at 200 nM Ca. The initial fluorescence values were 313 and 381 mV for 380 and 340 excitation, respectively, comprising 5% of the total fluorescence during maximal dialysis. By min 2 the raw fluorescence values had reached its plateau (6237 and 7550 mV for 380 and 340 excitation, respectively) and remained at those levels throughout the experiment. Note that the resulting ratio upon full dialysis at min 2 was 0.83, similar to the ratio for the 200 nM *in vitro* standard in Fig. 2 (triangles). In 15 cells the mean time required to reach full dialysis of the cell with solutions containing fura-2 and Ca clamped at concentrations from 100 to 1000 nM Ca was 3.69 $\pm$ 0.50 min, including a range of 1.3 to 6.9 min required.

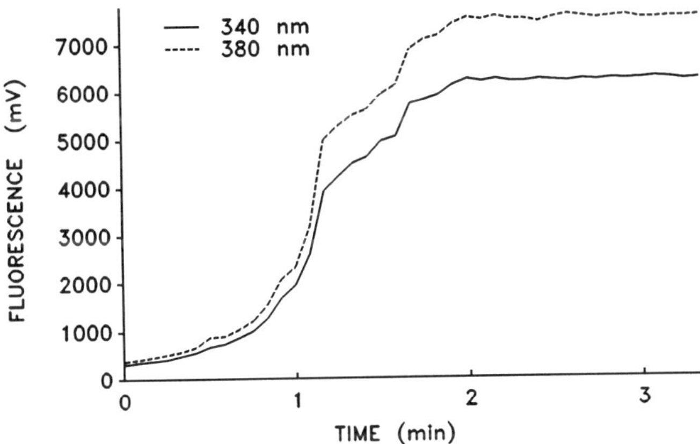

Figure 12. Time required for full dialysis of the pipette solution into a porcine smooth muscle cell. At min 0 access to the intracellular solution was gained and the raw emission fluorescence from 340 (solid line) and 380 (dashed line) excitation increased reaching a plateau within 2 min. Pipette solution was KCl,FURA,200 nM.

There are two obvious variables which might influence the time required for full dialysis of the cell. First, the size of the opening at the tip of the pipette; and second the volume of the cell's cytoplasm. We altered the diameter of the tip opening by changing temperature applied during the pulling of the micropipette. A relative measure of the tip diameter was provided by the pipette tip resistance which varied from 4.6 to 9.3 MOhms, a higher tip resistance indicating a smaller tip diameter. We found no statistical correlation between the micropipette tip resistance and the time required for full dialysis of the cell. There was a trend toward longer times required for dialysis with higher tip resistance (from 5 to 9 MOhms). However, 4 of 14 cells required 6 to 7 min for full dialysis using micropipettes with tip resistance of only 5 to 6 MOhms. We conclude that the time required for full dialysis is a variable that depends on multiple factors including the pipette tip resistance, the volume of the cell, and placement of the pipette on the cell. However, most coronary artery smooth muscle cells appear to be fully dialyzed with the pipette solution within 2 to 4.5 min after access to the cytosol has been gained.

Fig. 13 shows that clamping the Ca_i in the pipette to 400 nM for whole-cell recordings yielded a ratio similar to that shown in Fig. 2 for the *in vitro* calibration curve (triangles) and pentapotassium salt (circles) and caffeine exposure could not increase Ca-activated ionic currents or Ca_i. Therefore, fura-2 pentapotassium salt does behave in a single cell similarly to the *in vitro* calibrations and permeabilized cells and Ca_i in a single smooth muscle cells can be "clamped" with appropriated Ca-EGTA buffers with high buffer capacity (35, 58, 62). The fluorescence ratio in the cell averaged 1.78 $\pm$ 0.02 (X $\pm$ SE) for the 29 values displayed. "*In vitro*" ratios obtained were 1.83 $\pm$ 0.12 (N = 3). Another facet of the simultaneous measures is the ability to alter Ca_i buffering capacity with appropriate Ca-EGTA buffers in the pipette (See METHODS) (35). In another chapter (62) we have made extensive use of the whole-cell dialysis to clamp Ca_i at several different concentrations to determine the Ca_i-sensitivity of Ca-activated ion channels.

290

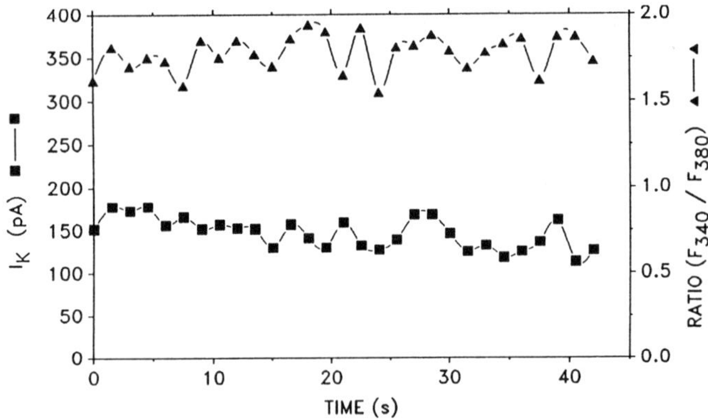

Figure 13. Simultaneous patch-clamp and fura-2 microfluorometry in smooth muscle cells: "clamping" of Ca_i with Ca-EGTA buffer. I_K (squares) and the fluorescence ratio (triangles) are shown for test depolarizations to 0 mV for only 200 ms, every 1.5 s (very similar to the second voltage step in Fig. 14A). The cell was exposed to 5×10^{-5} M caffeine starting at 0 s and continued for the entire duration shown here. The intracellular (pipette) solution was KCl,FURA,400 nM (See METHODS).

Relationship between Ca_i transients and activation of ion channels. The record plotted in Fig. 14A illustrates ionic currents and 3 different optical processor outputs during a typical experiment. The dashed line in the center represents zero current level. The composite output of the optical processor is shown at the bottom (PMT), with only the optical processor circuit output for the fluorescence resulting from 340 nm excitation (340 nm) superimposed on it. The zero volts level for these signals was the bottom of the PMT output waveform, which results from the opaque section of the wheel passing in front of the light beam. For clarity, the optical processor circuit output for the fluorescence resulting from 380 nm excitation is positioned in Fig. 14A using the dashed line as its zero level.

Data from our laboratory and other studies using smooth muscle cells (17, 23, 40, 53, 55, 56, 58, 60, 61, 67-69) (and those data in Figs. 3, 5, 7, and 8) indicate that elevation of Ca_i in tonic smooth muscle is most rapid when induced by release of Ca from the sarcoplasmic reticulum. Therefore, simultaneous patch-clamp and microphotometry studies were conducted during agonist-induced release of the sarcoplasmic reticulum Ca store. Relatively normal ionic equilibrium potentials of -80 mV for K, -30 mV for Cl, and +67 mV for Na (26) were set by the intracellular and extracellular solutions (See METHODS). The cells were held at -80 mV and brief depolarizations of 200 ms were given to a potential of -30 mV. The rationale was that if caffeine caused an increase in Ca_i and activated non-selective cation channels, the time course of the inward current would be similar to the increase in Ca_i and evident when the membrane potential was -80 mV. This was presented in more detail in another chapter (62). Also, a sensitive indicator of Ca_i directly under the sarcolemma would be shown by an increase in outward Ca-activated K current when the membrane potential was at -30 mV.

In Fig. 14B the $I_{K(Ca)}$ plotted is the value obtained at the end of the 200 ms test pulse when the $I_{K(Ca)}$ has plateaued. The depolarizations were applied every 1.5 s and data were acquired by the computer for 50 ms when the membrane potential was held at -80 mV prior to the depolarization and including the entire 200 ms depolarization as shown by the voltage protocol template in Fig. 14A; therefore, any change in fluorescence was noted during the pulses. The holding current at -80 mV was small (about 10 pA) and stable in the control condition. Depolarization to -30 mV elicited very little K current (Fig. 14A and B). Exposure to 5×10^{-5} M caffeine, however, increased both inward cation current ($I_{C(Ca)}$) and outward K current with identical time courses, peaking in about 4.5 s. The Ca-activated K current ($I_{K(Ca)}$) plotted in Fig. 14B was the value obtained after reaching a steady-state at near the end of the 200 ms test pulse. Note that Ca_i followed a similar time course, but was a larger amplitude than in Figs. 3 and 5 from fura-2/AM loaded cells. The inward holding current at -80 mV (circles) is increased concomitantly with the increase in

Ca$_i$. Evidence presented in another chapter indicates that some component of the Ca-activated inward current results from Na and/or Ca influx (62), and is thus considered a Ca-activated non-selective cation current ($I_{C(Ca)}$). Finally, on a technical note, the accuracy with which both the 340 and 380 nm sample-and-hold signals track the composite output of the optical processor is shown.

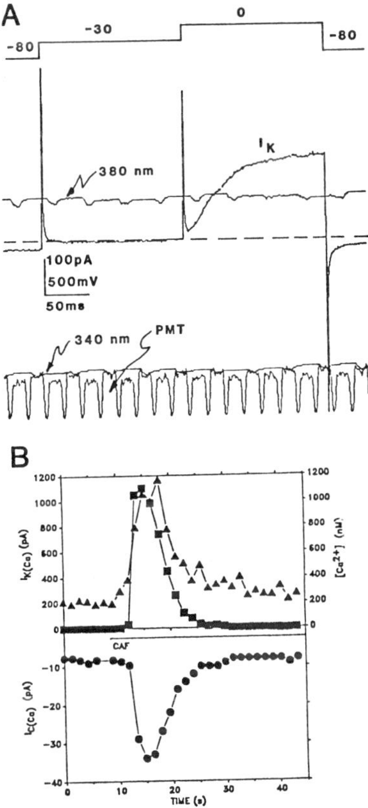

Figure 14. Simultaneous patch-clamp and fura-2 microfluorometry in smooth muscle cells. (A) Original record showing voltage template (top) and resulting whole-cell K currents (I_K). Intracellular (pipette) solution was KAsp,FURA and extracellular solution was PSS. (B) Caffeine-induced rapid Ca$_i$ transients and accompanying Ca-activated ionic currents. The cell was exposed to caffeine (CAF, 5 x 10^{-3} M) for the duration indicated by the horizontal line. Triangles = Ca$_i$ transient. Squares = Ca-activated K current ($I_{K(Ca)}$). Circles = inward holding current at -80 mV.

The data in Fig. 14 indicate a correlation between the time courses of Ca-activated ion channels and Ca$_i$, but plotting the data at 1.5 s intervals is not sufficient to describe accurately the time course. Therefore, we monitored the caffeine-induced increase in K current and Ca$_i$ with 50 ms time resolution. The cell was continuously clamped at 0 mV, combined with Ca$_i$ measurement at 50 ms intervals (instead of 1.5 s as in Fig. 14). The caffeine-induced increase in $I_{K(Ca)}$ <u>preceded</u> the increase in Ca$_i$ by 2 s. We interpret these data to indicate that Ca is released from the sarcoplasmic reticulum toward the sarcolemma. This "physiological localization" of Ca release is much more pronounced than found by Foskett et al. (15) for secretory cells. Simultaneous recordings of ionic currents and Ca$_i$ have provided a new dimension to the study of Ca$_i$ regulatory processes in smooth muscle. This example of simultaneous monitoring of ionic currents and dual wavelength fluorescence emission illustrates the strength and versatility of the system.

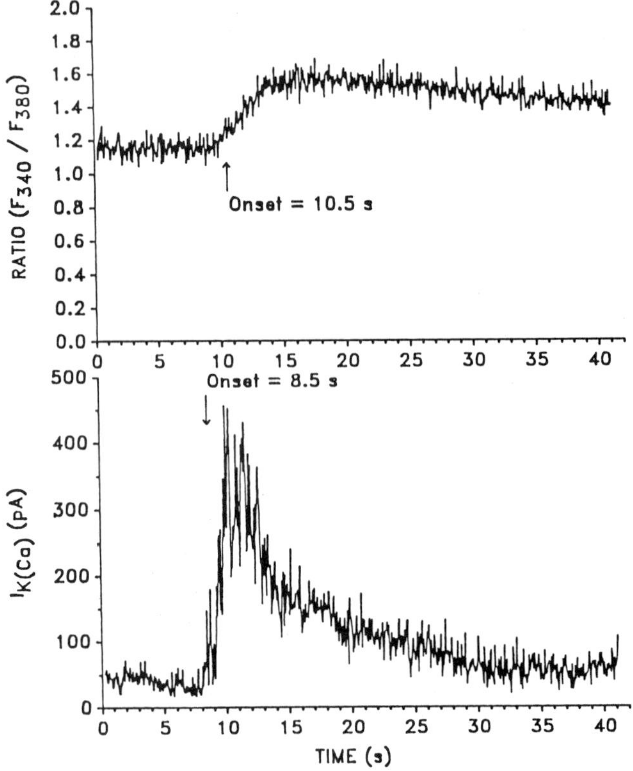

Figure 15. Simultaneous voltage-clamp measurement of Ca-activated K current ($I_{K(Ca)}$) and Ca_i (RATIO F_{340} / F_{380}) during exposure to caffeine. The SMC was exposed to CAF (5 x 10^{-3} M) at time 0 during a constant holding potential of 0 mV. The onset of the increase in $I_{K(Ca)}$ preceded the increase in the ratio by 2 s. Pipette solution was KAsp,FURA.

SUMMARY AND CONCLUSIONS

The experiments described here provide insight into the properties of fura-2 using the electronic and optical instrumentation described in the companion chapter (59). The experiments, although basic and at times tedious, are important to validate the use of fura-2 as a Ca_i indicator in smooth muscle cells. We have used this system for the past 3-4 years on over 4000 individual cells. We emphasize that many of the conclusions drawn from these experiments have application for other optical systems using fura-2, but the heterogeneity of different cell types must be appreciated and it is essential that similar experiments be conducted on every different preparation studied. This is especially appropriate given the heterogenous nature of the vascular system.

The main findings of this study are:

1) decreased Ca-sensitivity of fura-2/AM relative to the pentapotassium salt form of fura- 2 which can be corrected for by calibrating Ca_i under conditions which mimic intracellular;

2) buffering of Ca_i by fura-2 is minimal (cells still contract);

3) less than 3-4% of fura-2/AM is sequestered in intracellular organelles;

4) excessive loading with fura-2/AM resulted in higher initial fluorescence values than values with normal loading protocols. However, the higher values did not represent sequestration in intracellular organelles and Ca

concentrations calculated via 340/380 ratios were not different in cells excessively loaded vs. normal loading;

5)   photobleaching of fura-2 was not significant;

6)   ratio changes measured from groups of cells are indicative of the average change noted in single cells;

7)   fluorescence emission from the pipette during whole-cell recordings is negligible;

8)   Mn quenching of the fura-2 signal provides a means of separating changes in Ca_i due to Ca influx vs. internal release; and

9)   typically cells in whole-cell configuration are fully dialyzed with the pipette solution within 2 to 4.5 min after access to the cytosol is gained.

Further results which pertain specifically to the optical system on which measurements were made include:

1)   the ability to observe and record cell activity during the experiment;

2)   background subtraction and 340/380 division are completed on-line;

3)   time resolution of the system is adequate to monitor rapid changes in Ca_i in cardiac cells; and

4)   the ability to measure changes in other intracellular ions than Ca using different dyes. We believe the results presented will assist investigators working with fura-2 to appreciate the dyes limitations and use the dye under the most appropriate conditions.

ACKNOWLEDGEMENTS

This work was supported by a University of Missouri Alumni Fund for Faculty Development, Institutional Biomedical Support Grant RR 07053, the American Heart Association-Missouri Affiliate, Marion Merrell Dow, Inc., NIH HL41033 to M.S., a grant from the American Physical Therapy Association to L.S-B, and support from NIH Training Grant T-32 HL07094. The authors thank Charles Jorgenson of the Dalton Research Center for photography preparation and Laurel Bowman and Qicheng Hu for assistance with data collection and graphics. The authors will gladly share customized AxoBASIC 1.0 programs with licensed AxoBASIC users and Quattro Pro macros with authorized users.

REFERENCES

1.   Adams, D.J., J. Barakeh, R. Laskey, and C. Van Breemen, FASEB J. 3, 2389-2400 (1989).

2.   Almers, W. and E. Neher, FEBS Lett. 192, 13-18 (1985).

3.   Baylor, S.M., W.K. Chandler, and M.W. Marshall, J. Physiol. (Lond) 344, 625-666 (1983).

4.   Baylor, S.M. and S. Hollingworth, J. Physiol. (Lond) 403, 151-192 (1988).

5.   Becker, P.L. and F.S. Fay, Am. J. Physiol. 253, C613-C618 (1987).

6.   Becker, P.L., J.J. Singer, J.V. Walsh, Jr., and F.S. Fay, Science 244, 211-214 (1989).

7.   Bloch, K.D., S.P. Friedrich, M.-E. Lee, R.L. Eddy, T.B. Shows, and T. Quertermous, J. Biol. Chem. 264, 10851-10857 (1989).

8.   Cannell, M.B., J.R. Berlin, and W.J. Lederer, Science 238, 1419-1423 (1987).

9.   Cannell, M.B. and S.O. Sage, J. Physiol. (Lond) 419, 555-568 (1989).

10.   Cohen, L.B., B. Hille, and R.D. Keynes, J. Physiol. (Lond) 211, 495-515 (1970).

11.   Defeo, T.T., G.M. Briggs, and K.G. Morgan, Am. J. Physiol. 253, H1456-H1461 (1987).

12.   Defeo, T.T. and K.G. Morgan, Pflugers Arch. 404, 100-102 (1985).

13.   Defeo, T.T. and K.G. Morgan, Pflugers Arch. 406, 427-429 (1986).

14.   Fabiato, A. and F. Fabiato, J. Physiol. (Paris) 75, 463-505 (1979).

15.   Foskett, J.K., P.J. Gunter-Smith, J.E. Melvin, and R.J. Turner, Proc. Natl. Acad. Sci. USA 86, 167-171 (1989).

16.   Geisbuhler, T.P., D.A. Johnson, and M.J. Rovetto, Am. J. Physiol. 253, C645-C651 (1987).

17.   Goldman, W.F., S. Bova, and M.P. Blaustein, Cell Calcium 11, 221-231 (1990).

18.   Grynkiewicz, G., M. Poenie, and R.Y. Tsien, J. Biol. Chem. 260, 3440-3450 (1985).

19.   Hallam, T.J., R. Jacob, and J.E. Merritt, Biochem. J. 255, 179-184 (1988).

20.   Hallam, T.J., R. Jacob, and J.E. Merritt, Biochem. J. 259, 125-129 (1989).

21.   Hamill, O.P., A. Marty, E. Neher, B. Sakmann, and F.J. Sigworth, Pflugers Arch. 391, 85-100 (1981).

22.   Harootunian, A.T., J.P.Y. Kao, B.K. Eckert, and R.Y. Tsien, J. Biol. Chem. 264, 19458-19467 (1989).

23.   Himpens, B. and A.P. Somlyo, J. Physiol. (Lond) 395, 507-530 (1988).

24.   Iijima, K., L. Lin, A. Nasjletti, and M.S. Goligorsky, Am. J. Physiol. Cell Physiol. 260, C982-C992 (1991).

25.   Itoh, T., Y. Kubota, and H. Kuriyama, J. Physiol. (Lond) 397, 401-419 (1988).

26. Jones, A.W., in Content and fluxes of electrolytes. In, *Handbook of Physiology. The Cardiovascular System. Vol. II. Vascular Smooth Muscle*, D. F. Bohr, A. P. Somlyo, and H. V. Sparks, eds. (American Physiological Society, Bethesda 1980) pp. 253-299.
27. Kobayashi, S., H. Kanaide, and M. Nakamura, *Science* 229, 553-556 (1985).
28. Laughlin, M.H., M.E. Shaefer, and M. Sturek, (Submitted) (1991).
29. Molecular Probes, *Bioprobes* 9 (1988).
30. Mollard, P., P. Vacher, B. Dufy, B.P. Winiger, and W. Schlegel, *J. Biol. Chem.* 263, 19570-19576 (1988).
31. Monck, J.R., E.E. Reynolds, A.P. Thomas, and J.R. Williamson, *J. Biol. Chem.* 263, 4569-4575 (1988).
32. Moore, E.D.W., P.L. Becker, K.E. Fogarty, D.A. Williams, and F.S. Fay, *Cell Calcium* 11, 157-179 (1990).
33. Moore, E.D.W. and F.S. Fay, *Physiologist* 32, 147 (1989).(Abstract)
34. Murray, R.K. and M.I. Kotlikoff, *J. Physiol. (Lond.)* 435, 123-144 (1991).
35. Neher, E., *J. Physiol. (Lond.)* 395, 193-214 (1988).
36. Obye, P.K., M.H. Laughlin, and M. Sturek, (Submitted) (1991).
37. Oetliker, H., S.M. Baylor, and W.K. Chandler, *Nature* 257, 693-696 (1975).
38. Owens, G.K., A. Loeb, D. Gordon, and M.M. Thompson, *J. Cell Biol.* 102, 343-352 (1986).
39. Pallota, B.S., J.R. Helper, S.A. Oglesby, and T.K. Harden, *J. Gen. Physiol.* 89, 985-997 (1987).
40. Peeters, G.A., V. Hlady, J.H.B. Bridge, and W.H. Barry, *Am. J. Physiol.* 253, H1400-H1408 (1987).
41. Poenie, M., *Cell Calcium* 11, 85-91 (1990).
42. Rasmussen, H., P. Barrett, W. Zawalich, C. Isales, P. Stein, J. Smallwood, R. McCarthy, and W. Bollag, *Ann. N. Y. Acad. Sci.* 568, 73-80 (1989).
43. Rasmussen, H., H. Haller, Y. Takuwa, G. Kelley, and S. Park, *Prog. Clin. Biol. Res.* 327, 89-106 (1990).
44. Rembold, C.M., *J. Physiol. (Lond.)* 429, 77-94 (1990).
45. Rembold, C.M. and R. A. Murphy, *J. Cardiovasc. Pharmacol.* 12 Suppl. 5, S38-S42 (1988).
46. Reynolds, E.E. and G.R. Dubyak, *Biochem. Biophys. Res. Commun.* 136, 927-934 (1986).
47. Scanlon, M., D.A. Williams, and F.S. Fay, *J. Biol. Chem.* 262, 6308-6312 (1987).
48. Scarpa, A., in *Techniques in Cellular Physiology. Vol. 1, Part II, Techniques in Life Sciences, P127*, P.F. Baker, ed. (Amsterdam, Elsevier/North Holland 1982) pp. 1-40.
49. Schlegel, W., B.P. Winiger, P. Mollard, P. Vacher, F. Wuarin, G.R. Zahnd, C.B. Wollheim, and B. Dufy, *Nature* 329, 719-721 (1987).
50. Shepherd, J.T. and P.M. Vanhoutte, *J. Am. Coll. Cardiol.* 8, 50A-54A (1986).
51. Sick, T.J. and M. Rosenthal, *J. Neurosci. Methods* 28, 125-132 (1989).
52. Simonson, M.S., S. Wann, P. Mene , G.R. Dubyak, M. Kester, Y. Nakazato, J.R. Sedor, and M.J. Dunn, *J. Clin. Invest.* 83, 708-712 (1989).
53. Somlyo, A.P. and B. Himpens, *FASEB J.* 3, 2266-2276 (1989).
54. Spurgeon, H.A., M.D. Stern, G. Baartz, S. Raffaeli, R.G. Hansford, A. Talo, E.G. Lakatta, and M.C. Capogrossi, *Am. J. Physiol.* 258, H574-H586 (1990).
55. Stehno-Bittel, L., M.H. Laughlin, and M. Sturek, *Am. J. Physiol. Heart Circ. Physiol.* 259, H643-H647 (1990).
56. Stehno-Bittel, L., M.H. Laughlin, and M. Sturek, *J. Appl. Physiol.* in press (1991).
57. Stehno-Bittel, L. and M. Sturek, (Submitted) (1991).
58. Stehno-Bittel, L. and M. Sturek, (Submitted), (1991).
59. Sturek, M., W.M. Caldwell, D.A. Humphrey, and C. Wagner-Mann, in *Electrophysiology and Ion Channels of Vascular Smooth Muscle Cells and Endothelial Cells*, N. Sperelakis and H. Kuriyama, eds. (Elsevier, New York 1991) (in press).
60. Sturek, M., K. Kunda, and Q. Hu, J. Physiol. (Lond.) (in press) (1991).
61. Sturek, M., P. Smith, and L. Stehno-Bittel, in *Electrophysiology and Ion Channels of Vascular Smooth Muscle Cells and Endothelial Cells*, N. Sperelakis and H. Kuriyama, eds. (Elsevier, New York 1991) (in press).
62. Sturek, M., L. Stehno-Bittel, and P. Obye, in *Electrophysiology and Ion Channels of Vascular Smooth Muscle Cells and Endothelial Cells*, N. Sperelakis and H. Kuriyama, eds. (Elsevier, New York 1991) (in press).
63. Takamatsu, T. and W.G. Wier, *Cell Calcium* 11, 111-120 (1990).
64. Thayer, S.A., M. Sturek, and R.J. Miller, *Pflugers Arch.* 412, 216-223 (1988).
65. Tsien, R.Y., *Trends Neurosci.* 11, 419-424 (1988).
66. Van Dijk, A.M. and J.D. Laird, *Blood Vessels* 21, 267-278 (1984).
67. Wagner-Mann, C., Q. Hu, and M. Sturek, (Submitted) (1991).
68. Wagner-Mann, C.C., L. Bowman, and M. Sturek, *Am. J. Physiol. Cell Physiol.* 260, C763-C770 (1991).
69. Wagner-Mann, C.C. and M. Sturek, *Am. J. Physiol. Cell Physiol.* 260, C771-C777 (1991).
70. Williams, D.A., P.L. Becker, and F.S. Fay, *Science* 235, 1644-1648 (1987).
71. Williams, D.A. and F.S. Fay, *Cell Calcium* 11, 75-83 (1990).
72. Williams, D.A., K.E. Fogarty, R.Y. Tsien, and F.S. Fay, *Nature* 318, 558-561 (1985).
73. Winiger, P. and W. Schlegel, *Biochem. J.* 255, 161-167 (1988).
74. Yagi, S., P.L. Becker, and F.S. Fay, *Proc. Natl. Acad. Sci. USA* 85, 4109-4113 (1988).
75. Yamamoto, H. and C. Van Breemen, *J. Gen. Physiol.* 87, 369-389 (1986).

SECTION II

VASCULAR ENDOTHELIAL CELLS

Chapters 25 - 29

Ca^{2+} REGULATION OF VASCULAR SMOOTH MUSCLE AND RELEASE OF ENDOTHELIUM-DERIVED RELAXING FACTOR

HIDEAKI KARAKI

Department of Veterinary Pharmacology, Faculty of Agriculture,
The University of Tokyo, Bunkyo-ku, Tokyo 113, Japan

INTRODUCTION

Since Ebashi has established the pivotal role of Ca^{++} in the regulation of skeletal muscle contractility (for review see 15), correlation between cytosolic Ca^{++} levels ($[Ca^{++}]_i$) and muscle tension has been extensively studied by, for example, radioactive $^{45}Ca^{++}$ flux experiments. However, it is difficult to distinguish free Ca^{++} and bound Ca^{++} with this method, and resolution is not fast enough to follow the relatively rapid muscle contraction. Using the photoprotein, aequorin (7), and fluorescent Ca^{++} indicators such as fura-2 and indo-1 (20), it becomes possible to simultaneously measure $[Ca^{++}]_i$ and contraction in smooth muscle. The major emphasis of this review will be connected with our recent findings on the relationship among $[Ca^{++}]_i$, myosin light chain (MLC) phosphorylation and contractile tension in vascular smooth muscle. A discussion on the correlation between $[Ca^{++}]_i$ and release of endothelium-derived relaxing factor (EDRF) in vascular endothelial cells will be also presented.

Ca^{2+} REGULATION OF VASCULAR SMOOTH MUSCLE
Measurement of Cytosolic Ca^{2+} in Smooth Muscle

Method for simultaneous measurements of $[Ca^{++}]_i$ and muscle tension in vascular smooth muscle with fura-2 has been described by Ozaki *et al.* (60), Sato *et al.* (80) and Karaki et al. (39, 41). In brief, the thoracic aorta was isolated from a male Wistar rat (250-300 g), cut into spiral strips (1-2 mm in width and 5-7 mm in length) and placed in a normal physiological salt solution (PSS). Endothelium was removed by gently rubbing the intimal surface with a finger moistened with PSS. PSS contained (mM): NaCl 136.9, KCl 5.4, $CaCl_2$ 1.5, $MgCl_2$ 1.0, $NaHCO_3$ 23.8, ethylenediaminetetraacetic acid 0.01 and glucose 5.5. High K^+ solution was made by substituting NaCl with equimolar KCl. Ca^{++}-free solution was made by removing $CaCl_2$ and adding 0.5 mM ethyleneglycol bis(β-aminoethylether) N,N,N',N'-tetraacetic acid (EGTA). These solutions were saturated with 95% O_2 and 5% CO_2 mixture at 37°C and pH 7.4.

Acetoxymethyl ester of fura-2 (fura-2/AM) was dissolved in dimethyl sulfoxide, added to normal PSS to make final concentration of 5 μM and mixed vigorously by applying ultrasonic wave. Fluorometric determination showed that the effective concentration of fura-2/AM in this solution was approximately 1 μM. It was found that platelets and cultured smooth muscle cells, but not smooth muscle strip, was loaded with fura-2/AM using this solution. Centrifugation of this solution at 10,000 x g for 2 min decreased the effective concentration of fura-2/AM to approximately 70% and there was no detectable fura-2/AM in the supernatant after centrifugation at 50,000 x g for 20 min. This result indicates that fura-2/AM is insoluble in PSS and only a small amount disperses as particles of various size and most of the particles are so large that they are not able to enter the extracellular matrix of the smooth muscle tissue. So, we added a detergent, 0.02% Cremophor EL, followed by application of ultrasonic wave to solubilize fura-2/AM and it was found that centrifugation did not decrease the effective concentration of fura-2/AM. Using this solution, smooth muscle strips were loaded with enough amount of fura-2 to detect $[Ca^{++}]_i$ (for further detail see 2). Fura-2 loading changed neither muscle contraction nor phosphorylation of myosin light chain in vascular smooth muscle (19).

Muscle strips were treated with 5 μM fura-2/AM for 3-5 h at room temperature. After the fura-2 loading, muscle strips were washed with PSS for 20 min in a tissue bath at 37°C to remove unhydrolyzed fura-2/AM. Experiments were performed with a fluorimeter (CAF-100, JASCO, Tokyo, Japan) as shown in Fig. 1. The muscle strip was held horizontally in a temperature controlled, 7 ml volume organ bath. One end of the muscle strip was connected to force displacement transducer to monitor the mechanical activity. A part of the muscle strip was excited by light obtained from xenon high pressure lamp (75 W) equipped with rotating filter wheel (48 Hz) which contained 340 nm and 380 nm interference filters and the amount of 500 nm fluorescence induced by 340 nm excitation (F340) and that induced by 380 nm excitation (F380) was measured. The time constant of the optical channels was 0.25 sec.

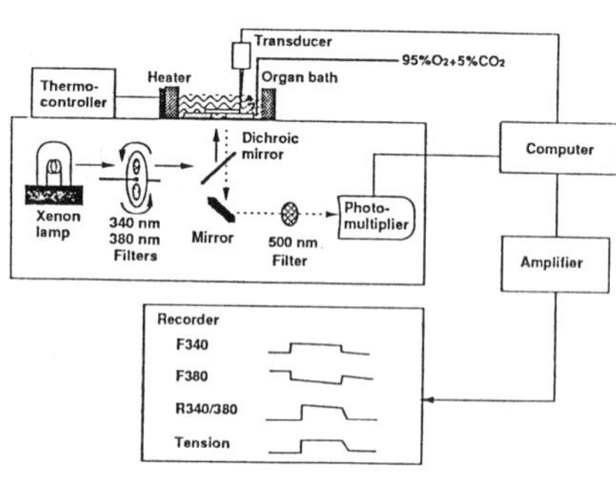

Figure 1. Block diagram of the fluorimeter designed to measure fluorescence in smooth muscle strip. The muscle strip was held horizontally in a temperature controlled organ bath. One end of the muscle strip was connected to a strain gauge transducer to monitor the mechanical activity. The excitation light was obtained from xenon high pressure lamp equipped with rotating filter wheel. The wheel contained 340 nm and 380 nm interference filters. Thus, muscle strip was excited by these two different wave length alternatively at 48 Hz. The 500 nm lights emitted from the muscle strip by 340 nm excitation (F340) and by 380 nm excitation (F380) were collected to photomultiplier and the ratio of these two fluorescence (R340/380) was calculated. Courtesy of Dr. Koichi Sato.

Absolute amount of $[Ca^{++}]_i$ was calculated by the method described by Scanlon *et al.* (82) and Malgaroli *et al.* (51). At the end of the experiment, the maximum and minimum levels of $[Ca^{++}]_i$ were determined using 10 μM ionomycin and 4 mM EGTA, respectively (80). Background fluorescence level was then measured in the presence of 4 mM $MnCl_2$ and 10 μM ionomycin. It was found that background fluorescence due to 340 nm or 380 nm excitation was approximately 45% and 32%, respectively, of the total fluorescence. F340 and F380 were corrected for background fluorescence, and the ratio of these two fluorescence (R340/380) was calculated. Because the intensity of 380 nm light was stronger than that of 340 nm light in our CAF-100 system, the maximum and the minimum R340/380 values (R_{max} and R_{min}, respectively) were smaller than those obtained with a conventional fluorimeter (FP-2060, JASCO) in which the intensities of two excitation lights are normalized. Furthermore, relative intensities of these two light changes due to the gradual degeneration of the filters. For this reason, R340/380 is used as an indicator of relative $[Ca^{++}]_i$ only in a same muscle strip. To use the R340/380 values obtained in different muscle strips using different CAF-100 systems as a $[Ca^{++}]_i$ indicator, it is necessary to compare these values with R_{max} and R_{min} (73).

The R_{max} obtained in a smooth muscle tissue with CAF-100 was smaller, whereas the R_{min} was larger than that obtained *in vitro*. To know the reason for this discrepancy, we examined the effects of cytoplasmic soluble proteins on fura-2 fluorescence (52). Ileal longitudinal smooth muscle was homogenized, centrifuged at 80,000 x g for 60 min, and the supernatant was dialyzed to remove contaminating inorganic ions and then freeze-dried. Using this method, 12.2 mg soluble protein/ml cell water was obtained. In the presence of Ca^{++}, these soluble proteins decreased F340 and increased F380, resulting in a decrease in the R_{max}. In contrast to this, these proteins decreased both F340 and F380 in the presence of EGTA, resulting in an increase in the R_{min}. Because spectrophotometric study (with UNIDEC-450 spectrophotometer, JASCO) showed that soluble proteins absorbed 340 nm light more strongly than 380 nm light, the decrease in R_{max} is at least partly attributable to this effect. However, the increase in R_{min} is not explained by the greater decrease in 340 nm light. These results support the suggestion that fura-2 binds to soluble proteins in cytoplasm, and its characteristics of fluorescence are changed (2, 43, 47, 52). In the presence of soluble proteins, however, R340/380 changed in proportion to the changes in Ca^{++} concentration, suggesting that fura-2 method is applicable to smooth muscle. We assumed the dissociation constant (K_d) of fura-2 for Ca^{++} to be 224 nM (20). However, since the K_d value in cytoplasm may be greater than this value which is obtained in the absence of proteins (47), calculated $[Ca^{++}]_i$ may be an underestimate.

Furthermore, smooth muscle has a background fluorescence which changes in accordance with $[Ca^{++}]_i$ (64) and this fluorescence may result in approximately 10% error in $[Ca^{++}]_i$ calculation (80). For these reasons, absolute $[Ca^{++}]/_i$ was not calculated in this experiment. Instead, R340/380 was used as an indicator of $[Ca^{++}]_i$ (60, 80), and the quantitative comparison of the $[Ca^{++}]_i$ was made by taking resting and high K^+-stimulated $[Ca^{++}]_i$ as 0% and 100%, respectively. Smooth muscle movements sometimes interfered with the fluorescence. The movement artifact was distinguishable from the Ca^{++} signal because muscle movements changed F340 and F380 to the same direction, whereas the changes in $[Ca^{++}]_i$ moved F340 and F380 to the opposite directions. Only the preparations in which F340 and F380 moved to the opposite directions were used in our experiments.

This method has been successively applied to other types of smooth muscle such as gastro-intestinal tract (24, 52, 61, 90), trachea (62), vas deferens (83) and uterus (77).

Ca^{++}-Tension Relationship

As shown in Fig. 2, high-K^+, norepinephrine (NE) and ionomycin induced sustained increase in $[Ca^{++}]_i$. Removing external Ca^{++}, sustained increments in $[Ca^{++}]_i$ was rapidly inhibited, suggesting that the increase in $[Ca^{++}]_i$ is attributable to Ca^{++} influx. In Ca^{++}-free solution, high K^+ was ineffective. However, NE induced transient increase in $[Ca^{++}]_i$ followed by transient contraction (Fig. 3). This result support the suggestion that the first transient increase in the $[Ca^{++}]_i$ induced by NE is attributable to Ca^{++} release from sarcoplasmic reticulum mediated by inositol-1,4,5-trisphosphate whereas the sustained contraction is due to Ca^{++} influx through Ca^{++} channels (35, 36). The second application of NE induced small sustained contraction without changing $[Ca^{++}]_i$. Mechanism of this contraction will be discussed later.

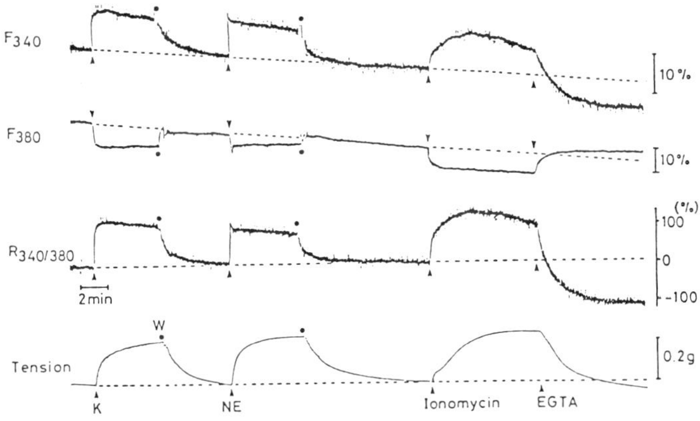

Figure 2. Changes in fura-2-Ca^{++} fluorescence measured simultaneously with muscle tension in a fura-2-loaded rat aorta stimulated by 72.7 mM KCl (K), 0.3 μM NE or 10 μM ionomycin. Changes in F340 and F380 are shown by relative value taking resting fluorescence level of the fura-2-loaded aorta as 100%. $[Ca^{++}]_i$ is shown by R340/380. Reproduced from Karaki (43) with permission.

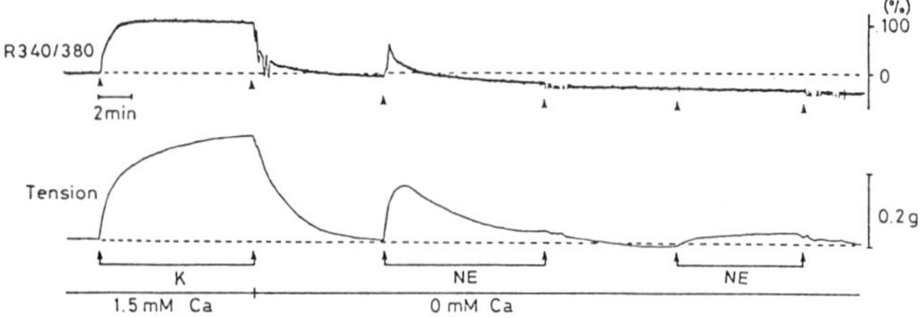

Figure 3. Effects of NE on $[Ca^{++}]_i$ (upper trace) and contraction (lower trace) in rat aorta in Ca^{++}-free solution. After stimulation with 72.7 mM KCl in the presence of 1.5 mM Ca^{++}, muscle was washed with Ca^{++}-free solution containing 0.5 mM EGTA. After the $[Ca^{++}]_i$ decreased to a resting level, 1 μM NE was added. After washing with Ca^{++}-free solution, NE was added again. Reproduced from Karaki *et al.* (40) with permission.

It has been reported using a photoprotein, aequorin, that receptor agonists cause initial large increase in $[Ca^{++}]_i$ which then declines to a very low level during sustained contraction in smooth muscle (54). The difference may be explained by the characteristics of aequorin luminescence which over-represents the region of high $[Ca^{++}]_i$ because its intensity corresponds to the 2.5th power of Ca^{++} concentration (8). Consequently, the presence of an area in which the Ca^{++} level is higher than average $[Ca^{++}]_i$ increases the total aequorin luminescence above that due to average $[Ca^{++}]_i$. Thus, it has been suggested that receptor-mediated Ca^{++} release produces a region of high Ca^{++} area which may cause the initial large and transient increase in aequorin signal (12, 70). Subsequent diffusion of this Ca^{++} in the cytoplasm establishes an uniform $[Ca^{++}]_i$. Simultaneously, opening of Ca^{++} channels increases the average $[Ca^{++}]_i$ which is detected as a steady level of aequorin signal. By contrast, fura-2 forms a simple 1:1 complex with Ca^{++} (20). Thus, fura-2 signals may represent the average $[Ca^{++}]_i$ (37, 84). MLC phosphorylation decreases more rapidly than $[Ca^{++}]_i$ during sustained contraction (14). This discrepancy does not necessarily indicate that fura-2 does not accurately measure $[Ca^{++}]_i$ because increase in $[Ca^{++}]_i$ may decrease MLC kinase activity to decrease MLC phosphorylation level, as discussed later.

As shown in Fig. 4, cumulative addition of KCl or NE-induced concentration-dependent increases in both muscle tension and $[Ca^{++}]_i$ and there was a positive correlation between $[Ca^{++}]_i$ and muscle tension. However, slope of the $[Ca^{++}]_i$-tension curve due to high K^+ is smaller than that due to NE, indicating that greater contraction is induced at a given $[Ca^{++}]_i$ in the presence of NE than in the presence of high K^+ (39, 80). Prostaglandin $F_{2\alpha}$ (PG), phorbol ester (63), endothelin-1 (ET) (78), serotonin (96), clonidine (91), thromboxane analogue and phenylephrine (25) showed similar Ca^{++}-sensitizing effects as NE.

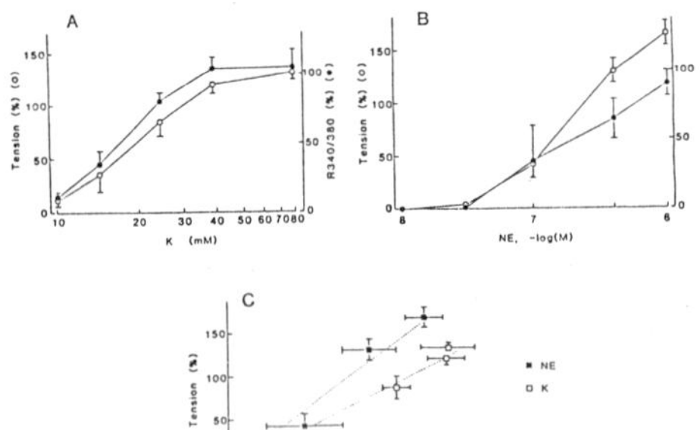

Figure 4. Effect of cumulative application of KCl (A) and NE (B) on muscle tension and $[Ca^{++}]_i$ (shown by R340/380). (C) $[Ca^{++}]_i$-tension relationship obtained with KCl or NE. Mean $\pm$ SEM of 4 to 6 experiments are shown. Reproduced from Sato *et al.* (80) with permission.

Mechanism of Ca^{++}-Sensitization

Since smooth muscle contraction is regulated by the increase in $[Ca^{++}]_i$ followed by MLC phosphorylation (23, 32, 33), Ca^{++} sensitization of contraction may be due to the modulation of the relationship between $[Ca^{++}]_i$ and MLC phosphorylation or MLC phosphorylation and tension development (38). In order to examine this possibility, effects of stimulants on MLC phosphorylation were examined.

Strips of rat aorta were frozen in acetone-dry ice. Muscle strips were crushed and mixed with 60 volumes of pyrophosphate buffer containing 100 mM NaP_2O_7, 5 mM EGTA, 2.5 mM dithiothreitol, 0.5 mM phenylmethyl-sulfonyl fluoride, 50 mM NaF, 0.6 M KI and 10% (w/v) glycerol at pH 8.8 (65). After a 5 min incubation at 0^oC, the mixture was centrifuged for 20 min at 5,000 x g. Glycerol and saturated sucrose were added to the supernatant to give final concentration of 30% and 20%, respectively. Sample was then subjected to pyrophosphate PAGE (PPi PAGE) for 16 hrs, as described by Takano-Ohmuro and Kohama (88, 89). Proteins were visualized by 0.1% coomassie brilliant blue R-250 staining. It has been reported that the increase in myosin motility measured with PPi PAGE is associated with the MLC phosphorylation which enhances the actin-activated myosin Mg^{++}-ATPase activity. Using PPi PAGE, therefore, it is possible to detect only the "active" MLC phosphorylation but not the "inactive" MLC phosphorylation which does not activates the actin-myosin interaction such as the phosphorylation induced by protein kinase C (87).

Fig. 5 shows the changes in $[Ca^{++}]_i$, MLC phosphorylation and muscle tension in rat aorta stimulated for 20 min with the maximally effective concentration of stimulant, that is, 72.7 mM KCl, 10 μM PG, 1 μM NE, 30 nM ET or 1 μM 12-deoxyphorbol 13-isobutyrate (DPB). All of these stimulants increased $[Ca^{++}]_i$, amount of phosphorylated MLC and muscle tension. However, there seems to be no correlation between $[Ca^{++}]_i$ and MLC phosphorylation or muscle tension. However, when the increments in MLC phosphorylation and muscle tension are normalized for a given increase in $[Ca^{++}]_i$ (Fig. 6), it is clearly shown that the efficiency of these stimulants to increase MLC phosphorylation and muscle tension at a given $[Ca^{++}]_i$ was dependent on stimulants; ET and DPB induced the greatest MLC phosphorylation and contraction followed by PG > NE > high K^+. Relationship between MLC phosphorylation and muscle tension stimulated by these stimulants is shown in Fig. 7. There is a curve-linear relationship between these two parameters supporting the suggestion that these stimulants do not modify the relationship between MLC phosphorylation and muscle tension (18, 22, 71). These results suggest that smooth muscle contraction is due to an increase in $[Ca^{++}]_i$ followed by an activation of MLC kinase and that receptor agonists and phorbol ester increase Ca^{++} sensitivity of MLC phosphorylation resulting in a greater contraction at a given $[Ca^{++}]_i$.

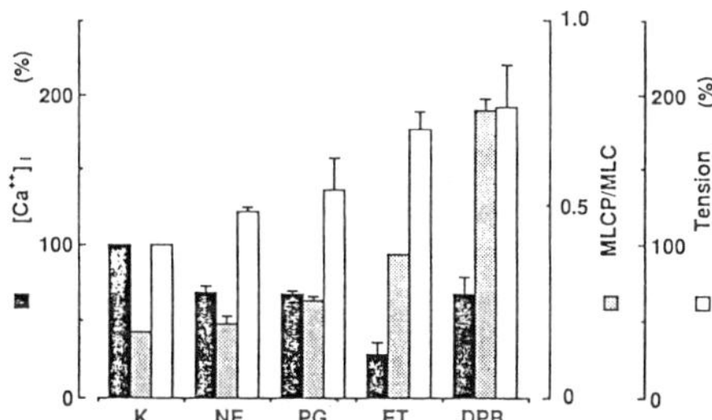

Figure 5. Changes in $[Ca^{++}]_i$, MLC phosphorylation and muscle tension in rat aorta induced by maximally effective concentrations of stimulants; 72.7 mM KCl, 1 μM NE, 10 μM prostaglandin $F_{2\alpha}$ (PG), 30 nM endothelin-1 (ET) and 1 μM DPB. $[Ca^{++}]_i$ and muscle tension stimulated by 72.7 mM KCl were taken as 100%. Each column represents mean $\pm$ SEM of 4-8 experiments.

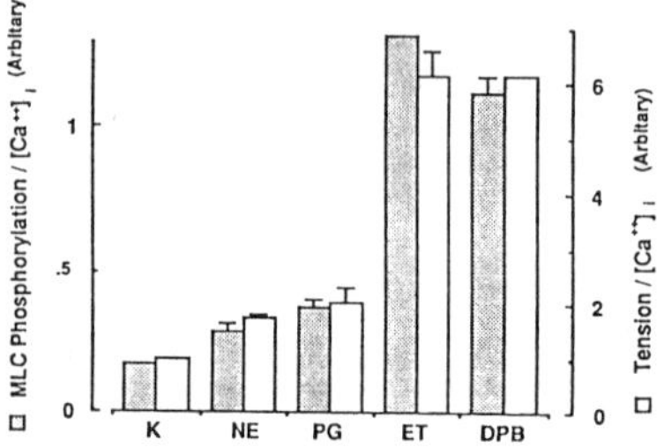

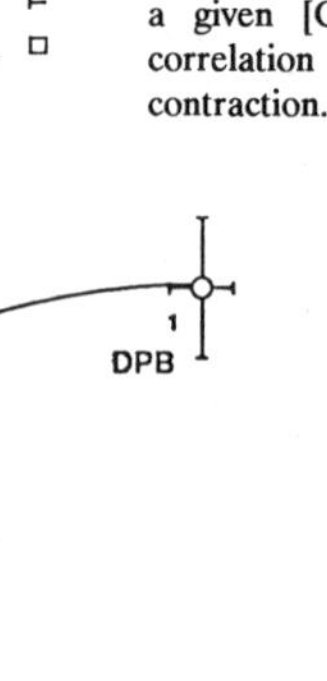

Figure 6. Changes in MLC phosphorylation and muscle tension per unit of $[Ca^{++}]_i$ increase in rat aorta. Data was adopted from Fig. 5. It is shown that receptor agonists and DPB induce greater MLC phosphorylation and contraction than KCl at a given $[Ca^{++}]_i$, and that there is a good correlation between MLC phosphorylation and contraction.

Figure 7. Correlation between MLC phosphorylation and muscle tension in rat aorta in the presence of 72.7 mM KCl, 1 μM NE, 10 μM prostaglandin $F_{2\alpha}$ (PG), 30 nM endothelin-1 (ET) and 1 μM DPB. 1: Normal PSS. 2: In the presence of 10 μM verapamil. 3: In the presence of 4 mM EGTA. Experiment was done as shown in Fig. 9. It is shown that there is a curve-linear correlation between MLC phosphorylation and muscle tension in normal PSS (1). In the presence of verapamil (2) or EGTA (3), MLC phosphorylation was almost completely inhibited whereas muscle tension was only partially inhibited, resulting in a greater contractions than those in normal PSS for a given increase in MLC phosphorylation.

There are two possible mechanisms for Ca^{++} sensitization (Fig. 8). Kubota *et al.* (48) and Kitazawa *et al.* (45) suggested that MLC phosphatase is inhibited by a mechanism coupled to GTP-binding protein resulting in a greater MLC phosphorylation at a given $[Ca^{++}]_i$. Alternative or additional mechanism has been reported by Stull *et al.* (85) and Tansey *et al.* (94, 95) in tracheal smooth muscle in which increase in $[Ca^{++}]_i$ not only activates MLC kinase but also Ca^{++}- and calmodulin-dependent protein kinase II which phosphorylates MLC kinase to inhibit its activity. Since receptor agonists uncouples this negative feedback loops, high K^+ more strongly decreases Ca^{++} sensitivity of MLC phosphorylation than receptor agonist (Fig. 8). In vascular smooth muscle, Gilbert *et al.* (19) also reported that Ca^{++} sensitivity of MLC kinase extracted from high K^+-treated tissues was lower than that extracted from unstimulated or histamine-treated tissues, supporting the above suggestion. These results also explain why MLC phosphorylation decreases during the sustained increase in $[Ca^{++}]_i$ (measured with fura-2) in the presence of various stimulants.

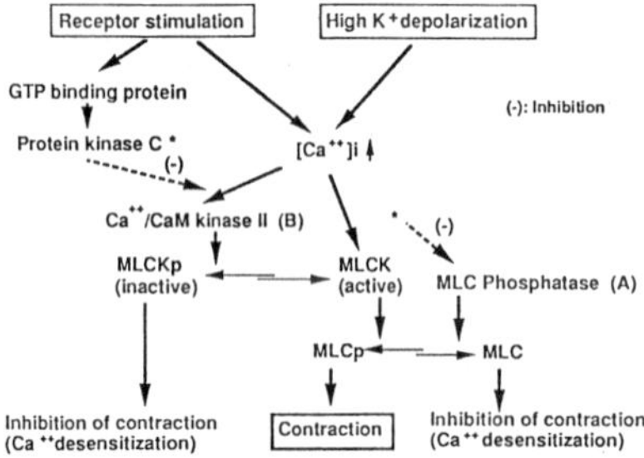

Figure 8. Possible mechanisms for Ca^{++} sensitization of contractile elements. **A:** Phosphorylated MLC is dephosphorylated by MLC phosphatase. **B:** Increase in $[Ca^{++}]_i$ activates not only MLC kinase but also Ca^{++}- and calmodulin-dependent protein kinase II. The latter kinase phosphorylates MLC kinase to decrease its activity. Receptor agonists inhibit these "Ca^{++}-desensitizing" mechanisms by activating GTP binding protein followed by activation of protein kinase C, resulting in the increase in Ca^{++} sensitivity of MLC phosphorylation.

Phorbol esters, at the concentration which did not change resting muscle tension or $[Ca^{++}]_i$, potentiated the high K^+-induced contraction with little effect on the high K^+-stimulated $[Ca^{++}]_i$ (unpublished observation). Since receptor agonists, but not high K^+, activate protein kinase C through formation of diacylglycerol (58), Ca^{++} sensitization induced by receptor agonists may be attributable to this mechanism.

Myosin Light Chain Phosphorylation-Independent Contraction

There are two methods to construct a $[Ca^{++}]_i$-tension curve. The first method is to cumulatively apply a stimulant, as shown in Fig. 4. Using this method, concentration of stimulant and $[Ca^{++}]_i$ are changed simultaneously. The second method is to change only $[Ca^{++}]_i$ by the addition of Ca^{++} channel blocker and/or EGTA in the presence of a fixed concentration of stimulant. Using this method, it is possible to examine more detailed effect of a stimulant on the $[Ca^{++}]_i$-tension relationship. Results of the latter type of experiment are shown in Fig. 9 and 10.

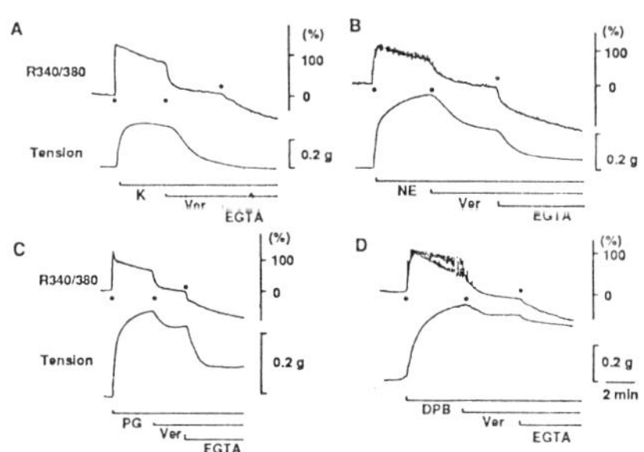

Figure 9. Effects of 72.7 mM KCl, 1 μM NE, 10 μM prostaglandin $F_{2\alpha}$ (PG) and 1 μM DPB on $[Ca^{++}]_i$ (upper trace) and muscle tension (lower trace) in rat aorta. Verapamil (Ver, 10 μM) and EGTA (4 mM) were sequentially added during the contraction. Reproduced from Ozaki et al. (62) with permission.

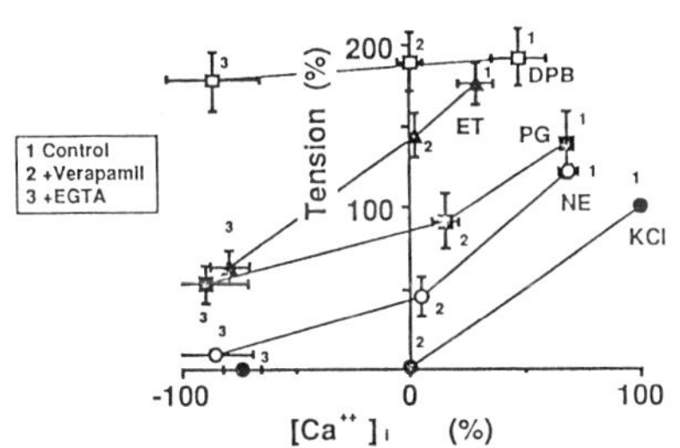

Figure 10. $[Ca^{++}]_i$-tension relationship in rat aorta obtained in the presence of 72.7 mM KCl, 1 μM NE, 10 μM prostaglandin $F_{2\alpha}$ (PG), 30 nM endothelin-1 (ET) and 1 μM DPB. Experiment was done as shown in Fig. 9. Each point represents mean $\pm$ SEM of 4-8 experiments. 100% represents high K^+-stimulated levels.

As shown in these figures, 10 μM verapamil and 4 mM EGTA were sequentially applied in the presence of the maximally effective concentrations of stimulants. Verapamil abolished the high K^+-induced increments in muscle tension and $[Ca^{++}]_i$, supporting the suggestion that high K^+-induced contraction is due to increase in Ca^{++} influx through the voltage-dependent Ca^{++} channels (35). The same concentration of verapamil strongly inhibited $[Ca^{++}]_i$ stimulated by NE, PG, ET or DPB, suggesting that these agonists increase $[Ca^{++}]_i$ mainly by opening of the voltage-dependent Ca^{++} channel. However, verapamil only partially inhibited the sustained contraction induced by these stimulants. Sequential addition of 4 mM EGTA further decreased $[Ca^{++}]_i$ to a level below a resting level. However, a portion of the contraction was not inhibited by EGTA. These results suggest that these stimulants shift $[Ca^{++}]_i$-tension relationship to the lower $[Ca^{++}]_i$ levels by increasing Ca^{++} sensitivity of contractile elements.

As shown in Table 1, 1 μM DPB increased $[Ca^{++}]_i$ followed by contraction (27, 42, 63). In other types of smooth muscle, however, DPB induced contraction without changing $[Ca^{++}]_i$. Other phorbol esters such as 12-deoxyphorbol 13-isobutyrate 20-acetate and phorbol 12, 13-dibutyrate showed similar effects as DPB (unpublished observation). These results suggest that increase in $[Ca^{++}]_i$ is not essential for the contraction induced by phorbol esters; activation of protein kinase C may increase Ca^{++} sensitivity of contractile elements to induce contraction at a resting $[Ca^{++}]_i$.

Table 1. Effect of 1 μM DPB on $[Ca^{++}]_i$ and muscle tension in smooth muscle (K. Sato, M. Mitsui, M. Hori, M. Tajimi, S.-C. Kwon, H. Ozaki and H. Karaki, unpublished).

Preparation	$[Ca^{++}]_i$	Tension
Rat		
Aorta	+ or ±	+
Carotid artery	+ or ±	+
Tail artery	−	+
Rabbit		
Aorta	−	+
Mesenteric artery	−	+
Canine trachea	−	+
Bovine trachea	−	+
Swine trachea	−	+

+: Increase. ±: Slight increase. −: No change.

In permeabilized smooth muscle with bacterial α-toxin (Fig. 11), NE (in the presence of GTP), GTP-γ-S and phorbol ester decrease the threshold Ca^{++} concentration for contraction. Similar results have been reported by Nishimura *et al.* (56, 57), Kitazawa *et al.* (46) and Anabuki *et al.* (5). ET showed similar effect (unpublished observation). These results also indicate that smooth muscle has an ability to develop contractile tension at or below a resting $[Ca^{++}]_i$ in the presence of receptor agonists and phorbol esters, and that activation of GTP binding protein followed by protein kinase C may be involved in this effect. In order to know the mechanism of this contraction, MLC phosphorylation was measured under the similar condition as that in Fig. 9.

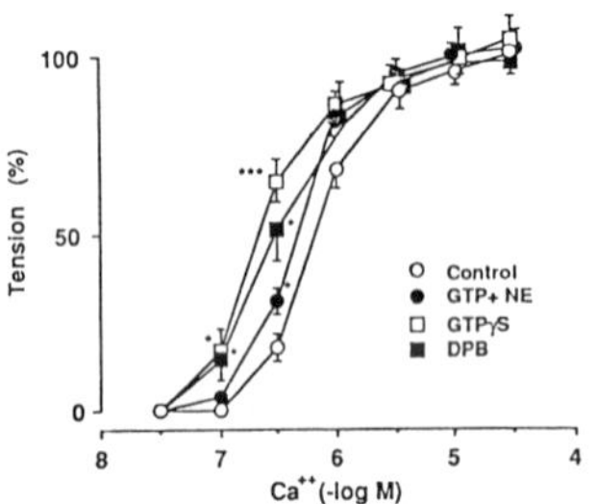

Figure 11. Ca^{++}-tension relationship in permeabilized rabbit mesenteric artery. Ca^{++} was cumulatively added in the absence or presence of 10 lM NE with 100 lM GTP, 10 μM GTP-γ-S or 1 lM DPB. Each point represents mean $\pm$ SEM of 4-8 experiments.

Fig. 12 shows the changes in MLC phosphorylation and $[Ca^{++}]_i$ in rat aorta. All of these stimulants increased $[Ca^{++}]_i$ and the amount of the phosphorylated MLC to various levels, as shown in Fig. 5. Subsequent addition of 10 μM verapamil decreased MLC phosphorylation and $[Ca^{++}]_i$ to or slightly above their respective resting levels. Sequential addition of 4 mM EGTA further decreased $[Ca^{++}]_i$. However, EGTA did not induce additional decrease in MLC phosphorylation which has already been at a resting level. Resulting $[Ca^{++}]_i$-MLC phosphorylation relationship indicates that MLC phosphorylation is dependent on the increase in $[Ca^{++}]_i$ above a resting level, and that the contraction obtained at or below a resting $[Ca^{++}]_i$ is not explained by MLC phosphorylation. Relationship between MLC phosphorylation and muscle tension is shown in Fig. 7. Contractions induced by receptor agonists and DPB in the presence of verapamil or EGTA were followed by no or only a small increase in MLC phosphorylation, resulting in a greater contraction at a given MLC phosphorylation than those obtained in the absence of these inhibitors. These results again suggest that there are MLC phosphorylation-dependent and independent mechanisms for regulation of contraction. It is also suggested that the contraction which is not dependent on MLC phosphorylation is highly sensitive to Ca^{++} and is regulated by $[Ca^{++}]_i$ below a resting level.

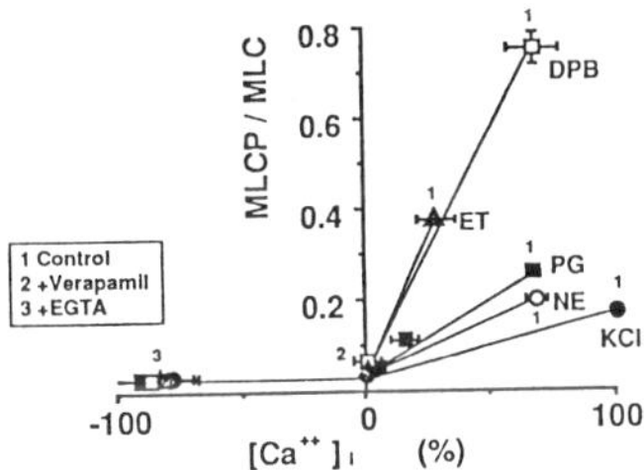

Figure 12. $[Ca^{++}]_i$-MLC phosphorylation relationship in rat aorta obtained in the presence of 72.7 mM KCl, 1 μM NE, 10 μM prostaglandin $F_{2\alpha}$ (PG), 30 nM endothelin-1 (ET) and 1 μM DPB. Experiment was done as shown in Fig. 9. Each point represents mean $\pm$ SEM of 4-8 experiments. 100% represents high K^+-stimulated $[Ca^{++}]_i$.

The EGTA-resistant portion of the contraction is not attributable to the initial MLC phosphorylation because addition of receptor agonists to a Ca^{++}-depleted muscle induced sustained contraction without increasing $[Ca^{++}]_i$ (see the second NE-induced contraction in Fig. 3) or MLC phosphorylation. Furthermore, an inhibitor of MLC kinase, ML-9 (30, 75), completely inhibited the contractions in the presence of external Ca^{++}, although the same concentration of ML-9 only partially inhibited the contractions in Ca^{++} free solution. These ML-9-resistant contractions were almost completely inhibited by non-selective kinase inhibitor, K252a and high concentrations of staurosporine. Measurements of stiffness and force recovery following a quick release suggested that the contraction induced in the absence of external Ca^{++} is due to cross-bridge cycling (27, 42). These results suggest that the contraction induced by receptor agonists in the absence of external Ca^{++} is regulated by a novel mechanism possibly phosphorylation by protein kinase C of protein(s) which is different from MLC.

The sustained contraction obtained in the presence of EGTA may be due either to a Ca^{++}-independent mechanism or to a mechanism which is dependent on a trace amount of Ca^{++} remaining in the cell. Since absolute $[Ca^{++}]_i$ was not calculated in the present experiment, the amount of Ca^{++} remaining in the cell is not known. In permeabilized mesenteric artery, NE, ET and DPB decreased threshold Ca^{++} level for contraction although these stimulants did not change the resting tone of the muscle in the absence of Ca^{++}. This result suggests that there is a trace amount of Ca^{++} remaining in the intact cells in the EGTA-containing solution and this Ca^{++} may activate the mechanism which is not dependent on MLC phosphorylation. Smooth muscle contraction which is not dependent on MLC phosphorylation has been reported in rat uterus stimulated with oxytocin in the absence of external Ca^{++} (59).

The order of potency of receptor agonists to induce a contraction in the absence of external Ca^{++} was same as that to increase Ca^{++}-sensitivity of contractile elements, that is, DPB > ET > PG > NE whereas high K^+ was ineffective (Fig. 9 and 10). The difference in the potency may be due to the ability of these stimulants to activate protein kinase C.

Summary

Although contraction of vascular smooth muscle is mainly due to an increase in $[Ca^{++}]_i$ followed by an increase in MLC phosphorylation, receptor agonists may increase Ca^{++} sensitivity of contractile elements by increasing Ca^{++} sensitivity of MLC phosphorylation when $[Ca^{++}]_i$ is higher than resting level. When $[Ca^{++}]_i$ is decreased to or below a resting level by the addition of verapamil and/or EGTA, receptor agonists still increase cross-bridge cycling by a mechanism which is highly sensitive to Ca^{++} but not dependent on the increase in MLC phosphorylation (Fig. 13).

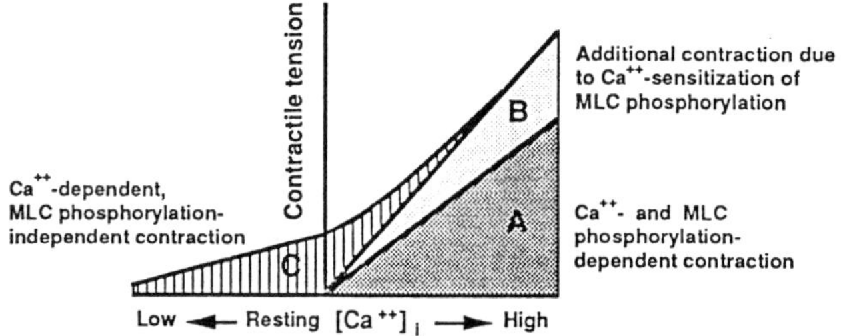

Figure 13. Mechanism of smooth muscle contraction. **A:** increase in $[Ca^{++}]_i$ induced by high K^+ or receptor agonists induces contraction which is dependent on MLC phosphorylation. **B:** Receptor agonists have an additional effect to increase Ca^{++} sensitivity of MLC phosphorylation (see Fig. 8) resulting in a greater contraction than high K^+ for a given increase in $[Ca^{++}]_i$. **C:** In the presence of verapamil or EGTA, $[Ca^{++}]_i$ and MLC phosphorylation returned to their respective resting levels although the contraction induced by receptor agonists was only partially inhibited. This portion of the contraction is regulated by $[Ca^{++}]_i$ but not by MLC phosphorylation.

EFFECTS OF VASODILATATORS

Ca^{++} Channel Blockers

As shown in Figs. 9 and 10, verapamil completely inhibited the $[Ca^{++}]_i$ and muscle tension stimulated by high K^+. Verapamil also strongly inhibited the $[Ca^{++}]_i$ stimulated by various receptor agonists suggesting that the pathway of receptor-mediated Ca^{++} influx is mainly the L type voltage-dependent Ca^{++} channels. However, a part of the $[Ca^{++}]_i$ stimulated by receptor agonists is sometimes not inhibited by verapamil (for example see Fig. 14) possibly because receptor agonists open also the non-selective cation channels which are permeable to Ca^{++} (6). Verapamil did not inhibit the transient increase in Ca^{++} and muscle tension stimulated by NE (Fig. 14) or caffeine in Ca^{++}-free solution (81) suggesting that verapamil does not inhibit Ca^{++} release (40). Similar results have been obtained with dihydropyridine Ca^{++} channel blockers such as nisoldipine (44), nifedipine, and felodipine (21). Since these dihydropyridines gradually decompose under ultraviolet light which is needed to stimulate fura-2, care must be taken to reduce the amount of excitation light when measuring $[Ca^{++}]_i$ with fura-2 in the presence of these dihydropyridines.

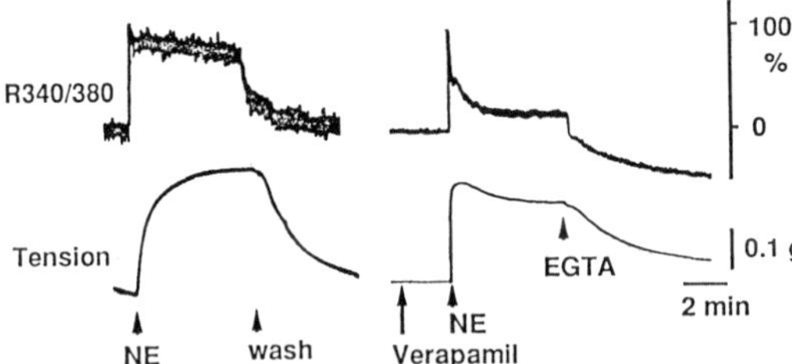

Figure 14. Effects of 1μM NE on $[Ca^{++}]_i$ (upper trace) and muscle tension (lower trace) in rat aorta. NE was added in the absence (left) or presence of 10 μM verapamil (right).

In spite of the strong inhibition of NE-stimulated $[Ca^{++}]_i$, verapamil only partially inhibited the NE-induced contraction. This is because NE increases not only $[Ca^{++}]_i$ but also Ca^{++} sensitivity of MLC phosphorylation and also activates a mechanism which is not dependent on MLC phosphorylation at a resting $[Ca^{++}]_i$ level, and that verapamil decreases the stimulated $[Ca^{++}]_i$ to a resting level without affecting other contractile mechanisms.

Cyclic AMP

In smooth muscle, cyclic AMP plays a role as a second messenger which decreases resting muscle tone and inhibits contraction induced by various agonists (9). An activator of adenylate cyclase, forskolin, and membrane permeable analogue of cyclic AMP, dibutyryl cyclic AMP (db-cAMP), selectively inhibited the NE-induced sustained contraction at lower concentrations whereas these inhibitors also inhibited the high K^+-induced contraction at higher concentrations (1-3, 49).

As for the mechanism of cyclic AMP-induced relaxation, a lower concentration of db-cAMP (10 μM, Fig. 15A) and forskolin (100 nM) inhibited NE-induced increase in both $[Ca^{++}]_i$ and muscle tension without changing the $[Ca^{++}]_i$-tension relationship. Increments in $[Ca^{++}]_i$ and muscle tension due to high K^+ were also inhibited without changing the $[Ca^{++}]_i$-tension relationship (Fig. 15B). However, NE-induced transient contraction obtained in a Ca^{++}-free solution was not inhibited. These results indicate that the inhibitory effect of lower concentrations of these inhibitors is attributable to the decrease in $[Ca^{++}]_i$. Since cyclic AMP does not change the membrane Ca^{++} pump activity in cultured vascular smooth muscle (17), decrease in $[Ca^{++}]_i$ may be due to inhibition of Ca^{++} channels and/or activation of Ca^{++} sequestration. Forskolin inhibits the NE-stimulated $^{45}Ca^{++}$ influx with little effect on the high K^+-stimulated influx (3), suggesting that low concentrations of these inhibitors relatively selectively inhibits receptor-linked Ca^{++} influx.

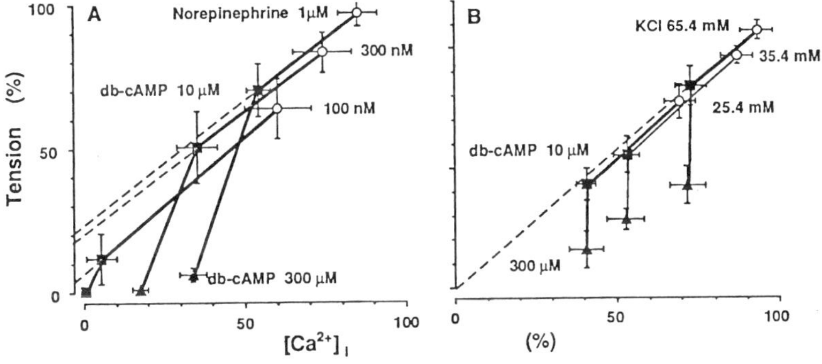

Figure 15. Effects of db-cAMP on $[Ca^{++}]_i$ and muscle tension in rat aorta stimulated by NE (A) or KCl (B). Db-cAMP was cumulatively added during the sustained contraction induced by NE (100 nM, 300 nM or 1 μM) or KCl (25.4 mM, 35.4 mM or 65.4 mM). Dotted lines indicate the effect of 10 μM verapamil added during the sustained contraction, as shown in Fig. 9. It is shown that 10 μM db-cAMP inhibited both $[Ca^{++}]_i$ and muscle tension, as is the case with verapamil. By contrast, 300 μM db-cAMP inhibited the contractions with only a small additional decrease in $[Ca^{++}]_i$, suggesting the decrease in Ca^{++} sensitivity. Furthermore, the contractions induced by higher concentrations of stimulants are less sensitive to db-cAMP. Each point represents mean $\pm$ SEM of 4 experiments

Higher concentrations of these inhibitors (300 μM db-cAMP and 1 μM forskolin) more strongly inhibited the contractions induced by NE or high K^+ with only a small additional decrease in $[Ca^{++}]_i$ (Fig. 15). In rabbit aorta, high concentration of forskolin inhibited high K^+-induced contraction with small decrease in $^{45}Ca^{++}$ uptake (3, 28). Cyclic AMP has been shown to inhibit contractile response in permeabilized smooth muscle through the activation of cyclic AMP-dependent protein kinase (55, 66). These results indicate that higher concentrations of these inhibitors inhibit contraction mainly by the decrease in Ca^{++} sensitivity of contractile elements. These inhibitors also inhibited the NE-induced transient contraction by the inhibition of Ca^{++} release. An inhibitor of type III phosphodiesterase, E-1020, showed similar inhibitory effect (86). Since cyclic AMP inhibits receptor-mediated increase in phosphatidyl inositol turnover (10, 31), stronger inhibitory effect of cyclic AMP on receptor-mediated contraction may be partly explained by this mechanism.

The results obtained with fura-2 were different from those obtained with aequorin; forskolin increased aequorin-Ca^{++} signal without increase in muscle tension in ferret portal vein (54) and in bovine trachea (93). Results with $^{45}Ca^{++}$ uptake experiments indicate that forskolin does not increase Ca^{++} influx (3, 28), suggesting that $[Ca^{++}]_i$ measured with aequorin represents the cellular sites which are different from those measured with fura-2 or $^{45}Ca^{++}$ uptake experiments.

The inhibitory effects of db-cAMP and forskolin, measured with IC_{50} values, were inversely proportional to the contractile tension induced by each stimulant. Plotting the IC_{50} values for these inhibitors against the $[Ca^{++}]_i$ stimulated by NE or high K^+, it was found that the inhibitory effect of cyclic AMP is stronger when the stimulated $[Ca^{++}]_i$ was lower irrespective of the type of stimulation. These results suggest that cAMP antagonizes the role of Ca++ in cytoplasm (1, 67).

It has been reported that cyclic AMP-dependent protein kinase phosphorylates MLC kinase, inhibits the activity of this kinase and thus inhibits contractile elements (4, 11). However, Stull *et al.* (85) reported that isoproterenol does not phosphorylate MLC kinase in tracheal smooth muscle suggesting that phosphorylation of other protein(s) is responsible for the cyclic AMP-induced decrease in Ca^{++} sensitivity.

Cyclic GMP

Low concentrations of an activator of guanylate cyclase, sodium nitroprusside, inhibited $[Ca^{++}]_i$ and muscle tension stimulated by high K^+ or NE with little effect on the slope of the $[Ca^{++}]_i$-tension curve. These results indicate that the decrease in $[Ca^{++}]_i$ is responsible for the inhibitory effect of low concentrations of sodium nitroprusside. Higher concentration of sodium nitroprusside strongly inhibited the contraction induced by NE and less strongly inhibited the contraction induced by high K^+. The increments in $[Ca^{++}]_i$ were also reduced, however to a lesser extent than the contractions, resulting in the decrease in slope of the $[Ca^{++}]_i$-tension curve (41). In canine coronary artery, nitroglycerin inhibited high K^+-induced contraction with little effect on $[Ca^{++}]_i$ (98). Cyclic GMP has been shown to inhibit contractile response in permeabilized smooth muscle through the activation of cyclic GMP-dependent protein kinase (55, 66). These results indicate that higher concentrations of sodium nitroprusside inhibits these contractions possibly by the decrease in $[Ca^{++}]_i$ and also by an increase in the threshold of $[Ca^{++}]_i$ for tension development.

Other Vasodilatators

Effects of various vasodilators on $[Ca^{++}]_i$ and contractile tension are summarized in Table 2. Ca^{++} channel blockers (21, 40, 44) but not K^+ channel openers (5) inhibited high K^+-stimulated Ca^{++} influx. By contrast, both Ca^{++} channel blockers and K^+ channel openers inhibited NE-induced contraction by decreasing $[Ca^{++}]_i$ resulted from direct and indirect inhibition of Ca^{++} channels, respectively. Various inhibitors such as TMB-8 (29), KT-362 (77) and ISF-2405 (53) inhibited the NE-induced contraction more strongly than $[Ca^{++}]_i$, suggesting that these compounds inhibit Ca^{++} sensitization. Furthermore, these inhibitors inhibited the portion of the NE-induced contraction which is not sensitive to verapamil, suggesting that the contraction which is not dependent on MLC phosphorylation is also inhibited by these inhibitors. Some inhibitors such as HA-1077 (92) and cytochalasin D (74) inhibited the contraction with little effect on $[Ca^{++}]_i$, suggesting that these inhibitors directly inhibits contractile elements.

Summary

Vasodilatators inhibit vascular smooth muscle contraction not only by decreasing $[Ca^{++}]_i$ but also by decreasing Ca^{++} sensitivity (69). Change in vascular smooth muscle contraction does not necessarily indicate change in $[Ca^{++}]_i$.

Table 2. Inhibitory effects of vasodilatators on $[Ca^{++}]i$ and Ca^{++} sensitivity in smooth muscle stimulated by KCl or NE.

Vasodilatator	$[Ca^{++}]i$		Ca^{++} sensitivity	Reference
	KCl	NE		
Verapamil	+	+	−	Karaki *et al.*, 1991
Nifedipine	+	+	−	Hagiwara *et al.*, 1991
Felodipine	+	+	−	Hagiwara *et al.*, 1991
Nisoldipine	+	+	−	Kim *et al.*, 1991
Pinacidil	−	+	−	Anabuki *et al.*, 1991
Forskolin	±	+	+	Abe & Karaki, 1990
Db-cAMP	±	+	+	Abe & Karaki, 1991
E-1020 a)	±	+	+	Tajimi *et al.*, 1991
Nitroprusside Na	±	+	+	Karaki *et al.*, 1988
Nitroglycerin	−	nd	+	Yanagisawa *et al.*, 1989
TMB-8 b)	+	+	+	Ishihara & Karaki, 1991
KT-362 c)	+	+	+	Sakata & Karaki, 1991
ISF 2405 d)	−	+	+	Mitsui *et al.*, 1991
Okadaic acid e)	−	+	+	Karaki *et al.*, 1990, Tansey *et al.*, 1991
HA-1077 f)	−	±	+	Takizawa *et al.*, 1991
Cytochalasin D	−	−	+	Saito *et al.*, 1991

+: decrease. ±: slight decrease. −: no effect. nd: not determined.
a) Inhibitor of type III phosphodiesterase. b) 8-(N,N-dimethyl-amino) octyl-3,4,5-trimethoxybenzoate. c) 5-[3-([2-(3,4-dimethoxy phenyl)ethyl]amino)-1-oxopropyl]-2,3,4,5-tetrahydro-1,5-benzo-thiazepine. d) Active metabolite of cadoralazine. e) Phosphatase inhibitor. f) Putative inhibitor of MLC kinase.

$C_A{}^{++}$ REGULATION OF EDRF RELEASE

It has been shown that various physiological and pharmacological stimuli release EDRF in blood vessels (16). In order to clarify the role of Ca^{++} in the release of EDRF, we developed a method to measure $[Ca^{++}]_i$ in endothelium of isolated vascular strip simultaneously with contractile tension (79).

Measurements of Endothelial Ca^{++} in a Vascular Strip

The thoracic aorta was isolated from a male Wistar rat (250-300 g), cut into a spiral strip (2 mm wide and 7 mm long) and placed in the normal PSS. Three types of muscle strips were prepared (Fig. 16): (1) strips with endothelium [E(++)]; (2) strips in which endothelium was removed only from a small area (approximately one third of the endothelial surface) from where fura-2-Ca^{++} fluorescence was detected [E(+-)]; and (3) strips without endothelium [E(--)]. $[Ca^{++}]_i$ was measured with fura-2. Muscle strips were treated with 5 μM fura-2/AM for 3-5 h at room temperature.

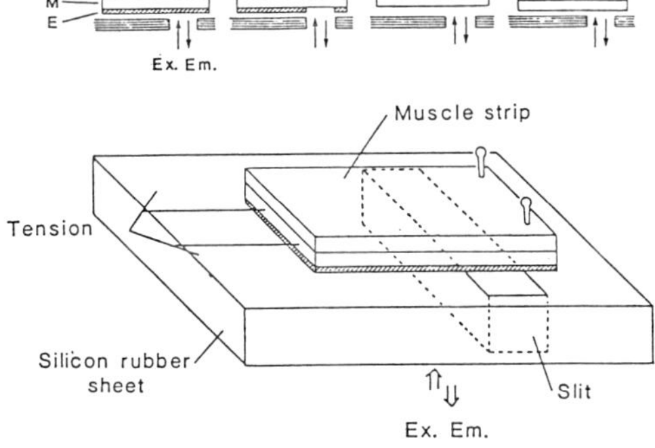

Figure 16. Schematic illustrations of the method for measuring $[Ca^{++}]_i$ in endothelium and/or smooth muscle in the aorta loaded with fura-2. E(++): preparation with endothelium; E(+-): endothelium was removed only from a portion (approximately one-third of the endothelial surface) from where the fura-2-Ca^{++} fluorescence was detected; E(--); all the endothelium was removed; Reversed; fluorescence was detected from adventitial surface of E(++). A, adventitia; M, tunica media; E, endothelium; Ex, excitation light; Em, emission light. Reproduced from Sato *et al.*, (79) with permission.

310

As shown in Fig. 16, the fura-2-loaded muscle strip was held horizontally on a silicon rubber sheet laid on the quartz bottom of an organ bath attached to a fluorimeter (CAF-100, Japan Spectroscopic, Tokyo, Japan). One end of the muscle strip was pinned to the silicon rubber sheet and another end was connected to a strain gauge transducer (Toyo Boldwin, Tokyo, Japan) to monitor the isometric tension. The muscle strip was excited alternatively by 340 nm and 380 nm light through a narrow slit of the sheet. The fluorescence emitted from the muscle strip was collected to a photomultiplier through a 500 nm filter and the ratio of the fluorescence due to excitation at 340 nm (F340) to that at 380 nm (F380) was calculated from successive illumination periods.

Cytosolic Ca^{++} in Endothelium

As shown in Fig. 16, fura-2-Ca^{++} fluorescence was detected from the endothelial surface. In E($+$-) and E($--$) preparations, therefore, fura-2-Ca^{++} fluorescence derives only from smooth muscle cells whereas the fluorescence derives from endothelial cells and also from smooth muscle cells in E($++$) preparation. In these three types of preparations, high K^+ and NE induced sustained increase in $[Ca^{++}]_i$ and muscle tension. By contrast, $3\,\mu$M carbachol (CCh) induced sustained increase in $[Ca^{++}]_i$ and a small relaxation in E($++$) preparation (Fig. 17A). However, CCh did not change $[Ca^{++}]_i$ and muscle tension in E($--$) preparation (Fig. 17B). In E($+$-) preparation, CCh significantly decreased the resting tension with no change in $[Ca^{++}]_i$. In E($++$) preparation, $30\,\mu$M histamine increased $[Ca^{++}]_i$ and slightly decreased the resting tone of the muscle although it was ineffective in E($+$-) and E($--$) preparations. These results indicate that high K^+ and NE increase $[Ca^{++}]_i$ in smooth muscle whereas CCh and histamine increases $[Ca^{++}]_i$ in endothelium but not in smooth muscle. In some experiments, the fluorescence was detected from adventitial surface of the E($++$) preparation ([Reversed] in Fig. 16). In the "reversed" preparation, 72.7 mM K^+ or $0.3\,\mu$M NE increased the $[Ca^{++}]_i$. CCh decreased the resting tone of this preparation. However, CCh induced only a slight increase in $[Ca^{++}]_i$. We also observed that the 500 nm light penetrates whole tissue of rat aorta whereas the lager portion of the 340 nm and 380 nm lights are absorbed. These results suggest that only a small portion of the fura-2-Ca^{++} fluorescence in endothelium (excited at 340 nm and 380 nm) is detected at the adventitial surface of the tissue. These results also support the suggestion that CCh dose not change $[Ca^{++}]_i$ in smooth muscle.

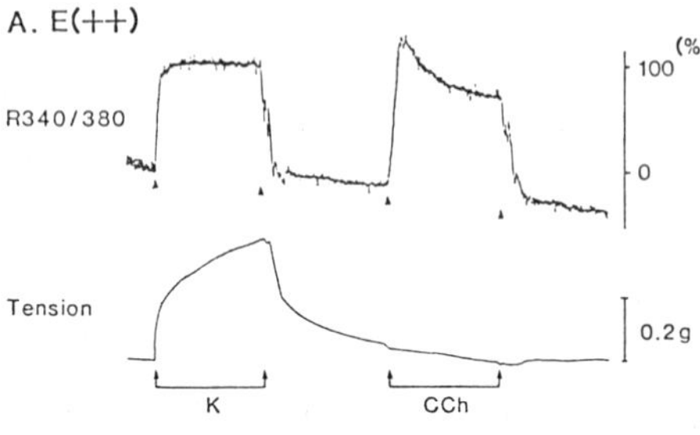

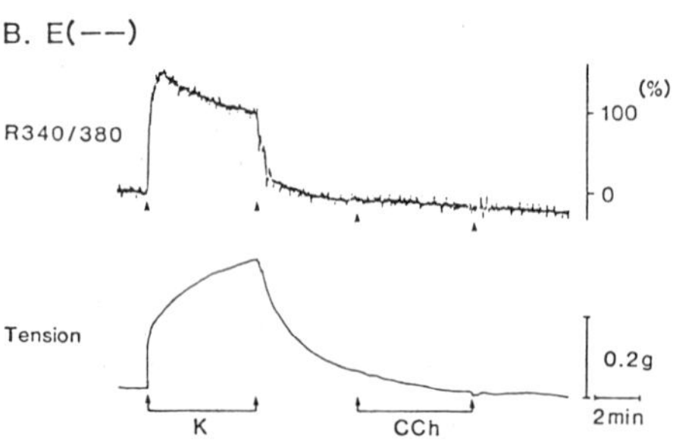

Figure 17. Effects of CCh on $[Ca^{++}]_i$ (upper trace) and contractile tension (lower trace) in E($++$) (A) and E($--$) (B). After the effects of 72.7 mM KCl (K) were examined, 3 μM CCh was added. Reproduced from Sato *et al.*, (79) with permission.

In Ca^{++}-free solution, 3 μM CCh induced a transient increase in $[Ca^{++}]_i$ and a decrease in muscle tension in $E(++)$ preparation but showed no effect in $E(--)$ preparation (Fig. 18). Histamine (30 μM) showed similar effect as CCh. By contrast, 1 μM NE transiently increased both $[Ca^{++}]_i$ and muscle tension in Ca^{++}-free solution in these two types of preparations which may be due to Ca^{++} release from smooth muscle cells (81). These results suggest that CCh and histamine induce only a transient increase in $[Ca^{++}]_i$ by releasing Ca^{++} from storage site in endothelial cells. These results also suggest that influx of extracellular Ca^{++} is essential for the sustained increase in $[Ca^{++}]_i$ in endothelium.

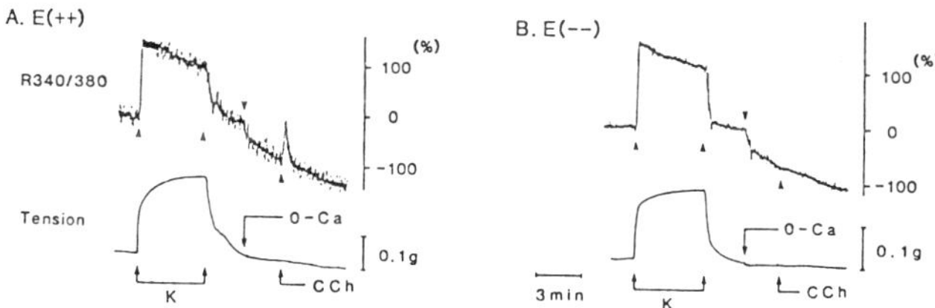

Figure 18. Effects of 3 μM CCh on $[Ca^{++}]_i$ (upper trace) and contractile tension (lower trace) in the absence of external Ca^{++} in $E(++)$ (A) and $E(--)$ (B). After treatment of the muscle strip with 72.7 mM KCl (K) solution for 5 min, external solution was replaced with Ca^{++}-free solution (with 0.5 mM EGTA), followed by an application of CCh. Reproduced from Sato *et al.*, (79) with permission.

Sustained increase in $[Ca^{++}]_i$ due to CCh and histamine was only partly inhibited by 10 μM verapamil suggesting that pathway of Ca^{++} influx is different from the voltage-dependent Ca^{++} channel. This result is consistent with the finding in cultured endothelial cells that a Ca^{++} channel activator, Bay-K-8644, does not increase $[Ca^{++}]_i$ (97) and that verapamil only partially inhibits the increase in $[Ca^{++}]_i$ (72).

In summary, CCh and histamine increased $[Ca^{++}]_i$ in endothelium mainly by an increase in Ca^{++} influx through a verapamil-insensitive pathway and partly by a release of Ca^{++}. The release of Ca^{++} may be mediated by generation of inositol 1,4,5-trisphosphate in endothelial cells (13, 50). Although the mechanism for Ca^{++} influx is not clarified as yet, activation of protein kinase C may be at least partly be responsible for this effect because phorbol esters release EDRF (76).

Ca^{++} Regulation of EDRF Release

As shown in Fig. 19, 0.3 μM NE induced sustained increase in $[Ca^{++}]_i$ and muscle tension in all the types of preparations. Addition of 3 μM CCh during the NE-induced sustained contraction caused a rapid increase in $[Ca^{++}]_i$ and a rapid relaxation of the contraction in $E(++)$ preparation. Sequential addition of 3 nM atropine decreased $[Ca^{++}]_i$ and increased muscle tension to the level before the addition of CCh (Fig. 19A). In $E(+-)$ preparation, CCh relaxed the NE-induced contraction and decreased the stimulated $[Ca^{++}]_i$ (Fig. 19B). By contrast, CCh had no effect on the muscle tension and $[Ca^{++}]_i$ stimulated by NE in $E(--)$ preparation (Fig. 19C). These results suggest that EDRF is released from the remaining portion of the endothelium in $E(+-)$ preparation. In "reversed" preparation, 3 μM CCh relaxed the NE-induced contraction with a decrease in $[Ca^{++}]_i$ (Fig. 19D). Histamine (30 μM) showed similar effects as CCh on NE-stimulated tension and $[Ca^{++}]_i$ in $E(++)$ and $E(--)$ preparations. Addition of CCh in the presence of high K^+ induced additional increase in $[Ca^{++}]_i$ with a small decrease in high K^+-induced contraction in $E(++)$ preparation although CCh was ineffective in $E(--)$ preparation. Fig. 20 shows the relationship between relaxation of NE-induced contraction and additional increase in $[Ca^{++}]_i$ due to CCh in $E(++)$ preparation. This figure shows that there is a positive correlation between the increase in $[Ca^{++}]_i$ in endothelium and release of EDRF.

In summary, it is suggested that the synthesis and/or release of EDRF is regulated by $[Ca^{++}]_i$ in endothelium.

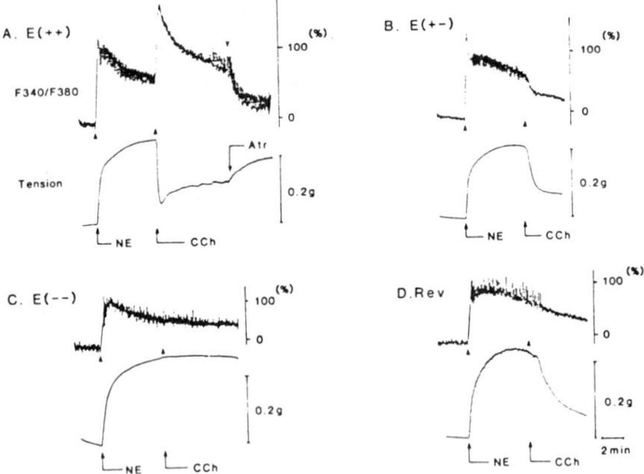

Figure 19. Effects of 3 μM CCh on $[Ca^{++}]_i$ (upper trace) and contractile tension (lower trace) in the presence of 0.3 μM NE in E(++) (A), E(+-) (B), E(--) (C), and reversed (D). In E(++), 3 nM atropine (Atr) was added after the application of CCh. Reproduced from Sato *et al.*, (79) with permission.

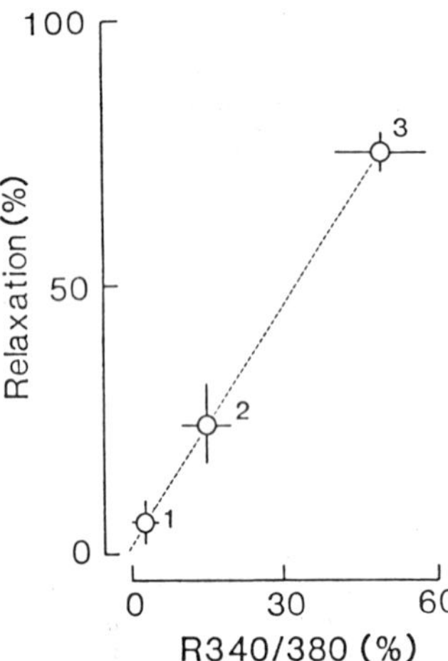

Figure 20. Relationship between the additional increase in $[Ca^{++}]_i$ (abscissa) and relaxation of contraction (ordinate) due to CCh in NE-stimulated E(++). CCh was added cumulatively during the contraction induced by 0.3 μM NE; 100% represents the level before the addition of CCh. The concentration of CCh is indicated by numbers (1; 10 nM; 2, 100 nM; 3, 1 μM). Each point represents mean of 4 experiments and S.E.M. is shown by vertical and horizontal bars. Reproduced from Sato *et al.*, (79) with permission.

Mechanism of EDRF-Induced Relaxation

EDRF decreases $[Ca^{++}]_i$ in smooth muscle cells. This result is consistent with the finding that acetylcholine-induced, endothelium-dependent relaxation of NE-contracted rabbit aorta is accompanied by a decrease in $^{45}Ca^{++}$ uptake (69). Comparison between the decrease in smooth muscle $[Ca^{++}]_i$ and relaxation in E(+-) preparation indicated that CCh decreased the NE-induced contraction by 67.6% whereas it inhibited the NE-stimulated $[Ca^{++}]_i$ only by 18.4%. This result suggests that the vasodilator effect of EDRF is not solely due to the decrease in $[Ca^{++}]_i$ but also to the inhibition of contractile elements. Vasodilatator effect of EDRF is at least partly due to activation of guanylate cyclase resulting in an increase in cyclic GMP content in smooth muscle cells (26, 68). Sodium nitroprusside, an activator of guanylate cyclase, inhibits NE-induced contraction in vascular smooth muscle by decreasing $[Ca^{++}]_i$ and also by decreasing Ca^{++} sensitivity of contractile elements (41, 55). These results support the suggestion that EDRF and sodium nitroprusside have similar mechanism of action.

In summary, it is suggested that EDRF released from endothelium acts on smooth muscle, deceases $[Ca^{++}]_i$ and Ca^{++} sensitivity of contractile elements and induces vasodilatation.

Summary

Major advantage of the present method is to measure the endothelial and/or smooth muscle $[Ca^{++}]_i$ simultaneously with muscle contraction to know the temporal as well as quantitative relationship between $[Ca^{++}]_i$ and function. In the present experiments, it was found that, when added in the presence of NE, CCh and histamine induced additional increase in $[Ca^{++}]_i$ which preceded the relaxation of the E(++) preparation. Furthermore, it was demonstrated that there is a positive correlation between the increase in $[Ca^{++}]_i$ and muscle relaxation. These results suggest that the CCh- or histamine-induced additional increase in $[Ca^{++}]_i$ is due to endothelial $[Ca^{++}]_i$ and that synthesis and/or release of EDRF is regulated by the amount of $[Ca^{++}]_i$ in the endothelial cells. EDRF may decease $[Ca^{++}]_i$ in the smooth muscle cells and also decrease Ca^{++} sensitivity of contractile elements resulting in vasodilatation.

CONCLUSION

Until several years ago, it was widely accepted that smooth muscle contraction is regulated by $[Ca^{++}]_i$. Measuring $[Ca^{++}]_i$ and muscle tension simultaneously with aequorin, Morgan and Morgan (26) first suggested that change in Ca^{++} sensitivity is also an important factor. Their suggestion was confirmed recently by the experiments using fura-2. It was also found that Ca^{++} sensitization of contractile elements is attributable to the increase in MLC phosphorylation at a given $[Ca^{++}]_i$. Furthermore, it was suggested that receptor-agonists stimulate a novel mechanism which is dependent on $[Ca^{++}]_i$ but not on MLC phosphorylation. Protein kinase C may be involved in this process. Release of EDRF is also regulated by $[Ca^{++}]_i$ although detailed mechanism remains to be elucidated.

ACKNOWLEDGEMENTS

Thanks are due to Drs. Masatoshi Hori, Minori Mitsui and Koichi Sato of our Department for their help in preparing the manuscript. Original work of this study was supported by the Grant-in-Aid for Scientific Research from the Ministry of Education, Science and Culture of Japan.

REFERENCES

1. Abe A. and H. Karaki, Folia Pharmacol. Japon. in press (1991).
2. Abe A. and H. Karaki, J. Pharmacol. Exp. Ther. 249, 895-900 (1989).
3. Abe A. and H. Karaki, Japan. J. Pharmacol. 46, 293-301 (1988).
4. Adelstein, R.S., M.A. Conti, D.R. Hathaway, and C.B. Klee, J. Biol. Chem. 253, 8347-8350 (1978).
5. Anabuki, J., M. Hori, H. Ozaki, I. Kato, and H. Karaki, Europ. J. Pharmacol. 190, 373-379 (1990).
6. Benham, C.D. and R.W. Tsien, Nature (Lond.) 328, 275-278 (1987).
7. Blinks, J.R., P.H. Mattingly, B.R. Jewell, M. VanLeewan, G.C. Harrer, and D.G. Allen, Methods Enzymol. 57, 292-328 (1978).
8. Blinks, J.R., Techniques Cell. Physiol 126, 1-38 (1982).
9. Bulbring, E. and T. Tomita, Pharmacol. Rev. 39, 49-96 (1987).
10. Campbell, M.D., S. Subramaniam, M.I. Kotolikoff, J.R. Williamson, and S.J. Fluharty, Mol. Pharmacol. 36, 282-284 (1990).
11. Conti, M.A. and R.S. Adelstein, J. Biol. Chem. 256, 3178-3181 (1981).

12. Defeo, T.T. and K.G. Morgan, J. Physiol. 369, 260-282 (1985).
13. Derian, C.K. and M.A. Moskowitz, J. Biol. Chem. 261, 3831-3837 (1986).
14. Dillon, P.F., M.O. Aksoy, S.P. Driska, and R.A. Murphy, Science 211, 495-497 (1981.).
15. Ebashi, S., Proc. R. Soc. London Ser. B. 207, 259-286 (1980).
16. Furchgott, R.F. and J.V. Zawadzki, Nature (Lond.) 288, 373-376 (1980).
17. Furukawa, K.-I., Y. Tawada, and M. Shigekawa, J. Biol. Chem. 263, 8058-8065 (1988).
18. Gerthoffer, W.T., K.A. Murphy, and S.J. Gunst, Am. J. Physiol. 26, C1062-C1068 (1989).
19. Gilbert, E.K., B.A. Weaver, and C.M. Rembold, FASEB J. (in press) (1991).
20. Grynkiewicz, G., M. Poenie, and R.Y. Tsien, J. Biol. Chem. 260, 3440-3450, (1985).
21. Hagiwara, S., M. Mitsui, and H. Karaki, Japan. J. Pharmacol. 55 (suppl. 1), 263P (1991).
22. Hai, C.-M. and R.A. Murphy, Am. J. Physiol. 255, C86-94 (1988).
23. Hartshorne, D.J. in: Physiology of the GI tract, L. R. Johnson ed. (Raven, New York 1987) pp.423-428.
24. Himpens, B. and A.P. Somlyo, J. Physiol. (Lond.) 395, 507-530 (1988).
25. Himpens, B., T. Kitazawa, and A.P. Somlyo, Pfluger's Arch. 417, 21-28 (1990).
26. Holzmann, S., J. Cyclic Nucleotide Res. 8, 409-419 (1982).
27. Hori, M., K. Sato, I. Kato, and H. Karaki, J. Muscle Res. Cell Motility 11, 355 (1990).
28. Hwang, K.S. and C. Van Breemen, Europ. J. Pharmacol. 134, 155-162 (1987).
29. Ishihara, H. and H. Karaki, Europ. J. Pharmacol. 197, 181-186 (1991).
30. Ishikawa, T., T. Chijiwa, H. Hagiwara, S. Mamiya, S. Saitoh, and H. Hidaka, J. Pharmacol. Exp. Ther. 33, 598-603 (1988).
31. Kaibuchi, K., Y. Takai, Y. Ogawa, S. Kimura, and Y. Nishizuka, Biochem. Biophys. Res. Commun. 104, 105-112 (1982).
32. Kamm, K.E. and J.T. Stull, Ann. Rev. Pharmacol. 25, 593-620 (1985).
33. Kamm, K.E. and J.T. Stull, Ann. Rev. Physiol. 51, 299-313 (1989).
34. Kamm, K.E., L.-C. Hsu, Y. Kubota, and J.T. Stull, J. Biol. Chem. 264, 21223-21229 (1989).
35. Karaki, H. and G.B. Weiss, Gastroenterology 87, 960-970 (1984).
36. Karaki, H. and G.B. Weiss, Life Sciences 42, 111-122 (1988).
37. Karaki, H., Acta Haematol. Japon. 52, 1506-1515 (1989).
38. Karaki, H., Am. J. Hypertension 3, 253S-256S (1990).
39. Karaki, H., K. Sato, and H. Ozaki, Europ. J. Pharmacol. 151, 325-328 (1988).
40. Karaki, H., K. Sato, and H. Ozaki, Japan. J. Pharmacol. 55, 35-42 (1991).
41. Karaki, H., K. Sato, H. Ozaki, and K. Murakami, Europ. J. Pharmacol. 156, 259-266 (1988).
42. Karaki, H., K. Sato, M. Hori, H. Ozaki, T. Ohyama, and K. Sakata, J. Muscle Res. Cell Motility 11, 355 (1990).
43. Karaki, H., Trend. Pharmacol. Sci. 10, 320-325 (1989).
44. Kim, B.-K., M. Mitsui, and H. Karaki, Japan. J. Pharmacol. 55 (suppl. 1), 189P (1991).
45. Kitazawa, T., B.D. Gaylinn, G.H. Denny, and A.P. Somlyo, J. Biol. Chem. 266, 1708-1715 (1991).
46. Kitazawa, T., S. Kobayashi, K. Horiuchi, A.V. Somlyo, and A.P. Somlyo, J. Biol. Chem. 264, 5339-5342 (1989).
47. Konishi, M., A. Olson, S. Hollingworth, and S.M. Baylor, Biophys. J. 54, 1089-1104 (1988).
48. Kubota, Y., K.E. Kamm, and J.T. Stull, Biophys. J. 57, 163a (1990).
49. Lincoln, T.M. and V. Fisher-Simpson, Europ. J. Pharmacol. 101, 17-27 (1983).
50. Lo, W.W.Y. and T.P.D. Fan, Biochem. Biophys. Res. Commun. 148, 47-53 (1987).
51. Malgalori, A., D. Milani, J. Meldolesi, and T. Pozzan, J. Cell Biol. 105, 2145-2155 (1987).
52. Mitsui, M. and H. Karaki, Am. J. Physiol. 258, C787-C793 (1990).
53. Mitsui, M., K. Nakao, T. Inukai, and H. Karaki, Europ. J. Pharmacol. 178, 171-177 (1990).
54. Morgan, J.P. and K.G. Morgan, J. Physiol. 357, 539-551 (1984).
55. Nishimura, J. and C. Van Breemen, Biochem. Biophys. Res. Commun. 163, 929-935,1989).
56. Nishimura, J., M. Kolber, and C. Van Breemen, Biochem. Biophys. Res. Commun. 157, 677-683 (1988).
57. Nishimura, J., R.A. Khall, J.P. Drenth, and C. Van Breemen, Am. J. Physiol. 259, H2-H8 (1990).
58. Nishizuka, Y., Nature 308, 693-698 (1984).
59. Oishi, K., H. Takano-Ohmuro, N. Minakami-Matsuo, O. Suga, H. Karibe, K. Kohama, and M.K. Uchida, Biochem. Biophys. Res. Comm. 176, 122-128 (1991).
60. Ozaki, H., K. Sato, T. Sato, and H. Karaki, Japan. J. Pharmacol. 45, 429-433 (1987).
61. Ozaki, H., R.J. Stevens, D.P. Blondfield, N.G. Publicover, and K.M. Sanders, Am. J. Physiol. 260, C917-C925 (1991 1991).
62. Ozaki, H., S.-C. Kwon, M. Tajimi, and H. Karaki, Pflugers Arch. 416, 351-359 (1990).
63. Ozaki, H., T. Ohyama, K. Sato, and H. Karaki, Japan. J. Pharmacol. 52, 509-512 (1990).
64. Ozaki, H., T. Satoh, H. Karaki, and Y. Ishida, J. Biol. Chem. 263, 14074-14079 (1988).
65. Persenchini, A., K.E. Kamm, and J.T. Stull, J. Biol. Chem. 261, 6293-6299 (1986).

66. Pfitzer, G., F. Hofmann, J. Di Salvo, and J.C. Ruegg, Pflugers Arch 401, 277-280 (1984).
67. Pfitzer, G., J.C. Ruegg, M. Zimmer, and F. Hoffmann, Pfluger's Arch. 405, 70-76 (1985).
68. Rapoport, R.M. and F. Murad, Circ. Res. 52, 352-357 (1983).
69. Ratz, P.H., M.M. Gleason, and S.F. Flaim, Circ. Res. 60, 31-38 (1987).
70. Rembold, C.M. and R.A. Murphy, Am. J. Physiol. 255, C719-C723 (1988).
71. Rembold, C.M., J. Physiol. 429, 77-94 (1990).
72. Rotrosen, D. and J.I. Gallin, J. Cell Biol. 103, 2379-2387 (1986).
73. Sada, T., H. Koike, M. Ikeda, K. Sato, H. Ozaki, and H. Karaki, Hypertension 16, 245-251 (1990).
74. Saito, S.,. M. Hori, K. Sato, H. Ozaki, and H. Karaki, Folia Pharmacol. Japon. 96, 14P (1990).
75. Saitoh, M., T. Ishikawa, S. Matsushima, M. Naka, and H. Hidaka, J. Biol. Chem. 262, 7796-7801 (1987).
76. Sakata, K. and H. Karaki, Europ. J. Pharmacol. 179, 207-210 (1990).
77. Sakata, K., and H. Karaki, Brit. J. Pharmacol. 102, 174-178 (1991).
78. Sakata, K., H. Ozaki, S.C. Kwon, and H. Karaki, Brit. J. Pharmacol. 98, 483-492 (1989).
79. Sato, K., H. Ozaki, and H. Karaki, J. Pharmacol. Exp. Ther. 255, 114-119 (1990).
80. Sato, K., H. Ozaki, and H. Karaki, J. Pharmacol. Exp. Ther. 246, 294-300 (1988).
81. Sato, K., H. Ozaki, and H. Karaki, Naunyn-Schmiedeberg's Arch. Pharmacol. 338, 443-448 (1988).
82. Scanlon, M., D.A. Williams, and S.F. Fay, J. Biol. Chem. 262, 6308-6312 (1987).
83. Shibata, S., N. Satake, M. Morikawa, S.-C. Kwon, H. Karaki, K. Kurahashi, T. Sawada, and I, Kodama, Europ. J. Pharmacol. 193, 1-7 (1991).
84. Somlyo, A.P. and B. Himpens, FASEB J. 3, 2266-2276 (1989.).
85. Stull, J.T., L.-C. Hsu, M.G. Tansey, and K.E. Kamm, J. Biol. Chem. 27, 16683-16690 (1990).
86. Tajimi, M., H. Ozaki, K. Sato, and H. Karaki, Naunyn-Schmiedeberg's Arch. Pharmacol. (in press) (1991).
87. Takano-Ohmuro, H. and K. Kohama in: Muscle and Motility 2, G. Marechal and U. Carraro, eds. (Intercept Ltd, Hampshire 1990) pp. 323-328.
88. Takano-Ohmuro, H. and K. Kohama, J. Biochem. 100, 1681-1684,1986).
89. Takano-Ohmuro, H. and K. Kohama, J. Biochem. 100, 259-268 (1986).
90. Takayanagi, I. and H. Ohtsuki, Japan. J. Pharmacol. 53, 525-528 (1990).
91. Takayanagi, I. and S. Onozuka, J. Pharmacobio-Dyn. 12, 781-786 (1989).
92. Takizawa, S., M. Hori, and H. Karaki, Japan. J. Pharmacol. 55 (suppl. 1), 263P (1991).
93. Takuwa, Y., N. Takuwa, and H. Rasmussen, J. Biol. Chem. 263, 762-768 (1988).
94. Tansey, M.G., M. Hori, H. Karaki, and J.T. Stull, FEBS Lett. 270, 219-221 (1991).
95. Tansey, M.G., Y. Kubota, R.A. Word, K.E. Kamm, and J.T. Stull, Biophys. J. 59, 232a (1991).
96. Thorin-Threscases, N., L. Oster, J. Atkinson, and C. Capdeville, Europ. J. Pharmacol. 179, 469-471 (1990).
97. Whitmer, W.R., J.S. Williams-Lawson, R.F. Highsmith, and A. Schwartz, Biochem. Biophys. Res. Commun. 154, 591-605 (1988).
98. Yanagisawa, T., M. Kawada, and N. Taira, Brit. J. Pharmacol. 98, 469-482 (1989.

ION CHANNELS AND REGULATION OF TRANSMEMBRANE Ca^{2+} INFLUX IN ENDOTHELIUM

BERND NILIUS

Medical Academy Erfurt, Institut of Medical Physiology, Nordhauser Str.74, O-5010 Erfurt, Germany

INTRODUCTION

Endothelial cells lining the inner surface of blood vessels act as a transducing surface for many physiological functions such as control of the contractile state of underlying smooth muscle cells, proliferation of these cells, modulation of function of white blood cells, platelets or constituents of plasma. Endothelial cells establish humoral communication via synthesis and release of endothelial active molecules such as proteins, e.g. platelet-derived growth factor and interleukins, peptides, e.g. endothelins, as well as smaller products, e.g. prostaglandins and especially endothelium-derived relaxation factor (EDRF) that is believed to play a key role in regulation of vascular tone. Other secreted vasoactive compounds (endothelium-derived contracting factors different from endothelins, endothelium-derived hyperpolarization factor) are not yet chemically characterized. EDRF is synthesised from the precursor L-arginine by a calcium-dependent NO-synthase (20) and released by a supposingly Ca-dependent mechanism. Thus, an increase in the intracellular Ca activity is therefore the trigger for EDRF synthesis and release. A similar mechanism has been described for another Ca-dependent enzyme (lyso-PAF-acetyl transferase) that triggers the synthesis of platelet-activating factor (PAF) in endothelial cells (13). There is now a large body of evidence that Ca-influx into vascular endothelial cells and the intracellular Ca-pattern is regulated by different voltage- and agonist-gated ion channels (for a review see (1 and 27)). After stimulation of endothelial cells by various vasoactive agonists, the functionally important sustained elevation of intracellular Ca depends on extracellular Ca and is supposingly mediated by a transmembrane Ca influx. The concerted action of several ion channels in regulation of intracellular Ca is the topic of this article.

AGONIST-INDUCED INTRACELLULAR Ca SIGNALS IN ENDOTHELIUM

Figure la shows a typical intracellular Ca signal in an endothelial cell induced by the vasoactive agonist histamine which is mediated by H_1-receptors. Concentrations below 1 μM induce oscillations of intracellular calcium. At higher concentrations, the initial peak of free intracellular calcium is followed by a long-lasting pedestal (15). It has been shown by several groups that the initial peak is generated by intracellular Ca release whereas the sustained increase is generated by a transmembrane Ca influx into the endothelial cell. The response shown in Figure la is similar for different agonists such as acetylcholine, ATP, bradykinin, and thrombin: oscillations at low concentrations, peak of Ca release followed by a pedestal at higher concentrations that is sometimes superimposed by calcium oscillations. The pedestal of Ca depends quantitatively on the extracellular concentration of calcium, disappears when cells are incubated in Ca free solutions or after application of EGTA, and can be blocked with lanthanum. Figure lb gives another example for an intracellular Ca transient evoked by application of 100 μM histamine that lasts several minutes. If no calcium is present in the extracellular solution, the increase in intracellular calcium is short and the long-lasting pedestal disappears (Figure lb (14)). The sustained increase in Ca^{++} may be of a special functional impact: EDRF is a short-living compound, therefore a mechanism that maintains a high intracellular Ca^{++} activity could prolong effects of EDRF. Because several different effects are triggered by an increase in the intracellular Ca^{++} activity, it is intriguing to ask whether the Ca-signal itself contains additional informations on the agonist coded in a specific Ca pattern.

Published 1991 by Elsevier Science Publishing Company, Inc.
Ion Channels of Vascular Smooth Muscle Cells and Endothelial Cells
Sperelakis and Kuriyama, Editors

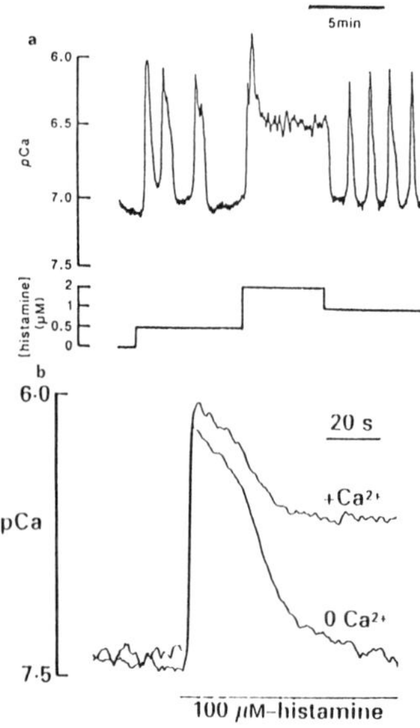

Figure 1. a: Oscillation of intracellular free calcium and peak response followed by a calcium pedestal evoked by histamine in endothelial cells from human umbilical vein. The response depends on the concentration of the agonist histamine. Calcium was measured with a FURA-2 AM fluorescenc probe (from 5 with kind permission of Nature and the authors). b: Histamine-stimulated influx of Ca^{++} in human umbilical vein endothelial cells. The intial peak is independent on the extracellular Ca-concentration. The following sustained plateau in intracellular calcium disappears by removing extracellular calcium. Thus, the maintained Ca response is mediated by transmembrane Ca influx (from 14 with permission).

MECHANICALLY-GATED ION CHANNELS

Endothelium is subjected to flow-related forces such as pressure (perpendicular to the surface) and shear stress (a tangential force that is generated by friction between blood and endothelial surface). Many effects of endothelial cells are related to blood flow: secretion of prostacyclin (PGI_2), EDRF, tissue plasminogen activator (TPA), cytoskeletal rearrangement, cell cycle entry, and others (see 8 for a review). Ion channels could be significantly involved in modulation of these responses. A potassium selective ion channel could act as a mechanosensor mediating a hyperpolarizing response due to shear stress even in the physiological range between 0.5 and 25 dyn/cm^2. Opening of this channel is believed to induce non-EDRF-dependent vasorelaxation via electrical coupling of endothelial cells (8, 24). Hyperpolarization of endothelial cells would also induce a hyperpolarization of smooth muscle cells which are coupled with the endothelium by low-resistance gap junctional pathways. Interestingly, regions with reduced shear stress (e.g. bifurcations of vessels) are favoured sites for appearance of atherosclerotic plaques.

Another mechanically gated channel is a stretch-dependent cation channel (17). Figure 2a shows openings of an ion channel induced by application of 10 to 20 mmHg negative pressure. Channel activation disappears when the suction is removed. In this example, the current is carried by sodium ions. If the pipette solution is changed to 110 mM Ca^{++} (Figure 2b), also stretch-induced channel openings can be recorded: currents are now carried by calcium ions. Thus, the stretch-activated channel is permeable for Ca^{++}. The current-voltage relationship is ohmic reflecting a single-channel conductance of 19 pS in isotonic calcium solutions (Figure 2c). The single-channel conductance in isotonic potassium solutions is 56 pS, for sodium being the charge carrier the conductance is 40 pS. The channel is more about six times more permeable for calcium than for sodium. A permeation ratio between P_{Ca}/P_{Na} = 1.2 to 8.4 could be measured (8, 17). Modulation of stretch-activated ion channels is now studied in several laboratories and seems to depend on changes in the cytoskeleton that focuses membrane distorsion from a large area to the site of a single ion channel.

The channel described is a possible candidate for permitting calcium to enter an endothelial cell under physiological conditions. However, the Ca influx will also critically depend on the transmembrane potential as well as the shape of the endothelial cell. Because endothelial cells are continuously exposed to blood flow, stretch-activated channels will open with a certain probability and allow calcium to enter the cell. Ca influx would then depend on the actual driving force that can be modulated via intracellular calcium itself as it will be shown later.

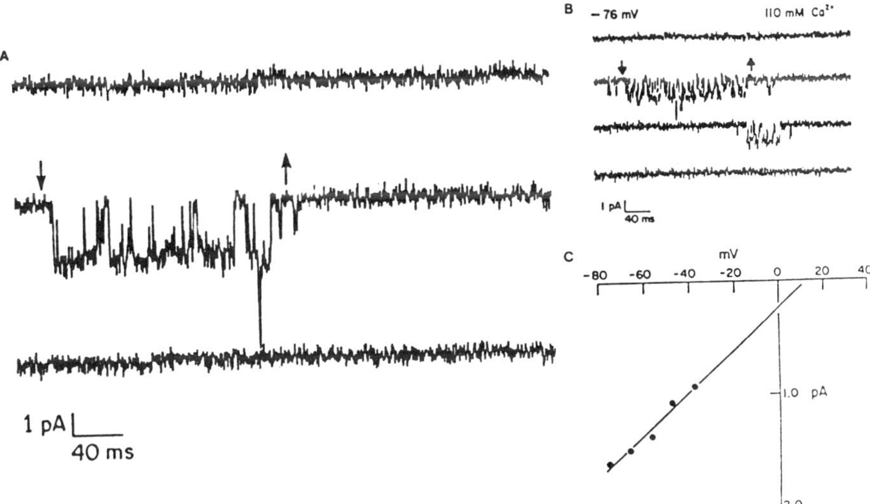

Figure 2. Stretch-activated ion channels in cultured endothelial cells from neonatal pig aorta. **A:** Opening of a non-selective cation channel in an inside-out patch by application of suction pulses (between the arrows) of 1.3 to 2.7 dyn/cm^2 through the pipette. The pipette contains 150 mM Na$^+$, the bath solution 150 mM K$^+$. Under this asymetrical condition single channel conductance is approximately 40 pS. **B:** The stretch activated channel is permeable for calcium. Between the arrows a suction pulse is applied and opens a channel with isotonic (110 mM) calcium in the pipette. **C:** The stretch-activated channel has an ohmic current-voltage relationship. Single channel conductance is 19 pS.(from 17 with kind permission of Nature and the authors).

VOLTAGE-GATED ION CHANNELS

In any tissue it is always very intriuging to ask for the existence of calcium channels because of the special functional significance of these channels. In most of the studies published on endothelium, there are statements on the absence of voltagegated Ca channels (e.g. (6, 22, 28), see 27 for a review). In endothelial cells dissociated from capillaries, however, voltage-gated Ca channels exist (2) and share some similarities with the classical L- and T-type calcium channels known in excitable tissues such as myocardium. In these cells, Ca channels similar to T-type channels have a conductance of approximately 10 pS. Ca transients evoked by short pulses of high potassium concentrations were sensitive to cadmium and amiloride which suggest a role of T-type Ca channels in the control of intracellular calcium because amiloride is know as an efficient blocker of T-type Ca-channels. Calcium channels similar to the L-type have a single-channel conductance of 16 pS and are sensitive to dihydropyridine calcium agonists. It is, however, difficult to imagine a functional role of the latter channels at the low resting membrane potential in most of endothelial cells.

Well documented is the existence of inwardly rectifying potassium channels in cultured endothelial cells. In most of excitable and non-excitable cells, these channels are responsible for stabilization of the resting membrane potential near to the K equilibrium potential, E_K. Inward currents can be measured at potentials negative to E_K, positive to E_K outward currents are almost blocked. Unlike to the cardiac inward rectifier, however, this block seems not to depend on intracellular Mg^{++}. With 150 mM K^+ in the pipette, the conductance of this channel is between 25 and 30 pS. Figure 3a shows an experiment in which the inward rectifier has been measured in a cell-attached patch. A linear voltage ramp (panel 3b) was used to record instantaneous currentvoltage relationships. No outward currents can be recorded. The channel is inwardly rectifying. At negative membrane potentials long openings can be observed. With 150 mM K^+ in the pipette the single-channel conductance is close to 27 pS. The flat current voltage relation was recorded with only 5.4 mM K^+ in the pipette (Figure 3c).

The extrapolated reversal potential was changed according to the external potassium concentration (change of 56 mV per K^+ decade, Figure 3d) indicating that the current is carried by potassium ions. It is worthwhile to mention that not all endothelial cells have an inward-rectifier potassium channel 17, 22, 27).

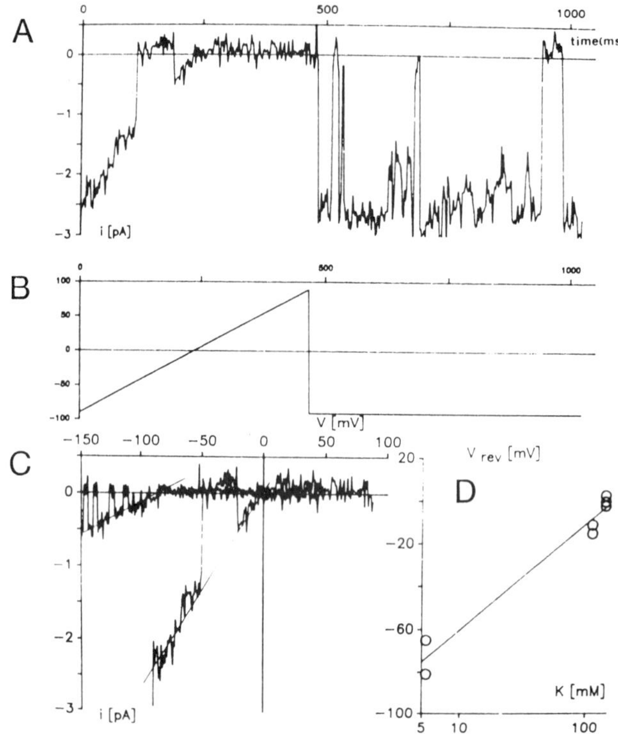

Figure 3. Inward rectifier potassium channels measured from endothelial cells of umbilical vein. Not in all cells of this preparation the channel could be observed. **A:** Single channel currents with 140 mM KCl in the pipette and the same concentration in the bath. Long openings can be observed at negative potentials, no outward currents can be detected referring to the inward retifying nature of the channel. **B:** Voltage protocol used, 450 ms ramp from -90 to +90 mV. **C:** Current-voltage, I-V, relationships for the inward rectifier channel at two different potassium concentrations in the pipette (5.4 mM, flat I-V curve; 140 mM, steep I-V curve). **D:** The panel shows the extrapolated reversal potentials plotted against the logarithm of the potassium concentration in the pipette. The straight line obtained from linear regression shows a slope of 56 mV per decade potassium concentration (1 kHz sampling, 1 kHz filter, cell attached patches, experiments B.N.).

one kind of a Ca-dependent potassium channel that could be involved in the regulation of Ca influx. This channel has a conductance of 40 pS and seems to be gated by an intracellular Ca release via activation of purinergic (P$_2$) receptors (26). The 40 pS channel is also activated by other agonists that induce Ca transients. Beside the 40 pS channel, a high-conductance channel with a conductance between 140 and 180 pS was observed that is also gated by intracellular calcium (9, 22).

These two types of channels belong to a class of potassium channels that could be made responsible for hyperpolarization of endothelial cells. A similar mechanism could also be provided by activation of chloride-selective ion channels that would hyperpolarize endothelial cells from low resting membrane potentials.

Until now, several outward ion currents activated by different agonists such as bradykinin, thrombin, platelet-activating factor, ATP, histamine and acetylcholine has been described at the level of whole-cell measurements and will not be reviewed in this context. Intracellular Ca release might be also involved in activation of these currents.

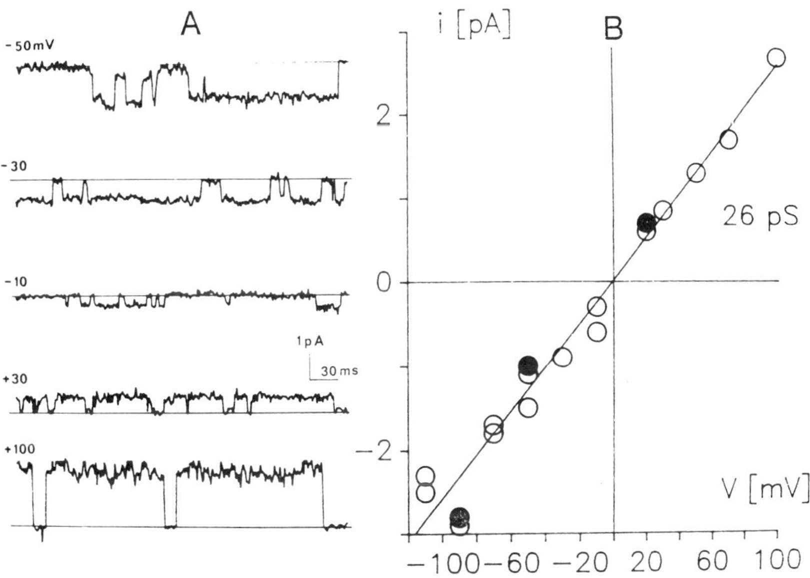

Figure 4. Histamine opens a non-selective cation channel in endothelial cells from human umbilical vein. **a:** Single channel recordings at different membrane potentials (5 μM histamine were applied to the bath outside the pipette to evoke the current; cell attached patch; 140 mM NaCl in the pipette, 140 M KCl in the bath; 1 kHz sampling rate, 500 Hz filter). **b:** Current-voltage relationship of the non-selective cation channel. The single channel conductance is 26 pS, the currents reverses near zero millivolts in asymmetrical solutions (points from two experiments, B.N.).

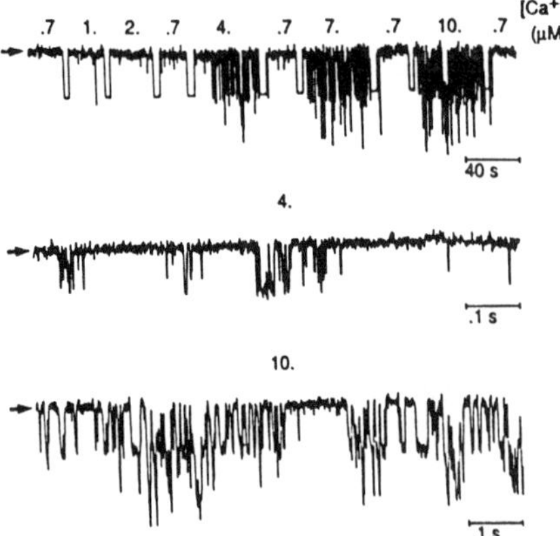

Figure 5. Calcium-activated-potassium channels measured in an inside-out patch from cultured bovine aortic endothelial cells. Numbers indicate the free intracellular calcium concentration in lM. The 40 pS channel is activated by an increase in the concentration of free calcium facing the inner side of the patch membrane (bath and pipette solution contain 212 mM K$^+$). This channel is an example for a single channel measurement of a low-conductance Ca-activated potassium channel in endothelium (from 28) with kind permission of Pflugers Archiv and the authors).

MEMBRANE TRANSPORTERS INVOLVED IN REGULATION OF CELL CALCIUM

Regulation of intracellular calcium seems to be influenced by membrane transporters. A Na/K pump has been demonstrated in bovine aortic endothelial cells and coronary endothelial cells. This pump is electrogenic and conributes to the membrane resting potential (7). Inhibition of the pump with ouabain causes a small depolarization of the cells. In endothelial cells, a cotransporter has been described (5). This cotransporter is stimulated by agonists such as bradykinin or directly by intracellular calcium. The stoichiometry is 1:2:2 for Na$^+$-K$^+$-Cl$^-$. The transporter is therefore electrogenic and could modulate the membrane potential. It is inhibited by diuretics.

A Na$^+$-Ca^{++} exchanger has also been discovered in endothelial cells (29). Any change in the intracellular sodium activity and in the membrane potential would therefore affect also calcium homeostasis. However, there are still conflicting results. At least, endothelial cells provide a pathway by which an increase in the intracellular sodium activity could result in an elevated intracellular calcium activity.

Calcium homeostasis is further influenced by the activity of a plasmalemmal Ca^{++}-Mg^{++}-dependent transport ATPase that provide sequestration of Ca out of the endothelial cell. Some agonists such as platelet activating factor (PAF) and histamine activate a calcium pump thereby modulating transmembrane Ca fluxes (3). However, special features of all these transporters are not yet clear.

Another pathway for intracellular signaling in endothelial cells acts via the Na$^+$-H$^+$ antiport. Activation of endothelial cells by thrombin results in a stimulation of PAF synthesis in these cells. In the activation pathway, stimulation of an Na$^+$-H$^+$ antiport results in alkalinization of the cells and an increase in the intracellular Ca^{++} activity that switches on the Cadependent lyso-PAF acetyltransferase (12). It is again intruiging to ask whether the link between activation of the Na$^+$-H$^+$ antiporter (resulting in an increased intracellular Na$^+$ activity) could be provided via the Na$^+$-Ca^{++} exchanger.

FUNCTIONAL ROLE OF ION CHANNELS IN ENDOTHELIUM

Signaling in endothelial cells is at least partially mediated by a transient increase of the intracellular calcium activity. Labile compounds like EDRF need a longlasting Ca signal to excert an efficient physiological response. The short intracellular Ca^{++}-release that appears as a Ca^{++}-peak is insufficient to trigger an EDRF-response. The calcium pedestal could be generated by a transmembrane influx of calcium. If a transmembrane pathway for calcium is available, the influx of Ca into the cells depend on the difference, E_m-E_{Ca}, between the membrane potential E_m (that is mainly determined by the potassium gradient and the potassium conductance) and the equilibrium potential for calcium, E_{Ca}. Any shift of E_m towards more negative values would increase the driving force for a transmembrane influx of calcium, and therefore induces an augmentation of the Ca^{++} influx. Candidates of ion channels for such an influx of calcium have been discussed in this chapter. At least the stretch-activated and agonist-gated ion channel in the surface membrane of endothelial cells can conduct calcium driven by its electrochemical gradient. It is also intruiging to speculate whether a sodium influx through stretch- or agonist-activated non-selective cation channels may be involved in the control of the transmembrane influx of calcium. If endothelial cells express functional sodium-calcium antiporters, opening of non-selective cation channels with a sufficiently high influx of sodium could induce a reduction of Ca^{++}-efflux or even cause a Ca^{++} influx mediated by this transporter.

A sufficient increase in intracellular calcium would also open Ca-activated potassium channels such as shown in Figure 5 or the high-conductance Ca-activated potassium channel. A vasoactive agonist that can release intracellular calcium would therefore necessarily gate Ca-activated potassium channels. One can assume that in any case of intracellular Ca release a coactivation of Ca-activated potassium channels should occur. Such an activation of potassium channels would stabilize the membrane potential at more negative values or even cause hyperpolarization in partially depolarized endothelial cells. It would be intruiging to ask whether in endothelial cells an even tighter (physical?) coupling between stretch- or agonist-gated ion channels with Caactivated potassium channels is realized than simply a coupling via free intracellular calcium.

If a depolarizing Ca-permeable cation channel such as the nonselective stretch- or agonist-activated channel is opened, the membrane potential would be clamped at negative values due to activation of the hyperpolarizing potassium channel. As a functional result, the driving force for calcium is increased. Therefore, concomitant activation of Ca-dependent potassium channels would be an efficient tool to control transmembrane Ca^{++} influx into an activated endothelial cell.

The following experimental findings strongly support this hypothesis that the membrane potential, regulated by potassium channels, play a major role in regulation of the transmembrane Ca-influx (1, 19, 21):
1)   decrease of the transmembrane potential by reducing the concentration gradient for potassium attenuates the agonistinduced long-lasting Ca-pedestal (e.g. bradykinin-evoked changes in the intracellular calcium activity are reduced by the elevation of extracellular potassium),
2)   agonist-induced Ca-influx is decreased by preincubation of endothelial cells with the potassium channel blocker tetraethylammonium (TEA) whereas the initial agonist-induced peak generated by intracellular Ca release is unchanged,
3)   the resting intracellular Ca-activity in cells stimulated with thimerosal (an agent that increase Ca fluxes into endothelial cell thereby causing long-lasting EDRF release) depends on the potassium gradient,
4)   in the presence of extracellular calcium, agonists induce long-lasting hyperpolarizations that are attenuated by changing to Ca free extracellular solutions.

Figure 6 summarizes the discussed cooperation between different channels and transporters in endothelial cells. Ion channels are depicted that could play a major role in the control of the intracellular calcium. Ca influx could be provided by the non-selective cation channel and the stretch-activated channel. If the agonist-gated channel would be really depend on the intracellular calcium itself thereby inducing a postive feedback, the necessary negative feedback possibly via PKC is tentatively drawn at the left hand side of the scheme (18). It is completely unclear how channels are gated by agonists (which second messengers?, which G-proteins?). The important regulation of Ca influx via activation of a Ca-dependent potassium channel is also included. The membrane transporters may play a modulatory role in the control of intracellular calcium.

The diversity of mechanisms to regulate intracellular calcium in endothelial cells (for a much more detailled review see 4) may provoke the following question: most of the effects on endothelial cell converge into an increased intracellular Ca^{++} activity, however, different cellular functions are triggered. Differences in the subcellular Ca- pattern induced by different receptors has recently been shown (10). It is intruiging to speculate whether the temporal and spatial pattern of the Ca signal in endothelium comprise a distinct functional information.

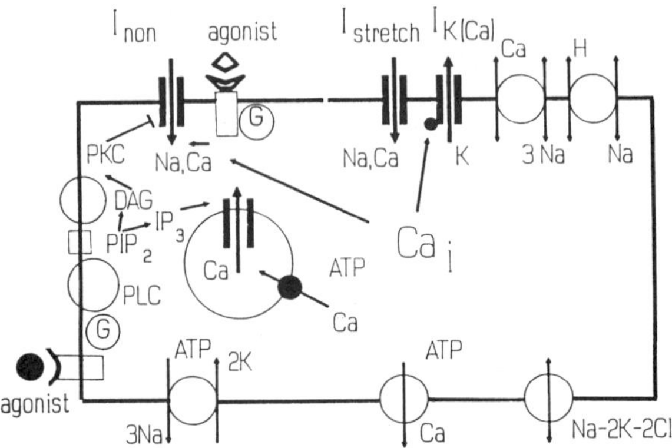

Figure 6. Functional scheme of the interaction of different ion channels in endothelial cells. Transmembrane Ca-influx is mediated by opening of at least two Ca-permeable ion channels (an agonist-gated non-selective cation channel, I_{non}, and a stretch-activated non-selective cation channel, $I_{stretch}$). The receptor for the agonist is assumed to be connected with a G-protein. The agonist can open the non-selective channel via an unknown second messenger (possibly Ca^{++} itself, see arrow) and can release Ca from intracellular stores of the endoplasmic reticulum (ER) possibly via IP_3. Ca entry through the agonist-gated and the stretch-activated channel as well as Ca release from ER can gate Ca-activated potassium channels, $I_{K(Ca)}$. The latter channel will clamp the cell at negative membrane potentials, and activation of the stretch-dependent channel will increase Ca. On the other hand, Ca_i will increase the open probability of the agonist-gated channel and provide a positive feedback. A negative feedback could be provided via activation of PKC. Other receptors for vasoactive agonists may stimulate via a G-protein phospholipase C (PLC). Hydrolysis of phosphatidyl-inositol-4,5-diphosphate (PIP_2) into diacylglycerol (DAG) and 1,4,5-inositol-triphosphte (IP_3) effects stimulation of protein kinase C (PKV) and a Ca release, respectively. PKC might excert a negative feedback on the transmembrane Ca influx as indicated by the bar. Five transmembrane transporters are shown. Double arrows always indicate a possible reverse in the direction of the transport. All mechanisms contribute to the regulation of intracellular calcium and generate a special concentration pattern in time and space.

CONCLUSION

Characterization of ion channels and membrane transporters in endothelial cells is still under way and by no means satisfying. There is some evidence that confirms the existence of mechanically- or agonist-gated calcium permeable ion-channels. Calcium-activated potassium channels would open by an increase in the intracellular Ca activity. An increase of a potassium conductance via Ca-dependent potassium channels would then efficiently clamp endothelial cells at negative membrane potentials or even induce hyperpolarization inspite of opening of depolarizing cation channels. Such mechanism may provide a finetuning of the transmembrane Ca-influx into endothelial cells by variation of the driving force for a transmembrane influx of calcium ions.

REFERENCES

1. Adams, D.J., J. Barakeh, R. Laskey, and C. van Breemen, FASEB J. 3, 2389-2400 (1990).
2. Bossu, J.-L., A. Feltz, and F. Tanz, FEBS Letters 255, 377-380 (1989).
3. Bregestovski, P. and U.S. Ryan, J. Mol. Cell. Cardiol. 21, Suppl. l, 103-108 (1989).
4. Bregestovski, P., A. Bakhramov, S. Moldobaeva, and K. Takeda, Br. J. Pharmacol. 95, 429-436 (1988).

5.   Brock, T.A., C. Brugnara, M. Canessa, and M.A. Gimbrone, Am. J. Physiol. 250, C888-C895 (1986).

6.   Colden-Stanfield, M., W.P. Schilling, A.K. Ritchie, S.G. Eskin, L.T. Navarro, and D.L. Kunze, Circ. Res. 61, 632-640 (1987).

7.   Daut, J., G. Mehrke, S. Nees, and W.H. Newman, J. Physiol. (Lond.) 402, 237-254 (1987).

8.   Davies, P.F., News in Physiol. Sci. 4, 22-25 (1989).

9.   Fichtner, H., U. Frobe, and M. Kohlhardt, J. Membrane Biol. 98, 125-133 (1987).

10.   Flavahan, N.A., H. Shimokawa, and P.M. Vanhoutte, J. Pharm. Exp. Therap. 256, 50-55 (1991).

11.   Flvahan N.A. and P.M. Vanhoutte, Blood Vessels 27, 218-229 (1990).

12.   Ghigo, D., F. Bussolino, G. Garbarinos, R. Heller, F. Turrini, G. Pescarmona, E.J. Cragoe, L. Pegararos, and A. Bosia, J. Biol. Chem. 263, 19437-19446 (1988).

13.   Gomez-Cambronero, J., P. Inarrea, F. Alonso, and M. Sanchez-Crespo, Biochem. J. 219, 419-424 (1984).

14.   Jacob R., J. Physiol. (Lond.) 421, 55-77 (1990).

15.   Jacob, R., J.E. Merritt, T.L. Hallam, and T.J. Rink, Nature 335, 4045 (1988).

16.   Johns, A., T.W. Lagetan, N.J. Lodge, U.S. Ryan, C. van Breemen, and D.J. Adams, Tissue & Cell 19, 733-745 (1987).

17.   Lansman, J.B., T.L. Hallam, and T.J. Rink, Nature 325, 881-813 (1987).

18.   Lechleiter, J., S. Girard, D. Clapham, and E. Peralta, Nature 350, 505-508 (1991).

19.   Luckhoff A.K. and R. Busse, Pflugers Arch. 416, 305-311 (1990).

20.   Mayer, B., K. Schmidt, P. Humbert, and E. Bohme, Biochem. Biophys. Res. Comm. 164, 678-685 (1989).

21.   Mehrke, G., U. Pohl, and J. Daut, J. Physiol. (Lond.) 435, in press (1991).

22.   Nilius B. and D. Riemann, Gen. Physiol. Biophys. 9, 89-112 (1990).

23.   Nilius, B., Pflugers Arch. 416, 609-611 (1990).

24.   Oleson, S.-P., D. Clapham, and P. Davies, Nature 331, 168-170 (1988).

25.   Ryan, U.S., J. Mol. Cell. Cardiol. 21, Suppl. 1, 85-90 (1989).

26.   Sauve, R., L. Parent, C. Simoneau, and G. Roy, Pflugers Arch. 412, 469-481 (1988).

27.   Takeda K. and M. Klepper, Blood Vessels 27, 169-183 (1990).

28.   Takeda, K., V. Schini, and H. Stoeckel, Pflugers Arch. 410, 385393 (1987).

29.   Winquist, R.J., P.B. Buntig, and T.L. Schonfield, J. Pharmacol. Exp. Ther. 235, 644-650 (1985).

CHLORIDE CURRENTS IN BOVINE PULMONARY ARTERY ENDOTHELIAL CELLS

MARK S. SHAPIRO* and THOMAS E. DeCOURSEY#

Department of Physiology, Rush Presbyterian St. Luke's Medical Center, 1653 W. Congress Parkway, Chicago, Illinois 60612

Abstract

Chloride currents in bovine pulmonary artery endothelial cells are described at the whole-cell and the single-channel level. In whole-cell experiments, an outwardly-rectifying Cl⁻ conductance is observed. This conductance has little voltage- or time-dependence at neutral external pH, but decays at positive potentials at acid pH. The relative permeability of Br⁻ is similar to that of Cl⁻ while MeSO₃⁻ is only half as permeant, estimated by reversal potential. The conductance of these anions corresponds approximately with their relative permeability. At least two types of Cl⁻ channels were detected, with conductances at positive potentials of approximately 77 pS and 300 pS. The smaller channel rectifies outwardly and has selectivity similar to that of whole-cell currents. The larger channel closes at large negative or positive potentials and resembles "mega-chloride" channels which have been described in a variety of cells. In addition, a pseudo-cation conductance is described which appears to arise from current flowing through gap junctions into neighboring cells, when endothelial cells grow to confluence. This conductance greatly complicates measurements on cells which are not isolated from other cells.

INTRODUCTION

Vascular endothelial cells, once considered to be a relatively inert structural component of blood vessels, have been implicated in the last decade in a number of physiological functions, one of the most important of which is to modulate the tone of the underlying vascular smooth muscle (for recent reviews see 12, 18). Not coincidentally, there has been a flurry of patch-clamp studies of the ion channels present in endothelial cells. This work has been reviewed recently (1, 31), and will not be described in detail here. A wide variety of channels and conductances have been reported, all of which to this point are either K^+ selective (5-7, 13, 21, 30, 33, 34, 39, 43, 45) or nonselective cation-permeable (3, 5, 21, 23, 30, 32). Several of these channels appear to be activatable by a variety of interventions (stretch or shear-stress, thrombin, acetylcholine, bradykinin, histamine, ATP) either directly or indirectly, some perhaps as a result of increased intracellular Ca^{2+} which accompanies agonist-stimulated endothelium-derived relaxing factor (EDRF) release. Chloride channels or currents have not been directly demonstrated, to our knowledge, in vascular endothelial cells, although recently evidence has been presented of Cl⁻ efflux in bovine pulmonary artery endothelial cells which can be blocked by Cl⁻ channel inhibitors (46). In rabbit corneal endothelium a K^+ conductance activated by external Cl⁻ has been reported (37). We describe here, both at the single-channel and whole-cell levels, Cl⁻ currents in bovine pulmonary artery endothelial cells. In addition, we illustrate an apparent macroscopic conductance which we attribute to current flowing across gap junctions into neighboring endothelial cells, rather than to a genuine membrane conductance in the cell under patch-clamp.

---

*Dr. Shapiro's present address is Department of Physiology and Biophysics, University of Washington, SJ-40, Seattle, WA 98195.

#Please direct all correspondence to T.E. DeCoursey.

Published 1991 by Elsevier Science Publishing Company, Inc.
Ion Channels of Vascular Smooth Muscle Cells and Endothelial Cells
Sperelakis and Kuriyama, Editors

328

METHODS

The preparation and culture conditions have been described previously (43). Briefly, bovine pulmonary artery endothelial cells were isolated by enzymatic digestion and maintained in culture. For voltage-clamp recording, cells were plated onto sterile glass cover-slip fragments. Standard "patch-clamp" techniques and configurations were used (17). Ringer's solution contained 160 mM Na^+, 4.5 mM K^+, 2 mM Ca^{2+}, 1 mM Mg^{2+}, and 5 mM HEPES buffer at *pH* 7.4. Other extracellular solutions, except where indicated, contained 2 mM Ca^{2+} and 5 mM HEPES buffer at *pH* 7.4, and other ionic constituents as indicated. The ionic composition of the pipette solution is described for each experimental condition, with the exception that 5 mM EGTA was always present to buffer Ca^{2+} and other divalent cations including those which might leach out of the pipette glass (9). Both pipette and bath solutions always included at least 4 mM Cl^- to avoid electrode polarization.

Details of the electrophysiological setups have also been described recently (10, 42). Experiments were done near 20° C.

RESULTS
Cl⁻ Currents are Present in Endothelial Cells

When bovine pulmonary artery endothelial cells are studied under appropriate ionic conditions, Cl^- permeability can be observed. In most of the experiments described in this chapter the pipette solution was K-free in order to minimize the contribution of K^+ currents; with large cations such as N-methyl-D-glucamine (NMG^+) or tetramethylammonium (TMA^+) in the pipette, outward currents are most likely carried by anions from the bath. A family of whole-cell voltage-clamp currents recorded in Ringer's solution is illustrated in Fig. 1*A*. The currents are largely time-independent, and the outward currents are larger than the inward currents. After nearly all the external Cl^- was replaced by methanesulfonate⁻ ($MeSO_3^-$), the outward currents were much smaller (Fig. 1*B*). Most of the outward current in Fig. 1*A* was therefore carried by Cl^- from the bath entering the cell. That some outward rectification remains in $MeSO_3^-$ solution suggests that the anion conductance in endothelial cells is not very selective, and that $MeSO_3^-$ can also carry current. Although this and evidence to be presented later indicate that the membrane of endothelial cells is permeable to anions other than Cl^-, this conductance will be called a Cl^- conductance because Cl^- is the main anion present under physiological conditions. The pipette solution in the experiment in Fig. 1 contained mainly $MeSO_3^-$ as the anion. Under these conditions of symmetrical anionic solutions in Fig. 1*B* there is still outward rectification, suggesting that outward rectification is a property of the anion conductance in these cells.

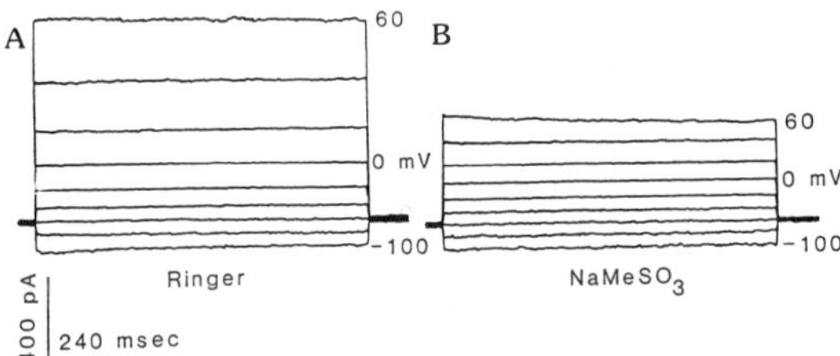

Figure 1. Whole-cell anion currents in an endothelial cell (*A*) in Ringer's solution (*see Methods*), and (*B*) in K^+-free NaMeSO₃ solution. Time-independent outward currents are carried mainly by Cl^- and are reduced but not abolished when Cl^- is replaced by $MeSO_3^-$. The small time-dependent current at -100 mV in Ringer's which is not visible in *B* after removal of K^+ is carried through inward rectifier channels. In both families, the cell was held at -60 mV and stepped to a range of potentials as indicated in 20 mV increments. Internal solution was TMAMeSO₃, *pH* 7.1. Calibration bars apply to both families of currents.

The most consistently detected ion channel in endothelial cells is the inwardly-rectifying K^+ channel (5-7, 21, 30, 34, 45). Just visible at -100 mV in Fig. 1*A* is a small time-dependent inward current which reflects current through inward rectifier channels. Increasing the external K^+ concentration resulted in larger inward currents at negative potentials which were first activated at a more positive potential range (*data not shown*). The inward rectifier currents disappear in Fig. 1*B* after the 4.5 mM K^+ present in Ringer's solution is replaced by Na^+. The K^+ dependence of these currents, and their voltage- and time-dependence (*not shown*) confirm the identity of these as inward rectifier currents, which can thus be detected with a complete absence of intracellular K^+. Inward rectifier currents can be seen in cardiac myocytes perfused with K^+-free solutions (28).

Block by Zn^{2+}. Identifying the selectivity of time-independent currents is difficult, especially if ion substitution does not abolish the current. It was assumed that the most likely interpretation of the inward currents in Fig. 1*B* is that they are carried by $MeSO_3^-$ through the same anion conductance responsible for the outward Cl^- currents. Indirect confirmation of this interpretation is shown in Fig. 2. In this experiment the pipette solution contained 60 mM Cl^- and 100 mM $MeSO_3^-$. Fig. 2*A* shows a family of Cl^- currents in another cell, and Fig. 2*B* shows the currents (at the same gain) elicited in this cell after addition to the bathing solution of $70\,\mu M$ Zn^{2+}, which blocks Cl^- currents in muscle (4). Both inward and outward currents are practically abolished. Washout of the Zn^{2+} resulted in gradual and partial recovery. The profound block by Zn^{2+} supports the idea that both Cl^- and $MeSO_3^-$ carry most of the time-independent membrane current under these conditions. Another possibility which cannot be ruled out at this time is that Zn^{2+} may also inhibit another conductance present under these conditions.

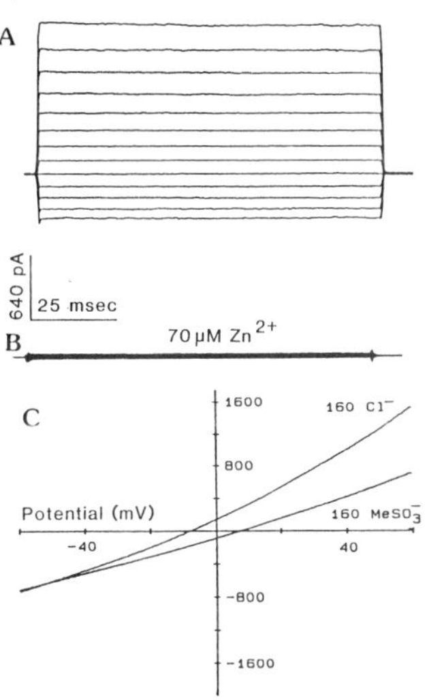

Figure 2. Whole-cell Cl^- currents. (A) Family of superimposed currents during 100 msec pulses from -60 mV to +60 mV every 10 mV, with 160 mM NMGCl in the bath. Holding potential -20 mV, 18^0 C. (B) Currents in the same cell over the same potential range after addition of $70\,\mu M$ Zn^{2+} to the bath, which abolishes the current. Calibrations apply to both *A* and *B*. Holding potential -10 mV, 17^0 C. (C) Currents recorded in the same cell during voltage ramps from 60 to -60 mV with 160 mM $NMGMeSO_3$ in the bath and then after the bath was changed to 160 mM NMGCl. Ramp speed -0.53 mV/msec, 19^0 C, filter 1 kHz, holding potential 0 mV, pipette 120 mM $NMGMeSO_3$ and 60 mM NMGCl.

Since the Cl^- currents in endothelial cells are generally time-independent, they can be studied efficiently by using voltage ramps. In Fig. 2*C* whole-cell currents in the same cell as in parts *A* and *B* of Fig. 2 are shown when the bath contained NMG^+ and Cl^- or $MeSO_3^-$. There is outward rectification in Cl^- which is decreased but not eliminated when $MeSO_3^-$ is in the bath. Since there was 60 mM Cl^- in the pipette, the persistence of outward rectification when the bath contains the less permeant $MeSO_3^-$ and only 4 mM Cl^- indicates that this conductance exhibits strong outward rectification.

Selectivity of the anion conductance. The reversal potential of the conductance under various ionic conditions can be determined from experiments like the one in Fig. 2*C*. These reversal potentials can be used to calculate the relative permeability of the endothelial cell membrane to various anions using the Goldman-Hodgkin-Katz equation. The results of such calculations are given in Table I. By this criterion, Br^- is as permeant as Cl^-, and $MeSO_3^-$ permeates anion-selective channels in endothelial cells about half as well as does Cl⁻. These data reinforce the idea that the outward currents in Fig. 1*B* were carried by $MeSO_3^-$ from the bath entering the cell, and that the inward currents were carried largely by $MeSO_3^-$ from the cell.

Table I. *Selectivity of Whole-cell Anion Conductance*

Ion	P_{rel}	S.D.	n
Cl⁻	≅1.0	-	-
Br⁻	1.02	0.07	3
$MeSO_3^-$	0.56	0.08	6

The permeability of an ion relative to Cl⁻, P_{rel}, was calculated using the Goldman-Hodgkin-Katz equation from the observed change in reversal potential after changing the bath from Cl⁻ to the test ion. *S.D.* is the standard deviation of the mean and *n* is the number of cells studied.

Effects of External pH on Cl⁻ Currents

The Cl⁻ conductance of skeletal muscle is profoundly affected by external *pH* (4, 16, 20, 35). Fig. 3 shows that *pH* affects Cl⁻ currents in endothelial cells. Identical voltage pulses were given to a cell bathed in Ringer's solution at three different *pH*. The current amplitude immediately after the depolarizing step is about the same at all three *pH*. This result indicates that the external *pH*, at least between 6 and 10, has little effect on the "resting" Cl⁻ conductance at the holding potential of -40 mV. At neutral or alkaline *pH* the outward Cl⁻ current during the test pulse to +80 mV is practically time-independent, but at acid *pH* the outward current decays substantially.

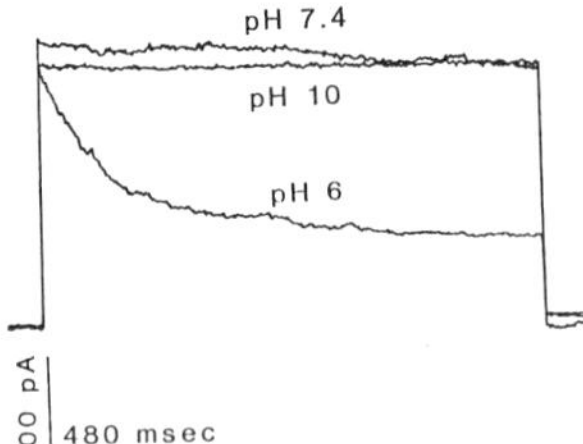

Figure 3. Effect of external *pH* on outward Cl⁻ currents. Identical voltage pulses to +80 mV were applied to a cell held at -40 mV, at the indicated *pH*. The bath contained Ringer's solution, with 5 mM MES buffer at *pH* 6 and CAPS buffer at *pH* 10.

Families of Cl⁻ currents at *pH* 7.4 and 6.0 from this experiment are shown in Fig. 4 (*A* and *B*, respectively). At *pH* 7.4 the currents are practically time-independent at all potentials studied, as was also illustrated in Figs. 1 and 2*A*. At *pH* 6 the currents are time-independent at negative potentials, but a slow inactivation-like decay appears first at +20 mV, which becomes faster and more complete at more positive potentials. Although this process superficially resembles inactivation, such as occurs in most K⁺ channels, the recovery time-course is more rapid at negative potentials than is the decay at positive potentials (*data not shown*). It might therefore be more appropriate to consider the decay as a deactivation process, with Cl⁻ channel activation being relatively rapid at negative potentials. Another possibility is that the decay in *pH* 6 reflects time- and voltage-dependent block of Cl⁻ channels by H⁺ (*cf.* 54). Note that there is a distinct negative conductance region in the steady-state current-voltage relation: the current at the end of the pulse is smaller at +40 or +60 mV than at +20 mV. The voltage-dependence and kinetics of Cl⁻ currents in endothelial cells at low *pH* closely resemble those of the volume-sensitive Cl⁻ conductance at neutral *pH* in tracheal epithelial cells (27) and in the T84 cell line (55).

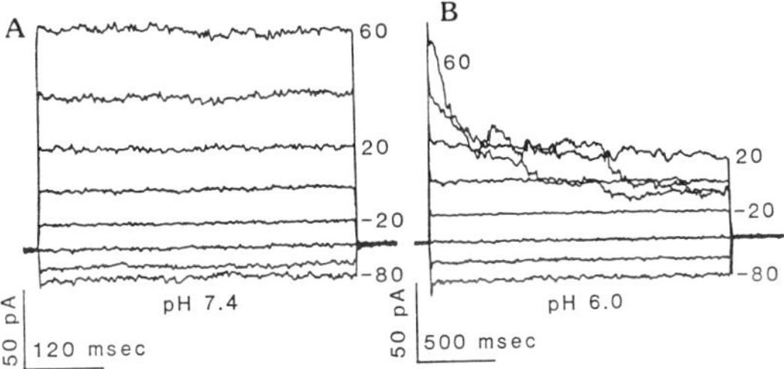

Figure 4. Voltage- and time-dependence of Cl⁻ currents at neutral and acid external *pH*. Both families were recorded in the same cell, but at different *pH*. The Cl⁻ currents at neutral *pH* are practically time-independent, but at acid *pH* they decay slowly during depolarizing pulses. Pipette TMAMeSO₃. Same cell as in Fig. 3.

Single-Channel Cl⁻ Currents

In on-cell patches a variety of single-channel currents can be seen in endothelial cells. All of the unitary currents described thus far are K^+ or cation-selective (see reviews 1, 31). Fig. 5 illustrates single anion-selective channel currents in an outside-out patch of endothelial cell membrane. This patch contained an anion channel which was studied both using voltage-ramps (Fig. 5A-C) and at constant voltages (Fig. 5D). With MeSO₃⁻ in the pipette (internal side of the membrane) and Br⁻ in the bath (external side) large outward currents carried by Br⁻ can be seen at +100 mV, and smaller but distinctly channel-like inward currents carried by MeSO₃⁻ can be seen in Fig. 5D. During voltage ramps the channel opened and closed often enough that closings and openings at all potentials could be collected and averaged to generate complete leak-subtracted open-channel current-voltage relations (Fig. 5A-C). Cl⁻ in the bath (Fig. 5A) carries large outward currents through the channel, and the unitary current rectifies strongly outwardly. The slope conductance at positive potentials is 77 pS, indicated with a line. The unitary conductance and strong outward rectification closely resemble the apical membrane Cl⁻ channel in tracheal epithelium (51). With Br⁻ in the bath (Fig. 5B) the outward current is at most a bit smaller than with Cl⁻ (note that in Cl⁻ the voltage range was extended farther), and the slope conductance is 18% smaller. In symmetrical MeSO₃⁻ solutions (Fig. 5C) the unitary current still rectifies outwardly, although the slope conductance at positive potentials is reduced to 23 pS. The selectivity of this Cl⁻ channel approximately corresponds with that of the whole-cell Cl⁻ currents in Table I, although it cannot be determined at this point what fraction of the macroscopic Cl⁻ conductance in endothelial cells is due to these channels.

"Mega" Cl⁻ channels. Larger anion-selective channels were also observed in endothelial cells. Fig. 6 illustrates whole-cell currents in a cell with a number of these channels. The bath contained 160 mM KCl, and the pipette NMGMeSO₃. From the amplitude of the transitions as channels open or close, the unitary conductance is roughly 300 pS for outward currents. These channels also carry inward current, although the conductance is lower and the gating transitions are harder to see. These large channels close at large positive or at large negative potentials, as is evident in the illustrated record. During depolarizing pulses the outward currents in this cell were similar and decayed markedly with time at neutral *pH*, whether the external cation was Na^+, K^+, or NMG^+, implicating Cl⁻ as the most probable current carrier. The selectivity of the channel for anions over cations was confirmed by diluting the bath solution by 50% with distilled water, which shifted the reversal potential reversibly by about 10 mV to more positive potentials. The outward current amplitude was decreased about 50% after diluting the bathing solution, with little change in the inward currents. The reduction of outward current and the shift in reversal potential are consistent with anion selectivity and a relative MeSO₃⁻ permeability about one-half that of Cl⁻.

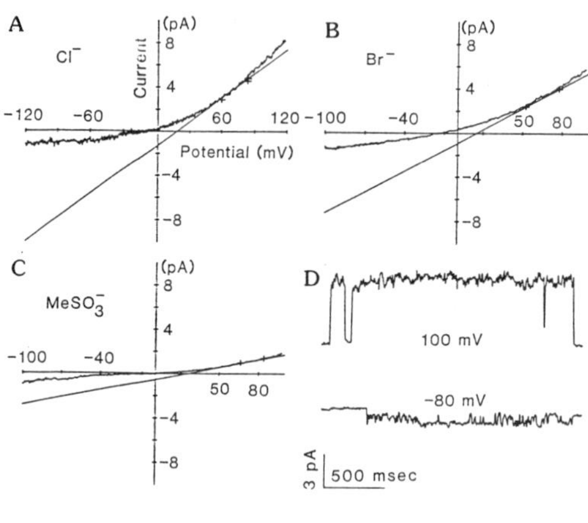

Figure 5. Unitary Cl⁻ channel currents in an excised outside-out patch. (*A-C*) Ensemble leak-subtracted open-channel current-voltage relations in the presence of 160 mM NaCl (*A*), 160 mM TEABr (tetraethylammonium bromide) (*B*), and 160 mM NaMeSO$_3$ (*C*). For each, a region of the current-voltage relations (between the cursors) *ca.* 67-87 mV positive to the reversal potential, V_{rev}, was fit by least-squares to obtain a slope conductance (g_s), the tangent shown in each figure. (*A*) g_s 77 pS, filter 2 kHz, ramp speed 2.1 mV/msec, V_{rev} -7 mV. (*B*) g_s 63 pS, filter 1 kHz, ramp speed 0.7 mV/msec, V_{rev} -12 mV. (*C*) g_s 23 pS, filter 1 kHz, ramp speed 0.7 mV/msec, V_{rev} 1 mV. (*D*) Two current records in 160 mM TEABr at 100 mV (*top*) and at -80 mV (*bottom*). Filter 100 Hz, sample interval 4.2 msec, pipette 160 KMeSO$_3$, 19° C.

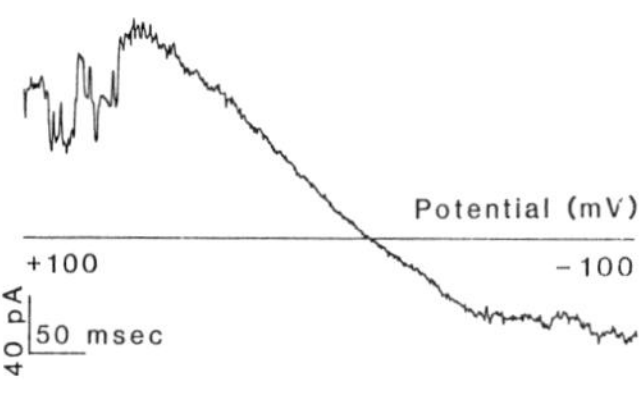

Figure 6. Large single-channel current events detectable at the whole-cell level. Whole-cell currents during a voltage ramp from +100 mV to -100 mV. The cell was held at -60 mV between ramps. The voltage-dependence of channel gating in this ramp is typical of that observed in many ramps, whether the voltage was ramped up or down. These large channels turn off at large positive or large negative potentials. The bath contained 160 mM KCl Ringer's, the pipette NMGMeSO$_3$, 20° C.

Pseudo-Cation Conductance

Most of the endothelial cells selected for study were spherical and separated from other cells on the cover slip. Occasionally cells which were in contact with other cells were studied. In some of these experiments an apparent conductance was observed which at first appeared to be K$^+$ selective. Fig. 7*A* shows whole-cell currents recorded during a voltage ramp from -120 mV to +80 mV. The bath contained Ringer's solution, the pipette CsMeSO$_3$. The current-voltage-relation is roughly linear and reverses near -80 mV. Changing the bath solution to isotonic K$^+$ Ringer's (*not shown*) or to isotonic Rb$^+$ Ringer's solution shifted the reversal potential to more positive potentials, suggesting that a K$^+$ selective conductance is involved. Reversal was more positive in K$^+$ Ringer's than in Rb$^+$ Ringer's, suggesting that Rb$^+$ is somewhat less permeant than is K$^+$. Replacing the Cl$^-$ in the bath with MeSO$_3$$^-$ had no distinct effect of either the reversal potential or the magnitude of these currents, making unlikely any significant contribution of Cl$^-$ to the conductance. However, when the bath solution was changed from Ringer's solution, with 4.5 mM K$^+$ and 160 mM Na$^+$, to K$^+$-free Na$^+$ Ringer's solution (Fig. 7*B*), the reversal potential shifted to substantially more positive potentials- in the opposite direction from what one would expect for a K$^+$ selective conductance! This phenomenon was observed in a number of experiments. The depolarizing shift in reversal potential was reversible when cells were exposed to K$^+$-free solutions briefly, but irreversible with longer exposures. This behavior is hard to interpret as a the result of a genuine K$^+$ conductance. Another peculiar aspect of these data is that the amplitude of conductance appears to be little affected, for either inward or outward currents, when the K$^+$ concentration is varied between 0 mM and 160 mM. Inward and outward currents were about the same magnitude, whether the bath contained K$^+$, Na$^+$, or NMG$^+$, and whether the cell contained Cs^+ or NMG^+. It is hard to imagine a K^+-selective conductance which conducts outward Cs^+ or NMG^+ currents as well as inward K^+ currents. These seemingly paradoxical results seem inconsistent with any particular selectivity, but may be explainable if the cells in which this conductance was observed were electrically coupled to other endothelial cells. If this were the case, then most of the observed current might pass through gap junctions into neighboring cells, rather than through the plasma membrane of the cell in contact with the pipette. This possibility is explored further in the *Discussion*.

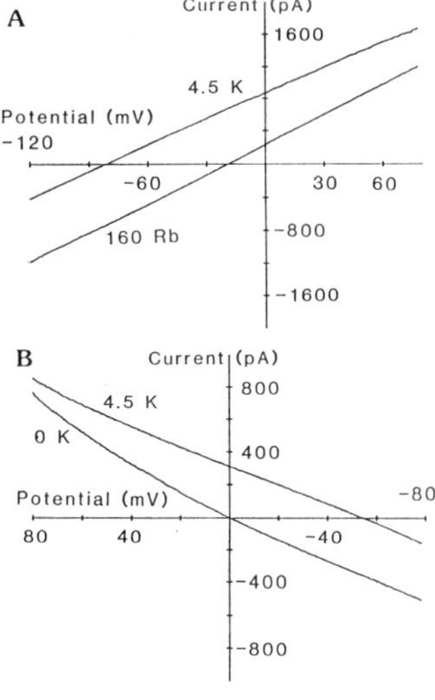

Figure 7. Whole-cell currents during voltage ramps in cells exhibiting the "pseudo-cation" conductance described in the text. (*A*) The cell was ramped from -120 mV to +80 mV at a rate of 0.8 mV/msec in a cell containing $CsMeSO_3$. The currents shown are in Ringer's solution with 4.5 mM K^+ and 160 mM Na^+ ("4.5 K"), and immediately after the bath was changed to 160 mM Rb^+ Ringer's. Note the large shift in reversal potential with minimal change in conductance. (*B*) This cell contained $NMGMeSO_3$ and was ramped from +80 mV to -80 mV at -0.64 mV/msec. In Ringer's solution the currents reverse near -60 mV, but when K^+-free Na^+ Ringer's solution ("0 K") was added the currents reversed near 0 mV, again with little change in slope conductance.

DISCUSSION
Chloride Conductance

The present results demonstrate the presence of a Cl^- conductance in endothelial cells. At the whole-cell level, Cl^- currents are usually time-independent and rectify outwardly. The membrane is not very selective, as Br^- permeates the membrane about as well as does Cl^-, and the large anion $MeSO_3^-$ also carries measurable current and has a relative permeability about half that of Cl^-. Whole-cell Cl^- currents are blocked completely by Zn^{2+}. Zn^{2+} has long been known to block the Cl^- conductance in frog skeletal muscle (19, 26), and also inhibits Cl^- currents at the single-channel level in both inside-out and outside-out muscle membrane patches (53).

Effects of pH. Acid *pH* inhibits the Cl^- conductance of frog, stingray, rat, and sunfish skeletal muscle (16, 20, 22, 35). At negative potentials Cl^- currents in frog muscle turn on with time at low *pH*, turn off with time at high *pH*, and are minimally time-dependent at neutral *pH* (25, 40, 49). The effects of low *pH* in endothelial cells were qualitatively similar, in that the conductance is on at negative potentials and turns off at positive potentials, but Cl^- currents in alkaline solutions were practically time-independent. Apparently protons either alter the behavior of the gating mechanism of Cl^- channels or produce a steep voltage- and time-dependent block.

Single Cl^- channel currents. Unitary Cl^- channels exhibiting strong outward rectification were observed in endothelial cells. These channels are about equally permeable to Br^- and Cl^- and carry $MeSO_3^-$ current less well. The selectivity, unitary conductance, and outward rectification of these channels closely resemble those of tracheal epithelial Cl^- channels (51), which are abnormally regulated in cystic fibrosis (14, 36, 50). The properties of these channels correspond fairly well with those of the macroscopic currents, but the extent to which they are responsible for whole-cell currents cannot be determined with certainty on the basis of existing data.

"Mega" Cl⁻ channels. Occasionally large-conductance (300-400 pS) anion-selective channels were seen in endothelial cells. These have a different voltage-dependence than the usual macroscopic Cl⁻ conductance. Even at neutral *pH* these channels turn off both at positive and negative potentials. This behavior is reminiscent of large-conductance Cl⁻ channels reported in a wide variety of cells (2. 29, 38, 40, 41, 44, 52).

Possible functions of a Cl⁻ conductance in endothelial cells. Several functional responses of vascular endothelial cells have been described which one may speculate require a Cl⁻ conductance. Ueda *et al.* (46) describe morphological changes which occur upon stimulation of bovine pulmonary artery endothelial cells with agents which increase cAMP, which are prevented by Cl⁻ channel inhibitors. Whether this inhibition is due to Cl⁻ efflux through channels, and whether the blockers used actually block Cl⁻ currents in these cells are questions which need to be explored. Histamine degradation by vascular endothelial cells involves an uptake step which appears to require extracellular Cl⁻ (15).

Pseudo-Cation Conductance

Some endothelial cells have what appears to be a large, linear conductance, which we suspect may be an artifact of electrical coupling between confluent cells. This phenomenon superficially resembles a K^+ conductance, but several types of evidence appear to rule out this possibility. The reversal potential varies non-monotonically with external K^+ concentration, and currents are about as large when other cations replace K^+ either in the bath or in the pipette solution. These results seem inconsistent with any particular selectivity, but may be explainable if the cells in which this conductance was observed were electrically coupled to other endothelial cells. This possibility was not directly tested, although large, slow, and multi-exponentially-decaying capacity transients were observed in some experiments. Endothelial cells in confluent monolayers in culture are known to form gap junctions with other endothelial cells, on the basis of transcellular diffusion of dyes and nucleotides, and also of electrotonic potential transfer as far as ten cell diameters away from the current-injecting electrode (24). Cooper et al. (8) have estimated the junctional resistance between lens epithelial cells to be 15.5 MΩ. Electrical coupling of this magnitude would greatly complicate any attempt to voltage-clamp confluent endothelial cells, because most of the current injected through the patch electrode would pass through gap junctions into neighboring cells, with only a small fraction actually passing through the surface membrane of the cell in contact with the pipette. Since neighboring cells will be more poorly voltage-clamped, the observed reversal potential of the current may approximately reflect the resting potential of the confluent cells, rather than a genuine reversal potential of the membrane conductance in the cell touching the pipette. Isotonic K^+ Ringer's or Rb^+ Ringer's would depolarize all of the endothelial cells on the cover slip, thus the apparent reversal potential will be close to 0 mV. Completely K^+-free bath solutions would eliminate inward rectifier currents by shifting the voltage-dependence of activation negatively, and consequently would also depolarize the monolayer. Eliminating inward rectifier currents by blocking with Ba^{2+} depolarizes endothelial cells (21), and also was found to shift the reversal potential of the "pseudo-cation" conductance to more positive potentials. This interpretation of the "pseudo-cation" conductance as reflecting mainly current through gap junctions also explains why the amplitude of the conductance does not change appreciably even when its reversal potential is shifted by varying the ionic conditions.

A somewhat similar phenomenon has been reported in human fibroblasts, in which the conductance was clearly demonstrated to be related to cell-to-cell coupling by its abolition when the cell was pulled away from its neighbors (11). Attempts to perform this procedure on confluent endothelial cells were not successful. If our interpretation of the "pseudo-cation" conductance is correct, and even if it is not but endothelial cells are electrically coupled, then this phenomenon comprises a significant danger in attempts to interpret data from confluent or near-confluent endothelial cells.

ACKNOWLEDGEMENTS

This work was supported by research grant HL37500 (TD), Research Career Development Award KO4-1928 (TD) from the National Institutes of Health.

REFERENCES

1. Adams, D.J., J. Barakeh, R. Laskey, and C. Van Breeman, FASEB J. 3,2389-2400 (1989).
2. Blatz A.L. and K.L. Magleby, Biophys. J. 43, 237-241 (1983).
3. Bregestovski, P., A. Bakhramov, S. Danilov, A. Moldobaeva, and K. Takeda, Brit. J. Pharmacol. 95, 429-436 (1988).
4. Bretag, A.H., Physiol. Rev. 67, 618-724 (1987).
5. Cannell, M.B. and S.O. Sage, J. Physiol. 419, 555-568 (1989).
6. Colden-Stanfield, M., W.P. Schilling, A.K. Ritchie, S.G. Eskin, L.T. Navarro, and D.L. Kunze, Circ. Res. 61, 632-640 (1987).
7. Colden-Stanfield, M., W.P. Schilling, L.D. Possani, and D.L. Kunze, J. Membrane Biol. 116, 227-238 (1990).
8. Cooper, K., J.L. Rae, and P. Gates, J. Membrane Biol. 111, 215-227 (1989).
9. Cota, G. and C.M. Armstrong, Biophys. J. 55, 107-109 (1988).
10. DeCoursey, T.E., J. Gen. Physiol. 95, 617-646 (1990).
11. Estacion, M., J. Physiol. 436, 579-601 (1991).
12. Fanburg, B.L., Chest 93,101S-105S (1988).
13. Fichtner, H., U. Frobe, R. Busse, and M. Kohlhardt, J. Membrane Biol. 98, 125-133 (1987).
14. Frizzell, R.A., T.I.N.S. 10, 190-193 (1987).
15. Haddock, R.C., P. Mack, S. Leal, and N.L. Baenziger, J. Biol. Chem. 265, 14395-14401 (1990).
16. Hagiwara, S. and K. Takahashi, J. Physiol. 238, 109-127 (1974).
17. Hamill, O.P., A. Marty, E. Neher, B. Sakmann, and F.J. Sigworth, Pflugers Arch. 391, 85-100 (1981).
18. Hassoun, P.M., B.L. Fanburg, and F. Junod in: The Lung: Scientific Foundations, R.G. Crystal, eds. (Raven Press, New York 1991) pp. 313-327.
19. Hutter O.F. and A.E. Warner, J. Physiol. 189, 445-460 (1967).
20. Hutter, O.F. and A.E. Warner, J. Physiol. 189, 403-425 (1967).
21. Johns, A., T.W. Lategan, N.J. Lodge, U.S. Ryan, C. Van Breeman, and D.J. Adams, Tissue and Cell 19, 733-745 (1987).
22. Klein, M.G., J. Exp. Biol. 114, 581-598 (1985).
23. Lansman, J.B., T.J. Hallam, and T.J. Rink, Nature 325, 811-813 (1987).
24. Larson D.M. and J.D. Sheridan, J. Cell. Biol. 92, 183-191 (1982).
25. Loo, D.D.F., J.G. McLarnon, and P.C. Vaughan, Can. J. Physiol. Pharmacol. 59, 7-13 (1981).
26. Mashima, H. and H. Washio, Jap. J. Physiol. 14, 535-550 (1964).
27. McCann, J.D., M. Li, and M.J. Welsh, J. Gen. Physiol. 94, 1015-1036 (1989).
28. Mitra, R.L. and M. Morad, J. Membrane Biol. 122, 33-42 (1991).
29. Nelson, D.J., J.M. Tang, and L.G. Palmer, J. Membrane Biol. 80, 81-89 (1984).
30. Nilius, B. and D. Riemann, Gen. Physiol. Biophys. 9, 89-112 (1990).
31. Nilius, B., N.I.P.S. 6,110-114 (1991).
32. Nilius, B., Pflugers Arch. 416, 609-611 (1990).
33. Olesen, S.P., D.E. Clapham, and P.F. Davies, Nature 331, 168-170 (1988).
34. Olesen, S.P., P.F. Davies, and D.E. Clapham, Circ. Res. 62, 1059-1064 (1988).
35. Palade, P.T. and R.L. Barchi, J. Gen. Physiol. 69, 325-342 (1977).
36. Quinton, P.M., FASEB J. 4, 2709-2717 (1990).
37. Rae, J.L., J. Dewey, K. Cooper, and P. Gates, J. Membrane Biol. 114, 29-36 (1990).
38. Ravesloot, J.H., R.J. Van Houten, D.L. Ypey, and P.J. Nijweide, J. Bone Mineral Res. 6, 355-363 (1991).
39. Sauve, R., L. Parent, C. Simoneau, and G. Roy, Pflugers Arch. 412, 469-481 (1988).
40. Schneider, G.T., D.I. Cook, P.W. Gage, and J.A. Young, Pflugers Arch. 404, 354-357 (1985).
41. Schwarze, W. and H.-A. Kolb, Pflugers Arch. 402, 281-291 (1984).
42. Shapiro, M.S. and T.E. DeCoursey, J. Gen. Physiol. 97, 1227-1250 (1991).
43. Silver, M.R. and T.E. DeCoursey, J. Gen. Physiol. 96, 109-133 (1990).
44. Soejima, M. and S. Kokubun, Pflugers Arch. 411, 304-311 (1988).
45. Takeda, K., V. Schini, and H. Stoeckel, Pflugers Arch 410, 385-393 (1987).
46. Ueda, S., S.-L. Lee, and B.L. Fanburg, Circ. Res. 66, 957-967 (1990).
47. Vaughan, P.C., J.G. McLarnon, and D.D.F. Loo, Can. J. Physiol. Pharmacol. 58, 999-1010 (1980).

48. Vigne, P., G. Champigny, R. Marsault, P. Barbry, C. Frelin, and M. Lazdunski, J. Biol. Chem. $\underline{264}$, 7663-7668 (1989).

49. Warner, A.E., J. Physiol. $\underline{227}$, 291-312 (1972).

50. Welsh, M.J., FASEB J. $\underline{4}$, 2718-2725 (1990).

51. Welsh, M.J., Pflugers Arch. $\underline{407}$, S116-S122 (1986).

52. Woll, K.H. and B. Neumcke, Pflugers Arch. $\underline{410}$, 641-647 (1987).

53. Woll, K.H., M.D. Leibowitz, B. Neumcke, and B. Hille, Pflugers Arch. $\underline{410}$, 632-640 (1987).

54. Woodhull, A.M., J. Gen. Physiol. $\underline{61}$, 687-708 (1973).

55. Worrell, R.T., A.G. Butt, W.H. Cliff, and R.A. Frizzell, Am. J. Physiol. $\underline{256}$, C1111-C1119 (1989).

MECHANO-SENSITIVE AND VOLUME-SENSITIVE ION TRANSPORT IN ENDOTHELIAL CELLS

W. CHARLES O'NEILL and BRIAN N. LING

Renal Division, Departments of Medicine and Physiology, Emory University School of Medicine and Veterans Administration Medical Center, Atlanta, Georgia 30322

INTRODUCTION

It has become apparent in recent years that endothelium is not merely a passive inner lining for the vascular system, but rather plays an active role in the regulation of vessel wall function and structure. In response to both neurohumoral and mechanical stimuli, endothelial cells alter their shape and cytoskeletal structure, and secrete factors that alter smooth muscle and platelet function (22, 24, 36, 37, 44, 46, 71, 75). The response of endothelium to mechanical stimuli is of particular interest since it may function as a transducer of dynamic changes in blood flow and act to modulate vasomotor tone. Endothelium is exposed to several forms of mechanical stress in vivo including hydrostatic pressure (force perpendicular to the endothelium), shear (force tangential to the endothelium), and stretch, all of which alter endothelial morphology and function (4, 33, 37, 60, 71). Recent attention has focused on the concept of "mechanotransduction," whereby external physical stimuli are perceived by cells and converted to biochemical signals (4, 77). Membrane deformation in multiple cell types, including endothelial cells, has been shown to affect the levels of intracellular second mesengers (cAMP, eicosanoids, phosphoinositols, and protein kinases) and to alter ionic properties (intracellular Na^+, K^+, H^+ and Ca^{2+} concentrations, and membrane potential) (77).

The mechanism by which mechanotransduction occurs is poorly understood, however considerable evidence suggests that changes in ion transport are an early event. The resulting changes in membrane potential or cytoplasmic Ca^{2+} levels may then signal subsequent biochemical events. The concept of mechanosensitive ion transporters has developed from two independent lines of research in non-endothelial cell types. Studies of mechanical effects on skeletal muscle led to the first demonstration of stretch-sensitive ion channels (62). Both stretch-activated and stretch-inactivated channels have been subsequently identified in a variety of cells (12, 49, 62, 77). Mechanosensitive ion conductances described to date include channels selective for K^+, Ca^{2+}, and Cl^-; as well as cation and anion nonselective channels (49). The second line of research has focused on the ability of cells to rapidly restore their volume after shrinkage (regulatory volume increase, RVI) or swelling (regulatory volume decrease, RVD) (12, 14, 23, 42, 49, 63). This was first characterized in avian red cells (31) and has now been observed in a wide variety of cells (14). In most cells, RVI is mediated either by Na-K-2Cl cotransport or by Na-H exchange (in conjunction with Cl/HCO_3 exchange); and RVD is mediated either by K^+ channels (with or without Cl^- channels) or by K-Cl cotransport. Although the mechanism by which cells sense changes in volume is not established, considerable evidence points to a mechanical signal. Studies in several cells have demonstrated that activation of volume-regulatory, ion transporters is independent of osmolarity, ionic strength, or intracellular ion concentrations (26, 38, 53, 54). Swelling-activated, K-Cl cotransport can be recreated in resealed erythrocyte ghosts (11, 55, 64), and RVD in several cell types is blocked by cytochalasins (16, 19). The activation of ion channels by both cell swelling and direct membrane stretching also strongly suggests that these two events are mechanistically linked (8, 66).

Investigation of mechanosensitive ion transport has recently extended to vascular endothelial cells, leading to the characterization of both mechano-sensitive and volume-sensitive ion transport proteins. This chapter will review the ion transport proteins present in vascular endothelial cells and focus on those that are regulated by physical stimuli. Possible mechanisms of ion transport regulation will be discussed and an integrated scheme for the functional role of these membrane transport proteins in endothelial cells will be presented.

Published 1991 by Elsevier Science Publishing Company, Inc.
Ion Channels of Vascular Smooth Muscle Cells and Endothelial Cells
Sperelakis and Kuriyama, Editors

MECHANO-SENSITIVE AND VOLUME-SENSITIVE ION TRANSPORTERS

Ion Channels

Potassium Channels. Multiple K^+-selective conductances, thought to be responsible for the regulation of membrane potential, have been described in vascular endothelial cells (1, 15, 58, 59, 68, 72), but only one has been shown to be mechanosensitive. In whole-cell patch clamp studies of bovine aortic endothelial cells (AEC) grown on the inside of glass capillary tubes and subjected to laminar flow, Olesen and associates (57) described an inwardly rectifying, K^+-selective current ($I_{K.S}$) that is activated by low levels of shear stress (0.2-17.0 dynes/cm^2). However, these investigators were unable to detect single K^+ channel events corresponding to this hyperpolarizing $I_{K.S}$ current in excised outside-out patches.

Characterization of the endothelial ion channels responsive to shear stress has been incomplete due to the obvious technical difficulties in applying patch clamp technology to the above model (33, 49, 57). This led us to examine the response of endothelial cells to another form of membrane stress caused by exposure to an anisotonic environment (62, 77). An increase in cellular volume of only 1% produces a membrane tension of 0.8 dynes/cm^2 (65). That mechanical activation of ion channels may also be involved in the regulation of cell volume homeostasis is not a new concept (12, 49, 63). Swelling-activated, K^+ and Cl$^-$ channels mediate regulatory volume decrease in several non-endothelial tissues via intracellular ion and water losses (42, 23).

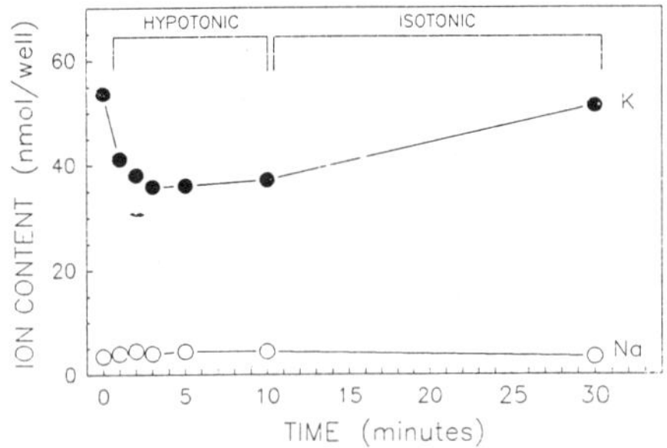

Figure 1. Bovine AEC intracellular ion content under anisotonic conditions.

When bovine AEC were placed in hypotonic medium (Fig. 1) we observed a prompt loss of intracellular K^+ K^+ consistent with regulatory volume decrease (39-41). Cell K^+ content was restored during subsequent reincubation in isotonic medium, indicating that the original K^+ loss with hypotonic exposure was not due to endothelial cell lysis. Although RVD has been previously demonstrated in endothelial cells (27, 47, 70), the specific ion transport mechanisms have not been identified. Using patch-clamp methodology, we have recently characterized two Ca^{2+}-dependent, K^+-selective channels (165 and 42 pS) that are biphasically activated when bovine AEC's are swollen (reducing bath osmolality from 285 to 215 mOsmol/kg H_2O), with a time course that corresponds to RVD (Fig. 2; Top Trace) (39-41). Outward K^+ current flow through these mechanosensitive channels thus appears to be responsible for the macroscopic K^+ efflux we observe in swollen endothelial cells.

Usually in the closed state under basal conditions (ie., isotonic media, T = 37° C, zero applied potential [V_{app} = 0 mV]), the open probability (P_O) for the 165 pS, K^+ channel increased dramatically when bovine AEC are placed in hypotonic media (Fig. 2; Top Trace and Fig. 3; Top Left) (39-41). Potassium channel stimulation was biphasic; a first phase (0-2 min.) of transient channel activation occurring within seconds of hypotonic bath exchange was followed by a delayed, but prolonged reactivation phase lasting over five minutes. The second reactivation phase was abolished by removal of extracellular Ca^{2+} from the hypotonic bath (Fig. 2; Bottom Trace). Under isotonic conditions, a channel with the same 165 pS conductance was occasionally observed. This channel was K^+ selective (P_K/P_{Na} = 25:1), and activated by depolarization (increasingly positive V_{app}) and/or increasing cytoplasmic Ca^{2+} in ripped-off patches. Fichtner and associates (15) have also described a 150 pS, K$^+$ channel in bovine AEC which was activated by depolarization and cytoplasmic Ca^{2+}, and rarely observed under isotonic conditions at resting membrane potential (2/55 patches). These characteristics are similar to swelling-activated "maxi-K" channels described by ourselves and others in a variety of mammalian cell types (32, 42, 73).

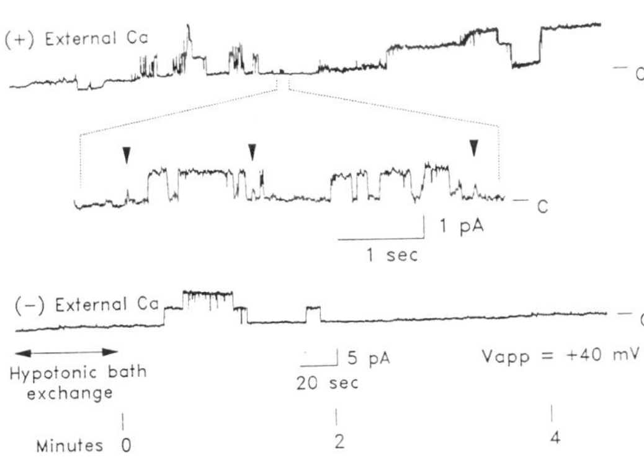

Figure 2. Ion channel activation by bovine AEC swelling. Top Trace) Hypotonic exposure, in the presence of extracellular Ca^{2+}. Biphasic activation of the 165 pS K$^+$ channel is depicted. Middle Trace) Enlargement of the Top Trace better illustrates activation of the 42 pS K$^+$ channel by cell swelling (larger amplitude events). The arrowheads point to still smaller amplitude current events representing the 28 pS NSCC. Bottom Trace) Hypotonic exposure in the absence of bath Ca^{2+}. Initial K$^+$ channel activation burst appears again, but the second reactivation phase is eliminated. Outward current is reflected by upward deflections. Horizontal bars mark the zero current level (C = closed state). Voltage (mV) represents the applied patch pipette voltage (V$_{app}$) displacement away from resting membrane potential (cell interior with respect to patch pipette interior). Orignial recording corner frequency was 2 KHz, sampling was performed at 5 KHz and traces are depicted without software filtering. External bath and pipettes contained 140 mM NaCl/5 mM KCL. Refer to references (39-41) for further details.

Flickering, brief openings of smaller amplitude, outward current events were also seen superimposed on the already mentioned "maxi-K" channel activity evoked by swelling-induced bovine AEC volume changes (Fig. 2; Top & Middle Traces) (39-41). This was found to represent biphasic activation of an inwardly rectifying,42 pS K$^+$ channel (Fig. 3, Top Right). Under isotonic conditions, single-channel records revealed a 40 pS K$^+$ selective (P$_K$/P$_{Na}$ = 64:1) channel which was also inwardly rectifying. Open probability for this latter channel was low (< 0.05) and independent of applied membrane potentials between -80 and +60 mV. In addition, this channel could be stimulated by increasing cytoplasmic Ca^{2+} or by exposure to exogenous bradykinin (200 nM). Bradykinin-induced activation of this 40 pS K$^+$ channel was biphasic, similar to the swelling-induced activation of the 42 pS K$^+$ channel. Other groups have described 40 pS K$^+$ selective (P$_K$/P$_{Na}$ > 100:1) channels in bovine AEC that are also inwardly rectifying, voltage independent, Ca^{2+} dependent, and biphasically activated by exogenous bradykinin and ATP (68, 72). As seen with swelling-induced, 165 and 42 pS K$^+$ channels, early activation of these 40 pS K$^+$ channels is independent of external Ca^{2+}, while the second reactivation phase is abolished by external Ca^{2+} removal. This bradykinin-activated and swelling-induced, 40-42 pS K$^+$ channel may represent the inwardly rectifying I$_{K,S}$ conductance that is induced by hemodynamic shear stress (57). Again, the membrane tension (0.8 dynes/cm^2 per 1% increase in cell volume) generated by our hypotonic exposure is in the same range known to activate both stretch-activated channels (65), and shear stress-activated I$_{K,S}$ (57). Physiologic shear stress on endothelial cells in large arteries ranges from 2-5 dynes/cm^2 (1 mmHg = 1.33 X 10^3 dynes/cm^2) under steady flow conditions, and can increase up to 20-50-fold in smaller vessels or under conditions of tubulent/pulsatile flow (33).

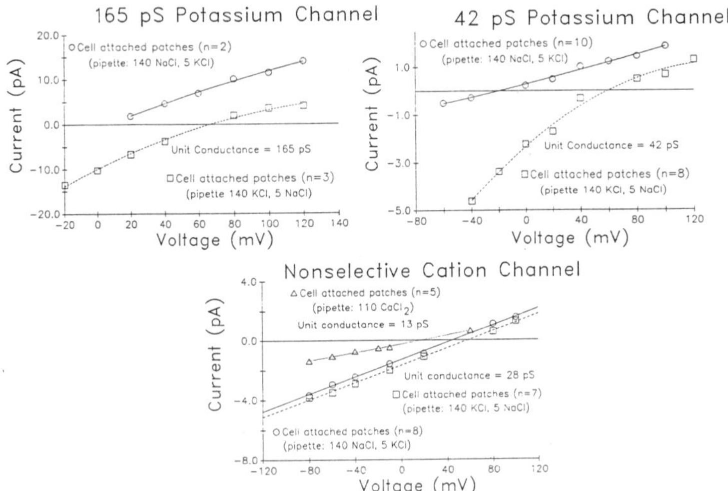

Figure 3. Current-voltage (I-V) curves for cell-attached patches. Top Left) 165 pS K^+ channel: Slope conductance (g) for inward current (pipette: 140 mM KCl/5 mM NaCl; squares, dashed line). Reversal potential (EREV) shifts negative with increasing external Na^+ (pipette: 140 mM NaCl/5 mM KCl; circles, solid line). Top Right) 42 pS K^+ channel: g for inwardly rectifying current (pipette: 140 mM KCl/5 mM NaCl). E_{rev} shifts negative with increasing external Na^+. Bottom) Nonselective Cation Channel: g = 28 pS for inward current (pipette: 140 mM KCl/5 mM NaCl). E_{rev} does not change significantly with increasing external Na^+. g = 13 pS for inward Ca^{2+} current (pipette: 110 mM CaCl₂; triangles, dotted line). Plots depict mean current measurements for n cell-attached patches.

Several other K^+ channel types are present in endothelial cells, but have features differentiating them from the mechanosensitive K^+ conductances described above. In bovine AEC there are inwardly rectifying K^+ channels with lower unitary conductances (25-26 pS) that are activated by hyperpolarization (58, 59). Finally, acetylcholine exposure hyperpolarizes bovine AEC via activation of a muscarinic receptor-linked K^+ conductance ($I_{K.ACh}$) (58) that is separate and additive to $I_{K.S}$ (57).

Non-selective Cation Channels. In addition to mechanical activation of 165 and 42 pS, Ca^{2+}-dependent K^+ channels, our group has also observed a 28 pS nonselective cation channel (NSCC) in response to bovine AEC swelling (Fig. 2; Top & Middle Traces and Fig. 3; Bottom) (39-41). More pertinent to our discussion of endothelial cell mechanotransduction and dependence on intracellular Ca^{2+} is the fact that this NSCC is capable of conducting inward Ca^{2+} current. When external (ie., intrapipette) cations were replaced with 110 mM CaCl₂ inward current (Ca^{2+} influx) was stimulated by bovine AEC swelling. Open probability for this latter 13 pS conductance was increased by membrane hyperpolarization. As noted above, initial activation of both 165 and 42 pS, Ca^{2+}-dependent K^+ channels in response to hypotonic exposure appears to be only partly dependent on external Ca^{2+} and presumably is also stimulated by cytoplasmic Ca^{2+} released from intracellular stores. In contrast, the delayed Ca^{2+}-dependent K^+ channel reactivation phase absolutely requires Ca^{2+} influx presumably occurring via Ca^{2+} permeable channels. This latter dependence on Ca^{2+} entry is also a feature of agonist-induced biphasic increases in cytosolic Ca^{2+} and biphasic activation of Ca^{2+}-dependent K^+ channels, as seen in endothelial cells with exogenous ATP, bradykinin and angiotensin II (48, 68, 69, 72).

We propose that the NSCC we have described may mediate Ca^{2+} influx into endothelial cells subjected to physical forces including cellular swelling and hemodynamic stress. NSCC's (16-56 pS) have been observed in a variety of endothelial cells (ie., human umbilical vein, porcine aorta, bovine aorta, and brain microvessels from rat and pig) and are activated by histamine, membrane stress or membrane hyperpolarization (1, 15, 34, 50, 76). Two of these NSCC's are also capable of conducting inward Ca^{2+} current with a wide range of conductances (8-19 pS) and P_{Ca}/P_{Na} permeability ratios (0.02-8.4) (34, 50). Membrane stretch caused by cell swelling, along with hyperpolarization from the swelling-induced, K^+ channel activation described above would promote Ca^{2+} entry via such NSCC's. Direct endothelial cell membrane deformation with a blunt micropipette induces both external Ca^{2+} entry and mobilization of intracellular Ca^{2+} stores (54). Vascular shear stress is associated with endothelial cell histamine release, membrane stretch, hyperpolarization, and K^+-selective channel activation, all of which would promote opening of NSCC's and influx of Ca^{2+} (33, 43, 46, 57). Increases in cytoplasmic Ca^{2+} have in fact been documented in vascular endothelial cells subjected to fluid shear stress (2). Similar stretch-activated, Ca^{2+} permeable NSCC's have also been described in non-endothelial tissues (ie., renal, lens, and choroid plexus epithelium) (12). Calcium influx through stretch-activated NSCC's has been proposed to be an early event in non-endothelial RVD, leading to subsequent activation of Ca^{2+}-dependent K^+ channels (12, 42, 73).

Our macroscopic measurements of endothelial ion composition changes with cellular swelling show a net decrease in intracellular cations largely due to potassium efflux. However, we also observe a small increase in intracellular Na^+ content (Fig. 1) (39-41). Given the physiologic gradient between extracellular and intracellular Na^+, activation of the 28 pS NSCC by swelling-induced hyperpolarization would favor inward Na^+ movement, along with Ca^{2+} entry and K^+ efflux. The contributions of other sodium-transporting mechanisms present in endothelial cells (ie., Na-K-2Cl cotransporter, Na-H and Na-Ca exchangers, Na-K-ATPase pump) to the observed swelling-induced changes in intracellular Na^+ are addressed in the following sections.

Calcium Channels. Alternative pathways for calcium entry into vascular endothelial cells have also been proposed (1). We have noted that both ligand-receptor interaction and mechanical stimuli can activate NSCC's permeable to calcium in endothelial cells (5, 25). Similarities in conductance, current-voltage relationship, voltage-dependence, and gating kinetics suggest that these agonist-evoked and mechanosensitive channels may be the same. In addition to these channels, basal ^{45}Ca and ^{22}Na entry into unstimulated endothelial cells also indicates the presence of leak pathways permeable to Ca^{2+} and Na^+ (1, 10, 25). However, there has been no evidence at a single-channel level that indicates the presence of classic voltage-dependent sodium or calcium channels in endothelial cells (74). Supporting this is the fact that both agonist-evoked increases in cytosolic Ca^{2+} and net inward current are insensitive to calcium channel blockers such as Bay-K-8644, verapamil, nimodipine and nitrendipine (9, 25, 74).

Chloride Channels. While volume-sensitive Cl^- channels are often associated with K^+ conductances in the maintainence of cell volume homeostasis in several non-endothelial cell types (14, 23, 42), there have been no published descriptions of chloride channels in endothelial cells. In endothelial cells, replacement of extracellular Cl^- with impermeant anions has no affect on resting membrane potential (25, 58). Certain nonselective channels in non-endothelial tissues are capable of conducting anions as well as cations (12). However, replacement of patch pipette (external) Cl^- with aspartate had no affect on the unitary conductance or reversal potential of nonselective cation channels in endothelial cells (34, 50).

Ion Channel Summary. Mechanosensitive endothelial ion channels that have been well characterized then include: Ca^{2+}-dependent K^+ channels and Ca^{2+}-permeable NSCC's activated by membrane stretching/cell swelling; and an inwardly rectifying K^+ conductance ($I_{K.S}$) activated by shear stress (1, 4, 34, 39-41, 57). A stretch-sensitive, 45 pS channel with a reversal potential of +20 mV has also been seen in porcine AEC cultures (1). However, voltage dependence, current-voltage relationship, and ion selectivity measurements for this latter channel have not been reported.

Na-K-2Cl Cotransport

Na-K-2Cl cotransport was first identified in vascular endothelial cells by Brock and coworkers (7) and found to comprise the majority of K^+ influx in arterial endothelial cells. This has been confirmed in subsequent studies by O'Donnell (51) and by Klein and O'Neill

(30). The abundance of cotransport in the first two studies is partly artifactual due to the use of ouabain, which causes cell shrinkage and secondary activation of cotransport (O'Neill and Klein, unpublished). Considerably less cotransport activity is observed in endothelial cells cultured from veins (7). Cotransport in aortic endothelial cells is modulated by several vasoactive compounds, suggesting that it may be important in endothelial modulation of smooth muscle tone. Cotransport is stimulated by bradykinin (7, 30, 51), vasopressin (7, 51), angiotensin II (51), and thrombin (O'Neill and Klein, unpublished). Removal of extracellular calcium or the addition of $LaCl_2$ to inhibit Ca^{2+} entry blocks the stimulation of cotransport by bradykinin (7), and treatment with bis-(o-aminophenoxy)-ethane-N,N,N',N'-tetraacetic acid (BAPTA) to chelate intracellular Ca^{2+} inhibits cotransport (Kartsonis and O'Neill - submitted), while treatment with the Ca^{2+} ionophores ionomycin or A23187 (7, 51, Kartsonis and O'Neill - submitted) increases cotransport. Thus cotransport is activated by intracellular Ca^{2+} and this is the mechanism of activation by bradykinin. The activation by the other vasoactive compounds has not been investigated but is likely mediated by intracellular Ca^{2+} as well. The addition of 8-Br-cAMP or agents that increase cyclic AMP (ie., norepinephrine, isoproterenol, forskolin) has been reported to inhibit cotransport (7, 51), but we have been unable to confirm this. Atrial natriuretic factor (ANF) has been reported to both stimulate (17) and inhibit (51) cotransport via activation of guanylyl cyclase. Other agents reported to inhibit cotransport in endothelial cells include histamine, acetylcholine, and phorbol esters (51).

[^{3}H]bumetanide, which binds to Na-K-2Cl cotransporters with high affinity (21) has been used to determine whether cotransport is regulated through changes in the density of cotransporters. Values of 230,000 (52) and 122,000 (30) specific binding sites per cell have been obtained under basal conditions in aortic endothelial cells. 8-Br-cAMP, 8-Br-cGMP, phorbol ester, norepinephrine, and ANF were found to reduce bumetanide binding and cotransport activity to the same degree, suggesting a decrease in the quantity of cotransporters (52). Stimulation of cotransport by bradykinin is due to an increase in the flux per binding site rather than to an increase in the quantity of cotransporters (30).

Since Na-K-2Cl cotransport mediates a net flux of ions, we have investigated its relationship to endothelial cell volume homeostasis (53). We found that hypertonic shrinkage of bovine AEC increased cotransport, measured as unidirectional bumetanide-sensitive K^+ influx (Fig. 4, solid circles). To ensure that this was due to a change in cell volume and not a change in osmolarity, ionic strength, or intracellular ion concentrations, we employed two other methods to shrink cells: 1) incubation in an isotonic Na^+-free, K^+-free medium which leads to KCl loss, and 2) preincubation in hypotonic medium which produces KCl loss and regulatory volume decrease. In each case the cells are shrunken when placed back in standard isotonic medium. As shown in Fig. 4, these methods produced even greater stimulation of cotransport than was seen in hypertonic medium. Conversely, hypotonic swelling inhibited cotransport, but this could have been due to dilution of transported ions. To avoid this problem, cells were preincubated in isotonic KCl medium, which produced rapid cell swelling due to uptake of KCl. This allowed cotransport to be measured using standard isotonic medium. A 27% increase in endothelial cell volume reduced cotransport to 22% of the baseline activity observed in cells of normal volume (53).

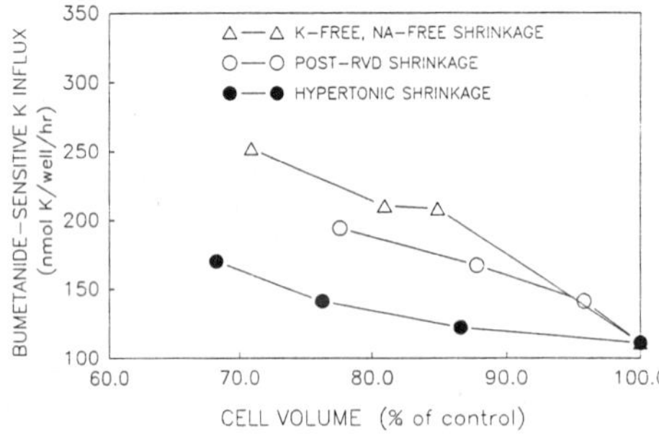

Figure 4. Bumetanide-sensitive K^+ influx in response to bovine AEC shrinkage.

The mechanism by which cell volume controls Na-K-2Cl cotransport in endothelial cells and other cells is not understood. We have found that okadaic acid, an inhibitor of protein phosphatases, stimulates cotransport to the level seen in shrunken cells, indicating that cotransport is regulated by phosphorylation (56). Using activators and inhibitors of known protein kinases (forskolin, 8-Br-cAMP, phorbol esters, staurosporine, and trifluoperazine), we have been unable to identify a kinase that may be involved.

Despite the fact that normal intracellular and extracellular ion concentrations would dictate a net influx of ions through cotransport, no net bumetanide-sensitive K^+ flux was observed in cells of normal volume. In normal cells, therefore, cotransport functions only in an exchange mode. This inhibition of net cotransport may be mediated by intracellular Cl^-, which has been shown to inhibit Na-K-2Cl cotransport in squid axons (6) and Ehrlich ascites cells (38). Whether the stimulation of cotransport in shrunken cells results in a net influx of ions with subsequent restoration of cell volume depends on the type of shrinkage. Net influx of ions does not occur in hypertonically shrunken cells, probably as a result of the increase in the intracellular Cl^- concentration (53). In isotonically shrunken cells, the intracellular Cl^- concentration is reduced due to KCl loss, and bumetanide-sensitive recovery of cell K+ and water is observed (Fig. 5). Thus, there is a dual effect of alterations in cellular volume on Na-K-2Cl cotransport in endothelial cells: 1) regulation by cell volume that is independent of intracellular Cl^- and is possibly mechanical in nature, and 2) inhibition by intracellular Cl^-, the concentration of which can vary with cellular volume.

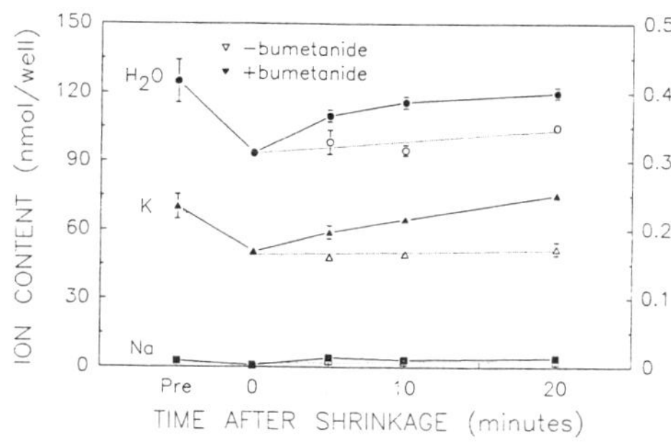

Figure 5. Dual effect of cell volume on Na-K-2Cl cotransport in bovine AEC.

Na-H Exchanger

Evidence for Na-H exchange in endothelial cells was first obtained in rat brain capillaries (3), and subsequent reports have documented its existence in bovine aortic (29), rabbit coronary microvascular (18), and human umbilical vein (13) endothelial cells. In addition to being activated by intracellular acidification, Na-H exchange in endothelial cells is also stimulated by ATP (29) and thrombin (18) suggesting that it is an important component of endothelial responses to these agonists. Ethylisopropylamiloride, a specific inhibitor of Na-H exchange, inhibited stimulation of prostaglandin release by thrombin, but not by ATP (18). The increase in intracellular Ca^{2+} concentration that occurs during alkalinization of endothelial cells (13) provides a potential link between Na-H exchange and endothelial agonists. As in other cells, transport is activated by hypertonic shrinkage (13) indicating that it may be a mechanosensitive transporter. The mechanism by which shrinkage activates Na-H exchange is unknown, but studies in other cells indicate that protein phosphorylation is involved (14). Activation of Na-H exchange in shrunken cells probably does not contribute to cell volume recovery since no regulatory volume increase occurs in hypertonically shrunken endothelial cells (53). It is not known whether other mechanical perturbations activate Na-H exchange in endothelial cells.

Na,K-ATPase Pump

Na-K-ATPase pump activity, identified as ouabain-sensitive transport, is present in vascular endothelial cells (7, 51). It participates indirectly in cell volume regulation by maintaining transcellular gradients for Na^+ and K^+, but it is not affected directly by changes in cell volume. This is also the case in cells other than endothelial cells (14). RVI in shrunken cells requires Na-K-ATPase pump activity to remove Na^+ that enters through Na-K-2Cl cotransport, and ouabain inhibits the reaccumulation of K^+ in shrunken cells (53). In normal cells, ouabain leads to cell shrinkage due to K^+ loss that is greater than Na^+ entry. Direct effects of mechanical forces on Na-K pump activity have not been reported.

ROLE OF MECHANO-SENSITIVE AND VOLUME-SENSITIVE ION TRANSPORTERS IN ENDOTHELIAL FUNCTION

There are at least two potential roles mechanically-activated ion transport may play in endothelium: transduction of mechanical stimuli into electrical and biochemical signals, and maintenance of cell volume and morphology. Mechanical stimuli in the form of increased blood flow and shear stress is associated with vasodilatation, and endothelial cell shape changes, hyperpolarization and release of endothelium-derived relaxing factor (EDRF), prostacyclin (PGI_2) and histamine (4, 33, 46, 61). Studies have shown that such flow-induced dilatation requires an intact endothelial layer (4). Shear stress stimulates EDRF release, whereas increased pulse pressure or transmural pressure (ie., myogenic forces) is inhibitory (60). In addition, shear stress increases intracellular Ca^{2+} (2), but there is no evidence for direct activation of classic voltage-dependent, Ca^{2+} channels by shear stress. Therefore, while it is known that EDRF and PGI2 are produced by endothelial cells in response to increases in cytoplasmic Ca^{2+} (33), how mechanical forces such as cell swelling and flow/shear stress induce such changes in intracellular ion composition has been poorly characterized.

Taken together the data summarized in this chapter suggests that physical stress in the form of swelling-induced membrane stretch results in macroscopic potassium efflux from endothelial cells through a series of local membrane-deliminated ion channel events. Hypotonically swollen bovine AEC exhibit biphasic activation of both 165 and 42 pS Ca^{2+}-dependent K^+ channels. Membrane stretch, along with hyperpolarization due to K^+ efflux promotes an electrical driving force favoring inward Ca^{2+} flow through Ca^{2+}-permeable NSCC's. This scheme for Ca^{2+}-dependent K^+ channel activation may also play a physiologic role in vascular endothelial cell responses to vasodilatory substances (ie., thrombin, bradykinin, histamine, platelet activating factor), changes in blood flow forces and pathologic cell volume abberations (ie., anisotonic shock; uremic Na^+,K^+-ATPase inhibitors; ischemia). Membrane potential changes resulting from Ca^{2+} influx would be stabilized by the second phase of Ca^{2+}-stimulated K^+ channel reactivation, permitting sustained Ca^{2+} Ca^{2+} Ca^{2+} Ca^{2+} Ca^{2+} Ca^{2+} entry and Ca^{2+} Ca^{2+}-dependent production of endothelium-derived relaxing factors (ie., EDRF and prostacyclin) (46). Bovine AEC K^+ efflux would also stabilize membrane depolarization resulting from any agonist or pathologic stimulus (ie., anisotonic swelling, endotoxic shock, ischemia) that increases Na^+ or Ca^{2+} entry, and the resulting vasodilatation would promote blood flow to ischemic tissues.

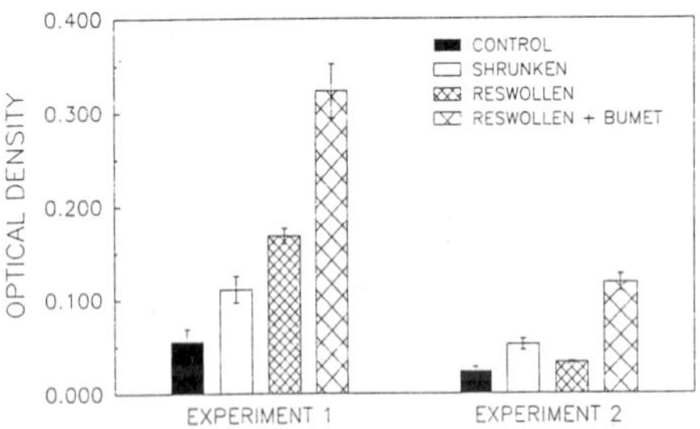

Figure 6. Protective effect of volume regulatory ion transport.

Regulation of endothelial cell volume may also be an important function of other mechanically-activated ion transporters. Cell contraction increases endothelial monolayer permeability by producing intracellular gaps (28, 35, 45). Hypertonic shrinkage of endothelial cells also increases monolayer permeability (20), presumably through he same mechanism. We have confirmed this finding and have demonstrated a protective effect of volume-regulatory transport. Bovine AEC were grown to confluence on collagen-coated microcarrier beads and permeability was measured by the uptake of albumin-bound Evan's blue dye into the beads (28). In experiment 1 (Fig. 6), shrinkage of the cells in Na$^+$-free, K$^+$-free medium to 67% of their original volume increased permeability (open bar). There was a further increase in permeability during a subsequent 20 minute period in standard medium to allow volume recovery (Fig. 6; narrow crosshatch). However, inclusion of bumetanide during this period to inhibit Na-K-2Cl cotransport and volume recovery resulted in substantially greater permeability (Fig. 6; wide crosshatch). In experiment 2 cells were shrunken to 80% of their original volume and the dye concentration was halved. In this case, permeability was restored during cell volume recovery, while inhibition of volume recovery resulted in a further increase in permeability. Although endothelial cells are not normally exposed to osmotic stress, other factors may dictate a need for cell volume regulation. Endothelial cells are constantly exposed to deforming forces such as shear and stretch, and rapid changes in cell volume may be necessary to preserve morphology and maintain permeability. Another potential role is the prevention of cell shrinkage during agonist stimulation. Several agonists increase intracellular Ca^{2+} with subsequent activation of K$^+$ channels. This would be expected to cause K$^+$ loss and cell shrinkage. We have found that exposure of bovine AEC to bradykinin in isotonic medium leads to a small loss of intracellular K$^+$ that is greatly augmented when Na-K-2Cl cotransport is inhibited with bumetanide (Fig. 7). Similar results were obtained with the calcium ionophore A23187. This suggests to us that endothelial cell volume is regulated by a balance between Na-K-2Cl cotransport and K$^+$ channels that is mediated by intracellular Ca^{2+}.

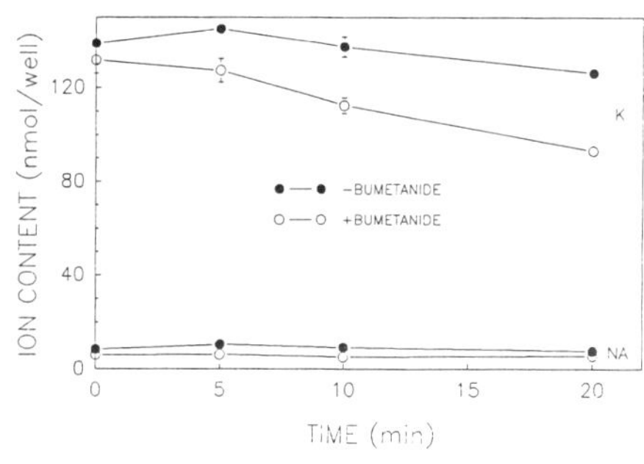

Figure 7. Cell volume regulation by coordinated response of Na-K-2Cl cotransport and K$^+$ channels.

ACKNOWLEDGEMENTS

Work from Dr. O'Neill's laboratory was supported by NIH grant DK 01643, National Kidney Foundation Georgia Affiliate Grant, and American Heart Association Georgia Affiliate Grant-In-Aid Award. Work from Dr. Ling's laboratory was supported by VA Merit Review Award and NIH Biomedical Research Support Grant S07 RR05364.

REFERENCES

1. Adams, D.J., J. Barakeh, R. Laskey, and C. Van Breemen, FASEB J. 3, 2389-2400 (1989).
2. Ando, J., T. Komatsuda, and A. Kamiya, In Vitro. Cell. and Develop. Biol. 24, 871-877 (1988).
3. Betz, A.L., J. Neurochem. 41, 1150-1157 (1983).
4. Bevan, J.A. and I. Laher, FASEB J. 5, 2267-2273 (1991).
5. Bregestovski, P., A. Bakhramov, S. Danilov, A. Moldobaeva, and K. Takeda, Br. J. Pharmacol. 95, 429-436 (1988).
6. Breitwieser, G.E., A.A. Altamirano, and J.M. Russell, Am. J. Physiol. 258, C749-C753 (1990).
7. Brock, T.A., C. Brugnara, M. Canessa, and Jr. Gimbrone, M.A., Am. J. Physiol. 250, C888-C895 (1986).
8. Christensen, O., Nature 330, 66-68 (1987).

9. Colden-Stanfield, M., W.P. Schilling, A.K. Ritchie, S.G. Eskin, L.T. Navarro, and D.L. Kunze, Circ. Res. 61, 632-640 (1987).

10. Danthuluri, N.R., M.I. Cybulsky, and T.A. Brock, Am. J. Physiol. 255, H1549-H1553 (1988).

11. Dunham, P.B. and P.J. Logue, Am. J. Physiol. 250, C578-C583 (1986).

12. Eaton, D.C., Y. Marunaka, and B.N. Ling in Membrane Transport in Biology, J. Schafer and G. Giebisch, eds. (Springer Verlag, New York 1991) (In press) (1991).

13. Escobales, N., E., Longo, Jr. Cragoe, E.J., N.R. Danthuluri, and T.A. Brock, Am. J. Physiol. 259, C640-C646 (1990).

14. Eveloff, J.L. and D.G. Warnock, Am. J. Physiol. 252, F1-F10 (1987).

15. Fichtner, H., U. Frobe, R. Busse, and M. Kohlhardt, J. Membr. Biol. 98, 125-133 (1987).

16. Foskett, J.K. and K.R. Spring, Am. J. Physiol. 248, C27-C36 (1985).

17. Fujita, T., H. Hagiwara, S. Ohuchi, M. Kozuka, M. Ishido, and S. Hirose, Biochem. Biophys. Res. Commun. 159, 734-740 (1989).

18. Gerritsen, M.E., C.A. Perry, T. Moatter, Jr. Cragoe, E.J., and M.S. Medow, Am. J. Physiol. 256, C831-C839 (1989).

19. Gilles, R., E. Delpire, C. Duchene, M. Cornet, and A. Pequeux, Comparative Biochemistry and Physiology 85A, 523-525 (1986).

20. Goligorsky, M.S., FEBS Lett. 240, 59-64 (1988).

21. Haas, M. and B. Forbush III, J. Biol. Chem. 261, 8434-8441 (1986).

22. Herman, I.M., A.M. Brant, V.S. Warty, J. Bonaccorso, E.C. Klein, R.L. Kormos, and H.S. Borovetz, J. Cell. Biol. 105, 291-302 (1987).

23. Hoffmann, E.K. and L.O. Simonsen, Physiol. Rev. 69, 315-382 (1989).

24. Ives, S.L., S.G. Eskin, and L.V. McIntire, In Vitro. Cell. & Develop. Biol. 22, 500-507 (1986).

25. Johns, A., T.W. Lategan, N.J. Lodge, U.S. Ryan, C. Van Breemen, and D.J. Adams, Tissue & Cell 19, 733-745 (1987).

26. Kaji, D., J. Gen. Physiol. 88, 719-738 (1986).

27. Kempski, O., M. Spatz, G. Valet, and A. Baethmann, J. Cell. Physiol. 123, 51-54 (1985).

28. Killackey, J.J.F., M.G. Johnston, and H.Z. Movat, Am. J. Pathol. 122, 50-61 (1986).

29. Kitazono, T., K. Takeshige, Jr. Cragoe, E.J., and S. Minakami, Biochem. Biophys. Res. Commun. 152, 1304-1309 (1988).

30. Klein J.D. and W.C. O'Neill, J. Biol. Chem. 265, 22238-22242 (1990).

31. Kregenow, F.M., Ann. Rev. Physiol. 43, 493-505 (1981).

32. Lang, F. and W. Rehwald, Physiol. Rev. (In Press) (1991).

33. Lansman, J.B., Nature 331, 481-482 (1988).

34. Lansman, J.B., T.J. Hallam, and T.J. Rink, Nature 235, 811-813 (1987).

35. Laposata, M., D.K. Dovarsky, and H.S. Shin, Blood 62, 549-556 (1983).

36. Levesque, M.J. and R.M. Nerem, J. Biochem. Engineering 107, 341-347 (1985).

37. Levesque, M.J., E.A. Sprague, C.J. Schwartz, and R.M. Nerem, Biotechnology Progress 5, 1-8 (1989).

38. Levinson, C., Biochim. Biophys. Acta 1021, 1-8 (1990).

39. Ling, B.N. and W.C. O'Neill, Am. J. Physiol. (In Press) (1991).

40. Ling, B.N. and W.C. O'Neill, Clin. Res. 39, 194A (Abstract) (1991).

41. Ling, B.N. and W.C. O'Neill, J. Am. Soc. Nephrol. (In Press) (1991).

42. Ling, B.N., C.L. Webster, and D.C. Eaton, Am. J. Physiol. (In Press) (1991).

43. Luckhoff, A. and R. Busse, Pflugers Arch. 416, 305-311 (1990).

44. Luscher, T.F., H.A. Bock, Z. Yang, and D. Diederich, Kidney Int. 39, 575-590 (1991).

45. Majno, G., S.M. Shea, and M. Leventhal, J. Cell. Biol. 42, 647-672 (1969).

46. Marsden, P.A., M.S. Goligorsky, and B.M. Brenner, J. Am. Soc. Nephrol. 1, 931-948 (1991).

47. Mazzoni, M.C., E. Lundgren, K-E. Arfors, and M. Intaglietta, J. Cell. Physiol. 140, 272-280 (1989).

48. Morgan-Boyd, R., J.M. Stewart, R.J. Vavarek, and A. Hassid, Am. J. Physiol. 253, C588-C598 (1987).

49. Morris, C.E., J. Membr. Biol. 113, 93-107 (1990).

50. Nilius, B., Pflugers Arch. 416, 609-611 (1990).

51. O'Donnell, M.E., Am. J. Physiol. 257, C36-C44 (1989).

52. O'Donnell, M.E., J. Biol. Chem. 264, 20326-20330 (1989).

53. O'Neill, W.C. and J.D. Klein, Am. J. Physiol. (In Press) (1991).

54. O'Neill, W.C., Am. J. Physiol. 253, C883-C888 (1987).

55. O'Neill, W.C., Am. J. Physiol. 256, C81-C88 (1989).

56. O'Neill, W.C., N. Kartsonis, and J.D. Klein, J. Am. Soc. Nephrol. (Abstract) (In Press) (1991).

57. Olesen, S-P., D.E. Clapham, and P.F. Davies, Nature 331, 168-170 (1988).

58. Olesen, S.-P., P.F. Davies, and D.E. Clapham, Circ. Res. 62, 1059-1064 (1988).

59. Piomelli, D., J.K.T. Wang, T.S. Sihra, A.C. Nairn, A.J. Czernik, and P. Greengard, Proc. Natl. Acad. Sci. USA 86, 8550-8554 (1989).

60. Rubanyi, G.M., A.D. Freay, K. Kauser, A. Johns, and D.R. Harder, Blood Vessels $\underline{27}$, 246-257 (1990).
61. Rubanyi, G.M., J.C. Romero, and P.M. Vanhoutte, Am. J. Physiol. $\underline{250}$, H1145-H1149 (1986).
62. Sachs, F., Fed. Proc. $\underline{46}$, 12-16 (1987).
63. Sachs, F., Membrane Biochemistry $\underline{6}$, 173-193 (1986).
64. Sachs, J.R., J. Gen. Physiol. $\underline{92}$, 685-711 (1988).
65. Sackin, H., Am. J. Physiol. $\underline{253}$, F1253-F1262 (1987).
66. Sackin, H., Proc Natl. Acad. Sci U.S.A. $\underline{86}$, 1731-1735 (1989).
67. Sato, M., M.J. Levesque, and R.M. Nerem, Arteriosclerosis $\underline{7}$, 276-286 (1987).
68. Sauve, R., L. Parent, C. Simoneau, and G. Roy, Pflugers Arch. $\underline{412}$, 469-481 (1988).
69. Schilling, W.P., L. Rajan, and E. Strobl-Jager, J. Biol. Chem. $\underline{264}$, 12838-12848 (1989).
70. Shepard, J.M., S.K. Goderie, N. Brzyski, P.J. Del Vecchio, A.B. Malik, and H.K. Kimelberg, J. Cell. Physiol. $\underline{133}$, 389-394 (1987).
71. Shirinsky, V.P., A.S. Antonov, K.G. Birukov, A.V. Sobolevsky, Y.A. Romanov, N.V. Kabaeva, G.N. Antonova, and V.N. Smirnov, J. Cell. Biol. $\underline{109}$, 331-339 (1989).
72. Stanfield-Colden, M., W.P. Schilling, L.D. Possani, and D.L. Kunze, J. Membr. Biol. $\underline{116}$, 227-238 (1990).
73. Sun, A.M., S.N. Saltzberg, D. Kikeri, and S.C. Hebert, Kidney Int. $\underline{38}$, 1019-1029 (1990).
74. Takeda, K., V. Schini, and H. Stoeckel, Pflugers Arch. $\underline{410}$, 385-393 (1987).
75. Vanhoutte, P.M., Nature $\underline{327}$, 459-460 (1987).
76. Vigne, P., G. Champigny, R. Marsault, P. Barbry, C. Frelin, and M. Lazdunski, J. Biol. Chem. $\underline{264}$, 7663-7668 (1989).
77. Watson, P.A., FASEB J. $\underline{5}$, 2013-2019 (1991).

IN VITRO MODELS OF VASCULAR ENDOTHELIAL CELL CALCIUM REGULATION

M. STUREK, P. SMITH, and L. STEHNO-BITTEL

Department of Physiology, School of Medicine, and Dalton Research Center, University of Missouri, Columbia, Missouri 65211

Abstract

We reviewed data suggesting that intracellular free Ca (Ca_i) regulation differs between endothelial cells in the intact artery and subcultured cells. We tested the hypothesis that Ca_i responses to K depolarization, the receptor agonists acetylcholine and bradykinin, and Na-Ca exchange in subcultured endothelial cells are depressed compared to Ca_i responses in freshly dispersed cells. Ca_i was measured with fura-2 microfluorometry in single cells from bovine or porcine left circumflex coronary artery. Three different cell preparations were compared: 1) cells freshly dispersed from the artery and used for Ca_i studies within 6 h; 2) subcultured monolayer cells used between 5-10 passages at confluency and; 3) subcultured cells dispersed from the monolayer with the same method as freshly dispersed cells. In all of the 3 bovine cell preparations (>250 cells) depolarization with 80 mM K did not increase Ca_i. A 1-min exposure to 10^{-7} M bradykinin increased Ca_i with 2 different time courses in freshly dispersed cells. A rapid, 4-fold increase in Ca_i was sustained >3-10 min in 29% of the cells and was dependent on extracellular Ca and the electrochemical driving force for Ca. In 71% of the cells the increase in Ca_i was transient, returning to resting levels in <2 min. In contrast, the monolayer and monolayer dispersed cells never showed a sustained response to bradykinin; only a transient, 4-fold increase in Ca_i. Acetylcholine (10^{-6} M) caused a transient, 2-fold increase in Ca_i in freshly dispersed cells, but never affected Ca_i in monolayer and monolayer dispersed cells. Cells freshly dispersed from porcine arteries showed a >4-fold increase in Ca_i when the Na gradient was reduced to 48% of normal, while subcultured cells showed a varied response with no Ca_i increase by the 7th passage. Data from these experiments that controlled for species, coronary artery segment, cell dispersion method, cell morphology, and Ca_i measurement methods indicate large differences in Ca_i regulation in coronary artery endothelial cell preparations. Species and arterial branch differences were also noted. These data suggest that freshly dispersed cells may be a more appropriate model for endothelial cell Ca_i regulation in the normal intact artery.

INTRODUCTION

The vascular endothelium is essential for normal regulation of vascular tone (10, 34). A wide variety of vasoactive agents cause the secretion of endothelium-derived relaxing factor (EDRF) and prostacyclin, which override the direct smooth muscle contracting effects of the vasoactive agents and relax the artery (7, 10, 24, 34). Release of EDRF and prostacyclin is dependent on the agonist-induced increase in intracellular free Ca (Ca_i) (15, 16, 20). Therefore, the ion transport mechanisms by which Ca_i is regulated in vascular endothelial cells have received great attention recently (1, 5). Furthermore, selective pharmacological modulation of endothelial cell Ca_i may be possible and ultimately of benefit therapeutically (24, 39).

Characterization of endothelial cell Ca_i regulatory mechanisms has been carried out using primarily subcultured cells (Ref. (1) for review). However, Ca_i dynamics in subcultured cells do not always parallel the endothelium-dependent and external Ca-dependent relaxtion of pre-contracted arteries (15, 20). Three main findings concerning endothelial cell Ca_i regulation are inconsistent with physiological endothelial-dependent relaxation of intact arteries. First, overwhelming direct evidence from ligand binding, voltage-clamp, and fura-2 studies convincingly shows the absence of voltage-gated Ca channels in subcultured vascular endothelial cells (1, 38). In contrast, there is indirect evidence from studies on arterial contraction suggesting that voltage-gated Ca channels exist in vascular endothelium (23, 24, 39). It is recognized that culture conditions or other physiological modulators could alter the expression of some Ca_i regulatory mechanisms (1,

Published 1991 by Elsevier Science Publishing Company, Inc.
Ion Channels of Vascular Smooth Muscle Cells and Endothelial Cells
Sperelakis and Kuriyama, Editors

8, 32). We have recently found with fura-2 measurements of Ca_i that dihydropyridine-sensitive Ca influx occurs in freshly dispersed cells from porcine coronary artery (30) and direct voltage-clamp recordings have shown transient-type Ca channels in freshly dispersed capillary endothelium (4). Second, acetylcholine-induced increases in Ca_i (6, 9) and EDRF release (7, 15, 20) have not been noted consistently in subcultured cells, while endothelium-dependent relaxation of arteries with an intact endothelium by receptor agonists such as acetylcholine has been widely noted (10, 34). Third, the physiological importance of Na-Ca exchange in endothelial cells is questionable because Na-dependent modulation of Ca_i in subcultured endothelial cells is apparent only when the cells have been highly Na-loaded (1, 2). In contrast, indirect evidence from contraction of arterial vessel segments shows that several Na-Ca exchange inhibitors alter vessel contractions in an endothelial-dependent manner (41).

Collectively, these data suggest that there are fundamental differences in Ca_i regulation between freshly dispersed and subcultured cells (4, 30), but we are aware of no studies which have <u>systematically</u> compared freshly dispersed and subcultured cells under close experimental control. One explanation for these discrepant results might be altered Ca_i regulation by K depolarization, receptor agonists, and Na-Ca exchange in subcultured endothelial cells. Therefore, the specific purpose of this manuscript was to review data indicating that subcultured endothelial cells from the coronary artery have depressed Ca_i responses compared to Ca_i responses in freshly dispersed cells. The stimuli include: 1) K depolarization; 2) the receptor agonists acetylcholine and bradykinin, and; 3) reduction of the transmembrane Na gradient. Experiments were controlled for species, coronary artery segment, species, cell dispersion procedure, cell morphology, and Ca_i measurement methodology, including ionic conditions, etc. We also briefly noted species (bovine vs. porcine) and artery (left anterior descending vs. right coronary) differences and emphasize the importance of the high specificity of vascular endothelial cell Ca_i regulation.

MATERIALS AND METHODS
Vascular Endothelial Cell Preparations

Three different preparations of vascular endothelial cells from coronary artery were used: 1) freshly dispersed cells were enzymatically harvested from bovine arteries within 2 d of obtaining the tissue and then were subsequently used for Ca_i studies within 6 h of dispersion (4, 19, 29); 2) subcultured monolayers of cells were obtained by culturing the freshly dispersed cells and passaging 5-10 times before being finally cultured on glass coverslips, thus providing cells similar to those used most widely in the literature for study of Ca_i regulation (1) and; 3) subcultured monolayer dispersed cells were obtained by harvesting the subcultured monolayer cells in a similar manner as for freshly dispersed cells.

Freshly dispersed cells. Endothelial cells were freshly dispersed from the left circumflex coronary arteries (except Fig. 7) of bovine or porcine hearts obtained from local abattoirs (19, 29, 30, 36). The surface of the heart was rinsed with 100 ml of sterile, physiological salt solution (PSS) consisting of (in mM): 2 $CaCl_2$, 10 glucose, 10 HEPES, 5 KCl, 1 $MgCl_2$, 138 NaCl, 200 U/ml penicillin, 200 μg/ml streptomycin (GIBCO, Grand Island, NY), adjusted to pH 7.4 with NaOH. The coronary artery was obtained within 30 minutes of the animals being humanely stunned and exsanguinated (as per Federal guidelines for the handling of food animals, USDA approval on file). The artery was cut open longitudinally while still in place on the heart at the abattoir and then dissected from the heart. The artery was rinsed again with 50 ml of sterile PSS and then immediately placed in a container with ice-cold cell culture media (abbreviated EH) similar to Eagle's Minimal Essential Media, which also contained HEPES as a pH buffer. The EH media consisted of (in mM): 2 $CaCl_2$, 135 NaCl, 1 $MgCl_2$, 5 KCl, 0.44 KH_2PO_4, 0.34 Na_2HPO_4, 2.6 $NaHCO_3$, 20 HEPES, 10 glucose, 100 U/ml penicillin, 100 μg/ml streptomycin. Several other components were also added to this buffer in the following dilutions (vol:vol): 0.02 amino acids, 0.01 vitamins, 0.002 phenol red, 2% horse serum (Hazelton, Lenexa, KS), pH 7.4 with NaOH. Amino acids, vitamins, and phenol red were obtained from Gibco. The artery was then transported to the laboratory on ice. The artery was cleaned of extraneous muscle and connective tissue using sterile techniques in a horizontal laminar air flow cell culture cabinet, while keeping the artery in a solution similar to the above, but containing only 0.5 mM $CaCl_2$ and no horse serum (termed Low Ca). Special care was taken not to damage the fragile endothelial cell layer; primarily, the lumenal surface was not pressed excessively on the petri dish during cleaning of the artery. Fresh dispersions of endothelial cells were always conducted within 2 d of dissection of the artery by first pinning a 1 to 1.5 cm segment of artery, lumen side up, onto the surface of silicone elastiomer in a small jar (30 ml) with 2 ml of Low Ca solution. After the artery was securely pinned the Low Ca was removed and then replaced by 2 ml of enzyme solution which contained the following components in Low Ca solution: 294 U/ml collagenase (CLS II, Worthington), 2 mg/ml bovine serum albumin (Fraction V, Sigma Chemical Co., St. Louis, MO), 1 mg/ml soybean trypsin inhibitor (Type I-S, Sigma), 0.4 mg/ml DNase I (Type IV, Sigma). With the adventitial side of the vessel facing down on the bottom of the jar the dispersion of excessive connective tissue and smooth muscle cells was minimized. The bottle was capped, tightly covered with plastic wrap, and placed in a water bath (37° C) with the vessel oriented longitudinally to the direction of the shaking movement of the bath, which was a rate of 100 strokes/min. The dispersion continued for 30-60 min and then the supernate fraction was removed. The supernate contained almost exclusively endothelial cells, easily identified morphologically by placing a drop on a dish and observing through a Nikon TMS microscope (Nikon, Inc., Garden City, NY). Few smooth muscle cells were seen unless the dispersion was continued for greater than 90 min. The fraction was then placed into a sterile 15 ml centrifuge and either loaded with fura-2 (See below) or used for monolayer cultures.

Subcultured monolayer cells. Cells were dispersed from a coronary artery segment as described above, except that the time of dispersion was usually limited to less than 30 min in order to further minimize contamination by smooth muscle cells. The cell-containing supernate was centrifuged (in a 15 ml tube) for 10 min at 1500 x g to form a pellet of cells and then the enzyme solution (supernate) was removed and replaced by 5 ml of growth media consisting of: Waymouth's MB media (Gibco, #430-1400EB), 20% (vol:vol) fetal bovine serum (Hyclone, Logan, Utah), 100 U/ml penicillin, and 100 μg/ml streptomycin. The cell suspension was placed into one 60 mm petri dish (Dow Corning, Midland, MI) and then allowed to grow in a 95% air, 5% CO_2 incubator at 37° C until reaching a confluent monolayer (primary culture). Cells were fed at 5 d intervals by removing about 1.5 ml of the media and replacing with growth media. The period of growth of this primary culture usually required about 7-14 d and at that time the plate was split (subcultured) to allow further growth. The resulting subcultured cells were referred to as first passage cells. Alternatively, the cells were frozen in liquid nitrogen for later use (See below).

Subculture of cells involved enzymatic removal of cells from the dish and placement of cells in growth media in new petri dishes. For secondary subculture (to yield first passage cells from primary cultures) the 5 ml of growth media was aspirated from the dish and saved as "conditioned media". The monolayer was rinsed once with 5 ml of phosphate buffered saline and incubated for 5 min at 37° C in 5 ml of Ca- and Mg-free phosphate buffered saline consisting of (in mM): 138 NaCl, 8 $Na_2HPO_4^O$ $7H_2O$, 2.7 KCl, 1.5 KH_2PO_4. Observation under the microsocope revealed that the cells often started to loosen from the dish at this point. The phosphate buffered saline was removed and 500 μl of 0.005% (wt:vol) trypsin + 5.3 x 10^{-5} M EDTA in phosphate buffered saline was added to the dish. The cells were again observed under the microscope to accurately determine the time at which the cells started to detach from the plate, usually 2-4 min. Once this occurred, 3 ml growth media was added to terminate the trypsinization. The supernate was aspirated through a 20 guage needle into a 3 ml syringe and then sprayed over the plate several times to remove loosened cells. Again, the supernate was aspirated and then cells were seeded at a density of 10,00-15,000 cells/cm^2 onto 3 100 mm diameter petri dishes containing 4 ml growth media plus 1 ml of conditioned media from the primary culture. Cloning of endothelial cells from the primary cultures was also used if necessary to establish a pure subcultured preparation of endothelial cells. If there were areas on the culture dishes where cells did not have the typical cobblestone morphology characteristic of endothelial cells present, i.e. the cells were spindle-shaped smooth muscle cells or fibroblasts (15, 35), then cloning rings were used to selectively trypsinize and remove only the cells having endothelial cell morphology. The edge of a cloning ring (1 cm diameter) was coated with silicone grease and then placed around the area of cells after the media has been removed from the dish. All steps involving rinsing in phosphate buffered saline, trypsinization, etc. were performed only on the 1 cm diameter area circumscribed by the cloning ring. The cells removed from different cloning areas were placed in separate 35 mm diameter petri dishes containing 2 ml of growth media and then placed in the incubator. If first passage (and subsequent passage) cells in 100 mm dishes did not become confluent in the typical 2-4 d period, then they were fed at 5 d intervals by removing about 6 ml of the media and replacing with growth media. For subculture of first passage cells and thereafter the procedure was the same, except that 1 100 mm dish was usually split into 5 100 mm dishes. For second and later passage cells, 10% rather than 20%, fetal bovine serum was used in the growth media. For Ca_i measurements cells were grown on 25 mm diameter coverslips (0.17 mm thickness, Fisher Scientific, St. Louis, MO) in a 35 mm culture dish until confluence, usually 3-7 d.

Long-term storage of cells involved freezing in liquid nitrogen. After a culture dish of cells were positively identified as endothelial cells (See below) they were harvested by trypsinization and treated as in subculturing. The differences were that only 1 ml of media containing 10% dimethylsulfoxide (vol:vol) was added to the cell pellet, giving a final cell number of 5 x 10^6 to 10^7 cells per 100 mm dish and 20% fetal bovine serum was included in the media. The cell suspension was placed in plastic freezing vials (Nunc, Inter Med, Thousand Oaks, CA) and frozen in liquid nitrogen (-198° C) to be used at a later date. Freezing was accomplished using a freezing tray which fit in the neck of a liquid nitrogen tank (Taylor-Wharton, Indianapolis, IN). The tray was adjusted to hold the freezing vials above the liquid nitrogen, thereby reducing the temperature of the cells more slowly from 23° C room temperature to -90° C in 20 min and then the cells were placed in the liquid nitrogen. Cells have retained viability up to 2 years using the liquid nitrogen storage conditions. Cells were brought out of the liquid nitrogen for use by partially thawing vials in a 37° C water bath until there was a frozen chunk of ice floating in the tube. The cells were then placed in a centrifuge tube containing 5 ml of pre-warmed (37° C) growth media and centrifuged for 5 min at 1500 x g. The supernate containing dimethylsulfoxide was aspirated and discarded and the cells were resuspended in 4 ml growth media and then placed in a 100 mm dish.

Subcultured monolayer dispersed cells. Dispersion of monolayer cells from the glass coverslips was essentially identical to the freshly dispersed cells (See above). The coverslip was placed in a vial and attached to the bottom by an inert waxy material. The enzyme solution (2 ml) was placed over the coverslip and the vial capped and incubated in the shaking water bath at 37° C for one h. The supernate was pipetted over the coverslip to remove loosened cells for use in fura-2 experiments.

Identification of endothelial cells. Three main factors were used for the positive identification of endothelial cells. First, the freshly dispersed endothelial cells exhibited a morphology similar to a cluster of grapes, in which the cells were completely spherical (See Results, Fig. 1A). Also, initially after dispersion the endothelial cells were often found in sheets strongly resembling the in situ appearance. Second, the endothelial cell was constantly observed during the experiment with a video monitor and no contraction or "blebbing" was noted that would be indicative of a smooth muscle cell (36). In other experiments on the smooth muscle cell fractions, those cells exhibited either shortening or blebbing. Third, the presence of receptors for acetylated low density lipoprotein (AcLDL) was verified in all 3 cell preparations by uptake of diI-AcLDL (1,1'-dioctadecyl-3,3,3',3'-tetramethylindocarbocyanine perchlorate, Biomedical Technologies, Stoughton, MA) (Figs. 3 and 4). Subcultured monolayer and subcultured dispersed cells were grown on glass coverslips and upon confluency were incubated for 4 h at 37° C in growth media containing 10 μg/ml DiI-AcLDL. The cells were rinsed in physiological salt solution free of DiI-AcLDL. The uptake was then visualized in the subcultured monolayer cells with rhodamine fluorescence optics (19, 29, 35) on a Nikon Diaphot microscope using a 40x oil immersion objective with a numerical aperture of 1.3. The subcultured dispersed cells were dispersed prior to fluorescence microscopy. The freshly dispersed cells were treated similarly to the subcultured dispersed cells, except that the entire arterial segment was incubated with DiI-AcLDL, then the cells were dispersed from the artery by the normal procedure (19, 29). All cells incoroporated the diI-AcLDL after incubation (See Results).

Fura-2 Microfluorometry of Ca_i

Electro-optical instrumentation. Measurements of Ca_i were made using the fluorescent indicator, fura-2 (11). The microfluorometry methods followed were similar to those described in great detail in other reports (33, 36). The major improvements of the methods include simultaneous monitoring of Ca_i and morphology of the cells (a qualitative not quantitative observation) and analog signal processing, which are briefly summarized here. To excite the fura-2, light from a 150W Xe arc lamp was passed via a liquid light guide (Oriel Corp., Stratford, CT) as a collimated beam through a circular interference filter wheel (Omega Optical, Brattleboro, VT). The filter wheel has been illustrated previously (3) and consisted of two sections that each comprised 144° of the circle; one of the filters provided 340 nm (10 nm HB) and the other provided 380 nm (10 nm HB) illumination. The wheel also had two opaque sections that each comprised 36° of the circle and separated the sections providing light transmission. This excitation filter wheel was mounted on an adjustable speed servo motor (B & B Electric, Indianapolis, IN) that rotated the wheel, thereby allowing the changing of the excitation wavelengths at rotational periods of 20, 50, 100, or 200 ms. These studies were conducted using a rotational period of 100 ms. Immediately after the interference filter wheel a collimating beam probe was placed to focus the light onto the end of another liquid light guide, which directed light to the epifluorescence illuminator of the Nikon Diaphot microscope. The light was reflected by a dichroic mirror (Nikon, DM 400) through a 40x oil immersion objective with a numerical aperture of 1.3 and the emitted fluorescence was selected for wavelengths >480 nm with a barrier filter. Recordings were defined spatially to primarily one endothelial cell by 3 procedures. First, the area of ultraviolet excitation was limited to an area of <100 μm diameter with an iris diaphragm in the epifluorescence illuminator of the microscope and placing the cell at the edge of the excitation area. Second, a rectangular aperture that had 4 sides which were independently adjustable (Ealing Electro-Optics, Holliston, MA) was inserted on the video port of the microscope and limited the area of measurement in the X-Y plane. Third, the area of measurement in the depth of the field was limited by adjusting the plane of focus appropriately. Fluorescence from neighboring cells may not have been completely eliminated by these measures, but was greatly minimized.

A reflective optocoupler monitored the position of the interference filters in the wheel and transmitted this information to the analog fluorescence signal processor. The processor was based on sample-and-hold circuitry, which demodulated the raw, oscillating fluorescence signal from the photomultiplier tube into two separate analog signals corresponding to 340 and 380 nm excitation and then were fed into separate channels of an analog-to-digital converter (Scientific Solutions, Inc., Solon, OH) and microcomputer equipped with the pClamp (Axon Instruments, Foster City, CA) data acquisition system. The sample interval in each channel of the A-D converter was 2 s, which was almost 10-fold more frequent than the most rapid change in Ca_i noted in these studies. Furthermore, although the sampling interval was 2 s, the data were also continuously displayed on an oscilloscope at the rotational period of 100 ms and, thus, any possible high frequency oscillation of Ca_i would have been noted. Select fluorescence values were then transferred into a spreadsheet program (Quattro, Borland, Inc.) for analysis and Sigmaplot (Jandel Scientific, Corte Madera, CA) for graphics. Background fluorescence was measured by moving to a cell-free area of the coverslip at the end of the experiment with both dispersed cell preparations. In the case of subcultured confluent monolayers of cells the coverslip was first wiped free of cells. None of the 3 endothelial cell preparations studied had autofluorescence significantly above the background fluorescence measured from the coverslip alone, which was ranged from 5-20% of the absolute fluorescence obtained from fura-2 loaded cells. None of the drugs used in these studies had intrinsic autofluorescence and did not change the autofluorescence of the cells. The background fluorescence was subtracted from the absolute fluorescence in off-line analysis before calculation of fluorescence ratios and Ca_i. The cell superfusion chamber was a machined plexiglass block, which utilized a 25 mm diameter glass coverslip as the bottom of the chamber (10 x 10 x 3 mm deep) and has been illustrated previously (33).

Qualitative observation of morphology simultaneously with fluorescence recordings was made possible by inserting onto the side port of the microscope a binocular head (Science Instrument Shop, University of Missouri) containing a dichroic mirror (Omega Optical, Inc.) that reflected the fura-2 fluorescence (peak of the emission is about 510 nm) to the photomultiplier. Simultaneously, additional illumination of the cell was provided by placing a 600 nm center wavelength/10 nm half-bandwidth interference filter in the transillumination light path. This narrow bandwidth light was then allowed to pass from the halogen lamp of the Diaphot through the dichroic mirror to a CCD camera (WV-D5010, Panasonic, Inc.), thus permitting observation of the cell on a video monitor during fura-2 microfluorometry. A 510 nm center wavelength/60 nm half-bandwidth interference filter was placed immediately in front of the photomultiplier to further insure no interference from the 600 nm illumination. This broad band filter allowed excellent throughput of fura-2 fluoresecence to the photomultiplier tube, but rejected 600 nm light as shown by background fluorescence of the same magnitude with or without simultaneous 600 nm illumination.

$[Ca^{2+}]_i$ vs. fluorescence ratio calibrations. In vitro $[Ca^{2+}]$ versus fluorescence ratio (F_{340} / F_{380}) calibrations were conducted for these experiments as previously described (36). Briefly, these calibrations validated the use of the fluorescence ratio as an estimate of Ca_i and consisted of the following procedures. Fura-2 pentapotassium salt (0.1 mM) was added to mock intracellular solutions consisting of (in mM): 126 KCl, 10 NaCl, 20 HEPES, 1 $MgCl_2$º $2H_2O$, 0.1 fura-2 pentapotassium salt, pH 7.1 with KOH. The mock intracellular solutions were adjusted to 10 different concentrations of free, ionized Ca ranging from 0-1500 nM, which are in the working range of the indicator and predicted Ca_i changes in these experiments (19, 29, 30). The $[Ca^{2+}]$ was achieved using appropriate ratios of H_2K_2EGTA and CaK_2EGTA as previously described (33, 36). Using the microfluorometry system described above, 2 μl droplets were placed on coverslips and fluorescence ratios were measured. This *in vitro* calibration was checked against "cytoplasmic" fluorescence ratios for fura-2 at $[Ca^{2+}]$ of 100 and 400 nM to determine whether fura-2, which had been loaded into the cells as fura-2 acetoxymethyl ester (fura-2/AM), was sufficiently de-esterified and, therefore, would provide ratios comparable to the *in vitro* calibration. For the "cytoplasmic" calibration subcultured dispersed endothelial cells were removed from 2 100 mm dishes with enzyme solution as described above, thus yielding 10 to 20 x 10^6 cells, which were loaded with 2.5 μM fura-2/AM (described below). The cell suspension was divided into 2 fractions of equal volume, centrifuged at 1500 x g for 5 min, and then the supernate was removed. The cell pellets were resuspended in 250 μl of calibration solutions containing 100 or 400 nM free $[Ca^{2+}]$, but no fura-2, and then mixed thoroughly, centrifuged, and the calibration solution supernate removed. The cell pellets were treated with 25 μl of calibration solutions containing saponin (200 μg/ml), which selectively permeabilizes the plasmalemma, thus releasing cytoplasmic fura-2 into the solution (42). Five droplets (2 μl) of each fraction were then placed on coverslips and the fluorescence ratios were determined with the microfluorometry system. Data from the *in vitro* calibration were fit to a second order polynomial with a non-linear least squares fit computer program (Sigmaplot, Jandel Scientific) and compared to the "cytoplasmic" calibration (Fig. 5).

Loading of cells with fura-2. The 3 different endothelial cell preparations were loaded with fura-2 by incubating them with 2.5 μM of the cell membrane-permeant acetoxymethylester (fura-2/AM, Molecular Probes, Pitchford, OR) for 30 min in a darkened incubator at 37° C. The vehicle for the fura-2/AM for freshly dispersed and subcultured dispersed cells was about 1.5 ml dispersion enzyme solution, which was brought up to 2 ml total in PSS (described above), except that penicillin and streptomycin were excluded and 0.2% bovine serum albumin (Fraction V, Sigma) was included. The cells were centrifuged for 5 min at 1500 x g, the supernate containing fura-2 removed, and the cells rinsed in 2 ml of pre-warmed PSS plus 0.2% albumin that was free of fura-2 for 30 min at 37° C in the dark. The cells were again centrifuged and the rinse solution removed and replaced by fresh PSS plus albumin, in which the cells remained until Ca_i measurements were conducted. The fura-2 loading procedure for subcultured monolayer cells was the same as for freshly dispersed and subcultured dispersed, except that the cells were exposed to solutions while remaining on the glass coverslip in the 35 mm dish and no centrifugation was required.

Solutions. The PSS (excluding penicillin, streptomycin, and albumin) was also used in experiments as extracellular fluid and in the additions of drugs, which were made directly. For experiments conducted in Ca-free medium (0 Ca), the buffer solution consisted of PSS without 2 mM $CaCl_2$, with an additional 2 mM NaCl, and with 10^{-5} or 10^{-3} M EGTA. A single endothelial cell at the perimeter of a cluster was selected for each experiment when studying freshly dispersed or subcultured dispersed cells. For subcultured monolayer cells almost any cell emitting a sufficiently large fluorescence signal and healthy morphology was chosen. A separate coverslip or droplet of cells was used for each experiment to eliminate any affect of prior exposure to drugs that may have affected subsequent responses. Each of the endothelial cells from which data were collected were superfused in PSS at a rate of 1-2 ml/min for a minimum of 5 minutes or until Ca_i was stable before data collection. Solutions could be exchanged completely with the superfusion system in 15-30 s. Acetylcholine (10^{-6} M, Sigma) or bradykinin (10^{-7} M, Sigma) additions to the solutions were made directly and depolarization with 80 mM K

(80K) was acheived by equimolar substitution of KCl for NaCl. The transmembrane Na gradient was decreased by substitution of 75 mM NaCl with LiCl, thus decreasing the extracellular to intracellular Na ratio from 14.3 to 6.8, a reduction to 48% of normal given an intracellular Na concentration of 10 mM. Ionomycin (Calbiochem, La Jolla, CA) was made as a stock solution of 10^{-2} M in 100% ethanol and then diluted in the appropriate buffer to a final concentration of 10^{-6} M. Experiments were conducted at room temperature (22-23° C).

RESULTS
Endothelial Cell Morphology

Freshly dispersed cells from bovine coronary artery most often displayed the "cluster of grapes" morphology after being loaded with fura-2 as shown in Fig. 1A. These cells had been centrifuged twice and vigorously pipetted. Immediately after the fraction was taken from the artery, prior to any centrifugation, the cells sometimes were flatter and sheetlike, more closely resembling the in vivo state (27). Cells loaded well with fura-2 via the fura-2/AM loading procedure, as shown by bright fluorescence of several cells in Fig. 1B, which is the same field of cells as in the brightfield micrograph in Fig. 1A. The lack of apparent fluorescence of some cells mainly reflects the different focal plane in which the cells were positioned. For example, the intense fluorescence of the cell located in the topmost region in Fig. 1B was associated with appropriate focus on the top left cell in the brightfield image (Fig. 1A). In contrast, the cell located in the top right region was out of focus under brightfield and had almost no fluorescence in the corresponding region in Fig. 1B. We used the focal plane of the cell as one means to restrict our area of Ca_i measurement primarily to one cell, e.g. the topmost cell in Fig. 1B.

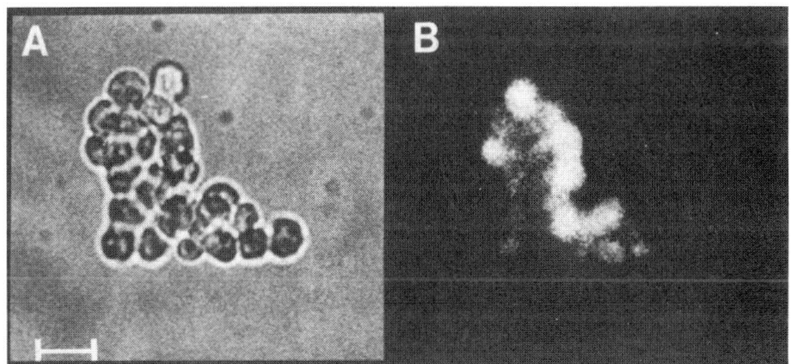

Figure 1. Freshly dispersed bovine endothelial cells. (A) Brightfield micrograph under oil immersion. Calibration bar = 8 μm. (B) Fluoresecence micrograph of cells loaded with fura-2/AM. Same field of cells as in A. Constant excitation was provided by 380 nm light.

Cells similar to those freshly dispersed in Fig. 1A were placed in long-term culture to yield the subcultured monolayer of cells in Fig. 2A that were grown to confluency between passages 5 and 10 on the glass coverslips used in all experiments on Ca_i. These cells displayed the classical "cobblestone" morphology under low magnification phase contrast optics (Fig. 2A). Higher magnification Nomarski differential interference contrast microscopy revealed the characteristic tight cell-to-cell coupling in most of the cells (Fig. 2B).

Positive identification of the cells as endothelial was evident by the uptake of diI-AcLDL in Fig. 3B. All cells incorporated the diI-AcLDL in localized regions of the cell after 4 h of incubation. A negative control was provided by the contractile smooth muscle cells occasionally found in the enzyme dispersion solution if the dispersion was continued for 1 1/2 to 2 h.

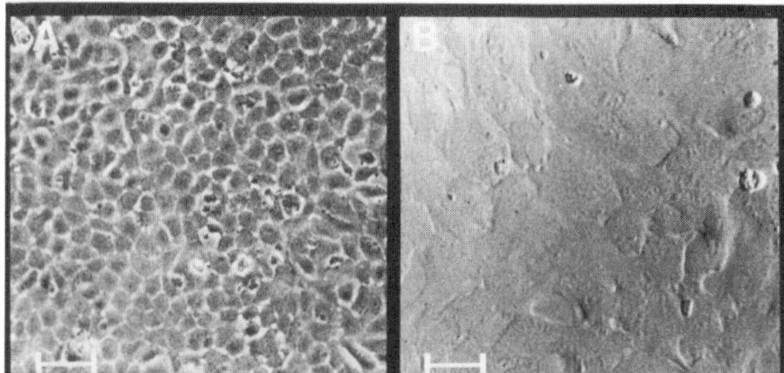

Figure 2. Subcultured monolayer of bovine endothelial cells. **(A)** Phase contrast micrograph of multiple passaged cells. Calibration bar = 30 lm. **(B)** Nomarski differential interference contrast micrograph of cells. Calibration bar = 10 μm.

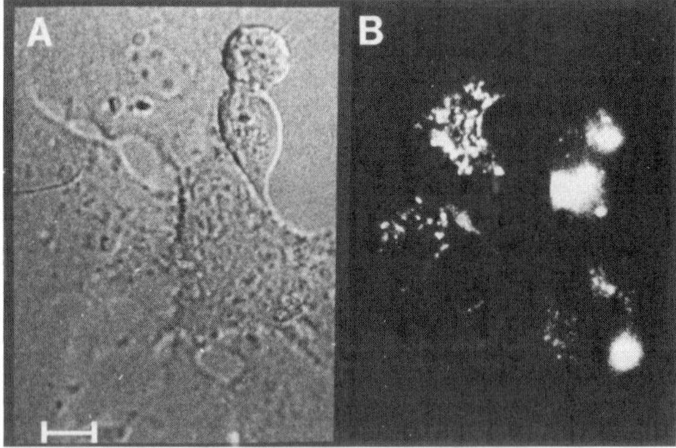

Figure 3. Endocytosis of DiI-AcLDL by bovine endothelial cells. **(A)** Brightfield micrograph of multiple passaged cells. Calibration bar = 4 μm. **(B)** Same field of cells as in A. Rhodamine fluorescence optics, using oil immersion objective.

The punctate fluorescence in the cytoplasmic area of the top cell is very typical and representative of the uptake of diI-AcLDL by endocytosis and resultant packaging into vesicles. Comparing the fluorescence micrograph in Fig. 3B with the brightfield image in Fig. 3A, it is clear that there was no transport into the nuclear region of the cell. Note also the 2 slightly rounded cells in the top right of Fig. 3A and the corresponding intense fluorescence, which was not punctate. The reason for this seemingly nonconfluent state of the cells is that the cells sometimes rounded as in Fig. 3A when coverslips were being mounted in the superfusion chamber for Ca_i measurements. These specific cells were not used for Ca_i measurements; instead, a cell such as the flat one on the left would have been used. All Ca_i measurements on subcultured cells were conducted on confluent monolayers such as those in Fig. 2, as verified by inspection of each coverslip while in cell culture media in the 35 mm dishes prior to fura-2 loading or mounting in the superfusion chamber. Cells similar to those in Figs. 2 and 3 were dispersed from their coverslips with the same methods as used to obtain cells in Fig. 1 and yielded the grapelike clusters in Fig. 4. The effect of depth of focus is also noted in Fig. 4B by the fluorescence of DiI-AcLDL being much more diffuse than in Fig. 3B. Assuming a near spherical morphology for these subcultured dispersed cells, the thickness of these cells is about 5-10 μm, compared to < 1-2 μm thickness of the subcultured monolayer.

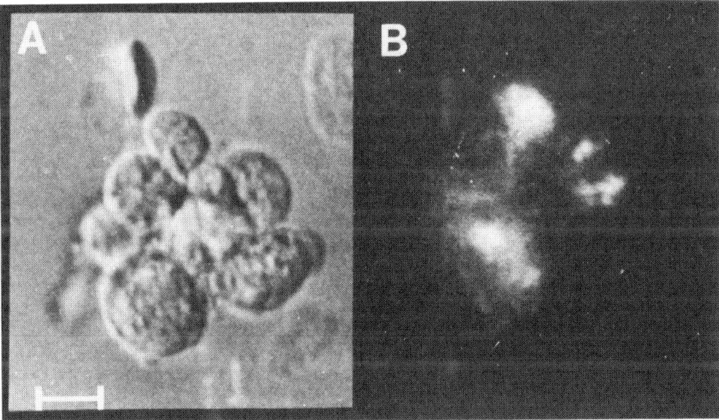

Figure 4. Subcultured bovine endothelial cells dispersed from monolayer. **(A)** Nomarski differential interference contrast micrograph of multiple passaged cells enzymatically dispersed from monolayer. Calibration bar = $8\,\mu$m. **(B)** Same field of cells as in A indicating endocytosis of DiI-AcLDL by cells. Rhodamine fluorescence optics, using oil immersion objective.

Fura-2 Fluorescence Ratio Versus $[Ca^{2+}]$ Calibration

Our hypothesis that Ca_i regulation was different in freshly dispersed versus cultured endothelial cells was absolutely dependent on the same dye handling by both cell preparations. If there was a non-specific depression of functions in the subcultured endothelial cells, then any differences noted in Ca_i regulation could have been attributed simply to altered cytoplasmic esterase activity or other nonspecific interferences with fura-2. Therefore, the Ca-sensitivity of fura-2, which had been trapped in the cytoplasm after loading via incorporation of fura-2/AM, was assessed by specifically permeabilizing the plasmalemma of bovine endothelial cells with saponin (42) to release the fura-2 and then bufferring the $[Ca^{2+}]$ in the lysate with Ca^0 EGTA calibration solutions. Other investigators have used Ca ionophores, such as ionomycin, to equilibrate Ca_i with the extracellular solution and, thereby, establish an "*in situ*" calibration using several $[Ca^{2+}]$ standards (21) or the R_{min} and R_{max} method commonly employed (11). However, the extremely variable response of cells to 2 x 10^{-5} M ionomycin (data not shown) prevented the reliable use on these single cells. Saponin lysis of a single cell was straightforward, but the dye was completely washed from the cell too rapidly to enable an equilibration of the extracellular and intracellular solutions and, thus, the use of cell suspensions was necessary. The *in vitro* calibration (circles, Fig. 5) was very similar to that found with a similar microfluorometry system (33) and other reports from this laboratory (28, 36), indicating a predictable relationship between the fluorescence ratio (F_{340} / F_{380}) and $[Ca^{2+}]$. Most importantly, the "cytoplasmic" calibration (squares, Fig. 5) was similar to the *in vitro* values in the range that most Ca_i values in this report lie; therefore, direct generalization from the *in vitro* calibration to fluorescence ratios obtained on single endothelial cells was appropriate in these studies. Freshly dispersed cells were not used because large enough numbers of cells could be generated. As the remaining results will indicate, depression of Ca_i responses usually occurred in subcultured preparations; therefore, if fura-2/AM hydrolysis were not sufficient and yielded an erroneous calibration, the error would be most apparent in the subcultured cells, not the freshly dispersed cells.

Acetylcholine and Bradykinin Regulation of Ca_i

Freshly dispersed cells. As indicated in the Methods, the maximum time resolution was the 2 s sampling interval of analog-to-digital converter, but data are plotted as mean $\pm$ SE only every 1/2 to 1 min for all group data. The resting values (baseline) for bovine endothelial cells were very stable as indicated by the SE lying within the area of the symbol at time points "baseline" and min 0 in Fig. 6. To acheive a stable resting Ca_i (baseline) it was important that the flow rate of the superfusion chamber was constant and fairly

358

laminar over the cells, because when the flow was pulsatile it caused the cell clusters to be shaken in the fluid and the Ca_i oscillated above basal levels (data not shown), consistent with the existence of mechanosensitive, Ca-permeable ion channels in vascular endothelial cells (14). Using a variety of protocols with several different orders of exposure to acetylcholine, bradykinin, and 80K, a remarkably bimodal response pattern was noted in the duration of the bradykinin-induced Ca_i increase in 34 total cells that were exposed to 10^{-7} M bradykinin for 1 min. The 10^{-7} M concentration of bradykinin chosen was maximal (data not shown). Pooling the data on the 34 cells in Fig. 6 shows a somewhat large variability in the bradykinin-induced increase in Ca_i relative to the baseline level of variability and suggested the possibility of two different cell populations that respond to bradykinin by eliciting either a transient or a sustained response to bradykinin. We determined the distribution of the Ca_i levels in the 34 cells 2 min after the end of exposure to bradykinin (min 3 in Fig. 6) and found that in 24 cells (71%) the Ca_i returned to within 2 standard deviations of the mean for the entire group of 34 cells. In contrast, Ca_i remained higher than 2 standard deviations above the mean ($\geq$ 120 nM) in the other 10 cells (29%). We termed a response that returned to baseline as "transient" and a response which did not return to baseline 2 min after the end of bradykinin exposure as "sustained" and plotted the responses in Fig. 6; subsequent analyses were conducted according to these response types. Since we found the extremely prolonged increase in Ca_i was elicited by only a 1-min exposure to 10^{-7} M bradykinin (up to 2 min exposures to bradykinin in preliminary studies) we did not attempt longer exposures to bradykinin because it was desirable to allow recovery of Ca_i to approximately resting levels and, thereby, provide a comparable baseline level for subsequent exposure to other agents. Accordingly, all experiments involved only a 1-min bradykinin exposure. This aspect of the protocols differed from almost all other reports on bradykinin regulation of Ca_i in vascular endothelial cells (8, 17, 20, 25). Similarly, for proper comparison to the effects of bradykinin, acetylcholine exposure was only 1 min in all studies.

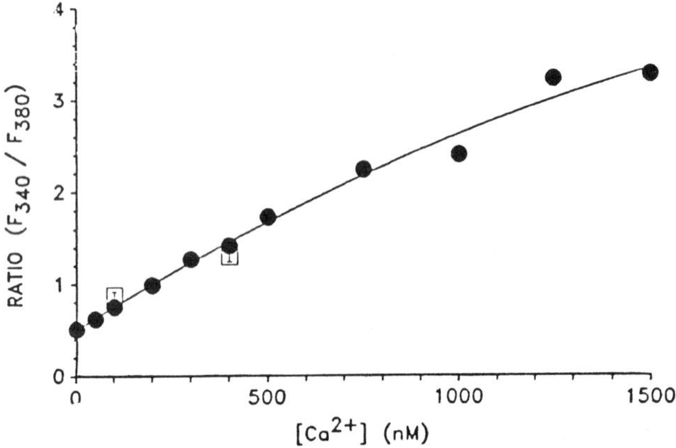

Figure 5. Fluorescence RATIO (F_{340} / F_{380}) versus [Ca^{2+}] calibration. *In vitro* calibration (See Methods) shown in circles fit by second order polynomial shown by line; n = mean of 3 measures per point; SE fall within the symbols. "Cytoplasmic" calibration shown in squares for [Ca^{2+}] of 100 and 400 nM; N = mean of 5 measures per point; SE shown within symbols.

Further characterization of the bradykinin-induced increase in Ca_i in freshly dispersed bovine endothelial cells involved estimation of the voltage-dependence of the sustained increase, since there is substantial evidence that the increase in Ca_i is mediated by Ca influx through a ligand-gated channel (13, 25). Accordingly, the bradykinin-induced sustained increase in Ca_i should be less when the electrochemical driving force for Ca entry is reduced by depolarization of the cells (25). As shown in Fig. 7A depolarization with 80K from min 2-4 did not elicit any increase in Ca_i above resting levels, a finding which was consistent in >250 bovine endothelial cells studied. This reliable finding suggests that no voltage-gated Ca channels exist in these bovine coronary endothelial cells (1, 8, 13, 32, 38).

These data are also consistent with our findings on cells from coronary artery of sedentary pigs, which show either no depolarization-induced increase in Ca_i until after exposure to bradykinin and return of Ca_i to resting level (30) or only a modest increase in Ca_i (19). In Fig. 7A the 80K solution was exchanged for a 2-min recovery in PSS and then a 1-min exposure to bradykinin elicited a 4-fold increase in Ca_i that was sustained for >3 min in 4 cells after return to normal PSS solution.

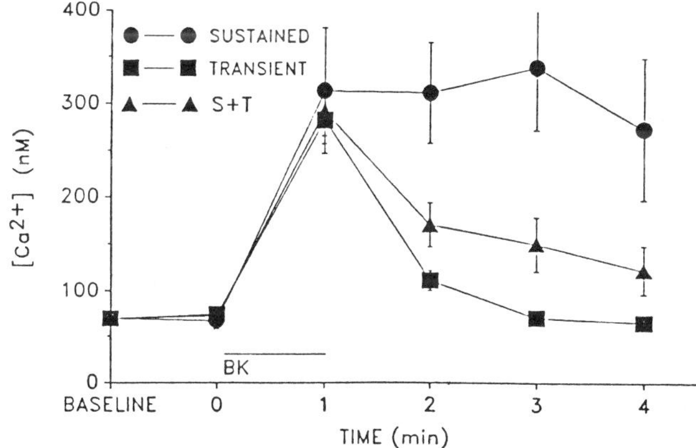

Figure 6. Freshly dispersed bovine endothelial cells: transient and sustained patterns of Ca_i response to bradykinin (BK). All cells were exposed to 10^{-7} M BK for 1 min indicated by the horizontal line. Values are mean $\pm$ standard error (SE) of all cells studied (sustained and transient responders, S+T, triangles) at each point; N = 34 cells/point. "Transient" responders (squares); N = 24 cells/point. "Sustained" responders (circles); N = 10 cells/point.

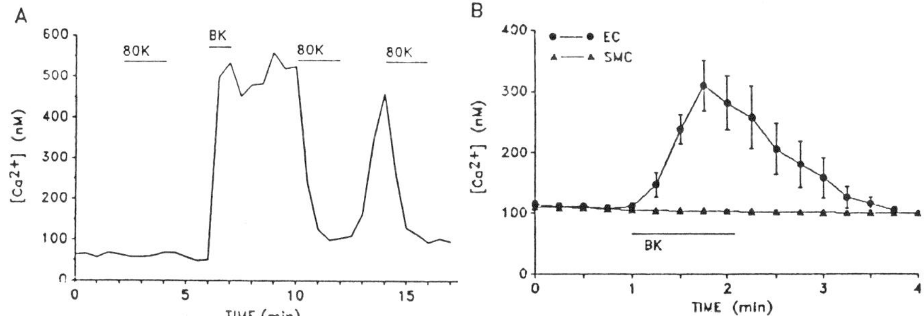

Figure 7. Freshly dispersed cells: effects of 10^{-7} M bradykinin (BK) in bovine and porcine cells. **(A)** Dependence of sustained Ca_i response to BK in bovine endothelial cells on electrochemical driving force for Ca entry; 80 mM K depolarization (80K). **(B)** Transient response to BK in porcine endothelial cells (EC, circles, N = 14; 5 cells from left circumflex and 9 cells from right coronary artery). No response to BK in porcine smooth muscle cells (SMC, triangles, N = 32).

In the bovine cells which showed a sustained phase of bradykinin-induced Ca_i increase, depolarization with 80K at min 10 decreased Ca_i to about 20% of the bradykinin-induced plateau increase in Ca_i. Subsequent repolarization of the cell by return to PSS (containing 5 mM K) then increased Ca_i, which was reversed again to only 20% of the plateau upon a second 80K-induced depolarization at min 14. The complete dependence of the bradykinin-induced sustained increase in Ca_i in bovine cells on extracellular Ca was shown by the transient response to bradykinin when extracellular Ca was removed and 10^{-5} M or 10^{-3} M EGTA added to Ca-free PSS (data not shown). The peak Ca_i response was

360

only about 50% of that in 2 mM extracellular Ca and always returned completely to baseline values by 2 min after the end of exposure to bradykinin (N = 15, data not shown). Also, pretreatment with EGTA probably did not deplete the intracellular store because after exposure to bradykinin (and the transient Ca_i response) 10^{-6} M ionomycin in 0 Ca solution also elicited a significant increase in Ca_i (data not shown).

Return of Ca_i to resting levels in bovine endothelial cells after exposure to bradykinin was also noted (transient responses, Figs. 6 and 8) and provided an ideal opportunity to test for the delayed potentiation of K depolarization-induced Ca influx by bradykinin, similar to that which we have noted in porcine coronary artery endothelial cells (30). There was no change in Ca_i after both 80K and milder depolarization with 30K at 4 min following the return of Ca_i to resting levels after the transient response to bradykinin in bovine endothelial cells (data not shown). These data suggest no potentiation of voltage-gated Ca channels in bovine endothelial cells by bradykinin. Fig. 7B shows a transient Ca_i response of porcine endothelial cells to a brief, 1-min exposure to bradykinin. The lack of a pronounced sustained Ca_i response to bradykinin indicates a subtle difference between endothelial cells of bovine and porcine species. In addition, even within the porcine species endothelial cells from the right coronary artery show a >4-fold longer sustained Ca_i response to bradykinin than cells from the left anterior descending coronary artery (19). Finally, note also the complete lack of Ca_i response to bradykinin in porcine smooth muscle cells in Fig. 7B, thus providing an unexpected tool to distinguish the cell types.

Subcultured cells vs. freshly dispersed cells. One advantage to using subcultured cells is the almost infinite number of cells that can be generated by multiple passages of cells harvested from only one artery. The potential problem, however, is that the resulting subcultures may be the product of selective culture of unusual cells from only one animal; therefore, Ca_i regulatory properties may not be representative of the larger animal population. To avoid such ambiguity subcultures for these studies were derived from arteries obtained from 6 different cows and data collection spanned 12 months, thereby minimizing the effects of any possible seasonal variability.

The effect of subculturing of bovine endothelial cells on Ca_i responses to bradykinin is shown in Fig. 8. Similar to the freshly dispersed cells (FD) 80K did not increase Ca_i in the subcultured monolayer cells (SM), thus indicating the absence of voltage-gated Ca channels. This is also indicative of the endothelial nature of these cells, because we have shown in several reports that 80K increases Ca_i in smooth muscle cells of the coronary artery (28, 36, 37). In contrast, the peak bradykinin-induced increase in Ca_i in subcultured monolayer cells was only about 75% of the peak Ca_i increase in freshly dispersed cells showing a transient Ca_i response. Another difference was the delay in the peak Ca_i increase in the monolayer cells.

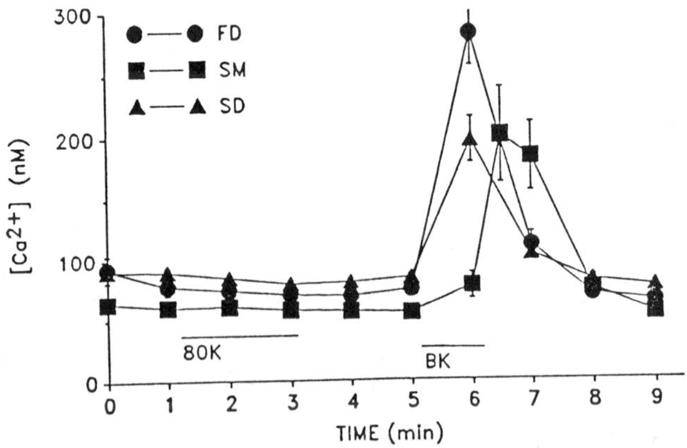

Figure 8. Ca_i responses to 80 mM K depolarization (80K) and 10^{-7} M bradykinin (BK) in bovine endothelial cells. Comparison of freshly dispersed (FD, circles, N = 24 from transient responses in Fig. 6), subcultured monolayer (SM, squares, N = 18), and subcultured dispersed (SD, triangles, N = 15) cell preparations.

The dispersion of subcultured cells from the glass coverslip provided a control for dispersion procedure and, more importantly, the morphology of the cells, which could have been responsible for diffusion limitations because of the monolayer morphology that allows primarily only one side to be exposed freely to the superfusate. The results are shown in Fig. 8 for the protocol which mimics that for freshly dispersed and subcultured monolayer cells. The subcultured dispersed cells (SD) were essentially the same as subcultured monolayer cells, except no delay in onset of bradykinin-induced Ca_i response (Fig. 8), thus indicating that the delay in the subcultured monolayer cells may have been due to slow diffusion of bradykinin to underside of cell in contact with the coverslip. A sustained Ca_i response to bradykinin similar to that in freshly dispersed cells in Fig. 6 was never noted in a total of 33 subcultured monolayer and subcultured dispersed bovine cells.

The effect of subculturing of bovine endothelial cells on Ca_i responses to acetylcholine is shown in Fig. 9, which summarizes the data from protocols very similar to that in Fig. 8. Similar to the freshly dispersed cells (FD) 80K again did not increase Ca_i in the subcultured monolayer cells (SM) or subcultured dispersed (SD) cells. These results strongly reinforce that finding that voltage-gated Ca channels are absent from bovine endothelial cells. In contrast, there was about a 2-fold acetylcholine-induced peak increase in Ca_i in freshly dispersed cells, while neither the monolayer (SM) nor subcultured dispersed (SD) cells showed a response to acetylcholine.

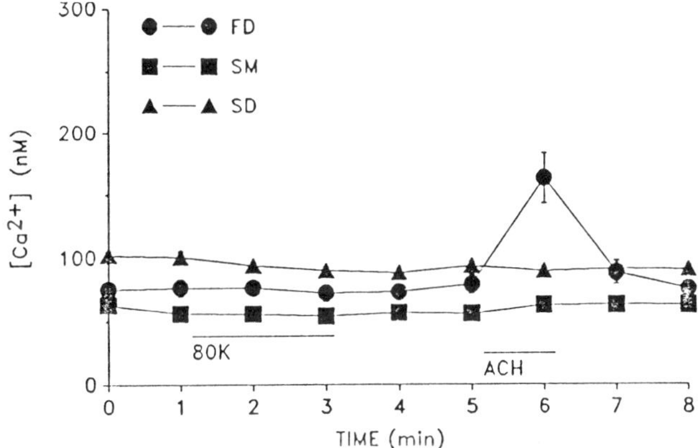

Figure 9. Ca_i responses to 80 mM K depolarization (80K) and 10^{-6} M acetylcholine (ACH) in bovine endothelial cells. Comparison of freshly dispersed (FD, circles, N = 18), subcultured monolayer (SM, squares, N = 18), and subcultured dispersed (SD, triangles, N = 15) cell preparations.

Regulation of Ca_i by the Transmembrane Na Gradient

Previous reports have shown that increases in Ca_i occur in vascular endothelial cells when the transmembrane Na gradient is decreased only when the cells were also Na-loaded (2). We also systematically compared freshly dispersed and subcultured endothelial cells from porcine coronary artery for affects of decreasing the Na gradient on Ca_i. Freshly dispersed cells show a greater than 4-fold increase in Ca_i when the Na gradient was reduced from the normal ratio of 14.3 to about 6.8 by replacement of 75 mM NaCl with LiCl in the extracellular solution (FD, Fig. 10). The magnitude of the Ca_i response was inversely related to the extracellular Na concentration (hence, the Na gradient) in freshly dispersed cells (29). As shown in Fig. 10 the Ca_i response of subcultured cells from the 5th passage was variable (29). Some cells showed a >3-fold increase in Ca_i when the Na gradient was decreased (5th - R), while other cells showed no response (5th - NR). The effect of decreasing the Na gradients, however, uniformly had no affect on Ca_i in all subcultured cells from the 7th passage (Fig. 10). These data strongly indicate that the Ca_i response was likely to be mediated by a Na-Ca exchange mechanism (12), which fails to remove Ca upon reduction of the Na gradient that normally is directed inward to drive the

362

extrusion of Ca from the cell. We do not have any data at this time to determine at what passage number the Na-Ca exchange mechanism begins to be altered.

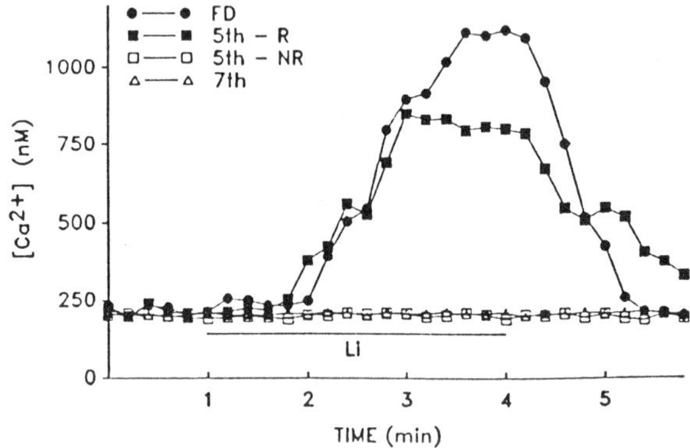

Figure 10. Ca_i responses to decreased transmembrane Na gradient in porcine endothelial cells. LiCl (Li) was substituted for NaCl. Comparison of freshly dispersed (FD, squares), 5th passage responders (5th - R), 5th passage non-responders (5th - NR), and 7th passage non-responding cells that were dispersed from the monolayer.

DISCUSSION

The main points of this manuscript are: 1) The literature indicates a variable presence of voltage-gated Ca influx in vascular endothelial cells, with the main source of variability being the cell preparation, e.g. subcultured vs. freshly dispersed (4, 30). Species differences may also exist, however, because all 3 bovine endothelial cell preparations clearly showed no sign of voltage-gated Ca influx. The expression of voltage-gated Ca influx may also be subject to great species variability and in vivo physiological state, as shown by augmentation of K depolarization-induced increases in Ca_i by both exercise training and by a pharmcological agent, Bay K 8644, in Yucatan miniature pigs (30). 2) Subculture of bovine coronary artery endothelial cells is associated with depression of the Ca_i responses of cells to the receptor agonists acetylcholine and bradykinin. 3) Subculture of porcine coronary artery endothelial cells is associated with the depression of Ca_i responses of cells to reductions in the transmembrane Na gradient, i.e. defects in Na-Ca exchange. We emphasize that in addition to the review of the literature suggesting differences in Ca_i regulation between freshly dispersed and cultured cells (1, 4, 8, 13, 23, 24, 30, 39, 41), our experiments systematically compared Ca_i responses of the different cell preparations with attention to species, coronary artery segment, species, cell dispersion procedure, cell morphology, and Ca_i measurement methodology, including ionic conditions, etc. The highly specific regulation of Ca_i in vascular endothelial cells must be recognized.

Although our data do clearly indicate a difference in Ca_i regulation between several *in vitro* endothelial cell models, the data do not indicate molecular signals for the long-term modulation of Ca_i regulation or precisely identify the ion transport alteration or molecular mechanism for the alterations, e.g. receptor, G protein coupling, etc. (26, 31). The microfluorometry methods we have used have permitted the study of freshly dispersed endothelial cells, since there fewer available than subcultured cells. Fortunately, the minimal number (300-1000) of freshly dispersed endothelial cells, as identified by uptake of DiI-AcLDL from a small segment of coronary artery, is adequate for our use of single cell microfluorometry methods (19, 29, 33, 36). Excellent possibilities exist for using these model cell preparations and methods for the technically somewhat straightforward pursuit of molecular characterization of these Ca_i regulatory mechanisms.

The general question which remains unresolved is What is an appropriate model (preparation) of vascular endothelial cells to represent *in situ* and/or *in vivo* states? Are subcultured endothelial cells a model for in vivo injury and regeneration as suggested by Ross (22)? Regenerating endothelium from atherosclerotic pig coronary artery that had undergone balloon denudation have impaired endothelium-dependent relaxation to bradykinin (27). This defect appeared to be a functional defect intrinsic to the cells, rather than lack of complete regeneration of cell number, because morphological analysis indicated a complete layer of cells on the artery. It is possible that vascular smooth muscle cells and other factors have a "trophic influence" on endothelial cells, which modulates the responsiveness of the endothelial cells to agonists, resulting in altered Ca_i dyanamics. Our data raise the intriguing question of whether cultured endothelial cells that are used in vascular grafts and/or seeded onto arteries in the intact animal for treatment of atherosclerosis or as a drug delivery system will assume a phenotype that is completely responsive to vasoactive agents (18, 40). Again, cellular and molecular characterization of the cytodifferentiation of endothelial cell Ca_i regulatory regulatory mechanisms is needed.

ACKNOWLEDGEMENTS

This work was supported by grants from the American Heart Association-Missouri Affiliate, Institutional Biomedical Support Grant RR07053, Marion, Merrill Dow Inc., and NIH HL41033 to M.S, a grant from the American Physical Therapy Association to LSB and support from NIH Training Grant T-32 HL07094. PS was a Dalton Research Center Cardiovascular Fellow. We acknowledge Laurel Bowman and Linda Rowland of the Core Vascular Cell Culture Facility of the Dalton Research Center for assistance in subculture of endothelial cells and Chuck Jorgenson for preparation of photographs.

REFERENCES

1. Adams, D.J., J. Barakeh, R. Laskey, and C. Van Breemen, FASEB J. 3, 2389-2400 (1989).
2. Adams, D.J., C. Van Breemen, M.B. Cannell, and S.O. Sage, J. Physiol. (Lond) 419, 192P (Abstract) (1989).
3. Becker, P.L. and F.S. Fay, Am. J. Physiol. 253, C613-C618 (1987).
4. Bossu, J.-L., A. Feltz, J.-L. Rodeau, and F. Tanzi, FEBS Lett. 255, 377-380 (1989).
5. Bregestovski, P.D. and U.S. Ryan, J. Mol. Cell. Cardiol. 21 Suppl. 1, 103-108 (1989).
6. Busse, R., H. Fichtner, A. Luckhoff, and M. Kohlhardt, Am. J. Physiol. 255, H965-H969 (1988).
7. Cocks, T.M., J.A. Angus, J.H. Campbell, and G.R. Campbell, J. Cell. Physiol. 123, 310-320 (1985).
8. Colden-Stanfield, M., W.P. Schilling, A.K. Ritchie, S.G. Eskin, Circ. Res. 61, 632-640 (1987).
9. Danthuluri, N.R., M.I. Cybulsky, and T.A. Brock, Am. J. Physiol. 255, H1549-H1553 (1988).
10. Furchgott, R.F. and P.M. Vanhoutte, FASEB J. 3, 2007-2018 (1989).
11. Grynkiewicz, G., M. Poenie, and R.Y. Tsien, J. Biol. Chem. 260, 3440-3450 (1985).
12. Hale, C.C. and R.S. Keller, Life Sci. 47, 9-15 (1990).
13. Johns, A., T.W. Lategan, N.J. Lodge, U.S. Ryan, C. Van Breemen, and D.J. Adams, Tissue Cell 19, 733-745 (1987).
14. Lansman, J.B., T.J. Hallam, and T.J. Rink, Nature 325, 811-813 (1987).
15. Loeb, A.L., N.J. Izzo, Jr., R.M. Johnson, J.C. Garrison, and M.J. Peach, Am. J. Cardiol. 62 Suppl. 11, 36G-40G (1988).
16. Luckhoff, A., U. Pohl, A. Mulsch, and R. Busse, Br. J. Pharmacol. 95, 189-196 (1988).
17. Luckhoff, A., R. Zeh, and R. Busse, Pflugers Arch. 412, 654-658 (1988).
18. Nabel, E.G., G. Plautz, F.M. Boyce, J.C. Stanley, and G.J. Nabel, Science 244, 1342-1344 (1989).
19. Obye, P.K., M.H. Laughlin, and M. Sturek, (Submitted).
20. Peach, M.J., H.A. Singer, N.J. Izzo, and A.L. Loeb, Am. J. Cardiol. 59, 35A-43A (1987).
21. Peeters, G.A., V. Hlady, J.H.B. Bridge, and W.H. Barry, Am. J. Physiol. 253, H1400-H1408 (1987).
22. Ross, R., J. Med. 314, 488-500 (1986).
23. Rubanyi, G.M., A. Schwartz, and P.M. Vanhoutte, Eur. J. Pharmacol. 117, 143-144 (1985).
24. Rubyani, G.M. and P.M. Vanhoutte, Ann. N. Y. Acad. Sci. 522, 226-233 (1988).
25. Schilling, W.P., Am. J. Physiol. 257, H778-H784 (1989).
26. Shimokawa, H., N.A. Flavahan, and P.M. Vanhoutte, Circ. Res. 65, 740-753 (1989).
27. Shimokawa, H. and P.M. Vanhoutte, Circ. Res. 64, 900-914 (1989).
28. Stehno-Bittel, L., M.H. Laughlin, and M. Sturek, J. Appl. Physiol. (In press).
29. Stehno-Bittel, L. and M. Sturek, (Submitted).
30. Sturek, M., L. Stehno-Bittel, L. Bowman, C. L. Oltman, and M.H. Laughlin, (Submitted).
31. Sung, C.-P., A.J. Arleth, K. Shikano, B. Zabko-Potapovich, Biochem. Pharmacol. 38, 696-699 (1989).
32. Takeda, K., V. Schini, and H. Stoeckel, Pflugers Arch. 410, 385-393 (1987).
33. Thayer, S.A., M. Sturek, and R.J. Miller, Pflugers Arch. 412, 216-223 (1988).

34.   Vanhoutte, P.M., Hypertension 13, 658-667 (1989).
35.   Voyta, J.C., D.P. Via, C.E. Butterfield, and B.R. Zetter, J. Cell Biol. 99, 2034-2040 (1984).
36.   Wagner-Mann, C.C., L. Bowman, and M. Sturek, Am. J. Physiol. Cell Physiol. 260, C763-C770 (1991).
37.   Wagner-Mann, C.C. and M. Sturek, Am. J. Physiol. Cell Physiol. 260, C771-C777 (1991).
38.   Whitmer, K.R., J.S. Williams-Lawson, R.F. Highsmith, and A. Schwartz, Biochem. Biophys. Res. Commun. 154, 591-605 (1988).
39.   Williams, J.S., Y.H. Baik, W.J. Koch, and A. Schwartz, J. Pharmacol. Exp. Ther. 241, 379-386 (1987).
40.   Wilson, J.M., L.K. Birinyi, R.N. Salomon, P. Libby, A.D. Callow, and R.C. Mulligan, Science 244, 1344-1346 (1989).
41.   Winquist, R.J., L.A. Heaney, A.A. Wallace, E.P. Baskin, R.B. Stein, M.L. Garcia, and G.J. Kaczorowski, J. Pharmacol. Exp. Ther. 248, 149-156 (1989).
42.   Yamamoto, H. and C. Van Breemen, J. Gen. Physiol. 87, 369-389 (1986).

INDEX